AF896359

Interdisciplinary Cancer Research

Volume 6

Series Editor
Nima Rezaei
Department of Clinical Immunology, Karolinska Institutet, Stockholm, Sweden

Cancer Immunology Project (CIP), Universal Scientific Education and Research Network (USERN), Stockholm, Sweden

Editorial Board Members
Rodrigo Aguiar, Universidade Federal de São Paulo, São Paulo, Brazil
Atif A. Ahmed, University of Missouri–Kansas City, Kansas City, Missouri, USA
Maria R. Ambrosio, University of Siena, Siena, Italy
Mehmet Artac, Department of Medical Oncology, Necmettin Erbakan University, Konya, Türkiye
Tanya N. Augustine, Faculty of Health Sciences, University of the Witwatersrand, Johannesburg, South Africa
Rolf Bambauer, Institute for Blood Purification, Homburg, Germany
Ajaz Ahmad Bhat, Division of Translational Medicine, Sidra Medical and Research Center, Doha, Qatar
Luca Bertolaccini, European Institute of Oncology, Milan, Italy
Chiara Bianchini, University Hospital of Ferrara, Ferrara, Italy
Milena Cavic, Institute of Oncology and Radiology of Serbia, Belgrade, Serbia
Sakti Chakrabarti, Medical College of Wisconsin, Milwaukee, USA
William C. S. Cho, Department of Clinical Oncology, Queen Elizabeth Hospital, Kowloon, Hong Kong
Anna M. Czarnecka, Maria Sklodowska-Curie National Research Institute of Oncology, Warsaw, Poland
Cátia Domingues, University of Coimbra, Coimbra, Portugal
A. Emre Eşkazan, Istanbul University-Cerrahpaşa, Istanbul, Türkiye
Jawad Fares, Northwestern University, Chicago, Illinois, USA
Carlos E. Fonseca Alves, São Paulo State University, São Paulo, Brazil
Pascaline Fru, University of the Witwatersrand, Johannesburg, South Africa
Jessica Da Gama Duarte, Olivia Newton-John Cancer Research Institute, Heidelberg, Australia
Mónica C. García, Universidad Nacional de Córdoba, Córdoba, Argentina
Melissa A. H. Gener, Children's Mercy Hospital, Kansas City, Missouri, USA

José Antonio Estrada Guadarrama, Universidad Autónoma del Estado de México, Toluca, Mexico
Kristian M. Hargadon, Gilmer Hall, Hargadon Laboratory, Hampden–Sydney College, Hampden Sydney, Virginia, USA
Paul Holvoet, Catholic University of Leuven, Leuven, Belgium
Vladimir Jurisic, Faculty of Medical Sciences, University of Kragujevac, Kragujevac, Serbia
Yearul Kabir, University of Dhaka, Dhaka, Bangladesh
Theodora Katsila, National Hellenic Research Foundation, Athens, Greece
Jorg Kleeff, Martin-Luther-University Halle-Wittenberg, Halle, Germany
Chao Liang, Hong Kong Baptist University, Hong Kong, Hong Kong
Mei Lan Tan, Universiti Sains Malaysia, Pulau Pinang, Malaysia
Weijie Li, Children's Mercy Hospital, Kansas City, Missouri, USA
Sonia Prado López, Institute of Solid State Electronics, Technische Universität Wien, Vienna, Austria
Muzafar A. Macha, Islamic University of Science and Technology, Awantipora, India
Natalia Malara, Magna Graecia University, Catanzaro, Italy
Adile Orhan, University of Copenhagen, Copenhagen, Denmark
Heriberto Prado-Garcia, National Institute of Respiratory Diseases "Ismael Cosío Villegas", Mexico City, Mexico
Judith Pérez-Velázquez, Helmholtz Zentrum München, Munich, Germany
Wafaa M. Rashed, Children's Cancer Hospital, Cairo, Egypt
Francesca Sanguedolce, University of Foggia, Foggia, Italy
Rosalinda Sorrentino, University of Salerno, Fisciano, Italy
Irina Zh. Shubina, N. N. Blokhin National Medical Research Center of Oncology, Moscow, Russia
Heloisa Sobreiro Selistre de Araujo, Universidade Federal de São Carlos, Sao Carlos, Brazil
Ana Isabel Torres-Suárez, Universidad Complutense de Madrid, Madrid, Spain
Jakub Włodarczyk, Medical University of Lodz, Lodz, Poland
Joe Poh Sheng Yeong, Singapore General Hospital, Singapore, Singapore
Marta A. Toscano, Hospital de Endocrinología y Metabolismo Dr. Arturo Oñativia, Salta, Argentina
Tak-Wah Wong, National Cheng Kung University Medical Center, Tainan, Taiwan
Jun Yin, Central China Normal University, Wuhan, China
Bin Yu, Zhengzhou University, Zhengzhou, China

The "Interdisciplinary Cancer Research" series publishes comprehensive volumes on different cancers and presents the most updated and peer-reviewed articles on human cancers. Over the past decade, increased cancer research has greatly improved our understanding of the nature of cancerous cells which has led to the development of more effective therapeutic strategies to treat cancers. This translational series is of special value to researchers and practitioners working on cell biology, immunology, hematology, biochemistry, genetics, oncology and related fields.

Nima Rezaei
Editor

Breast Cancer Genetics, Immunology, and Immunotherapy: An Interdisciplinary Approach

Springer

Editor
Nima Rezaei
Department of Clinical Immunology
Karolinska Institutet
Stockholm, Sweden

Research Center for Immunodeficiencies,
Children's Medical Center
Tehran University of Medical Sciences
Tehran, Iran

ISSN 2731-4561 ISSN 2731-457X (electronic)
Interdisciplinary Cancer Research
ISBN 978-3-031-65807-5 ISBN 978-3-031-65808-2 (eBook)
https://doi.org/10.1007/978-3-031-65808-2

© The Editor(s) (if applicable) and The Author(s), under exclusive license to Springer Nature Switzerland AG 2024
This work is subject to copyright. All rights are solely and exclusively licensed by the Publisher, whether the whole or part of the material is concerned, specifically the rights of translation, reprinting, reuse of illustrations, recitation, broadcasting, reproduction on microfilms or in any other physical way, and transmission or information storage and retrieval, electronic adaptation, computer software, or by similar or dissimilar methodology now known or hereafter developed.
The use of general descriptive names, registered names, trademarks, service marks, etc. in this publication does not imply, even in the absence of a specific statement, that such names are exempt from the relevant protective laws and regulations and therefore free for general use.
The publisher, the authors and the editors are safe to assume that the advice and information in this book are believed to be true and accurate at the date of publication. Neither the publisher nor the authors or the editors give a warranty, expressed or implied, with respect to the material contained herein or for any errors or omissions that may have been made. The publisher remains neutral with regard to jurisdictional claims in published maps and institutional affiliations.

This Springer imprint is published by the registered company Springer Nature Switzerland AG
The registered company address is: Gewerbestrasse 11, 6330 Cham, Switzerland

If disposing of this product, please recycle the paper.

Preface

It has been estimated that about 5–10 percent of breast cancers are hereditary. Meanwhile mutation in the BRCA1 or BRCA2 gene are the most common cause of hereditary breast cancer. Triple-negative breast cancer accounts for about 10–15 percent of breast cancers; it refers to the cancer cells that do not have estrogen or progesterone receptors (ER or PR) and also do not make HER2 protein. Immunotherapy is helpful to treat early-stage and metastatic triple-negative breast cancer.

The Interdisciplinary Cancer Research series publishes comprehensive volumes on different cancers. It plans to present the most updated and peer-reviewed interdisciplinary chapters on cancers. This interdisciplinary book series is of special value to researchers and practitioners working on cell biology, immunology, hematology, biochemistry, genetics, oncology and related fields. This is the main concept of Cancer Immunology Project (CIP) and Network of Immunity in Infection, Malignancy and Autoimmunity (NIIMA), which are two active interest groups of the Universal Scientific Education and Research Network (USERN).

The sixth volume of the book entitled *Breast Cancer Genetics, Immunology, and Immunotherapy: An Interdisciplinary Approach* starts with diagnostic, prognostic, and treatment role of miRNAs in triple-negative breast cancer. Approaches to breast cancer immunology and immunotherapy are discussed in Chapter 2. Molecular genetics of breast cancer are the subject of Chapter 3, while the role of interleukins, chemokine receptor, and tumor-infiltrated lymphocytes-TILS in breast cancer is explained in Chapters 4, 5, and 6, respectively. Multidrug-resistant and metastatic breast cancer are presented in Chapters 7, 8, and 9. The other Chapters (10–13) present targeting myeloid determinants, SiRNA loaded in drug delivery nanosystems, nucleic acid and drug dual agent nanoplatforms, and oncolytic virotherapy in treatment of breast cancer.

I hope that this interdisciplinary book will be comprehensible, cogent, and of special value for researchers, oncologists, geneticists, and immunologists who wish to extend their knowledge on breast cancer.

Tehran, Iran — Nima Rezaei

Contents

About the Editor

Nima Rezaei MD, PhD, gained his medical degree (MD) from Tehran University of Medical Sciences and subsequently obtained an MSc in Molecular and Genetic Medicine and a PhD in Clinical Immunology and Human Genetics from the University of Sheffield, UK. He also spent a short-term fellowship of Pediatric Clinical Immunology and Bone Marrow Transplantation in the Newcastle General Hospital. Professor Rezaei is now the Full Professor of Immunology and Vice Dean of Research and Technologies, School of Medicine, Tehran University of Medical Sciences, and the co-founder and Head of the Research Center for Immunodeficiencies. He is also the Founder of Universal Scientific Education and Research Network (USERN). Professor Rezaei has already been the Director of more than two hundreds research projects and has designed and participated in several international collaborative projects. Professor Rezaei is the editor, editorial assistant, or editorial board member of more than 50 international journals. He has edited more than one hundred international books, has presented more than a thousand lectures/posters in congresses/meetings, and has published more than 1500 scientific papers in international journals.

Role of miRNAs in Triple-Negative Breast Cancer: Diagnosis, Prognosis, and Treatment

Alireza Bagheri, Mahsa Zargar, Dorsa Safari, Zahra Aminzade, Amir Hossein Loghman, and Nima Rezaei

Abstract

MicroRNAs (miRNAs) are small noncoding RNAs that regulate cellular normal function and physiology. There is increasing evidence indicating that miRNAs regulate pathways involved in cancer initiation and progression. Recently, the expression patterns of miRNAs have been used as biomarkers in diagnosing

A. Bagheri
Network of Interdisciplinarity in Neonates and Infants (NINI), Universal Scientific Education and Research Network (USERN), Tehran, Iran

Department of Genetics, Faculty of Basic Sciences, Shahrekord University, Shahrekord, Iran

M. Zargar
Network of Interdisciplinarity in Neonates and Infants (NINI), Universal Scientific Education and Research Network (USERN), Tehran, Iran

Department of Medical Genetics, School of Medicine, Tehran University of Medical Sciences, Tehran, Iran

D. Safari
Network of Interdisciplinarity in Neonates and Infants (NINI), Universal Scientific Education and Research Network (USERN), Tehran, Iran

Tehran Medical Sciences Branch, Islamic Azad University, Tehran, Iran

Z. Aminzade
Network of Interdisciplinarity in Neonates and Infants (NINI), Universal Scientific Education and Research Network (USERN), Tehran, Iran

Student Research Committee, School of Medicine, Shahid Beheshti University of Medical Sciences, Tehran, Iran

Endometriosis Research Center, Iran University of Medical Sciences, Tehran, Iran

A. H. Loghman
Network of Interdisciplinarity in Neonates and Infants (NINI), Universal Scientific Education and Research Network (USERN), Tehran, Iran

© The Author(s), under exclusive license to Springer Nature Switzerland AG 2024
Interdisciplinary Cancer Research, https://doi.org/10.1007/16833_2024_222
Published online: 29 May 2024

various cancers including breast cancer. MiRNAs are highly stable and can be found in various human samples including blood, serum, urine, and tissues. Also, the correlation between clinicopathological features and the level of expression makes them more suitable biomarkers. Triple-negative breast cancer (TNBC) is the most malignant subtype of breast cancer. The lack of estrogen receptor (ER), progesterone receptor (PR), and human epidermal growth factor receptor 2 (HER2) receptors has made TNBC treatment more challenging. Increasing the expression of miRNAs with oncogenic activities (OncomiRs) and decreasing the expression of miRNAs with tumor-suppressing features are related to TNBC pathophysiology. Investigating miRNAs relating to TNBC development could lead to finding novel ways in diagnosis, prognosis, and treatment of TNBC; thus, this study aims to review the mechanisms and roles of miRNAs in TNBC.

Keywords

Breast neoplasms · MicroRNAs · Triple-negative breast neoplasms

1 Introduction

Roughly, one third of cancer-related deaths in women are attributed to breast cancer (BC), as reported by a study (Siegel et al. 2019). Based on individual features, it was divided into five categories: progesterone receptor positive (PR+), estrogen receptor positive (ER+), human epidermal growth factor receptor 2 amplification (HER2+), and estrogen and progesterone receptor negative (ER/PR−). ER and PR negative (−) include two subtypes, “basal-like” and “normal-like” (Xu et al. 2020), which were previously called triple-negative breast cancer (TNBC) (Borri and Granaglia 2021). Approximately, 15–20% of all breast cancer cases are classified as TNBC (Morris et al. 2007). It is the most malignant subtype of breast carcinoma and has a shorter survival time than the other subtypes (Dent et al. 2007). It is often observed in premenopausal women under 40 years of age (Morris et al. 2007). It is correlated with poor prognosis, elevated mortality rate, and heightened metastasis (Xu et al. 2020). The lack of specific receptors like ER, PR, and HER2 makes this agent ineffective for targeted treatment and hormone therapy. Therefore, it is a great therapeutic challenge, and it is important to find diagnostic or therapeutic biomarkers

N. Rezaei (✉)
Network of Interdisciplinarity in Neonates and Infants (NINI), Universal Scientific Education and Research Network (USERN), Tehran, Iran

Research Center for Immunodeficiencies, Pediatrics Center of Excellence, Children’s Medical Center Hospital, Tehran University of Medical Sciences, Tehran, Iran

Department of Immunology, School of Medicine, Tehran University of Medical Science, Tehran, Iran
e-mail: rezaei_nima@tums.ac.ir

for TNBC (Xu et al. 2020). Several papers investigated the role of microRNAs (miRNAs) in regulating molecular pathways in TNBC (Borri and Granaglia 2021).

MicroRNAs are small noncoding RNAs, 18 and 22 nucleotides in length, that play a role in the pathogenesis of many diseases, including cancers. They can be encoded by exons, introns, exon/intron junctions, or through their own genes (Xu et al. 2020). Following transcription by RNA polymerase (type II or III), Drosha and DiGeorge syndrome critical region in gene 8 (DCR8) cleave the primary miRNA (pri-miRNA) to generate a precursor miRNA (pre-miRNA) with a hairpin-shaped structure. Transporters then facilitate the movement of pre-miRNAs from the nucleus to the cytoplasm, where they become mature miRNAs (Xu et al. 2020). miRNAs are involved in a range of biological activities that are important for tumorigenesis (Hata and Kashima 2016). Based on their expression patterns, miRNAs are divided into oncogenic or suppressor categories. Overexpressed miRNAs, known as oncomiRs, belong to the oncogenic category, whereas underexpressed miRNAs are categorized as tumor suppressors (Svoronos et al. 2016). miRNAs can degrade or prevent translation of their target mRNAs by interacting with 5′ untranslated region (UTR), 3′ UTR, or coding regions (Lytle et al. 2007). High stability and effective identification in body fluids, targeting several genes simultaneously, and the association between the level of expression and clinicopathological characteristics have made them ideal biomarkers, especially in TNBCs, which are devoid of predictive markers and possible targets for treatment (Kunej et al. 2014; Matamala et al. 2015; Piasecka et al. 2018; Wang et al. 2019a, b, c, d; Xu et al. 2020).

In this study, our goal was to explore the mechanisms and roles of miRNAs in TNBC and advance our comprehension of their significance in diagnosing, predicting outcomes, and treating this specific type of cancer.

2 Role of miRNAs to TNBC

By targeting mRNAs, miRNAs play a role in multiple facets of TNBC growth and maturation, such as cellular proliferation, invasion of tumors and migration, stimulation of new blood vessel formation, escape from immunosurveillance, and chemoresistance. miRNAs involved in these mechanisms can act either as oncogenic miRNAs promoting cancer development or as suppressor miRNAs impeding it. Subsequent sections delve into the exploration of these miRNAs (Fu et al. 2023).

miR-20a is coded by mir-17-92 cluster, and its upregulation in breast cancer, especially in the TNBC subtype, indicates its oncogenic role. This increased miR-20a level is linked to tumor growth. Some studies have indicated that miR-20a is involved in modulating key pathways. For example, miR-20a-5p promotes TNBC cell invasion and migration by targeting the gene *RUNX3*. In addition, miR-20a in TNBC cells controls vital markers, including CDH1, fibronectin, and N-cadherin, by controlling the expression of Twist-1. The loss of CDH1 is a critical event in the process of epithelial-mesenchymal transition (EMT). In contrast, silencing the expression of TGFBR2 terminated the expression of TGF-β, thus inhibiting the mesenchymal-epithelial transition (MET) (Xu et al. 2020). Numerous investigations have revealed that miR-21 shows increased expression in breast cancer compared to that in healthy tissue, and this overexpression indicates that this miRNA probably has an oncogenic function. Researchers have identified several target genes, including *LZTFL1*, which promote cell proliferation and metastasis. In 2017, a study showed that miR-21 can increase proliferation and invasion in MDA-MB-468 by targeting and downregulating tumor suppressors such as PTEN (Xu et al. 2020). In a study performed by Chen et al. (2018a, b), the upregulation of miR-25-3p in TNBC cells was found to suppress the expression of BTG2, thereby activating the AKT and ERK-MAPK signaling pathways (Chen et al. 2018a, b). miR-26a-5p upregulation promotes PTX cytotoxicity in vitro and in vivo (Li et al. 2021). miR-29c normally exerts its growth inhibitory role by directly binding to TGIF2, CREB5, and AKT3. Nevertheless, in TNBC, there is a reduction in miR-29c, leading to a loss of this inhibitory effect (Bhardwaj et al. 2017). miR-30a-5p acts as a suppressor of tumor activity in TNBC and suppresses proliferation, movement, and infiltration by targeting a variety of genes and pathways. The decrease in miR-30a-5p presence in TNBC increases β3 integrin (responsible for cell adhesion), and this increase in expression stimulates Erk/Ets-1 signaling, consequently influencing the development and progression of cancer cells, and metastatic phenotype (Li et al. 2016). The expression of miR-33b diminishes in breast cancer, and this reduction is linked to an increase in lymph node metastasis. miR-33b inhibits self-renewal, proliferation, and cell infiltration of breast cancer cells by focusing on HMGA2, SALL4, and Twist1 (Lin et al. 2015). miR-34a exhibits tumor suppressor properties in both laboratory conditions and in living organisms. Maintaining the expression of this miRNA by targeting oncogene *c-SRC* in TNBC cells prevents cell growth and infiltration, activates senescence, and increases responsiveness to dasatinib (Adams et al. 2016). Additionally, *miR-93* is one of the genes that is significantly more active in TNBC compared to non-TNBC. The elevation of this miRNA in MCF-7 cell lines promoted their proliferation, invasion, and metastasis potential (Hu et al. 2015). In TNBC tissues and cells, the expression of miR-101 is significantly reduced. The suppression of tumor progression and the promotion of apoptosis and increased sensitivity to paclitaxel are achieved by targeting and inhibiting MCL-1 expression by this miRNA (Liu et al. 2015a, b). Upregulation of MiR-105/93-3p by targeting SFRP1 contributes to TNBC cell resistance to cisplatin and stemness (Li et al. 2017a, b, c). miR-124 is downregulated in TNBC cells and tissues and results in the inhibition of proliferation, metastasis, and EMT by targeting the 3' UTR of ZEB2 mRNA (Ji et al. 2019). Low levels of miR-125b, functioning as a tumor suppressor, are linked to poor prognosis and chemoresistance. Knockdown of MAP2K7, as one of the targets of miR-125b, inhibits EMT in Hs578T (TNBC cells) (Xu et al. 2020). Also, miR-125b modulates the Wnt/β-catenin pathway by targeting the 3′ UTR of APC and inhibits EMT, proliferation, and invasion (Nie et al. 2019). MiR-126-3p serves as a tumor suppressor in TNBC cells, inhibiting proliferation, invasion, migration, and angiogenesis by targeting RGS3 mRNA (Hong et al. 2019). miR-129-5p is downregulated in breast cancer tissues and adriamycin-resistant breast cancer cells (MDA-MB-231/ADR and MCF-7/ADR). By targeting SOX2, miR-129-5p reduces adriamycin resistance

(Zeng et al. 2018). miR-130a is strongly downregulated in TNBC tissues compared to non-TNBC tissues. Increased expression of miR-130a suppresses the migration and invasion of TNBC cells by targeting the 3′ UTR of FOSL1 mRNA and suppressing ZO-1 inhibition (Chen et al. 2018a, b). The expression of miR-134 in triple-negative breast cancer (TNBC) is noticeably reduced. Elevating miR-134 expression has shown to be effective in diminishing the expression of STAT5B, Hsp90, and Bcl-2. This reduction leads to decreased proliferation and aggressiveness while also enhancing drug sensitivity, especially in promoting cisplatin-induced apoptosis (O'Brien et al. 2015). Furthermore, the upregulation of miR-142-3p has been linked to heightened sensitivity to PTX treatment in breast cancer. By targeting GNB2 and activating the AKT-mTOR pathway, miR-142-3p serves as a valuable component in improving treatment outcomes (Shi et al. 2023). Studies have emphasized the significance of miR-145 in TNBC, showcasing its decreased expression compared to non-TNBC. This downregulation is closely associated with tumor cell growth, invasion, and metastasis. Notably, miR-145 achieves its effects in TNBC by targeting genes involved in EMT, such as *ARF6*, which plays a role in regulating tumor cell invasion. *ARF6* affects cell-cell adhesion by changing the location of CDH1 (Eades et al. 2015). Additionally, the induction of miR-145-5p expression has displayed inhibitory effects on cell proliferation while inducing apoptosis and increasing sensitivity to cisplatin therapy in MDA-MB-231 cells (García-García et al. 2019). Upregulated miR-149 inhibits metastasis by targeting the cerebrospinal fluid (CSF) (Sánchez-González et al. 2020). In 2019, Bolandghamat Pour et al. showed that miR-154 reduced cell viability and increased sensitivity to doxorubicin by targeting and inhibiting nicotinamide adenine dinucleotide (NAD) (Bolandghamat Pour et al. 2019). The increase in ABCC3 under the influence of miR-181b-2-3p increased the sensitivity of doxorubicin-resistant TNBC cells (Zeng et al. 2020). miR-182-5p upregulates the expression of FBXW7, leading to inhibition of the TLR4/NF-κB pathway. Consequently, this process enhances the proliferation and invasion of TNBC cells (Wu et al. 2020a, b). Chen et al. (2018a, b) found that miR-199a-5p has a suppressive impact on TNBC progression, as it modulates the genes involved in the epithelial mesenchymal transition, such as *ZEB1* and *CDH1*, thereby inhibiting proliferation, migration, and invasion (Xu et al. 2020). The miR-200 family, including miR-200a, miR-200b, and miR-200c, exhibits decreased expression in TNBC cells and tissues. These members target genes such as *CASP-3, CDH1, EPHA2, TWIST, XIAP, and ZEB1/2*, thereby inhibiting proliferation, migration, and EMT and inducing apoptosis. For example, sustained expression of miR-200b suppresses TNBC metastasis by regulating the *ARHGAP18* and *Rho A* genes (Xu et al. 2020). Knockdown of miR-200 leads to the reversal of properties inhibited by cancer stem cells (CSC) and paclitaxel resistance induced by Jagged2 knockdown (Li et al. 2018). miR-203 demonstrates downregulation in TNBC. Induction of this miRNA targets 3′-UTR of BIRC5 and LASP mRNAs and inhibits cell proliferation and migration (Wang et al. 2012). Downregulation of miR-206 is observed in TNBC cell lines and tumor samples. In its normal state, miR-206 acts as a suppressor, targeting CORO1C, hence inhibiting proliferation and invasion and arresting cell cycle in the G1/S phase (Wang et al.

2014). miR-211 experiences significant downregulation in breast cancer. Re-expression of miR-211 results in the suppression of growth, migration, invasion, and cell cycle by targeting the CDC25B 3′ UTR mRNA (Song and Zhao 2015). The excessive presence of microRNA-223 controls the pathway involving mitochondria and reactive oxygen species (ROS) by focusing on HAX-1, a protein that prevents cell death and is found in mitochondria. This regulation enhances the receptiveness of TNBC stem cells to apoptosis induced by tumor necrosis factor (TNF)-related apoptosis-inducing ligand (TRAIL) (Sun et al. 2016). The miRNA known as miR-296-5p, which has tumor-suppressive properties, plays a role in the development of resistance to paclitaxel in TNBC cells. Induced miR-296-5p targets the BOK mRNA, suppressing proliferation, invasion, and migration while impairing paclitaxel-induced apoptosis (Onyeagucha et al. 2016). miR-342-3p decreases viability, migration, and proliferation by targeting MCT1 and increasing changes in lactate efflux (Hidalgo-Miranda et al. 2016). Prior research has indicated a downregulation of miR-361-5p in breast cancer. Overexpressing miR-361-5p suppresses the invasion and migration of TNBC cells by targeting RQCD1 and inhibiting EGFR/PI3K/Akt signaling pathway, as demonstrated in other studies (Han et al. 2019). miR-378 is also downregulated in breast cancer. It suppresses proliferation, invasion, and aggressive phenotypes by targeting the 3′ UTR of Runx1 mRNA (Browne et al. 2016). Inducing the expression of miR-384 by targeting ACVR1 mRNA inhibits migration, proliferation, and breast cancer progression (Wang et al. 2018). miR-423-5p upregulation is effective in dichlorodiammine platinum (Cisplatin) (DDP) resistance. Thus, TNBC cells are resistant to DDP change sensitivity of other breast cancer cells to miR-423-5p-dependent exosomes (Wang et al. 2019a, b, c, d). miR-455-3p, among the overexpressed miRNAs within TNBC cell lines, promotes invasion, cell proliferation, and metastasis potential by targeting the 3′ UTR of tumor suppressor EI24 (Li et al. 2017a, b, c). By targeting TNKS2, miR-490-3p blocks β-catenin signaling, leading to the suppression of growth and invasion in TNBCs (Jia et al. 2016). miR-496 acts as a tumor suppressor, restraining proliferation and invasion in TNBC cells by targeting Del-1 mRNA. Therefore, the downregulation of miR-496 expression can contribute to the progression of TNBC cells (Lee et al. 2021). miR-574-5p attenuates the rate of proliferation, invasion, and epithelial-mesenchymal transition in TNBC cells by targeting SOX2 and BCL11A mRNAs, leading to the inhibition of SKIL/TAZ/CTGF axis (Zhang et al. 2020). miR-613 acts as a tumor suppressive whose inducible expression by targeting the Daam1/RhoA signaling pathway inhibits migration and invasion in TNBC (Xiong et al. 2018). MiR-655 demonstrates its tumor-suppressive effect in TNBC by suppressing EMT by directly inhibiting the expression of PRRX1 (Lv et al. 2016). In 2019, Wang et al. stated that a reduction in the activity of miR-890 may be linked to the increase in the expression of CD147, which promotes malignant transformation in TNBC cells (Wang et al. 2019a, b, c, d). miR-1,296 is downregulated in both TNBC cells and tissues. When this microRNA's expression is triggered to target the 3′ UTR of CCND1 mRNA, it hinders cell proliferation and reduces sensitivity to cisplatin in TNBC cells (Phan et al. 2016). MiR-1301 is overexpressed in TNBC cells and tissues and functions by binding to the 3′ UTR

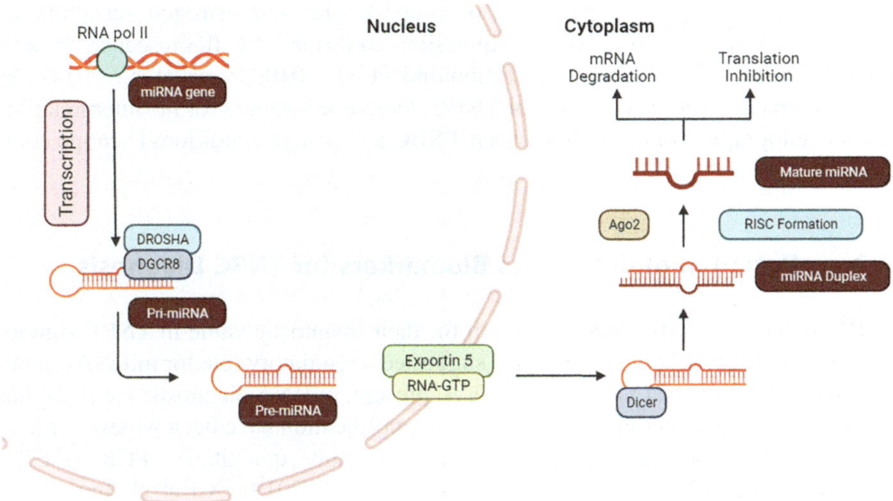

Fig. 1 miRNA synthesis and mechanism of action
The transcription of RNA polymerase II of the miRNA gene starts the synthesis of miRNA. Drosha and the DiGeorge syndrome critical region in gene 8 (*DGCR8*) processes the resultant primary miRNA (pri-miRNA) to create pre-miRNA, which is then carried to the cytoplasm by the nuclear transport receptor exportin-5, which is reliant on RanGTP. To create the miRNA-miRNA duplex, Dicer cuts the pre-miRNA in the cytoplasm. RNA-induced silencing complex (RISC) assembly and pre-miRNA processing are mediated by Argonaute 2 (Ago2). While the other strand of the duplex is broken down, one strand of the duplex stays mature miRNA (miRNA). Mature miRNAs have the ability to either promote target mRNA degradation or inhibit target mRNA translation once they are ready

of EZH2 mRNA, leading to enhanced proliferation, migration, and colony formation (Wu et al. 2018). In the context of TNBC, miR-4306 serves as a tumor-suppressive microRNA. It inhibits cell proliferation, migration, invasion, and angiogenesis by directly targeting SIX1/Cdc42/VEGFA and deactivating their associated signaling pathways (Zhao et al. 2019). The upregulation of miR-5195-3p, by targeting EIF4A2, enhances the sensitivity of TNBC cells and tissues to paclitaxel (Liu et al. 2019) (Fig. 1).

3 Diagnosis

3.1 Current Methods for TNBC Diagnosis

Today, TNBC is detected based on a combination of imaging modalities and immunohistochemistry (IHC). Both of these methods rely on the technician, besides taking a great deal of time. Therefore, new diagnostic methods, including blood-based liquid sampling, PET scan, nano-biosensors, analytical data tools, and digital polymerase chain reaction (PCR) have emerged (Dass et al. 2021). TNBC is conventionally diagnosed by $<1\%$ of progesterone and estrogen receptors in IHC with absent or low HER2 expression confirmed by fluorescence in situ hybridizations (FISH) technique (Hammond et al. 2010; Perez et al. 2014). In terms of imaging for the detection of TNBC, there is a potential for misinterpretation due to radiographic similarities between TNBC and benign conditions (Schopp et al. 2023).

3.2 Potential of miRNAs as Biomarkers for TNBC Diagnosis

miRNAs have recently gained attention for their diagnostic value in TNBC tumors (Sabit et al. 2021). Many authors have suggested a regulatory role for miRNAs in the proliferation of TNBC (Yan et al. 2023). At present, miRNA diagnostic methods that are based on nanomaterials and nucleic acid amplification have been widely used, as well as conventional techniques concluding real-time quantitative PCR (qPCR), microarray technology, and northern blot (Ye et al. 2019). Xu et al. found that sponging of the miR-671-5p and miR-665 functions in the circAR-E2E4-miR-665-STAT3 axis can be used in the diagnosis of TNBC (Xu et al. 2023). Zhong et al. designed an enhanced electrochemiluminescence (ECL) biosensing method using Au nanoparticles/zinc-metal organic framework heterostructure with magnetic spheres/gold shell, providing a highly sensitive identification of miRNA-522 in TNBC tumor tissues (Zhong et al. 2023). Wang et al. also reported that using a novel nano-biosensor based on Cu nanoclusters, even low expression levels of miRNA-585–3p in TNBC tumor tissues are detectable (Wang et al. 2023). Downregulation of miRNA-200c in TNBC patients would be promising in the detection of TNBC tumors (Niedźwiecki et al. 2018). A study by Berber et al. shows that the underexpression of miRNA-200c and miRNA-205 in TNBC patients would also predict lymph node metastasis (Berber et al. 2014). Regarding the miRNA 200 family (miRNA-429, miRNA-141, miRNA-200a, miRNA-200c, and miRNA-200b), other members, such as miR-141, have been found to take part in the malignant transformation of breast tumor cells as a result of interaction with circular RNAs (Guz et al. 2023). Upregulation of both miRNA-105 and miRNA-93-3p activates the signaling pathway of Wnt/β-catenin and leads to metastasis, cell stemness, resistance to chemotherapy, and poor survival; overall, these biomarkers are suitable for early or late detection of TNBC (Li et al. 2017a, b, c). A meta-analysis in 2021 demonstrated that overexpression of miR-93, miR-19a, miR-210, and miR-19b in TNBC is significantly correlated with worse overall survival (Qattan et al. 2021). A combination of miR-21 and miR-210 levels in breast cancer cell lines has been used to identify TNBC (Mao et al. 2023).

Presenting in serum, blood, and urine, as well as tissue samples, miRNAs could be detected using minimally invasive techniques (Kahraman et al. 2018). Ritter et al. discovered that the expression levels of miRNA-17, miRNA-19b, and miRNA-30b in the serum of TNBC patients are associated with early response to neoadjuvant chemotherapy and may be used as a diagnostic biomarker for post-therapeutic screening in TNBC patients (Ritter et al. 2020). A longitudinal study of

146 TNBC patients, who had been followed up by a median duration of 68 months, demonstrated that the upregulation of miRNA-27a, miRNA-206, and miRNA-214 were independently related to diminished patient survival (Liu et al. 2023). Shin et al. observed lower expression of miRNA-21, miRNA-16, and miRNA-199a-5p in patients with TNBC, while only miRNA-199a-5p was related to the disease stage and would be a potential biomarker for early diagnosis of TNBC (Shin et al. 2015). Overexpression of miR-10b, miR-9, and miR-17-5p has been shown as diagnostic biomarker for TNBC (Malla et al. 2019). Kahraman et al. identified a seven-miRNA signature for early diagnosis of basal-like TNBC with 74.2% specificity, 83.8% sensitivity, and an area under the receiver operating characteristic (ROC) curve (AUC) of 0.814 (Kahraman et al. 2018). Naorem et al. discovered a six-miRNA meta-signature in 133 TNBC samples which was highly dysregulated and could provide promising biomarkers for TNBC diagnosis with a great accuracy. Downregulation was observed in two miRNAs (miRNA-449a and miRNA-190b), while four miRNAs were overexpressed (miRNA-18a-5p, miRNA-135b-5p, miRNA-522-3p, and miRNA-9-5p) (Naorem et al. 2019). A review study by Chakkaravarthi et al. indicated that promising miRNA candidates for the detection of TNBC include miRNA-9–1–5p, miRNA-155, miRNA-429, miRNA-182–5p, miRNA-200a, miRNA-200b, miRNA-195, miRNA-145–5p, miRNA-506, and miRNA-22–3p (Chakkaravarthi et al. 2023). Wijesinghe et al. developed a highly sensitive fluorescence sensor for the detection of miR-342-3p in TNBC cells (Wijesinghe et al. 2022). Several suggested miRNAs for TNBC diagnosis are indexed in Table 1.

3.3 Exosomal miRNAs as a Biomarker for TNBC

Based on the International Society for Extracellular Vesicles guidelines, extracellular vesicles (EVs) are defined as cell-derived particles that are enclosed with a lipid bilayer and cannot replicate. EVs are subdivided into exosomes (endosome originated), microvesicles (cell membrane originated), and many other particle types (Théry et al. 2018). Exosomal miRNAs, which are miRNA-containing exosomes, have a crucial role in cellular communication in cancers, particularly TNBC (Encarnación-Medina et al. 2023). Accordingly, recognition of exo-miRNAs might be superior to circulatory cell-free miRNAs (Stevic et al. 2018). Several methods were applied for exo-miRNA detection in body fluids, such as next-generation sequencing (NGS) (Wu et al. 2020a, b), TaqMan microRNA array, as well as RT-qPCR. Some exo-miRNAs related to TNBC diagnosis are organized in Table 2. Considering TNBC tumor characteristics, exo-miR-374 significantly correlated with higher tumor size (Stevic et al. 2018). Todorova et al. demonstrated that TNBC patients have lower expression of exosomal miR-182 compared with other breast carcinoma subtypes, which also predicted a non-complete pathological response of TNBC to neoadjuvant chemotherapy (Todorova et al. 2022). High levels of exosomal miR-222 have been reported in TNBC and luminal B subtypes compared to luminal A (Lee et al. 2019, Rodríguez-Martínez et al. 2019).

Table 1 Diagnostic miRNAs in TNBC patients

Author, year	No. of patients with TNBC	Patient sample	Overexpressed miRNA	Underexpressed miRNA	Measurement technique	References
Liu, 2023	146	Tumor tissue	miR-27a, miR-206, and miR-214	–	qRT-PCR	Liu et al. (2023)
Triantafyllou, 2022	8	Urine	–	miR18a, miR19b, miR30b, miR-222b, miR-320c	qRT-PCR	Triantafyllou et al. (2022)
		Serum	let7a, let7e, miR21B	miR15a, miR17, miR18a, miR19b, miR30b		
Qattan, 2021	36	Plasma	miR19a, miR19b, miR93, miR25, miR22 and miR210	MiR199a	qRT-PCR	Qattan et al. (2021)
Niedźwiecki, 2018	9	Serum	–	miR 200c	qRT-PCR	Niedźwiecki et al. (2018)
Kahraman, 2018	21	Blood	miR-126-5p, miR-144-3p, miR-144-5p, miR-301a-3p, miR-101-3p, miR-126-3p, miR-664b-5p	–	Microarray, qRT-PCR	Kahraman et al. (2018)
Li, 2017	74	Plasma	miR93 and miR105	–	qRT-PCR	Li et al. (2017)
Shin, 2015	5	Plasma	–	miR-16, miR-21 miR-199a-5p	Microarray, qRT-PCR	Shin et al. (2015)

Table 2 Extracellular vesicle miRNAs associated with TNBC

Author, year	Type of extracellular vesicle	No. of patients with TNBC	No. of validation TNBC patients	Patient sample	Overexpressed exo-miRNA	Underexpressed exo-miRNA	Measurement technique	References
Todorova, 2022	Exosome	6	–	Plasma	–	mir-182	NGS	Todorova et al. (2022)
Wu, 2020	Exosome	6	20	Plasma	miR-148a-5p, miR-210a-3p, miR-200a-5p, miR-378a-3p, miR-7,110-5p, miR-483-5p	miR-150-5p, miR-92b-3p	NGS, RT-PCR	Wu et al. (2020a, b)
Stevic, 2018	Exosome	15	224	Plasma	miR-376c, miR-335, miR-382, miR-433, miR-628, miR-422a	miRNA-365, miRNA-27b, miRNA-27a, miRNA-16, miRNA-143	TaqMan MicroRNA Assays	Stevic et al. (2018)

3.4 Challenges and Limitations in miRNA-Based Diagnosis

miRNA signaling pathways are considerably influenced by neoadjuvant chemotherapy. Therefore, it might predict the response to neoadjuvant chemotherapy (Kahraman et al. 2018). Low quantities of miRNAs in circulation may require large entities of samples, while some authors have overcome this limitation by isolating miRNAs by an enrichment step before profiling the miRNA expression (Hamam et al. 2016). The use of ethylenediamine tetraacetic acid (EDTA), serving as an anticoagulant in blood sampling tubes, significantly alters miRNA expression with intense effect over prolonged exposure (Leidinger et al. 2015). Conventional methods for detecting miRNAs such as microarray allow for a high-throughput diagnosis of miRNAs while being limited by poor specificity in cases of similar sequences (Ye et al. 2019). Even newly developed methods to detect miRNAs require additional amplification methods to increase their sensitivity due to the nature of miRNAs which are low concentrated in samples of patients susceptible to degradation, short in sequence, and similar within the same miRNA family (Wijesinghe et al. 2022).

4 Prognosis

4.1 RNA Expression as Prognostic Biomarkers in TNBC

MicroRNAs, also known as noncoding RNAs, are tiny (18–21 nucleotides) that have the ability to modify posttranscriptionally the expression of genes. Compared to mRNA, miRNAs have the following benefits for expression analysis: They are useful for therapeutic reasons, particularly in regard to TNBC diagnosis and therapy, because (1) they are tiny and hence more stable and (2) they can be retrieved from frozen tissue, formalin-fixed paraffin-embedded tissues, as well as blood, with little to no degradation. MicroRNAs have been associated with TNBC, three receptors (ER, PR, and HER2), and/or metastasis in prognostic studies. (Mathe et al. 2015). Breast cancer exhibits heterogeneity, with several molecular subgroups within the illness. Globally, TNBC is the second most prevalent subtype of BC in women, representing 15–20% of all BCs (Kumar et al. 2012). Numerous research in recent years have demonstrated that miRNAs are expressed differently in BC (Nassar et al. 2020; Wu and Chu 2022). In TNBC, information on RNA and microRNA (miRNA) biomarkers is changing. Here are a few recently published works in this field. MiRNAs are circulating, noncoding RNA molecules that range in length from 17 to 27 nucleotides. They control the posttranscriptional expression of genes, including tumor suppressors and oncogenes, which are involved in the oncogenic pathway (Paszek et al. 2017). Using high-throughput sequencing methods, 28,000 mature miRNAs have been found. MiRNAs have the potential to be explored as prognostic biomarkers in noninvasive samples including blood, serum, and urine because of their stability. Research has demonstrated that dysregulated miRNAs have a role in BC carcinogenesis. Particular miRNA signatures seem to be predictive for triple-negative breast cancer (Yang et al. 2015; Paszek et al. 2017; Sugita et al. 2019).

4.2 Predicting Disease Outcomes Using miRNAs

For instance, it has been demonstrated that in TNBC patients, reduced expression of miR-155 is predictive of a bad outcome, whereas higher expression of miR-21, miR-27a/b, miR-210, and miR-454 was linked to a shorter overall survival (OS). Similar to this, shorter disease-free survival (DFS) was associated with higher levels of miR-454 and lower expression of miR-374a/b. It was discovered that additional miRNA panels were linked to chemoresistance; TNBC tissue samples from patients who did not react to neoadjuvant chemotherapy had higher expression of miR-181a. A pilot study that used bloodborne miRNA signatures from 21 basal-like TNBC cases that received neoadjuvant therapy also revealed 321 deregulated miRNAs, including miR-34a, when comparing expression before and after treatment, and it discovered that complete responders tended to have higher levels of miRNA following platinum-based chemotherapy (Kahraman et al. 2018).

Furthermore, a pCR in TNBC may be predicted by certain miRNA profiles, as the GeparSixto study showed (Stevic et al. 2018). Transcripts longer than 200 nucleotides, known as long noncoding RNAs (lncRNAs), may not be translated into proteins. They carry out regulatory tasks in several cancer biology hallmarks, just as miRNAs. Many cancer forms, including TNBC, have abnormalities in their lncRNAs. TNBC tissues have been revealed to have higher levels of the lncRNA known as highly upregulated in liver cancer (HULC), which has been linked to worse clinical outcomes. Additionally, new preclinical research has identified MALAT1 as a possible biomarker in TNBC, aiding in the prediction of prognosis and metastasis. MALAT1 is a highly conserved long noncoding RNA (Sukumar et al. 2021). In patients with TNBC, lower expression of miR-155 indicated a poor overall survival, but higher expression of miR-21, miR-27a/b, miR-210, and miR-454 was linked to a shorter overall survival (Kleivi Sahlberg et al. 2015; Thakur et al. 2016; Liu et al. 2017). Similarly, reduced disease-free survival was associated with higher levels of miR-454 and decreased expression of miR-374a/b/5p (Cascione et al. 2013; Cao et al. 2016). Particularly for BC, miR-155 seems to be a good predictive tool in various cancer types. MiR-155 is generally regarded as an oncogenic miR that increases the aggressiveness of breast cancer, angiogenesis, and tumor development (Grimaldi et al. 2020). One of the first oncomiRs to be found to be overexpressed in breast cancer was miR-21, which has been linked to lymph node metastases, an advanced clinical stage, and a poor patient prognosis (Yan et al. 2008). One of the miRNAs that is markedly elevated in TNBC is miR-27a/b, and its high expression has been linked to a poor overall patient survival rate (Gong et al. 2021). Tumor progression and a poorer prognosis are highly correlated with lower expression of miR-34a/34b/34c (Yu et al. 2014). TNBC has a high expression level of miR-210, and a worse prognosis for the condition has been linked to this overexpression (Bar et al. 2017) (Table 3).

Table 3 miRNAs which are predictive or prognostic

MiRNA	Result	Predictive/ prognostic	References
miR-27b-3p	When down in TNBC is linked to a favorable prognosis	Prognostic	Liu et al. (2015a, b)
miR-374b-5p, miR-218-5p, miR-126-3p	When up in TNBC is linked to a favorable prognosis	Prognostic	Liu et al. (2015a, b)
miR-10b-5p, miR-451a, miR-125b-5p, miR-31-5p, miR-195-5p, miR-130a-3p	Downregulated in TNBC/ associated with chemoresistance	Predictive	Ouyang et al. (2014)
miR-155-5p, miR-21-3p, miR-181a-5p, miR-181b-5p, miR-183-5p	Upregulated in TNBC/ associated with chemoresistance	Predictive	Ouyang et al. (2014)
miR-342, miR-150	miRNAs for a favorable outcome	Prognostic	Buffa et al. (2011)
miR-27b, miR-210, miR-144	miRNAs linked to a bad outcome	Prognostic	Buffa et al. (2011)

5 Treatment

5.1 Current Treatment Option for TNBC: Different Strategies for TNBC Therapeutics

Among all breast cancer subtypes, TNBC is uniquely known as the most heterogeneous and extremely aggressive phenotype of this malignancy, with the worst outcomes and the lowest survival rate. From a therapeutic standpoint, TNBC treatment faces several challenges due to the lack of estrogen receptor (ER), progesterone (PR), and human epidermal growth factor receptors (HER2), high molecular heterogeneity, poor prognosis, advanced stages at the time of diagnosis, high probability of distant metastasis, and early locoregional recurrence (Qattan 2020).

Since TNBC patients do not benefit from endocrine and HER2-targeted therapies, the primary treatment options are currently limited to chemotherapy, surgery, radiation, and immunotherapy. To decide on the choice of treatment or a combination of therapeutic methods in TNBC, clinical and pathological parameters such as patient age, tumor stage, presence or absence of metastasis, tumor histology and molecular subtype, and tumor grade should be taken into consideration (Obidiro et al. 2023).

5.2 Surgical Management

Over the last three decades, removing the tumor through surgery has been one of the key approaches in treating breast cancer including TNBC. Surgical techniques have considerably evolved, since the first introduction of radical mastectomy in 1882 by

William Halsted. Vast advancements in early detection, radiotherapy, neoadjuvant chemotherapy, etc. significantly changed the timing and extent of modern surgery compared to its early forms, leading to a major shift toward breast-conserving surgery (BCS) (including lumpectomy and quadrantectomy) and oncoplastic surgery instead of simple or radical mastectomies (Halsted et al. 2014).

Several studies and long-term trials have compared the effectiveness and clinical benefits of radical mastectomy and breast-conserving surgery (BCS) in treating TNBC. The findings show that patients who undergo BCS have similar long-term survival rates as those who undergo radical mastectomy. Locoregional treatment of breast cancer by BCS alongside radiation does not lead to a worse prognosis than radical mastectomy, and in the early stages of TNBC, it may also result in better overall survival and breast cancer–specific survival outcomes (Saifi et al. 2022; Obidiro et al. 2023). However, a major challenge in treating TNBC, regardless of the surgical approach, is the high risk of local and distant recurrence. While BCS, including postoperative radiotherapy, may raise concerns about local failure, current data implicates that it is not associated with a higher frequency of early local recurrence compared to mastectomy. In fact, in some cases, early local recurrence may be lower in TNBC who undergo BCS compared to non-TNBC. Additionally, more extensive surgical interventions do not necessarily reduce the likelihood of tumor recurrence (Di Leone et al. 2023).

Organ preservation surgery is becoming increasingly popular among breast cancer patients due to its ability to provide more favorable cosmetic results and positively impact the patient's overall quality of life. As a result, it is crucial for the breast surgeon to carefully assess and select the appropriate treatment method for each patient, taking into account various factors such as age, tumor size, tumor stage, nodal status, lymphovascular invasion, etc., in order to achieve the best possible oncological outcome (Kaviani et al. 2013).

5.3 Radiation

When considering locoregional therapy in TNBC, the importance of radiotherapy (RT) for optimal long-term outcomes is well established. In the setting of both mastectomy and BCS, post-mastectomy RT is accompanied by reduced risk of recurrence and increased overall survival. Based on current guidelines and protocols, adjuvant RT should be offered to all patients after BCS since it significantly improves survival benefits and local control after BCS (de Faria Bessa and Marta 2022; Obidiro et al. 2023).

5.4 Chemotherapy

Despite the poor prognosis and lower survival rate, TNBC is highly responsive to chemotherapy (CT), known as the TNBC paradox. This phenomenon makes cytotoxic chemotherapy the main standard therapeutic option in both early and locally advanced TNBC. Common chemotherapy regimens in adjuvant and neoadjuvant chemotherapy (NACT) settings include anthracyclines as DNA-intercalating agents and topoisomerase II blockers (e.g., doxorubicin), alkylators (e.g., cyclophosphamide), an anti-microtubule agent taxane (e.g., paclitaxel), and anti-metabolite fluorouracil (5-FU). However, no standard regimen is currently available for metastatic or relapsed disease. Platinum-based NACT could also be considered in BReast CAncer gene (BRCA)-mutated TNBC and metastatic patients to obtain more favorable outcomes. Capecitabine as adjuvant CT could also be beneficial in patients who did not achieve pathological complete response (PCR) (defined as the complete disappearance of invasive cancer in the breast and the absence of tumor in the axillary lymph nodes) after NACT (Obidiro et al. 2023).

For many years, NACT has been the primary standard for managing TNBC, particularly early-stage disease. The shrinkage and downstaging of the tumor and axillary lesions following NACT lead to better response to BCS and reduced need for axillary resection. TNBC patients are more likely to achieve PCR after NACT than non-TNBC patients, leading to improved overall survival and disease-free survival. Definitive surgery combined with adjuvant chemotherapy also results in improved overall survival and lower risk of recurrent disease (Yin et al. 2022; Obidiro et al. 2023).

5.5 Targeted Therapy

The FDA has recently approved poly (ADP-ribose) polymerase inhibitors (PARPis) as a new class of drug for treating locally advanced and metastatic breast cancer harboring BRCA1/2 mutations. PARP inhibitors cause the accumulation of unrepaired DNA in cells lacking functional BRCA, leading to the initiation of apoptosis and cell death processes. Olaparib and talazoparib are PARP inhibitors clinically approved for use as monotherapy in advanced and metastatic TNBC (Obidiro et al. 2023).

Although TNBC does not respond to targeted therapies used in other subtypes of breast cancer due to the absence of ER, PR, and HER2, there are other potential targets such as androgen receptors (AR) and glucocorticoid receptors (GR) in TNBC. Approximately, 10–35% of TNBC express AR, particularly the luminal androgen receptor (LAR) subtype. Based on the results of recent trials, targeting AR is accompanied by clinical benefits and might be a potential treatment strategy (Won and Spruck 2020).

In addition to hormone receptors, epidermal growth factor receptor (EGFR) and vascular endothelial growth factor (VEGF) receptors are promising targets for TNBC therapy. Both proteins play roles in cell proliferation, differentiation, angiogenesis, metastasis, and apoptosis and are being studied in several trials as target receptors for TNBC (Li et al. 2022).

5.6 Immunotherapy

From an immunotherapeutic perspective, TNBC is more immunologic than other subtypes of breast cancers with more tumor-infiltrating lymphocytes (TILs), higher tumor mutational burden (TMB), and elevated level of programmed cell death-ligand 1 (PD-L1) expression. Atezolizumab is an FDA-approved immune checkpoint inhibitor targeting PD-L1 on the surface of TNBC cells. Ongoing research is exploring the effectiveness and safety of other monoclonal antibodies targeting PD-1 and its ligand, PD-L1 in TNBC management. Additionally, antibody-drug conjugates (ADC) and P13/AKT/mTOR inhibitors are under study for TNBC therapy (Li et al. 2022; Obidiro et al. 2023).

5.7 Therapeutic Potential of Targeting miRNAs in TNBC

As previously noted, dysregulation of miRNA is well recognized in numerous diseases and cancers including TNBC. Concerning cancer, miRNA molecules can be categorized into two distinct groups based on their established functions: cancer-promoting miRNAs, called onco-miRNAs, and cancer-inhibiting miRNAs, referred to as tumor suppressors. Downregulation of tumor-suppressive miRNAs and overexpression of onco-miRNAs are recognized factors in various stages of tumor progression including tumor initiation, cell proliferation, immune cell infiltration, invasive behavior, metastasis, and the resistance of neoplastic cells to chemotherapeutic drugs (Qattan 2020). With miRNAs influencing different stages of tumorigenesis, miRNA-based therapies have become a dynamic research area, aiming to uncover novel targets for advancing cancer treatment including TNBC. miRNA-based therapies rely on two primary strategies: the use of miRNA antagonists or miRNA or non-miRNA targets to reduce the expression or level of once miRNAs and restore the compromised function of tumor suppressor miRNAs (Tang et al. 2019).

Several investigations demonstrated that targeting miRNAs in both tumor cells and tumor stroma potentially interferes with cell division and differentiation and tumor progression. For instance, miRNA-29c is a miRNA with anti-proliferative and anti-clonogenic effect in normal cells; its function is lost in TNBC tumorigenesis. Targeting miRNA-29c and its associated downstream signaling pathway could be a promising approach in the treatment of TNBC at preneoplastic (Bhardwaj et al. 2017). Similarly, miRNA-544 overexpression in TNBC is associated with suppressed tumor growth and impaired cell proliferation and migration in both in vitro and in vivo models (Grimaldi et al. 2021). Modifying the expression of miRNAs not only affects tumor cells but also has a diverse impact on tumor microenvironment (TME) elements such as endothelial cells, tumor-associated fibroblasts, tumor-infiltrating immune cells, etc. (Grimaldi et al. 2021). Increased level of miRNA-142-5p is associated with hypoxia-induced lysyl oxidase (LOX) in TNBC stroma which is strongly related to poor survival and chemotherapy resistance in TNBC patients (Saatci et al. 2020). miRNA-34a expression targets the immune component of TNBC TME by suppressing interleukin-6 (IL-6) production and macrophage cell differentiation (Qattan 2020). miRNA-based therapeutic approach can also control tumor aggressiveness by limiting cell migration and metastasis. Tumor suppressor miRNA-34a, which is downregulated in TNBC, relatively targets p53 and its signaling pathway and inhibits metastasis. miRNA-31 is also an anti-metastatic miRNA that is downregulated in TNBC. miR-146a-5p, miR-26a, miR-136, and miR-136 are additional miRNAs with known inhibitory effects in TNBC metastasis (Tang et al. 2019). Conversely, miRNA-181a overexpression is associated with a more aggressive form of TNBC with higher tumor grade and positive lymph node metastasis (Sabit et al. 2021).

TNBC radiosensitivity and chemosensitivity also appear to be under the control of specific miRNAs. miRNA-302 showed a promising effect as a potential radiotherapy sensitizer in radioresistant TNBC, in both in vitro and in vivo settings (Liang et al. 2013). Studies also demonstrated that miRNA-155−5p, miRNA-21-3p, miRNA-181a-5p, miRNA-181b-5p, and miRNA-183−5p are upregulated in Doxorubicin resistant TNBC, while the expression of other miRNAs such as miRNA-10b-5p, miRNA-129-5p, miRNA-451a, miRNA-125b-5p, miRNA-200c, miRNA-31−5p, and miRNA-195−5p is significantly reduced (Sabit et al. 2021; Fu et al. 2023). Since chemoresistance is the leading cause of treatment failure in TNBC, extensive research is needed to identify the optimal miRNA targets to effectively mimic or reverse chemoresistance and overcome this obstacle.

Considering the extensive advancements in genomics and RNA-based technologies, assessment of miRNA expression abnormalities in each TNBC patient, and focus on distinctive disruptions of each individual can open up a new intriguing area in precision oncology (Fu et al. 2023). It should be noted that although miRNAs can be considered targets in various aspects of TNBC treatment, due to the high instability, low specificity, short half-life, and rapid metabolization through the kidney, selecting an appropriate delivery system into the cancer tissue is highly crucial (Baumann and Winkler 2014).

5.8 Clinical Trials and Research on miRNA-Based Treatments

Despite the involvement and presence of miRNA abnormal expression in all stages of TNBC initiation and progression, the development of miRNA-based therapeutics is still in the preclinical stages, and the clinical research exploring the effects of miRNA-based therapeutic in cancer is in phase I and II of trials (Wang et al. 2020). Most of the clinical trials studying human subjects focus on miRNA profile changes during cancer treatment to utilize miRNA as a predictive element for treatment response and patient prognosis (Kahraman et al. 2018; Wang et al. 2019a, b, c, d). Ongoing clinical trials investigating miRNA-based drug design for TNBC and other malignancies are mainly focused on four major domains including miRNA mimetics, small interfering RNA (siRNA), anti-miRNA oligonucleotide (AMO) inhibitors, and anti-sense oligonucleotides (ASO) (Sabit et al. 2021).

miRNA-34a is one of the main targets in antitumor miRNA-based therapies with favorable outcomes in both in vitro and in vivo experiments. When miRNA-34a is replaced in TNBC cells, it can promote tumor growth inhibition and downregulated oncogenic signaling pathways in both TNBC cell lines and models (Adams et al. 2016). The co-administration of miRNA-34a and doxorubicin using nanoparticles demonstrates a synergistic tumor suppression effect in both in vitro and in vivo environments (Deng et al. 2014). Furthermore, delayed tumor growth and impaired cell motility and invasion were observed when miR-544 and miR-603 were reintroduced to the TNBC cell lines and mice xenograft tumor models. Among miRNAs with oncogenic roles, miRNA-21 has been extensively studied in the anti-miRNA therapeutic approaches. The knockdown of miRNA-21 plays a regulatory role in signaling pathways connected to both tumor growth and apoptosis (Wang et al. 2020). Considering the significant heterogenicity of TNBC and the challenges along with its treatment course, integrating miRNA-based therapies into conventional treatments appears to be promising. Therefore, advancing the studies beyond phase II is crucial.

6 Challenges and Future Directions

Cancer treatment paradigms will undergo a transformation with the use of precision medicine. Remarkably, treating malignancies that are genetically related can bring this ambition closer to reality. Triple-negative breast cancer (TNBC) is a good candidate for precision medicine because of its heterogeneous character.

The human genome project made it possible to uncover genetic variants linked to illness and to better understand the structure and function of human genes. Genetic technology has come a long way since the human genome project, and there has been constant development toward more potent and affordable methods (Alvarez-Erviti et al. 2011).

The advent of the chip-based microarray made it possible to conduct genome-wide association studies for millions of SNPs and early gene expression profiling investigations. However, the majority of non-SNP applications were later superseded by RNA-derived cDNA and genomic DNA NGS (Biswas et al. 2019).

With the understanding of illness genetics, doctors may make better judgments on patient treatment, resulting in higher quality care and fewer pointless tests or procedures (Chaudhary et al. 2020).

Precision medicine has made strides, but regular clinical deployment still faces several challenges. The huge number of variations in the human genome that are of unknown relevance and the challenge of obtaining large enough sample numbers to examine their potential roles are major obstacles to the therapeutic use of genomic data (Dioufa et al. 2017).

The lack of facilities and training in clinical settings is another obstacle to the development of precision medicine (Gong et al. 2019).

Finally, before incorporating genetic data into the clinical context entirely, difficulties related to cost and reimbursement must be resolved (Haffty et al. 2006).

Creating a technological infrastructure, implementing outcome-based payment guidelines, educating and promoting staff in clinical settings, and encouraging patient participation in precision medicine are some ways to overcome these obstacles. Precision medicine has a bright future, but before it is completely incorporated into the healthcare system, it must overcome a number of challenges (Higginbotham et al. 2016).

A very variable and aggressive form of breast cancer is called triple-negative breast cancer (TNBC). More people die from TNBC than from other forms of breast cancer, and the disease recurs more frequently, particularly in the first 5 years of life (Mollah and Varamini 2021).

Although there are a few targeted treatments available at the moment for TNBC, chemotherapy is still the backbone of treatment, with immunotherapy being a more recent and significant addition (Perou et al. 2000; Poggio et al. 2019; Piombino et al. 2021).

A major step forward would be to optimize TNBC therapy based on genetic and maybe immunological characteristics.

6.1 Recently Licensed TNBC Treatments by the FDA

For most TNBC patients, chemotherapy and surgery continue to be the mainstay of treatment. A few kinds of medications, including antibody drug conjugates (ADC), PARP inhibitors (PARPis), and immune checkpoint blockers (ICBs), have, nonetheless, shown to be clearly beneficial in TNBC patients and have even been approved by the FDA in some situations. However, response rates to the new drugs are not constant, and a significant effort in precision oncology will be directed at forecasting the response to these novel classes of therapies (Hossain et al. 2021).

Lipid bilayers are assembled into nanoscale extracellular vesicles known as exosomes. Exosomes are derived from many kinds of donor cells and include a complex variety of materials, such as proteins and several kinds of nucleic acids (DNA, siRNA, LnRNA, and miRNA). Once the exosomes attach to recipient cells, they release their contents, which carry out their biological purposes. Exosomes have been implicated in several aspects of TNBC carcinogenesis, such as immune evasion, treatment resistance, angiogenesis, tumor start, and metastasis. Exosomes, on the other hand, may be useful indicators for TNBC diagnosis, tracking, and treatment. Even more intriguingly, exosomes can be used to deliver drugs at the nanoscale with a focus on TNBC (Weaver et al. 2022) (Table 4).

Given TNBC's molecular heterogeneity and frequent multiclonality, precision combination treatments appear to have a higher chance of success than monotherapy approaches. Promising outcomes have been shown when regular chemotherapy or other targeted drugs are used in conjunction with targeted agents.

Clinical trials that are more accurate, adaptable, and based on biomarkers will be essential to the development of TNBC developing therapies. With somatic and, sometimes, germline genomics driving the rapid development of targeted therapy and immunotherapy, we may reasonably expect more accurate, effective, and less toxic treatment of TNBC soon. To fully realize the potential of precision medicine in TNBC, more research is essential.

Table 4 Clinical application of exosomes in TNBC

Targets	Application	Source	Aim	References
Circulating exosomal miRNAs	TNBC	TNBC patients	As indicators to forecast the effectiveness of TNBC therapy as diagnostic tools	Todorova et al. (2022)
Exosomal miRNAs which may modulate chemoresistance	TNBC	TNBC patients	Biomarkers of pembrolizumab response in TNBC patients	Stevic et al. (2018)
Profiles of exosomal miRNA	TNBC	Patients with TNBC receiving neoadjuvant chemotherapy	Establishing biomarkers	Sueta et al. (2021)

7 Conclusion

This review concludes by delving into the complex terrain of triple-negative breast cancer (TNBC) and emphasizing the critical role that microRNAs (miRNAs) play in TNBC diagnosis, prognosis, and treatment. By thoroughly examining miRNA expression patterns, signaling pathways, and their effects on TNBC development, our analysis highlights the importance of these small RNA molecules in the complex regulatory network of gene expression associated with TNBC.

The diagnostic landscape, described in this chapter, highlights the potential of miRNAs as biomarkers and considers the difficulties and existing approaches in TNBC diagnosis. miRNAs emerge as major factors in determining the future of TNBC diagnostics, providing a sophisticated knowledge of the illness at the molecular level, despite the obstacles and restrictions in their use.

MiRNAs demonstrate their prognostic potential, offering important insights into the course of illness. The ramifications for treatment choices and patient care highlight the revolutionary potential of miRNA-based prognostic instruments, offering a paradigm change in the customization of therapeutic strategies for specific patients.

The therapy section explores the therapeutic potential of targeting miRNAs in addition to going over the current approaches for TNBC therapies. A new age in TNBC therapy appears to be approaching, one that promises more tailored and efficient interventions, as we observe the explosion of clinical trials and research devoted to miRNA-based treatments.

But even as we move closer to these bright futures, it is important to recognize that there are still obstacles in the way of miRNA research for TNBC. By highlighting the obstacles that still need to be surmounted, this chapter opens the door for more study and creativity. Emerging technology and promising advancements provide a look into an exciting future where miRNAs might be a key player in revolutionizing TNBC treatment.

Essentially, this analysis provides a thorough road map that clarifies the many roles that miRNAs play in TNBC. It is impossible to exaggerate the significance of miRNAs in the setting of TNBC nor to deny their potential influence on the direction of TNBC treatment in the future. The quest to understand the complexity of TNBC continues as we stand on the brink of a new era in cancer research, driven by the promise of miRNAs. Our ultimate goal is to translate these discoveries into real advances in patient outcomes.

Acknowledgments We would like to thank Dr. Noosha Samieefar for her valuable comments on our chapter.

References

Adams BD, Wali VB, Cheng CJ, Inukai S, Booth CJ, Agarwal S, Rimm DL, Győrffy B, Santarpia L, Pusztai L, Saltzman WM, Slack FJ (2016) miR-34a silences c-SRC to attenuate tumor growth in triple-negative breast cancer. Cancer Res 76:927–939

Alvarez-Erviti L, Seow Y, Yin H, Betts C, Lakhal S, Wood MJ (2011) Delivery of siRNA to the mouse brain by systemic injection of targeted exosomes. Nat Biotechnol 29:341–345

Bar I, Merhi A, Abdel-Sater F, Ben Addi A, Sollennita S, Canon J-L, Delrée P (2017) The MicroRNA miR-210 is expressed by cancer cells but also by the tumor microenvironment in triple-negative breast cancer. J Histochem Cytochem 65:335–346

Baumann V, Winkler J (2014) miRNA-based therapies: strategies and delivery platforms for oligonucleotide and non-oligonucleotide agents. Future Med Chem 6:1967–1984

Berber U, Yilmaz I, Narli G, Haholu A, Kucukodaci Z, Demirel D (2014) miR-205 and miR-200c: predictive micro RNAs for lymph node metastasis in triple negative breast cancer. J Breast Cancer 17:143–148

Bhardwaj A, Singh H, Rajapakshe K, Tachibana K, Ganesan N, Pan Y, Gunaratne PH, Coarfa C, Bedrosian I (2017) Regulation of miRNA-29c and its downstream pathways in preneoplastic progression of triple-negative breast cancer. Oncotarget 8:19645–19660

Biswas S, Mandal G, Roy Chowdhury S, Purohit S, Payne KK, Anadon C, Gupta A, Swanson P, Yu X, Conejo-Garcia JR (2019) Exosomes produced by mesenchymal stem cells drive differentiation of myeloid cells into immunosuppressive M2-polarized macrophages in breast cancer. J Immunol 203:3447–3460

Bolandghamat Pour Z, Nourbakhsh M, Mousavizadeh K, Madjd Z, Ghorbanhosseini SS, Abdolvahabi Z, Hesari Z, Ezzati Mobasser S (2019) Suppression of nicotinamide phosphoribosyltransferase expression by miR-154 reduces the viability of breast cancer cells and increases their susceptibility to doxorubicin. BMC Cancer 19:1–13

Borri F, Granaglia A (2021) Pathology of triple negative breast cancer. In: Seminars in cancer biology. Elsevier, London

Browne G, Dragon JA, Hong D, Messier TL, Gordon JA, Farina NH, Boyd JR, VanOudenhove JJ, Perez AW, Zaidi SK (2016) MicroRNA-378-mediated suppression of Runx1 alleviates the aggressive phenotype of triple-negative MDA-MB-231 human breast cancer cells. Tumour Biol 37:8825–8839

Buffa FM, Camps C, Winchester L, Snell CE, Gee HE, Sheldon H, Taylor M, Harris AL, Ragoussis J (2011) microRNA-associated progression pathways and potential therapeutic targets identified by integrated mRNA and microRNA expression profiling in breast cancer. Cancer Res 71: 5635–5645

Cao Z-G, Li J-J, Yao L, Huang Y-N, Liu Y-R, Hu X, Song C-G, Shao Z-M (2016) High expression of microRNA-454 is associated with poor prognosis in triple-negative breast cancer. Oncotarget 7:64900

Cascione L, Gasparini P, Lovat F, Carasi S, Pulvirenti A, Ferro A, Alder H, He G, Vecchione A, Croce CM (2013) Integrated microRNA and mRNA signatures associated with survival in triple negative breast cancer. PLoS One 8:e55910

Chakkaravarthi K, Ramesh R, Palaniyandi T, Baskar G, Viswanathan S, Wahab MRA, Surendran H, Ravi M, Sivaji A (2023) Prospectives of mirna gene signaling pathway in triple-negative breast cancer. Pathol Res Pract 248:154658

Chaudhary P, Gibbs LD, Maji S, Lewis CM, Suzuki S, Vishwanatha JK (2020) Serum exosomal-annexin A2 is associated with African-American triple-negative breast cancer and promotes angiogenesis. Breast Cancer Res 22:1–15

Chen H, Pan H, Qian Y, Zhou W, Liu X (2018a) MiR-25-3p promotes the proliferation of triple negative breast cancer by targeting BTG2. Mol Cancer 17:1–11

Chen X, Zhao M, Huang J, Li Y, Wang S, Harrington CA, Qian DZ, Sun XX, Dai MS (2018b) MicroRNA-130a suppresses breast cancer cell migration and invasion by targeting FOSL1 and upregulating ZO-1. J Cell Biochem 119:4945–4956

Dass SA, Tan KL, Selva Rajan R, Mokhtar NF, Mohd Adzmi ER, Wan Abdul Rahman WF, Tengku Din TADA-A, Balakrishnan V (2021) Triple negative breast cancer: a review of present and future diagnostic modalities. Medicina 57:62

de Faria BJ, Marta GN (2022) Triple-negative breast cancer and radiation therapy. Rep Pract Oncol Radiother 27:545–551

Deng X, Cao M, Zhang J, Hu K, Yin Z, Zhou Z, Xiao X, Yang Y, Sheng W, Wu Y (2014) Hyaluronic acid-chitosan nanoparticles for co-delivery of MiR-34a and doxorubicin in therapy against triple negative breast cancer. Biomaterials 35:4333–4344

Dent R, Trudeau M, Pritchard KI, Hanna WM, Kahn HK, Sawka CA, Lickley LA, Rawlinson E, Sun P, Narod SA (2007) Triple-negative breast cancer: clinical features and patterns of recurrence. Clin Cancer Res 13:4429–4434

Di Leone A, Franco A, Zotta F, Scardina L, Sicignano M, Di Guglielmo E, Castagnetta V, Magno S, Terribile D, Sanchez AM, Franceschini G, Masetti R (2023) Local treatment of triple-negative breast cancer: is mastectomy superior to breast-conserving surgery? J Pers Med 13

Dioufa N, Clark AM, Ma B, Beckwitt CH, Wells A (2017) Bi-directional exosome-driven inter-communication between the hepatic niche and cancer cells. Mol Cancer 16:1–14

Eades G, Wolfson B, Zhang Y, Li Q, Yao Y, Zhou Q (2015) lincRNA-RoR and miR-145 regulate invasion in triple-negative breast cancer via targeting ARF6. Mol Cancer Res 13:330–338

Encarnación-Medina J, Godoy L, Matta J, Ortiz-Sánchez C (2023) Identification of Exo-miRNAs: a summary of the efforts in translational studies involving triple-negative breast cancer. Cells 12

Fu Y, Yang Q, Yang H, Zhang X (2023) New progress in the role of microRNAs in the diagnosis and prognosis of triple negative breast cancer. Front Mol Biosci 10:1162463

García-García F, Salinas-Vera YM, García-Vázquez R, Marchat LA, Rodríguez-Cuevas S, López-González JS, Carlos-Reyes Á, Ramos-Payán R, Aguilar-Medina M, Pérez-Plasencia C (2019) miR-145-5p is associated with pathological complete response to neoadjuvant chemotherapy and impairs cell proliferation by targeting TGFβR2 in breast cancer. Oncol Rep 41:3527–3534

Gong C, Tian J, Wang Z, Gao Y, Wu X, Ding X, Qiang L, Li G, Han Z, Yuan Y (2019) Functional exosome-mediated co-delivery of doxorubicin and hydrophobically modified microRNA 159 for triple-negative breast cancer therapy. J Nanobiotechnol 17:1–18

Gong W, Zhu C, Liu Y, Muckenhuber A, Bronger H, Scorilas A, Kiechle M, Dorn J, Magdolen V, Dreyer T (2021) Elevated levels of both microRNA 378 (miR-378) and kallikrein-related peptidase 4 (KLK4) mRNA are associated with an unfavorable prognosis in triple-negative breast cancer. Am J Transl Res 13:1594

Grimaldi AM, Nuzzo S, Condorelli G, Salvatore M, Incoronato M (2020) Prognostic and clinicopathological significance of MiR-155 in breast cancer: a systematic review. Int J Mol Sci 21:5834

Grimaldi AM, Salvatore M, Incoronato M (2021) miRNA-based therapeutics in breast cancer: a systematic review. Front Oncol 11:668464

Guz M, Jeleniewicz W, Cybulski M (2023) Interactions between circRNAs and miR-141 in cancer: from pathogenesis to diagnosis and therapy. Int J Mol Sci 24:11861

Haffty BG, Yang Q, Reiss M, Kearney T, Higgins SA, Weidhaas J, Harris L, Hait W, Toppmeyer D (2006) Locoregional relapse and distant metastasis in conservatively managed triple negative early-stage breast cancer. J Clin Oncol 24:5652–5657

Halsted CP, Benson JR, Jatoi I (2014) A historical account of breast cancer surgery: beware of local recurrence but be not radical. Future Oncol 10:1649–1657

Hamam R, Ali AM, Alsaleh KA, Kassem M, Alfayez M, Aldahmash A, Alajez NM (2016) microRNA expression profiling on individual breast cancer patients identifies novel panel of circulating microRNA for early detection. Sci Rep 6:25997

Hammond MEH, Hayes DF, Dowsett M, Allred DC, Hagerty KL, Badve S, Fitzgibbons PL, Francis G, Goldstein NS, Hayes M (2010) American Society of Clinical Oncology/College of American Pathologists guideline recommendations for immunohistochemical testing of estrogen and progesterone receptors in breast cancer (unabridged version). Arch Pathol Lab Med 134:e48–e72

Han J, Yu J, Li J, Guo M, Song J, Zhou X (2019) Overexpression of miR-361-5p in triple-negative breast cancer (TNBC) inhibits migration and invasion by targeting RQCD1 and inhibiting the EGFR/PI3K/Akt pathway. Bosn J Basic Med Sci 19:52

Hata A, Kashima R (2016) Dysregulation of microRNA biogenesis machinery in cancer. Crit Rev Biochem Mol Biol 51:121–134

Hidalgo-Miranda A, Romero-Cordoba S, Rodriguez-Cuevas S, Rebollar-Vega R, Tagliabue E, Iorio M, D'Ippolito E, Baroni S (2016) A microRNA signature identifies subtypes of triple-negative breast cancer and reveals MIR-342-3P as regulator of a lactate metabolic pathway. In: Human genomics. Biomed Central, London

Higginbotham JN, Zhang Q, Jeppesen DK, Scott AM, Manning HC, Ochieng J, Franklin JL, Coffey RJ (2016) Identification and characterization of EGF receptor in individual exosomes by fluorescence-activated vesicle sorting. J Extracell Vesicles 5:29254

Hong Z, Hong C, Ma B, Wang Q, Zhang X, Li L, Wang C, Chen D (2019) MicroRNA-126-3p inhibits the proliferation, migration, invasion, and angiogenesis of triple-negative breast cancer cells by targeting RGS3. Oncol Rep 42:1569–1579

Hossain F, Majumder S, David J, Miele L (2021) Precision medicine and triple-negative breast cancer: current landscape and future directions. Cancers 13:3739

Hu J, Xu J, Wu Y, Chen Q, Zheng W, Lu X, Zhou C, Jiao D (2015) Identification of microRNA-93 as a functional dysregulated miRNA in triple-negative breast cancer. Tumour Biol 36:251–258

Ji H, Sang M, Liu F, Ai N, Geng C (2019) miR-124 regulates EMT based on ZEB2 target to inhibit invasion and metastasis in triple-negative breast cancer. Pathol Res Pract 215:697–704

Jia Z, Liu Y, Gao Q, Han Y, Zhang G, Xu S, Cheng K, Zou W (2016) miR-490-3p inhibits the growth and invasiveness in triple-negative breast cancer by repressing the expression of TNKS2. Gene 593:41–47

Kahraman M, Röske A, Laufer T, Fehlmann T, Backes C, Kern F, Kohlhaas J, Schrörs H, Saiz A, Zabler C, Ludwig N, Fasching PA, Strick R, Rübner M, Beckmann MW, Meese E, Keller A, Schrauder MG (2018) MicroRNA in diagnosis and therapy monitoring of early-stage triple-negative breast cancer. Sci Rep 8:11584

Kaviani A, Sodagari N, Sheikhbahaei S, Eslami V, Hafezi-Nejad N, Safavi A, Noparast M, Fitoussi A (2013) From radical mastectomy to breast-conserving therapy and oncoplastic breast surgery: a narrative review comparing oncological result, cosmetic outcome, quality of life, and health economy. ISRN Oncol 2013:742462

Kleivi Sahlberg K, Bottai G, Naume B, Burwinkel B, Calin GA, Børresen-Dale A-L, Santarpia L (2015) A serum microRNA signature predicts tumor relapse and survival in triple-negative breast cancer patients. Clin Cancer Res 21:1207–1214

Kumar R, Sharma A, Tiwari RK (2012) Application of microarray in breast cancer: an overview. J Pharm Bioallied Sci 4:21

Kunej T, Obsteter J, Pogacar Z, Horvat S, Calin GA (2014) The decalog of long non-coding RNA involvement in cancer diagnosis and monitoring. Crit Rev Clin Lab Sci 51:344–357

Lee JU, Kim WH, Lee HS, Park KH, Sim SJ (2019) Quantitative and specific detection of exosomal miRNAs for accurate diagnosis of breast cancer using a surface-enhanced Raman scattering sensor based on plasmonic head-flocked gold nanopillars. Small 15:1804968

Lee SJ, Jeong J-H, Lee J, Park HY, Jung JH, Kang J, Kim EA, Park NJ-Y, Park J-Y, Lee IH (2021) MicroRNA-496 inhibits triple negative breast cancer cell proliferation by targeting Del-1. Medicine 100

Leidinger P, Backes C, Rheinheimer S, Keller A, Meese E (2015) Towards clinical applications of blood-borne miRNA signatures: the influence of the anticoagulant EDTA on miRNA abundance. PLoS One 10:e0143321

Li W, Liu C, Zhao C, Zhai L, Lv S (2016) Downregulation of β3 integrin by miR-30a-5p modulates cell adhesion and invasion by interrupting Erk/Ets-1 network in triple-negative breast cancer corrigendum in/10.3892/ijo. 2021.5169. Int J Oncol 48:1155–1164

Li H-Y, Liang J-L, Kuo Y-L, Lee H-H, Calkins MJ, Chang H-T, Lin F-C, Chen Y-C, Hsu T-I, Hsiao M (2017a) miR-105/93-3p promotes chemoresistance and circulating miR-105/93-3p acts as a diagnostic biomarker for triple negative breast cancer. Breast Cancer Res 19:1–14

Li HY, Liang JL, Kuo YL, Lee HH, Calkins MJ, Chang HT, Lin FC, Chen YC, Hsu TI, Hsiao M, Ger LP, Lu PJ (2017b) miR-105/93-3p promotes chemoresistance and circulating miR-105/93-3p acts as a diagnostic biomarker for triple negative breast cancer. Breast Cancer Res 19:133

Li Z, Meng Q, Pan A, Wu X, Cui J, Wang Y, Li L (2017c) MicroRNA-455-3p promotes invasion and migration in triple negative breast cancer by targeting tumor suppressor EI24. Oncotarget 8:19455

Li C, Miao K, Chen Y, Liu L, Zhao G, Lin M, Jiang C (2018) Jagged2 promotes cancer stem cell properties of triple negative breast cancer cells and paclitaxel resistance via regulating microRNA-200. Eur Rev Med Pharmacol Sci 22:6008–6014

Li P-P, Li R-G, Huang Y-Q, Lu J-P, Zhang W-J, Wang Z-Y (2021) LncRNA OTUD6B-AS1 promotes paclitaxel resistance in triple negative breast cancer by regulation of miR-26a-5p/MTDH pathway-mediated autophagy and genomic instability. Aging (Albany NY) 13:24171

Li Y, Zhang H, Merkher Y, Chen L, Liu N, Leonov S, Chen Y (2022) Recent advances in therapeutic strategies for triple-negative breast cancer. J Hematol Oncol 15:121

Liang Z, Ahn J, Guo D, Votaw JR, Shim H (2013) MicroRNA-302 replacement therapy sensitizes breast cancer cells to ionizing radiation. Pharm Res 30:1008–1016

Lin Y, Liu AY, Fan C, Zheng H, Li Y, Zhang C, Wu S, Yu D, Huang Z, Liu F (2015) MicroRNA-33b inhibits breast cancer metastasis by targeting HMGA2, SALL4 and Twist1. Sci Rep 5:9995

Liu Y, Cai Q, Bao P-P, Su Y, Cai H, Wu J, Ye F, Guo X, Zheng W, Zheng Y (2015a) Tumor tissue microRNA expression in association with triple-negative breast cancer outcomes. Breast Cancer Res Treat 152:183–191

Liu X, Tang H, Chen J, Song C, Yang L, Liu P, Wang N, Xie X, Lin X, Xie X (2015b) MicroRNA-101 inhibits cell progression and increases paclitaxel sensitivity by suppressing MCL-1 expression in human triple-negative breast cancer. Oncotarget 6:20070

Liu C, Zhang S, Wang Q, Zhang X (2017) Tumor suppressor miR-1 inhibits tumor growth and metastasis by simultaneously targeting multiple genes. Oncotarget 8:42043

Liu M, Gong C, Xu R, Chen Y, Wang X (2019) MicroRNA-5195-3p enhances the chemosensitivity of triple-negative breast cancer to paclitaxel by downregulating EIF4A2. Cell Mol Biol Lett 24:1–11

Liu Y, Gong W, Panoutsopoulou K, Singer-Cornelius T, Augustin K, Bronger H, Kiechle M, Dorn J, Scorilas A, Avgeris M, Magdolen V, Dreyer T (2023) Association of high miR-27a, miR-206, and miR-214 expression with poor patient prognosis and increased chemoresistance in triple-negative breast cancer. Am J Cancer Res 13:2471–2487

Lv ZD, Kong B, Liu XP, Jin LY, Dong Q, Li FN, Wang HB (2016) miR-655 suppresses epithelial-to-mesenchymal transition by targeting Prrx1 in triple-negative breast cancer. J Cell Mol Med 20:864–873

Lytle JR, Yario TA, Steitz JA (2007) Target mRNAs are repressed as efficiently by microRNA-binding sites in the 5′ UTR as in the 3′ UTR. Proc Natl Acad Sci 104:9667–9672

Malla RR, Kumari S, Gavara MM, Badana AK, Gugalavath S, Kumar DKG, Rokkam P (2019) A perspective on the diagnostics, prognostics, and therapeutics of microRNAs of triple-negative breast cancer. Biophys Rev 11:227–234

Mao H, Cao Y, Zou Z, Xia J, Zhao J (2023) An enzyme-powered microRNA discriminator for the subtype-specific diagnosis of breast cancer. Chem Sci 14:2097–2106

Matamala N, Vargas MT, Gonzalez-Campora R, Minambres R, Arias JI, Menendez P, Andres-Leon E, Gomez-Lopez G, Yanowsky K, Calvete-Candenas J (2015) Tumor microRNA expression profiling identifies circulating microRNAs for early breast cancer detection. Clin Chem 61:1098–1106

Mathe A, Scott RJ, Avery-Kiejda KA (2015) MiRNAs and other epigenetic changes as biomarkers in triple negative breast cancer. Int J Mol Sci 16:28347–28376

Mollah F, Varamini P (2021) Overcoming therapy resistance and relapse in TNBC: emerging technologies to target breast cancer-associated fibroblasts. Biomedicines 9:1921

Morris GJ, Naidu S, Topham AK, Guiles F, Xu Y, McCue P, Schwartz GF, Park PK, Rosenberg AL, Brill K (2007) Differences in breast carcinoma characteristics in newly diagnosed African–American and Caucasian patients: a single-institution compilation compared with the National Cancer Institute's surveillance, epidemiology, and end results database. Cancer 110:876–884

Naorem LD, Muthaiyan M, Venkatesan A (2019) Identification of dysregulated miRNAs in triple negative breast cancer: a meta-analysis approach. J Cell Physiol 234:11768–11779

Nassar FJ, Chamandi G, Tfaily MA, Zgheib NK, Nasr R (2020) Peripheral blood-based biopsy for breast cancer risk prediction and early detection. Front Med 7:28

Nie J, Jiang H-C, Zhou Y-C, Jiang B, He W-J, Wang Y-F, Dong J (2019) MiR-125b regulates the proliferation and metastasis of triple negative breast cancer cells via the Wnt/β-catenin pathway and EMT. Biosci Biotechnol Biochem 83:1062–1071

Niedźwiecki S, Piekarski J, Szymańska B, Pawłowska Z, Jeziorski A (2018) Serum levels of circulating miRNA-21, miRNA-10b and miRNA-200c in triple-negative breast cancer patients. Ginekol Pol 89:415–420

Obidiro O, Battogtokh G, Akala EO (2023) Triple negative breast cancer treatment options and limitations: future outlook. Pharmaceutics 15

O'Brien K, Lowry MC, Corcoran C, Martinez VG, Daly M, Rani S, Gallagher WM, Radomski MW, MacLeod RA, O'Driscoll L (2015) miR-134 in extracellular vesicles reduces triple-negative breast cancer aggression and increases drug sensitivity. Oncotarget 6:32774

Onyeagucha B, Rajamanickam S, Subbarayalu P, Bansal B, Bansal H, Chang Y, Timilsina S, Abdelfaltah N, Eedunuri V, Rao M (2016) Abstract P2-03-04: down-regulation of Bcl2-related ovarian killer (BOK) by miR-296-5p protects breast cancer cells from paclitaxel-induced apoptosis. Cancer Res 76:P2-03-04-P02-03-04

Ouyang M, Li Y, Ye S, Ma J, Lu L, Lv W, Chang G, Li X, Li Q, Wang S (2014) MicroRNA profiling implies new markers of chemoresistance of triple-negative breast cancer. PLoS One 9:e96228

Paszek S, Gabło N, Barnaś E, Szybka M, Morawiec J, Kołacińska A, Zawlik I (2017) Dysregulation of microRNAs in triple-negative breast cancer. Ginekol Pol 88:530–536

Perez EA, Cortés J, Gonzalez-Angulo AM, Bartlett JM (2014) HER2 testing: current status and future directions. Cancer Treat Rev 40:276–284

Perou CM, Sørlie T, Eisen MB, Van De Rijn M, Jeffrey SS, Rees CA, Pollack JR, Ross DT, Johnsen H, Akslen LA (2000) Molecular portraits of human breast tumours. Nature 406: 747–752

Phan B, Majid S, Ursu S, de Semir D, Nosrati M, Bezrookove V, Kashani-Sabet M, Dar AA (2016) Tumor suppressor role of microRNA-1296 in triple-negative breast cancer. Oncotarget 7:19519

Piasecka D, Braun M, Kordek R, Sadej R, Romanska H (2018) MicroRNAs in regulation of triple-negative breast cancer progression. J Cancer Res Clin Oncol 144:1401–1411

Piombino C, Mastrolia I, Omarini C, Candini O, Dominici M, Piacentini F, Toss A (2021) The role of exosomes in breast cancer diagnosis. Biomedicines 9:312

Poggio M, Hu T, Pai C-C, Chu B, Belair CD, Chang A, Montabana E, Lang UE, Fu Q, Fong L (2019) Suppression of exosomal PD-L1 induces systemic anti-tumor immunity and memory. Cell 177: 414–427. e413

Qattan A (2020) Novel miRNA targets and therapies in the triple-negative breast cancer microenvironment: an emerging hope for a challenging disease. Int J Mol Sci 21

Qattan A, Al-Tweigeri T, Alkhayal W, Suleman K, Tulbah A, Amer S (2021) Clinical identification of dysregulated circulating microRNAs and their implication in drug response in triple negative breast cancer (TNBC) by target gene network and meta-analysis. Genes (Basel) 12

Ritter A, Hirschfeld M, Berner K, Rücker G, Jäger M, Weiss D, Medl M, Nöthling C, Gassner S, Asberger J, Erbes T (2020) Circulating non-coding RNA-biomarker potential in neoadjuvant chemotherapy of triple negative breast cancer? Int J Oncol 56:47–68

Rodríguez-Martínez A, de Miguel-Pérez D, Ortega FG, García-Puche JL, Robles-Fernández I, Exposito J, Martorell-Marugan J, Carmona-Sáez P, Garrido-Navas MC, Rolfo C (2019) Exosomal miRNA profile as complementary tool in the diagnostic and prediction of treatment response in localized breast cancer under neoadjuvant chemotherapy. Breast Cancer Res 21:1–9

Saatci O, Kaymak A, Raza U, Ersan PG, Akbulut O, Banister CE, Sikirzhytski V, Tokat UM, Aykut G, Ansari SA, Dogan HT, Dogan M, Jandaghi P, Isik A, Gundogdu F, Kosemehmetoglu K, Dizdar O, Aksoy S, Akyol A, Uner A, Buckhaults PJ, Riazalhosseini Y, Sahin O (2020) Targeting lysyl oxidase (LOX) overcomes chemotherapy resistance in triple negative breast cancer. Nat Commun 11:2416

Sabit H, Cevik E, Tombuloglu H, Abdel-Ghany S, Tombuloglu G, Esteller M (2021) Triple negative breast cancer in the era of miRNA. Crit Rev Oncol Hematol 157:103196

Saifi O, Chahrour MA, Li Z, Hoballah J, Panoff J, Vallow LA, Zeidan YH (2022) Is breast conservation superior to mastectomy in early stage triple negative breast cancer? Breast 62: 144–151

Sánchez-González I, Bobien A, Molnar C, Schmid S, Strotbek M, Boerries M, Busch H, Olayioye MA (2020) miR-149 suppresses breast cancer metastasis by blocking paracrine interactions with macrophages. Cancer Res 80:1330–1341

Schopp JG, Polat DS, Arjmandi F, Hayes JC, Ahn RW, Sullivan K, Sahoo S, Porembka JH (2023) Imaging challenges in diagnosing triple-negative breast cancer. Radiographics 43:e230027

Shi Y, Wang J, Tao S, Zhang S, Mao L, Shi X, Wang W, Cheng C, Shi Y, Yang Q (2023) miR-142-3p improves paclitaxel sensitivity in resistant breast cancer by inhibiting autophagy through the GNB2-AKT-mTOR pathway. Cell Signal 103:110566

Shin VY, Siu JM, Cheuk I, Ng EK, Kwong A (2015) Circulating cell-free miRNAs as biomarker for triple-negative breast cancer. Br J Cancer 112:1751–1759

Siegel RL, Miller KD, Jemal A (2019) Cancer statistics, 2019. CA Cancer J Clin 69:7–34

Song G-Q, Zhao Y (2015) MicroRNA-211, a direct negative regulator of CDC25B expression, inhibits triple-negative breast cancer cells' growth and migration. Tumour Biol 36:5001–5009

Stevic I, Müller V, Weber K, Fasching PA, Karn T, Marmé F, Schem C, Stickeler E, Denkert C, van Mackelenbergh M (2018) Specific microRNA signatures in exosomes of triple-negative and HER2-positive breast cancer patients undergoing neoadjuvant therapy within the GeparSixto trial. BMC Med 16:1–16

Sueta A, Fujiki Y, Goto-Yamaguchi L, Tomiguchi M, Yamamoto-Ibusuki M, Iwase H, Yamamoto Y (2021) Exosomal miRNA profiles of triple-negative breast cancer in neoadjuvant treatment. Oncol Lett 22:1–10

Sugita BM, Pereira SR, de Almeida RC, Gill M, Mahajan A, Duttargi A, Kirolikar S, Fadda P, de Lima RS, Urban CA (2019) Integrated copy number and miRNA expression analysis in triple negative breast cancer of Latin American patients. Oncotarget 10:6184

Sukumar J, Gast K, Quiroga D, Lustberg M, Williams N (2021) Triple-negative breast cancer: promising prognostic biomarkers currently in development. Expert Rev Anticancer Ther 21: 135–148

Sun X, Li Y, Zheng M, Zuo W, Zheng W (2016) MicroRNA-223 increases the sensitivity of triple-negative breast cancer stem cells to TRAIL-induced apoptosis by targeting HAX-1. PLoS One 11:e0162754

Svoronos AA, Engelman DM, Slack FJ (2016) OncomiR or tumor suppressor? The duplicity of microRNAs in cancer. Cancer Res 76:3666–3670

Tang Q, Ouyang H, He D, Yu C, Tang G (2019) MicroRNA-based potential diagnostic, prognostic and therapeutic applications in triple-negative breast cancer. Artif Cells Nanomed Biotechnol 47:2800–2809

Thakur S, Grover RK, Gupta S, Yadav AK, Das BC (2016) Identification of specific miRNA signature in paired sera and tissue samples of Indian women with triple negative breast cancer. PLoS One 11:e0158946

Théry C, Witwer KW, Aikawa E, Alcaraz MJ, Anderson JD, Andriantsitohaina R, Antoniou A, Arab T, Archer F, Atkin-Smith GK (2018) Minimal information for studies of extracellular vesicles 2018 (MISEV2018): a position statement of the International Society for Extracellular Vesicles and update of the MISEV2014 guidelines. J Extracell Vesicles 7:1535750

Todorova VK, Byrum SD, Gies AJ, Haynie C, Smith H, Reyna NS, Makhoul I (2022) Circulating exosomal microRNAs as predictive biomarkers of neoadjuvant chemotherapy response in breast cancer. Curr Oncol 29:613–630

Triantafyllou A, Dovrolis N, Zografos E, Theodoropoulos C, Zografos GC, Michalopoulos NV, Gazouli M (2022) Circulating miRNA expression profiling in breast cancer molecular subtypes: applying machine learning analysis in bioinformatics. Cancer Diagn Prognosis 2:739

Wang C, Zheng X, Shen C, Shi Y (2012) MicroRNA-203 suppresses cell proliferation and migration by targeting BIRC5 and LASP1 in human triple-negative breast cancer cells. J Exp Clin Cancer Res 31:1–8

Wang J, Tsouko E, Jonsson P, Bergh J, Hartman J, Aydogdu E, Williams C (2014) miR-206 inhibits cell migration through direct targeting of the actin-binding protein coronin 1C in triple-negative breast cancer. Mol Oncol 8:1690–1702

Wang Y, Zhang Z, Wang J (2018) MicroRNA-384 inhibits the progression of breast cancer by targeting ACVR1. Oncol Rep 39:2563–2574

Wang D-Y, Gendoo DM, Ben-David Y, Woodgett JR, Zacksenhaus E (2019a) A subgroup of microRNAs defines PTEN-deficient, triple-negative breast cancer patients with poorest prognosis and alterations in RB1, MYC, and Wnt signaling. Breast Cancer Res 21:1–13

Wang DY, Gendoo DMA, Ben-David Y, Woodgett JR, Zacksenhaus E (2019b) A subgroup of microRNAs defines PTEN-deficient, triple-negative breast cancer patients with poorest prognosis and alterations in RB1, MYC, and Wnt signaling. Breast Cancer Res 21:18

Wang C, Xu C, Niu R, Hu G, Gu Z, Zhuang Z (2019c) MiR-890 inhibits proliferation and invasion and induces apoptosis in triple-negative breast cancer cells by targeting CD147. BMC Cancer 19:1–11

Wang B, Zhang Y, Ye M, Wu J, Ma L, Chen H (2019d) Cisplatin-resistant MDA-MB-231 cell-derived exosomes increase the resistance of recipient cells in an exosomal miR-423-5p-dependent manner. Curr Drug Metab 20:804–814

Wang F, Zuroske T, Watts JK (2020) RNA therapeutics on the rise. Nat Rev Drug Discov 19: 441–442

Wang D, Nie Y, Wang P, Ma Q (2023) In situ synthesis of Cu nanoclusters/CeO(2) nanorod as aggregated induced ECL probe for triple-negative breast cancer detection. Talanta 258:124400

Weaver JW, Zhang J, Rojas J, Musich PR, Yao Z, Jiang Y (2022) The application of exosomes in the treatment of triple-negative breast cancer. Front Mol Biosci 9:1022725

Wijesinghe KM, Kanak MA, Harrell JC, Dhakal S (2022) Single-molecule sensor for high-confidence detection of miRNA. ACS Sens 7:1086–1094

Won KA, Spruck C (2020) Triple-negative breast cancer therapy: current and future perspectives (review). Int J Oncol 57:1245–1261

Wu H-J, Chu P-Y (2022) Current and developing liquid biopsy techniques for breast cancer. Cancers 14:2052

Wu Q, Chen Z, Zhang G, Zhou W, Peng Y, Liu R, Chen C, Feng J (2018) EZH2 induces the expression of miR-1301 as a negative feedback control mechanism in triple negative breast cancer. Acta Biochim Biophys Sin 50:693–700

Wu X, Chen H, Wu M, Peng S, Zhang L (2020a) Downregulation of miR-182-5p inhibits the proliferation and invasion of triple-negative breast cancer cells through regulating TLR4/NF-κB pathway activity by targeting FBXW7. Ann Transl Med 8

Wu H, Wang Q, Zhong H, Li L, Zhang Q, Huang Q, Yu Z (2020b) Differentially expressed microRNAs in exosomes of patients with breast cancer revealed by next-generation sequencing. Oncol Rep 43:240–250

Xiong H, Yan T, Zhang W, Shi F, Jiang X, Wang X, Li S, Chen Y, Chen C, Zhu Y (2018) miR-613 inhibits cell migration and invasion by downregulating Daam1 in triple-negative breast cancer. Cell Signal 44:33–42

Xu J, Wu K-j, Jia Q-j, Ding X-f (2020) Roles of miRNA and lncRNA in triple-negative breast cancer. J Zhejiang Univ Sci B 21:673

Xu H, Fang M, Zuo B, Yao W, Ren J, Zhang Y (2023) circAR-E2E4-miR-665-STAT3 axis is a potential regulatory network in triple-negative breast cancer. Heliyon 9:e12654

Yan L-X, Huang X-F, Shao Q, Huang M-Y, Deng L, Wu Q-L, Zeng Y-X, Shao J-Y (2008) MicroRNA miR-21 overexpression in human breast cancer is associated with advanced clinical stage, lymph node metastasis and patient poor prognosis. RNA 14:2348–2360

Yan H, Huang W, Chen C, Zhang X, Zhu K, Yuan J (2023) MiR-133a/CD47 axis is a novel prognostic biomarker to promote triple negative breast cancer progression. Pathol Res Pract 244:154400

Yang F, Zhang W, Shen Y, Guan X (2015) Identification of dysregulated microRNAs in triple-negative breast cancer. Int J Oncol 46:927–932

Ye J, Xu M, Tian X, Cai S, Zeng S (2019) Research advances in the detection of miRNA. J Pharm Anal 9:217–226

Yin J, Zhu C, Wang G, Gu J (2022) Treatment for triple-negative breast cancer: an umbrella review of meta-analyses. Int J Gen Med 15:5901–5914

Yu H, Li H, Qian H, Jiao X, Zhu X, Jiang X, Dai G, Huang J (2014) Upregulation of miR-301a correlates with poor prognosis in triple-negative breast cancer. Med Oncol 31:1–6

Zeng H, Wang L, Wang J, Chen T, Li H, Zhang K, Chen J, Zhen S, Tuluhong D, Li J (2018) microRNA-129-5p suppresses Adriamycin resistance in breast cancer by targeting SOX2. Arch Biochem Biophys 651:52–60

Zeng C, Fan D, Xu Y, Li X, Yuan J, Yang Q, Zhou X, Lu J, Zhang C, Han J (2020) Curcumol enhances the sensitivity of doxorubicin in triple-negative breast cancer via regulating the miR-181b-2-3p-ABCC3 axis. Biochem Pharmacol 174:113795

Zhang KJ, Hu Y, Luo N, Li X, Chen FY, Yuan JQ, Guo L (2020) miR-574-5p attenuates proliferation, migration and EMT in triple-negative breast cancer cells by targeting BCL11A and SOX2 to inhibit the SKIL/TAZ/CTGF axis. Int J Oncol 56:1240–1251

Zhao Z, Li L, Du P, Ma L, Zhang W, Zheng L, Lan B, Zhang B, Ma F, Xu B (2019) Transcriptional downregulation of miR-4306 serves as a new therapeutic target for triple negative breast cancer. Theranostics 9:1401

Zhong W, Zhang Y, Zhao H, Liang Z, Shi J, Ma Q (2023) High electrochemical active Au-NP/2D zinc-metal organic frameworks heterostructure-based ECL sensor for the miRNA-522 detection in triple negative breast cancer. Talanta 265:124875

Approaches to Breast Cancer Immunology and Immunotherapy

Anh D. Le, Huu-Thinh Nguyen, and Sao-Mai Dam

Abstract

Breast cancer is a common disease affecting women globally. There is growing interest in using the body's immune system to fight breast cancer due to the limitations and side effects of conventional treatments, such as chemotherapy, surgery, and radiation therapy. Abnormal or malignant cells are quickly identified and destroyed by the immune system. However, cancer cells can hide from the immune system and avoid being attacked. Immunotherapy appears to be a promising treatment option. This treatment boosts the body's immunological response to fight cancer cells. CAR-T cell (chimeric antigen receptor T cell) therapy and other checkpoint inhibitors may be used. Vaccines against cancer can potentially unleash the body's defenses against breast cancer. Research is ongoing on the role of the microbiome in breast cancer and the immune response. The composition of the microbiome can influence immune function and cancer development. Scientists are interested in understanding how the microbiome, immune system, and breast cancer cells interact in order to develop new prevention and therapeutic strategies for the disease. Utilizing the immune system to fight breast cancer is a promising new research area with numerous potential benefits.

Keywords

Breast cancer · Cancer vaccines · CAR-T cell therapy · Checkpoint inhibitors · Immunology · Immunotherapy

A. D. Le · S.-M. Dam (✉)
Industrial University of Ho Chi Minh City, Ho Chi Minh City, Vietnam
e-mail: damsaomai@iuh.edu.vn

H.-T. Nguyen
University of Medicine and Pharmacy in Ho Chi Minh City, Ho Chi Minh City, Vietnam

© The Author(s), under exclusive license to Springer Nature Switzerland AG 2023

Interdisciplinary Cancer Research, https://doi.org/10.1007/16833_2023_178
Published online: 15 August 2023

1 Introduction

Breast cancer is a complex and heterogeneous disease that affects a large number of women around the globe. It is the most prevalent cancer diagnosed in females. According to the World Health Organization, there were 2.26 million breast cancer cases in 2020, including 685 thousand fatalities (WHO 2022). Unfortunately, breast cancer affects low- and middle-income countries disproportionately and remains the primary cause of deaths due to cancer in women. In high-income countries, the 5-year survival rates for breast cancer exceed 90%, while in India, it is only 66%, and in South Africa, it is as low as 40% (WHO 2023). Traditional therapies, including chemotherapy, radiation therapy, and surgery, have proven helpful in some situations; however, they can have adverse impacts and may not be effective for all patients. As a result, an increasing number of people are interested in harnessing the immune system to target and destroy breast cancer cells.

The immune system is essential to the body's defense mechanism, identifying and eliminating abnormal or cancerous cells. Cancer cells, however, have learned to use various strategies to avoid being recognized and destroyed by the immune system. This allows them to grow and spread throughout the body, making cancer a challenging disease to treat. Immune editing is one of these techniques, which entails choosing cancer cells that are less immunogenic, or less likely to be identified and killed by the immune system. Understanding immune editing is critical for the development of new cancer treatments that boost the immune system's ability to detect and fight cancer cells. Researchers want to develop more effective cancer treatments by targeting immune editing, which can prevent cancer cells from evading the body's immune system and potentially lead to improved results for cancer patients (Schreiber et al. 2011). Researchers have been working to better understand the mechanism by which cancer cells evade the immune system and how to boost the immune system's ability to eliminate tumor cells. One approach involves immunotherapy, which entails the use of drugs or other agents to enhance the immune response against cancer cells. Immunotherapy has revolutionized cancer treatment and is effective in various cancer types, including breast cancer.

Breast cancer immunotherapy has taken many forms, such as checkpoint inhibitors, CAR-T cell treatments, and vaccines for cancer. Checkpoint inhibitors block cancer cells' immune system-evading checkpoint molecules. Pembrolizumab is one of the immune checkpoint inhibitor. In metastatic triple-negative breast cancer patients, inhibiting the programmed death 1 (PD-1) receptor on T cells with pembrolizumab increases the immune response against cancer cells (Schmid et al. 2022). During CAR-T cell therapy, a patient's T cells are modified to better recognize and destroy cancer cells. T cells are genetically altered to express a chimeric antigen receptor (CAR) that targets a specific antigen produced in cancer cells in CAR-T cell therapy. CAR-T cell therapy has shown considerable promise in the treatment of HER2 (human epidermal growth factor receptor 2)-positive recurrent or refractory breast cancer (Yang et al. 2022).

In addition to checkpoint inhibitors and CAR-T cell treatments, research has been conducted on cancer vaccines to enhance the immune system's ability to target breast cancer cells. Cancer vaccines stimulate the immune system by injecting specific proteins prevalent in cancer cells. This improves the immune system's ability to detect and eliminate cancer cells. The immunogenicity of vaccine formulations can be improved in a number of ways. Autologous, allogeneic, and dendritic cell-based immunizations are examples of such approaches. More examples include vaccinations based on tumor-associated peptides or proteins, DNA vaccines, heat shock proteins, and recombinant technology with viral or bacterial vectors (Shumway et al. 2009).

Despite immunotherapy potential, not all patients react to these treatments, and resistance to immunotherapy remains a major obstacle for those dealing with breast cancer. Combining immunotherapy with additional therapies like chemotherapy and radiation is one strategy researchers are exploring to overcome this resistance. Combining multiple treatment modalities may help enhance immunotherapy's effectiveness and provide better outcomes for breast cancer patients who do not respond to immunotherapy alone. Further research is necessary to identify new targets and approaches to overcome immunotherapy resistance and improve breast cancer treatment's effectiveness. Researchers want to enhance patient outcomes and give more effective treatment choices for patients with breast cancer by elucidating the reasons of resistance to immunotherapy and creating innovative techniques to overcome it.

2 Checkpoint Inhibitors

Immunotherapy in the form of checkpoint inhibitors works by inhibiting proteins on immune cells. These safeguards are natural defense mechanisms of the body to stop the immune system from attacking healthy tissue. However, they can be used by cancer cells to escape the body's defenses and spread. By inhibiting these checkpoints, checkpoint inhibitors improve the immune system's ability to recognize and kill cancer cells.

Blocking immune checkpoints is one of the most hopeful ways to get the immune system to fight cancer. Immune checkpoints are immune system pathways that prevent the immune system from attacking itself. They are important for keeping self-tolerance and controlling the length and strength of immune responses in peripheral tissues to limit damage to other tissues (Pardoll 2012). Specifically, tumors use immune checkpoint pathways to evade destruction by T cells that are trained to recognize tumor antigens. Many of these immunological checkpoints can be easily inhibited by antibodies or modified by recombinant ligands or receptors since they are triggered by ligand-receptor interactions.

In this group of immune therapeutics, the approval of the CTLA4 (cytotoxic T-lymphocyte-associated 4) antibodies was the first to be granted by the FDA (Food and Drug Administration). Preliminary clinical evidence using inhibitors of several immune checkpoint proteins, such as PD-1, on the other hand, shows broad and diverse choices for boosting antitumor immunity, perhaps resulting in clinical responses that last for an extended period (Pardoll 2012).

PD-1 inhibitors are one form of checkpoint inhibitor that has been developed specifically for breast cancer. T cells include a protein known as PD-1, and these cells are part of the immune system. PD-1 aids in the prevention of these cells attacking healthy cells. Some breast cancer cells, however, can create PD-L1 (Programmed Death Ligand 1), a protein that attaches to PD-1 and stops the immune system from attacking the cancer cells. To better direct the immune system toward targeting and eliminating cancer cells, PD-1 inhibitors block the connection between PD-1 and PD-L1. In their study, Brahmer et al. (2012) found that blocking the interaction between the PD-1 protein and its ligand PD-L1 can enhance immune function and exhibit anti-tumor activity. In a multicenter phase I trial, patients with selected advanced cancers were administered intravenous anti-PD-L1 antibodies at escalating doses every 14 days for up to 16 cycles. Of the 207 patients who received the antibody, 9% experienced Grade 3 or 4 toxic effects (Brahmer et al. 2012). A few patients showed a complete or partial objective response, with some reactions lasting over a year. Patients with advanced tumors such as non-small-cell lung cancer, melanoma, and renal-cell cancer saw objective response rates between 6% and 17% and disease stability rates between 12% and 41% after 24 weeks of antibody-induced tumor shrinkage and disease stabilization (NCT00729664 2015).

PD-1 inhibitors are immunotherapy that blocks the PD-1 protein found in T cells. This facilitates the immune system's ability to spot and destroy cancer cells. While PD-1 inhibitors effectively treat some types of cancer, including melanoma and lung cancer, their role in breast cancer treatment is still being studied. However, outcomes from recent clinical studies using PD-1 inhibitors in specific subsets of breast cancer, particularly triple-negative breast cancer (TNBC) and HER2-negative breast cancer, have been proven to be encouraging. Triple-negative breast cancer (TNBC) is an aggressive type of cancer that does not respond well to traditional chemotherapy or hormone therapy, making it an incredibly challenging disease to treat. PD-1 inhibitors have shown some success in treating TNBC alone or in combination with chemotherapy.

A subtype of breast cancer known as triple-negative breast cancer, or TNBC, is characterized by the absence of expression of the estrogen receptor, the progesterone receptor, and the human epidermal growth factor receptor 2. In treating several cancers, including TNBC, immunotherapy with immune checkpoint inhibitors has demonstrated some encouraging results. Pembrolizumab, an inhibitor of programmed cell death protein 1 (PD-1), was tested for its therapeutic value and safety in a study conducted by Nanda et al. (2016) on patients diagnosed with advanced TNBC. The KEYNOTE-012 research was a phase Ib clinical trial carried out on 32 women with advanced TNBC whose tumors were examined to determine whether or not they expressed PD-L1. PD-L1 is a protein often overexpressed in cancer cells and inhibits the immune response. Pembrolizumab was given intravenously at 10 milligrams per kilogram every 2 weeks. According to the study's findings, 58.6% of the patients had tumors that tested positive for PD-L1. In this study, the median number of doses given out was five, and the most common adverse effects noticed were minor. These negative effects included tiredness, arthralgia, nausea, and myalgia (Nanda et al. 2016). One patient died directly from the treatment; five individuals sustained grade 3 or higher toxicity. For the 27 patients tested for anticancer activity, the overall response rate was 18.5%, and the median time it took for patients to respond was 17.9 weeks. It was impossible to determine a median value for the duration of response, which ranged from 15.0 to 47.3 weeks. Results of this research suggest that pembrolizumab may have a safety profile that is acceptable, as well as therapeutic activity, in patients with advanced TNBC who have undergone a number of different treatments in the past. In conclusion, immunotherapy with pembrolizumab could be a promising therapeutic option for TNBC, which currently lacks targeted treatments (NCT02447003 2020).

The use of the checkpoint inhibitor pembrolizumab in managing metastatic triple-negative breast cancer has been authorized. Complete pathological response (pCR) was achieved in considerably more patients undergoing definitive surgery for early-stage triple-negative breast cancer when pembrolizumab was added to neoadjuvant chemotherapy in phase III clinical studies (Schmid et al. 2022). On the other hand, preliminary results about event-free survival (EFS) were not shared until quite recently. In this trial, pembrolizumab or a placebo was given as neoadjuvant therapy to 1174 patients with newly diagnosed stage II or III triple-negative breast cancer. Every 3 weeks, patients were given four cycles of pembrolizumab or a placebo. Then, patients received either doxorubicin-cyclophosphamide or epirubicin-cyclophosphamide in combination with pembrolizumab or a placebo for a total of four cycles of treatment. Adjuvant pembrolizumab or placebo was administered to patients every 3 weeks for nine cycles after their final operation. The primary endpoints were pCR and EFS. In the fourth planned interim analysis, the median time spent following up was 39.1 months. Estimated event-free survival (EFS) at 36 months was higher in the pembrolizumab-chemotherapy group (84.5% vs. 76.8%; hazard ratio for event or death, 0.63; $P < 0.001$) than in the placebo-chemotherapy group (76.8%). Clinical study results showed that patients who received neoadjuvant pembrolizumab in addition to chemotherapy and adjuvant pembrolizumab after surgery had considerably longer EFS than those with neoadjuvant chemotherapy alone. The gold standard, neoadjuvant chemotherapy, was used as a comparison. In addition, the incidence of adverse events was highest during the neoadjuvant phase, and this data was compared to the previously established safety qualifications of pembrolizumab and chemotherapy. These findings illuminate the potential of pembrolizumab in treating early triple-negative breast cancer and open the way for future research to optimize treatment regimens for those suffering from this problematic disease (Schmid et al. 2022).

In the KEYNOTE-012 trial, patients with PD-L1+ metastatic triple-negative breast cancer (TNBC) were included. The researchers observed that pembrolizumab, an inhibitor of the programmed cell death protein 1 (PD-1), demonstrated long-lasting effectiveness while maintaining a tolerable level of safety. To further investigate the efficacy and safety of pembrolizumab in this population, cohort A of KEYNOTE-086 was conducted. The study enrolled 170 female patients with centrally confirmed metastatic TNBC who had previously been treated with at least one course of chemotherapy. Those patients had an Eastern Cooperative Oncology

Group Performance Status (ECOG PS) of 0-1. Patients were given 200 mg of pembrolizumab every 3 weeks for up to 24 months. The duration of the response (DOR), disease control rate (DCR), progression-free survival (PFS), and overall survival (OS) were secondary goals.

In contrast, the overall response rate (ORR) and safety were the primary endpoints. Nine patients were still receiving pembrolizumab after a median follow-up of 10.9 months. 60% of patients experienced treatment-related adverse events; 12% were classified as severe. Adverse events connected to treatment accounted for the discontinuation of care for 4% of patients. No matter how strongly PD-L1 was expressed, the total response rate was just 5%, although 8% of patients had their condition under control. The median duration of response was 6.3 months, with 63% of responders still not experiencing disease progression at the time of data cutoff. The median progression-free survival and overall survival were 2.0 months and 8.9 months, respectively, with an overall survival rate of 69% after 6 months. The ORR was considerably lower in patients with unfavorable prognostic features, according to the study. This trial showed that in a subset of patients with extensively pretreated metastatic TNBC, pembrolizumab monotherapy was safe and produced durable responses. These patients are the focus of monotherapy and combination therapy randomized clinical trials that are now underway (Schmid et al. 2018).

Another study by Kagihara et al. (2020) showed the long-term clinical benefits of atezolizumab, another PD-1 inhibitor, in patients with metastatic TNBC. Cancer cells are more easily attacked by the immune system after atezolizumab, an immune checkpoint inhibitor, inhibits the connection between PD-L1 and PD-1. Nab-paclitaxel is a type of chemotherapy combined with atezolizumab to enhance its efficacy. This study uses patients with metastatic TNBC to investigate this combination's pharmacology and toxicological profile. It has been demonstrated that the combination of atezolizumab with nab-paclitaxel, administered in the first-line scenario to patients with metastatic TNBC, results in a significant extension of progression-free survival in both the intention-to-treat and PD-L1-positive categories (Kagihara et al. 2020). The FDA approved this combination for use in patients with PD-L1-positive, metastatic TNBC due to the IMpassion 130 study. The analysis results support continuing studies into potential new immunotherapy combinations for TNBC. Developing predictive biomarkers and immunotherapy in early-stage TNBC patients can potentially improve survival rates for this high-risk patient population. The combination has demonstrated efficacy and a manageable toxicity profile in clinical trials. Ongoing research in this area could lead to improved treatments and better outcomes for patients with TNBC.

In the other clinical trial, Schmid et al. (2018) found that atezolizumab, combined with nab-paclitaxel, effectively treated advanced triple-negative breast cancer. In this phase III trial, researchers investigated the efficacy of combining nanoparticle albumin-bound (nab)-paclitaxel with atezolizumab for treating untreated metastatic TNBC. In the study, 451 patients diagnosed with untreated metastatic triple-negative breast cancer (TNBC) were offered the choice to receive either atezolizumab in combination with nab-paclitaxel or a placebo in combination with nab-paclitaxel as their treatment regimen. The treatment was continued until the patient's condition deteriorated too much or until the harmful effects became intolerable. The critical success indicators were progression-free survival and overall survival. If the results were statistically significant, both were tested in the intent-to-treat group first, then in the PD-L1 positive cohort. In the intention-to-treat population and the PD-L1-positive subset, atezolizumab plus nab-paclitaxel was observed to prolong progression-free survival in patients with metastatic TNBC (Schmid et al. 2018). In the intention-to-treat analysis, progression-free survival was 7.2 months with atezolizumab and nab-paclitaxel, but only 5.5 months with placebo and nab-paclitaxel. The median progression-free survival for patients with PD-L1-positive tumors was 7.5 months with atezolizumab plus nab-paclitaxel and 5 months with placebo plus nab-paclitaxel. With atezolizumab plus nab-paclitaxel, the median overall survival was 21.3 months, while with placebo plus nab-paclitaxel, the median overall survival was 17.6 months.

Patients with tumors that were positive for PD-L1 had a median overall survival time of 25.0 months, while patients with tumors that were negative for PD-L1 had a median overall survival time of 15.5 months. The study also found that adverse effects occurred in a manner consistent with the safety profiles already established for each medicine and that no adverse effects had previously been unknown. When atezolizumab was given with nab-paclitaxel, 15.9% of patients had side effects that made them stop treatment. On the other hand, patients who got a placebo plus nab-paclitaxel had 8.2% of bad things happen that made them stop treatment. With the combination of atezolizumab and nab-paclitaxel, patients with untreated metastatic TNBC who have tumors that are positive for PD-L1 now have a treatment choice that is effective and well tolerated (NCT02425891 2022). The study suggests that this combination could improve outcomes for patients with TNBC, a breast cancer subtype lacking effective targeted treatments.

While these studies show promising results for PD-1 inhibitors in specific subsets of breast cancer, it is essential to note that not all breast cancer patients will benefit from these treatments (Kagihara et al. 2020; Nanda et al. 2016; Schmid et al. 2018, 2022) and there may be risks associated with their use, such as autoimmune reactions or side effects from the immune system being activated. So, just like with any other cancer treatment, patients need to talk to their healthcare provider about their choices and decide based on their situation.

Another checkpoint inhibitor developed for breast cancer is a CTLA-4 (cytotoxic T-lymphocyte-associated antigen 4) inhibitor. CTLA-4 inhibitors are a type of immunotherapy used to treat various types of cancer. CTLA-4 is a protein found on T cells and helps regulate their activity. They work by blocking CTLA-4, a molecule that helps regulate the immune system, which can enhance the body's natural ability to fight cancer (O'Donnell et al. 2019). By blocking CTLA-4, CTLA-4 inhibitors can help increase T cell activity and enhance the immune response against cancer cells. Nivolumab and ipilimumab were tested in tandem in individuals with advanced melanoma in a phase I experiment. The experiment results showed that the combination therapy effectively shrank tumors in many patients with manageable side effects. The results suggest that combining these two therapies could provide a new treatment option for patients with advanced melanoma (Wolchok et al. 2013).

Regarding breast cancer, there is some evidence that CTLA-4 inhibitors may be effective in treating certain types of the disease. Specifically, clinical trials have shown that a CTLA-4 inhibitor called ipilimumab, combined with chemotherapy, can improve outcomes for patients with TNBC (Nanda et al. 2016). TNBC is a particularly aggressive form of breast cancer that is difficult to treat with traditional therapies, representing a promising development. While CTLA-4 inhibitors have shown promise in treating breast cancer, not all breast tumors are the same, and further study is needed to discover which patients will benefit the most.

CTLA-4 inhibitors, like other cancer medications, might have undesirable side effects. Fatigue, diarrhea, nausea, and skin rash are the most frequently reported adverse reactions of ipilimumab. More serious adverse effects, such as inflammation of the lungs, liver, or colon, have been reported seldom. Patients who are thinking about undergoing this treatment should talk to their doctor about the advantages and disadvantages.

Clinical trials have shown that checkpoint inhibitors can effectively treat certain types of breast cancer. Patients with metastatic TNBC treated with the PD-1 inhibitor pembrolizumab had improved progression-free survival, for instance (Adams et al. 2017). Similarly, the CTLA-4 inhibitor ipilimumab has been shown to improve overall survival in patients with advanced breast cancer that has not responded to other treatments (Wolchok et al. 2013). Another trial confirmed the efficacy of the PD-1 inhibitor nivolumab in conjunction with the CTLA-4 inhibitor ipilimumab in treating patients with metastatic breast cancer (Nanda et al. 2016).

Nivolumab with ipilimumab as a combination therapy for the treatment of metastatic breast cancer requires additional investigation to evaluate whether or not it is successful. It is important to note that this combination therapy can also have significant side effects and should only be used under the supervision of a qualified healthcare provider.

It is important to note that checkpoint inhibitors may not be effective for all patients with breast cancer and can be associated with side effects. These side effects can vary in severity and may include fatigue, skin rash, diarrhea, and inflammation of organs such as the lungs or liver. Therefore, it is crucial for patients considering checkpoint inhibitor therapy to have open discussions with their healthcare provider regarding the potential risks and benefits involved. Such discussions can help patients make informed decisions about their treatment options and ensure that the potential benefits outweigh the potential risks in their specific case (Larkin et al. 2015).

In conclusion, checkpoint inhibitors are immunotherapy for breast cancer that targets proteins that modulate the immune response. Although they have demonstrated promise in clinical trials, they are ineffective for all patients and can cause undesirable adverse effects. Further research is needed to understand better how checkpoint inhibitors can treat breast cancer and identify patients most likely to benefit from this therapy.

3 CAR-T Cell Therapy

Triple-negative breast cancer (TNBC) is a difficult-to-treat breast cancer subtype, as it does not express the commonly targeted receptors. Traditional treatment options, such as surgery, radiotherapy, and chemotherapy, often provide minimal clinical benefit or result in disease recurrence. Chimeric antigen receptor T cell (CAR-T) therapy has emerged as a promising treatment option for various hematologic malignancies.

CAR-T cell therapy is an immunotherapy that genetically modifies a patient's T cells to recognize and destroy cancer cells. This is done to treat patients who have advanced forms of cancer. CAR-T therapy utilizes genetically modified T lymphocytes to identify and eliminate cancer cells expressing specific tumor-associated antigens (TAAs). The T cells are modified to express a chimeric antigen receptor (CAR) that recognizes a particular protein on the surface of cancer cells. After being reinfused into the patient, CAR-T cells have the ability to kill cancer cells that express a target protein by selectively binding to and destroying those cells.

Although CAR-T cell therapy has proven effective in treating certain forms of blood malignancies, including leukemia and lymphoma, the FDA has not yet approved its use in breast cancer treatment (Nasiri et al. 2022). CAR-T cell therapy is now being researched and tested in ongoing clinical trials for breast cancer treatment, and preliminary findings have been encouraging (Corti et al. 2022; Dees et al. 2020; Maalej et al. 2023).

CAR-T cell therapy is a potentially effective treatment for breast cancer, and it is now being researched on a wide variety of potential targets in both preclinical and clinical trials. Among the targets, there have been 19 antigens that have been investigated in preclinical studies, and 12 specific antigens have moved on to clinical trials to be tested for their safety and effectiveness (Y.-H. Yang et al. 2022). The overexpression of the human epidermal growth factor receptor-2 (HER2) is a significant characteristic observed in approximately 20–30% of breast cancer patients. This overexpression has made HER2 a crucial target for gene therapy in breast cancer. Researchers studying breast cancer have devoted considerable attention to studying HER2, recognizing its importance in the development and progression of the disease. The understanding of HER2 overexpression has paved the way for targeted therapies aimed at inhibiting HER2 signaling and improving outcomes for patients with HER2-positive breast cancer. There are ongoing clinical trials for HER2-CAR-T cell therapy and multi-target CAR-T cell therapy for HER2-positive solid tumors, including breast cancer. HER2 is the receptor tyrosine kinase that is addressed with CAR-T cell therapy in breast cancer more often than any other receptor tyrosine kinase (NCT03696030 2018; NCT03740256 2023; NCT04430595 2020; NCT04650451 2020).

In 2003, the discovery of breast cancer stem cells marked a significant breakthrough in understanding breast cancer's initiation, progression, and treatment resistance. Ganglioside GD2 was identified as a potential marker of breast cancer stem cells in 2012, and targeting therapies for GD2 may improve survival and cure rates for breast cancer patients. CD44v6 is an adhesion molecule found on the surface of cells that plays a crucial role in promoting tumor cell invasion, metastasis, and degradation of the extracellular matrix. Overexpression of CD44v6 has been linked to breast cancer invasion and metastasis, as well as tumor size, staging, and lymph node metastasis. As a result, CD44v6 has become an essential target for breast cancer treatment. CAR-T cells are genetically engineered T lymphocytes designed to target specific tumor-associated antigens (TAAs). HER2-, GD2-, and CD44v6-specific CAR-T cells have shown promise in improving immunotherapy treatment, preventing tumor cells from evading treatment, and achieving long-term disease relief. This approach holds great potential in improving the treatment and prognosis of breast cancer patients (NCT04430595 2020).

A clinical trial was conducted by Shanghai PerHum Therapeutics in the year 2020 to explore the efficacy, safety, and tolerability of CAR-modified autologous T cells in patients with HER2-positive solid tumors, such as breast cancer. The trial aimed to ascertain the maximum dose of the medication that could be safely tolerated and evaluate the participants regarding their response rate, progression-free survival, and overall survival. During this time, the University of Pennsylvania started two clinical studies for a treatment called cMET-CAR-T cell therapy in breast cancer patients (NCT04511871 2023). Unfortunately, one was ended due to financial concerns (NCT03060356 2020), while the other showed promising results (NCT01837602 2018) with intratumoral injection of cMET-CAR-T cells. cMET stands for chemo-resistant epidermal growth factor receptor. Another phase I clinical trial is being carried out to examine the potential side effects of the ROR1-CAR-T cell treatment and the optimum dose for treating ROR1-positive malignancies. These malignancies include breast cancers in stage IV and TNBC with metastases (NCT02706392 2021). These clinical trials demonstrate CAR-T cell therapy's potential in treating breast cancer and emphasize the significance of identifying appropriate target antigens for effective immunotherapy.

Despite the advancements in preclinical studies targeting 12 cell surface proteins in breast cancer, only six have progressed to clinical trials. Mesothelin-targeted CAR-T cell therapy is now participating in three phase I clinical trials to study potential adverse effects (NCT02580747 2018; NCT02792114 2023), whereas the remaining CARs are participating in phases I and II clinical trials to evaluate their efficacy and safety (NCT02414269 2015). The purpose of these clinical trials is to assess the practicability, safety, and effectiveness of using CAR-T cell therapy to treat patients with breast cancer. Due to its overexpression in various solid tumors, including breast cancer, mesothelin has been identified as a potential target for CAR-T cell therapy. This is because mesothelin is a glycoprotein found on the cell surface. Clinical trials for mesothelin-targeted CAR-T cell therapy have shown promising results, improved overall survival rates, and tolerable side effects. With ongoing trials, there is hope that CAR-T cell therapy will become a new therapeutic option for breast cancer patients, especially those with metastatic or refractory disease. However, further studies are required to determine the optimal target antigen and dosing regimens to improve the efficacy of CAR-T cell therapy in breast cancer.

MUC1, a transmembrane glycoprotein often overexpressed and aberrantly glycosylated in breast cancer, has become a promising target for CAR-T cell therapy.

The MUC1-targeted CAR-T cell treatment being investigated at present is now the subject of clinical trials, which are being done to assess the safety and efficacy of the treatment. The trials have different targets: the first (NCT04020575 2020) is investigating the different domains of the cleaved form of MUC1 at the City of Hope Medical Center in California, while the second (NCT04025216 2019) is targeting the aberrantly glycated form of MUC1 and is being conducted at various universities and research centers in the United States. The third trial (NCT02587689 2015) targets the entire MUC1 molecule and is undertaken by PersonGen Biomedicine (Suzhou) Co., Ltd in China. By providing critical insights into the safety and efficacy of MUC1-targeted CAR-T cell therapy, these clinical studies have the potential to discover new and effective medicines for breast cancer, a difficult-to-treat illness.

The Shenzhen Geno-Immune Medical Institute plans to conduct a phase I/II clinical trial (NCT04427449 2020) in the years 2020–2023 to investigate the efficacy and safety of using CD44v6-specific CAR-T cell therapy to treat patients who have CD44v6-positive cancers. The trial aims to examine the feasibility, safety, and effectiveness of 4SCAR-CD44v6 T cell therapy in targeting a variety of cancers, as CD44v6 is expressed in many types of cancers. T cells taken from patients or donors will be genetically modified with lentiviral CAR vectors to target CD44v6-positive cancer cells. The genetically modified T cells will be delivered to patients intravenously. The trial will evaluate the feasibility, safety, and efficacy of CD44v6 CAR-T cell therapy in various cancers, including stomach, breast, prostate, multiple myeloma, and lymphoma, that is CD44v6-positive. Objective responses (complete response and partial response) will be assessed using Response Evaluation Criteria in Solid Tumors (RECIST). The trial will also examine the functionality and persistence of 4SCAR-CD44v6 T cells in patients.

The West China Hospital at Sichuan University is conducting a phase I clinical trial for third-generation EpCAM-CAR-T cells to evaluate their safety and efficacy (NCT02915445) from 2016 to 2023. Patients suffering from solid tumors, such as nasopharyngeal carcinoma, breast cancer, stomach cancer, and other types, are sought out for participation in this research trial. Epithelial cell adhesion molecule, also known as EpCAM, has been linked to a worse prognosis and increased tumor dissemination and invasion. On the other hand, various medicines targeting EpCAM benefit cancer patients. The primary purpose of this research is to ascertain whether or not EpCAM CAR-T cells, which are engineered to include a chimeric antigen receptor (CAR-T) capable of recognizing EpCAM, are risk-free. The effectiveness of the treatment will also be evaluated in this trial using the Response Evaluation Criteria in Solid Tumors (RECIST). Lentiviral transduction of novel CAR genes from the second generation creates EpCAM CAR-T cells. EpCAM CAR-T cells are administered in dose-escalating cohorts to patients with nasopharyngeal carcinoma, breast cancer, and other EpCAM-positive solid tumors. This is done to determine the most significant acceptable dose and evaluate adverse effects while simultaneously identifying whether or not the intervention is helpful to these patient populations.

Only patients with high expression levels of EpCAM, as determined by staining of tumor tissue specimens, will be included in the study. Many CD3 T cells (at least

2×107) will be isolated from the patient's blood, activated using CD3 and CD28 antibodies, and transduced with lentivirus encoding the EpCAM CAR gene. EpCAM CAR-T cells will be allowed to proliferate 10–100 times before being frozen or infused, depending on their availability. If the white blood cells and lymphocyte levels are normal, patients will be given cyclophosphamide for preconditioning to prepare them for lymphodepletion. Infusion of T cells will begin at the lowest level and escalate if proven safe. Patients will have a cardiogram taken of them while receiving the infusion, and blood will be drawn both before and after the infusion to assess whether or not CAR-T cells and cytokines are present. In addition, the standard imaging studies will be carried out. Patients given CAR-T cells will be monitored for at least 15 years to see whether or not this treatment has any adverse consequences in the long run (NCT02915445 2016).

Compared to receptor tyrosine kinases (RTKs) and cell surface proteins, several other targets have reached clinical trials. Currently, only a few therapies have advanced to the clinical trial stage. In this group, two chimeric antigen receptor (CAR)-T cell treatments that are specific to carcinoembryonic antigen (CEA) are now undergoing phase I clinical trials (NCT02349724 2014; NCT03682744 2018), and one is in phase I and II clinical trials (NCT04348643 2020). In addition, two clinical trials in phase I are being conducted to examine CAR-T cell treatment's safety, effectiveness, and tolerance. These clinical trials target NKG2D (NCT04107142 2019) and GD2 (NCT03635632 2019).

The research study (NCT02349724 2014) conducted by the Southwest Hospital of Third Military Medical University in China aimed to evaluate the safety and efficacy of chimeric antigen receptor (CAR) T cells targeting carcinoembryonic antigen (CEA) in patients with CEA-positive solid tumors, including gastric, lung, pancreatic, breast, and colorectal cancer. CEA is a protein overexpressed in various cancer cells and used as a cancer diagnosis and treatment biomarker. The study aimed to assess the potential of CAR T cells to target and kill cancer cells expressing CEA. The preclinical studies showed that CAR T cells targeting CEA had effective killing activity against CEA-positive cancer cells, indicating that CAR T cells have the potential to be a promising treatment option for CEA-positive solid tumors. The study involved administering CAR T cells to patients with advanced-stage CEA-positive cancer who had exhausted all standard treatment options. The primary objective was to determine the safety of CEA-targeted CAR T cells and identify the appropriate dosage of CAR T cells to be infused. The study also aimed to evaluate the efficacy of CAR T cells in killing cancer cells, tumor targeting ability, and disease status after treatment. The researchers used Common Toxicity Criteria for Adverse Effects (CTCAE) version 4.0 to determine the toxicity profile of the CAR T cells. The results showed that CAR T cell therapy was generally well-tolerated and had manageable adverse effects. The research showed encouraging results, suggesting that CAR T cell therapy targeting CEA is a treatment option that is both safe and effective for patients who have solid tumors that test positive for CEA. These findings offer promise for developing new cancer treatments that are more effective than those already available.

Patients at the Rutgers Cancer Institute of New Jersey who had CEA-expressing adenocarcinoma peritoneal metastases or malignant ascites participated in an open-label, dose-escalating phase I clinical trial between 2018 and 2021. The purpose of the trial was to evaluate the safety and efficacy of intraperitoneal chimeric antigen receptor T (CAR-T) cell infusions targeting carcinoembryonic antigen (CEA) (NCT03682744 2018). Peripheral blood mononuclear cells were extracted from the patients via leukapheresis, and T cells were re-engineered to express CEA-specific CARs. After being developed in culture, the CAR-T cells were later administered to the patients through intraperitoneal injection at dosages of cells that had been previously calculated. This process was repeated several times. One anti-CEA CAR-T dose was supposed to be given to each participant in the trial, and the investigator was supposed to decide whether or not to provide any subsequent cycles. At the time of the CAR-T infusion, on the final day of treatment, and during the third reporting interval, biopsies of the peritoneum and the tumor were obtained. These samples were collected during the therapy period. The study's primary objective was to assess the safety of intraperitoneal CAR-T infusions and determine the appropriate dosage of CAR-T cells. In addition, this research aimed to determine whether or not patients with CEA-expressing adenocarcinoma peritoneal metastases or malignant ascites could benefit from treatment with CAR-T cells. According to the research findings, intraperitoneal administration of anti-CEA CAR-T cells was typically well-tolerated and displayed tolerable side effects. In addition, the study showed that intraperitoneal CAR-T infusions targeting CEA could induce tumor regression and improve the clinical outcomes of patients with CEA-expressing adenocarcinoma peritoneal metastases or malignant ascites. The study demonstrated the potential of CAR-T cell therapy to treat CEA-expressing solid tumors and highlighted the importance of intraperitoneal delivery of CAR-T cells for peritoneal carcinomatosis. The study provides critical data on the safety and efficacy of CAR-T cell therapy and may help refine the use of these treatments in cancer patients.

(NCT04348643 2020) is a significant phase I and II clinical trial conducted at Chongqing University Cancer Hospital, China, evaluating the safety and efficacy of chimeric antigen receptor (CAR)-T cell therapy targeting carcinoembryonic antigen (CEA) for patients with relapsed/refractory CEA+ Cancer. The study aims to determine the recommended dosage and infusion plan for CAR-T cell therapy targeting CEA. The single-arm study evaluates the efficacy and safety of CEA-targeted CAR-T cell therapy, and CAR-T cell reinfusion is carried out 1–3 times. CEA-CAR-T cells are administered intravenously. The study's primary objective is to identify the recommended dosage and infusion plan for CAR-T cell therapy targeting CEA. The study provides critical data on the efficacy and safety of CAR-T cell therapy for cancer treatment, particularly for relapsed/refractory CEA+ Cancer patients. The study is significant in advancing cancer treatments and refining the use of CAR-T cell therapy targeting CEA. In addition, the study's outcomes may provide valuable insights into developing more targeted and effective cancer therapies in the future.

A phase I clinical trial (NCT04107142 2019) was conducted at Landmark Medical Centre in Johor Bahru, Johor, Malaysia, from 2019 to 2021. The CAR-T cell therapy's safety, effectiveness, and tolerability in targeting NKG2D for relapsed or refractory solid tumors were subjected to the research study's evaluation. It was an open-label, single-center, dose-escalation study to determine the safety and tolerability of haploidentical/allogeneic NKG2DL-targeting chimeric antigen receptor-grafted Gamma Delta (γδ) T Cells (CTM-N2D) in persons with various types of recurrent or refractory solid malignancies. The research was designed to examine the safety and tolerability of the treatment. The trial was designed to investigate the safety and tolerability of the treatment.

The main goals of the phase I study were to find out if haploidentical or allogeneic chimeric antigen receptor-grafted γδ T cells that target NKG2DL were safe, how well they worked, and what the safe amount was. This treatment was to be administered four times weekly to patients who had relapsed or could not respond to treatment for solid tumors. CTM-N2D-101 is a dose-escalation study carried out during phase I of the research project, as stated in the comprehensive synopsis of the research endeavor. This research aims to determine whether or not it is possible to manufacture CTM-N2D at one of three target dose levels, the range of which is from 3×10^8 to 3×10^9 per infusion, while simultaneously reducing the likelihood of unwanted side effects. Following the delivery of four doses every week, a 3 + 3 design was implemented to determine the safest and most successful manner of treating the patient. The analysis of the incidence of dose-limiting toxicity (DLT) was the method that was used to achieve this goal. The safe regimen will be used in more phase II tests for certain diseases. In this study, triple-negative breast cancer was the disease of interest. Adoptive cell transfer of NKG2DL-targeting chimeric antigen receptor-grafted Gamma Delta T cells was the treatment for this condition. The study's outcomes may provide valuable insights into developing more effective and targeted cancer therapies in the future.

The other phase I clinical trial (NCT03635632 2019) is being conducted at Houston Methodist Hospital in Houston, Texas, United States, from 2019 through 2038. The purpose of this study is to find out if CAR-T cell therapy that targets GD2 is safe, successful, and well tolerated by people with neuroblastoma, sarcoma, uveal melanoma, breast cancer, or any other type of cancer that expresses GD2. Because there is no typical treatment for this type of cancer, people who have it are asked to participate in a gene transfer research study using immune cells called T cells. The use of antibodies and T cells in the battle against cancer has demonstrated some success. Still, it has not yet proven to be effective enough to cure the majority of patients.

T cells engineered with a chimeric antigen receptor (CAR) derived from an antibody that detects GD2 were administered to 11 neuroblastoma patients in a study that was conducted in the past. After around 2 weeks, the cells, which had grown for a while, started to die off despite increasing. Because of this, a new gene called C7R was introduced into GD2 T cells so that they could have a consistent supply of cytokines that aid in their survival and allow them to live for a more extended period. In addition, administering chemotherapy before the infusion of T cells, a process known as lymphodepletion, can extend the time that the T cells remain in the body and potentially make them more efficient at killing cancer cells.

The main goals of this study are to figure out how much GD2-C7R T cells can be used safely, how long they can be found in the blood, and what effect they have on cancer. The study's outcomes may provide insights into developing more effective and targeted cancer therapies in the future. Please be aware that the GD2-C7R T cells are not yet officially recognized as safe and effective by the FDA. The study is conducted hoping to pave the way for future approvals.

In recent years, CAR-T cell therapy has emerged as a potentially effective treatment for several forms of cancer, including breast cancer. Although CAR-T cell treatment for breast cancer has been the subject of numerous clinical trials, only a tiny fraction of trial outcomes have been made public. Understanding the safety and efficacy of CAR-T cell therapy in breast cancer is essential, and ongoing clinical trials targeting different antigens will be critical in advancing our knowledge in this area. Low penetration of CAR-T cells into tumors is one of the critical issues that must be overcome to successfully apply CAR-T cell therapy in clinical settings to treat breast cancer. In addition, breast tumors are characterized by a microenvironment that suppresses the immune system, tumor heterogeneity, and the possibility of on-target and off-tumor adverse effects, all of which present additional obstacles. Various strategies have been proposed to address these challenges, such as engineering CAR-T cells to secrete cytokines, developing bispecific CAR-T cells, gene editing techniques, and combinatorial therapies. In addition, it is required to conduct clinical trials using CAR-T cells that target a variety of antigens to increase the number of treatment choices accessible to patients. Ongoing trials, such as the ones targeting NKG2D and GD2, are particularly promising. However, more research is needed to determine CAR-T cell therapy's long-term safety and efficacy in breast cancer treatment.

In conclusion, CAR-T cell therapy holds enormous promise to transform breast cancer treatment, but it also faces substantial obstacles. Continuing clinical trials to examine CAR-T cell therapy's safety and efficacy in treating breast cancer is critical, as is addressing these issues to improve clinical results. However, CAR-T cell therapy is a growing field, and the promising results shown in breast cancer treatment are encouraging.

4 Breast Cancer Vaccines

Significant shifts have occurred over the previous several years in the adjuvant setting of breast cancer treatment. Immunotherapy is an effective adjuvant treatment for breast cancer that has recently joined the ranks of regular medical procedures such as chemotherapy, radiation, hormonal manipulation, and surgical tumor removal. Additionally, passive immunotherapeutic drugs like trastuzumab have seen a widespread application for treating HER-2/neu overexpressing breast cancer. On the other hand, vaccine therapy in the setting of metastatic disease has not yet demonstrated clinical impact in phase III testing. This could be because of the increased immunosuppressive effects that are present in the microenvironment of the tumor.

Treating patients in the adjuvant setting who have common illness residual and a high risk of relapse may, on the other hand, give a greater possibility of clinical benefit. For humans, vaccines against breast cancer are immunotherapy that helps the body's immune system find and kill breast cancer cells. Their research has conducted clinical trials on several breast cancer vaccines, including peptide, DNA, whole-cell, and dendritic cell vaccines. Nevertheless, finding effective vaccinations against breast cancer has proven difficult, and their utility in a context where the disease has spread to other organs is still restricted. The tumor's microenvironment is critical to both breast cancer's progression and vaccination therapy's success. The ineffectiveness of vaccination therapy can be attributed to several variables, including the absence of co-stimulatory molecules, the activation of cytotoxic T-lymphocyte antigen-4 (CTLA-4), an increase in the number of T regulatory cells, and the generation of soluble immunosuppressive substances by the tumor (Shumway et al. 2009).

There has been a shift toward treating patients with limited residual disease and significant relapse risk in an adjuvant setting. This is because immunological systems that tumors use to evade treatment are less formidable in this situation. This strategy may have a better probability of producing positive clinical results. This review summarizes the most recent research on creating breast cancer vaccines for use in adjuvant settings, particularly on vaccine formulation and antigen selection. Future research should continue to investigate the potential of vaccine therapy in breast cancer treatment, with a particular emphasis on optimizing vaccine design and identifying novel strategies to overcome immune evasion mechanisms in the tumor microenvironment.

Peptide vaccines use short fragments of proteins called peptides specific to breast cancer cells. When the vaccine is injected into the patient, it stimulates the immune system to produce T cells that can recognize and attack breast cancer cells. Peptide vaccines are a potent type of neoadjuvant immunotherapy that can precisely target tumor-associated antigens (TAAs) expressed by cancer cells. For example, breast cancer cells express MUC1, HER2/neu, and CEA proteins. Peptide vaccines can specifically target these cancer-cell-expressed TAAs. Peptides derived from these TAAs can elicit T cell responses, which are essential for effective antitumor immunity (Nicolás-Morales et al. 2022). Therefore, selecting peptides that can accommodate the natural variations in human leukocyte antigen (HLA) genes is crucial to ensure optimal T cell responses.

Peptide vaccines have several advantages over other therapies and vaccines, including low cost, antigen purity, and the induction of both humoral and cellular immune responses. Promising clinical trials of peptide-based vaccines, such as NeuVax-E75 and DPX-0907, have shown their efficacy in breast, ovarian, and prostate cancer. NeuVax-E75 targets the HER2/neu and GM-CSF epitopes, while DPX-0907 targets HLA-A2-TAAs expressed in multiple types of cancer (Peres et al. 2015).

Peptide-based vaccines targeting HER2, such as the E75 and GP2 vaccines, have been developed and are currently under clinical research for patients with breast cancer. A systematic review and meta-analysis of 24 clinical studies retrieved from various databases, including Cochrane Library, PubMed, Medline, Embase, and Web of Science, evaluated the clinical efficacy and safety of E75 and GP2 vaccines in breast cancer patients (You et al. 2021). The study comprised 1704 patients who were immunized and 1248 patients who were not. The meta-analysis found that the E75 vaccination improved clinical outcomes such as delayed-type hypersensitivity reaction, change in CD8+ T cell counts, recurrence rate, and disease-free survival rate with few adverse effects. The overall survival rate, however, did not differ significantly between the vaccinated and control groups.

In contrast, the GP2 vaccine showed a significant difference in the change in CD8+ T cell numbers before and after injection, suggesting that it could elicit a robust immune response. However, more clinical trials are necessary to confirm its clinical efficacy. This study provides important insights into developing and optimizing peptide-based vaccines for breast cancer immunotherapy. The results support the efficacy and safety of the E75 vaccine and provide a foundation for future research to improve the clinical efficacy of the GP2 vaccine.

Between 2013 and 2018, a research trial (NCT01570036 2013) was conducted at 29 sites in the United States to investigate the effectiveness and safety of combination immunotherapy using Herceptin and the HER2 vaccine NeuVax in preventing breast cancer recurrences. The study specifically targeted breast cancer cases with low (1+) or intermediate (2+) levels of HER2 expression, and included patients with either node-positive (NP) or node-negative (NN) status. This multi-center, prospective, randomized, single-blind, placebo-controlled phase II trial enrolled disease-free individuals who had completed standard care treatment. Participants who tested positive for HLA-A2+, HLA-A3+, HLA-A24+, or HLA-A26+ received the E75 vaccine in combination with GM-CSF after undergoing screening and HLA typing. This combination immunotherapy was tested for its efficacy and safety in avoiding recurrences in the target patient population. The trial attempted to boost the immune response against HER2-expressing cancer cells by combining Herceptin, a targeted therapy for HER2-positive breast cancer, with NeuVax. E75 and GM-CSF were added to boost CD8+ T cell response. Enrolled patients will be randomized at random to either Herceptin plus NeuVax vaccination or Herceptin with GM-CSF alone, with the primary effectiveness goal being a comparison of disease-free survival (DFS) at 24 months across treatment groups. As a secondary aim, the trial will assess DFS at 36 months and describe immunological responses to the vaccine, which will be linked with clinical benefit. The primary safety issue to be addressed in the trial is demonstrating that combination HER2-directed therapy does not result in additive cardiac toxicity. In addition, the trial aims to demonstrate that a variety of Herceptin and NeuVax vaccinations can prevent recurrences in patients with NP or NN breast cancer whose tumors express low or intermediate levels of HER2. The trial is critical because it will assess the efficiency of combination immunotherapy with Herceptin and NeuVax vaccine in avoiding recurrences in a specific patient cohort, thereby helping the development of more effective and tailored breast cancer treatments.

The clinical trial (NCT01730118 2013), conducted at the National Institutes of Health Clinical Center in Maryland, United States, from 2013 to 2019, aims to evaluate the safety and efficacy of a personalized cancer vaccine, AdHER2, in individuals with HER2-expressing tumors. Nearly one-third of breast cancers and several other types of malignancies, such as colon, prostate, and non-small cell lung cancer, contain HER2, a protein in tumors. The vaccine is an adenoviral transduced autologous human epidermal growth factor receptor/neu dendritic cell immunization manufactured using an individual's immune cells. The vaccine is made specifically for the individual. Participants must be at least 18 years old and have HER2-expressing malignancies to participate in the trial.

A screening process will be performed on the participants, which will involve a physical exam, a medical history review, blood and urine sample collection, and imaging studies. An apheresis technique will be performed to harvest the immune cells to manufacture the vaccine. The vaccination will be given in five doses spread out over the trial at weeks 0, 4, 8, 16, and 24. To evaluate the vaccine's efficacy and safety, all participants will undergo rigorous testing, including comprehensive physical examinations, frequent blood tests, and imaging scans. The study's primary objective is to test the safety and effectiveness of the AdHER2 vaccination. The personalized cancer vaccine approach could pave the way for developing more effective and targeted cancer treatments. The clinical study is critical because it looks into how safe and effective a personalized cancer vaccine is at making the immune system better able to spot and fight cancer cells. Additionally, the trial targets a specific patient population with tumors expressing HER2 protein.

Peptide vaccines offer increased specificity and reduced toxicity compared to other therapies and vaccines. Future research aims to develop degenerate multi-epitope immunogens against multiple TAA targets, although this remains a challenge for many breast cancers. In addition, further studies are needed to optimize peptide vaccine formulations and identify additional TAA targets for breast cancer immunotherapy.

DNA vaccines also have emerged as a promising immunotherapeutic option in Breast cancer therapy. DNA vaccines are designed to elicit an immune response by introducing genetic material encoding a specific antigen into the host's cells. Upon uptake, the antigen is produced, processed, and presented to the immune system, resulting in an immune response against the targeted cancer cells.

DNA vaccines offer advantages over traditional vaccines, including cost-effectiveness and long-lasting protection through repeated administration. Yang et al. also reported in 2014 that clinical tests of plasmid DNA vaccines showed that they were safe and triggered a broad as well as particular immune reaction (Yang et al. 2014). However, the immunogenicity of DNA vaccines is limited because immunological tolerance against endogenous self-antigens in tumors causes a failure to produce robust antigen-specific cellular immune responses. This results in the DNA vaccine's inability to prevent the development of tumors.

To combat breast cancer, researchers have looked at using DNA vaccines to attack the disease. Researchers are currently conducting clinical trials to determine the efficacy of DNA vaccines in treating breast cancer. These researches aim to learn more about the efficacy of DNA vaccines in inducing an immune response against breast cancer cells, as well as their safety, tolerability, and immunogenicity – two possible ways to boost the immunological response prompted by the vaccine or treatment. The first way is to improve the efficacy of DNA vaccines by selecting the optimal antigen(s) to include in the plasmid DNA and ensuring that they act correctly. This method is carried out to improve the performance of DNA vaccines. The second approach is to combine DNA vaccines with other medicines that could improve their efficiency by reducing immunosuppression in the tumor's microenvironment or raising the number of immune cells or their activity (Lopes et al. 2019).

Utilizing DNA vaccines in breast cancer treatment holds great promise and could potentially revolutionize the field of cancer therapy. In breast cancer, DNA vaccines contain small pieces of DNA that encode breast cancer proteins. When injected into the patient, the DNA is taken up by cells in the body and used to produce breast cancer proteins, which then stimulate the immune system to produce T cells that can attack breast cancer cells. The immunogenicity and effectiveness of DNA vaccines can be optimized by carefully selecting the antigen(s) used in the vaccine and combining it with other complementary therapies. By doing so, clinical outcomes for breast cancer patients can be improved, offering hope for those suffering from this disease.

Mizutani et al. (2004) presented the results of a study in which they studied the efficacy of two oral DNA-based vaccinations in suppressing angiogenesis and protecting against the formation of breast cancer metastases in animal models. The study was conducted by investigating the effectiveness of these vaccines in inhibiting angiogenesis and providing protection against the development of breast cancer metastases. According to research, the vaccines were delivered by utilizing a strain of Salmonella typhimurium that had been weakened and targeted novel antigens. These antigens consisted of things like the transcription factor Fos-related antigen 1, also known as Fra-1, and endoglin, also known as CD105. According to the study's findings, both vaccinations caused a reduction in the angiogenesis that occurred inside the vasculature of breast cancers. This effectively reduced the growth of breast tumors and their ability to metastasis to other parts of the body. In addition, it has been demonstrated that the vaccines can overcome peripheral tolerance by stimulating a robust cell-mediated immune response against the tumor's self-antigens. This was revealed after the vaccinations were hypothesized to break peripheral tolerance (Mizutani et al. 2004).

DNA-based vaccines have become available as a result of the discovery that tumor-associated antigens (TAAs), which have a low immunogenic potential but can be expressed in highly immunogenic forms by antigen-presenting cells, can serve as targets for the immune system to attack in the battle against cancer. Cohen (2000) successfully developed DNA-based vaccinations that treat cancer in mice effectively. He did this by separating the genomic DNA from spontaneous breast cancer that developed in a C3H/He animal and then transfecting it into a mouse fibroblast cell line (LM). Before transfection, the fibroblasts were changed so that they would produce allogeneic MHC-determinants (H-2K(b)) and release IL-2, making them immunogenic in a general way. Following this step, the breast tumor DNA and a plasmid known as pHyg that conferred resistance to the antibiotic hygromycin were co-transfected into the cells. Cell colonies resistant to hygromycin were combined and tested to see whether they had immunotherapeutic potential against tumor formation in mice. Mice who carried tumors and were immunized with cells transfected with a foreign gene lived noticeably longer than mice who carried tumors and were not immunized. Both mice with syngeneic melanoma introduced into their bodies and animals treated with fibroblasts that made IL-2 and transfected with DNA from melanoma cells exhibited favorable benefits. The immunity, unique to the tumor from which the DNA was extracted, was brought on by CD8+ T cells responsible for its production. These results illustrate the potential of DNA-based vaccines in inducing an immune response against cancer cells, specifically in breast and melanoma cancers. However, clinical studies have shown that these vaccinations usually have limited therapeutic effects because the tumor weakens the immune system. Therefore, choosing and optimizing the best antigen(s) to be introduced into the plasmid DNA and combining DNA vaccines with complementary therapies are the two strategies to increase the vaccine-induced immune response and treatment efficacy. These techniques aid in reducing immune system suppression in the tumor's microenvironment or boosting immune cell activity and quantity. In conclusion, the study by Cohen (2000) demonstrates the potential of DNA-based vaccines to fight cancer. It has paved the way for further research into improving the immunogenicity of DNA vaccines and exploring their potential in combination with other therapies to enhance their activity against cancer (Cohen 2000).

Patients who had been diagnosed with advanced-stage ERBB2-positive breast cancer were participated in a clinical experiment that was done by Disis et al. (2023). The purpose of this trial was to evaluate the immunogenicity and safety of a plasmid DNA vaccine that encoded the intracellular domain (ICD) of ERBB2. This trial was a non-randomized phase I clinical trial. The major goal of the study was to evaluate whether or not the vaccination was successful in producing an immunological response. The study included 66 patients who received three vaccine doses at different concentrations (10, 100, and 500 μg). The vaccine was given intradermally alongside soluble granulocyte-macrophage colony-stimulating factor during each of the three immunizations, which served as an adjuvant. The vaccine's safety and ability to induce an immune response were the primary focus of the experiment. According to the data, the bulk of the side effects caused by the vaccine were grade 1 or 2 and did not exhibit any significant variations between the dosage groups. At most time points, patients in the 100 μg and 500 μg arms had a more significant immune response to ERBB2 ICD type 1 than patients in the 10 μg arm. Furthermore, the highest vaccine dose was substantially related to a higher leftover DNA incidence at the injection site. The study also found that patients who had DNA persistence at week 16 had significantly lower average ERBB2 ICD immunity at later time points following immunization completion than patients who did not have DNA persistence (Disis et al. 2023).

The majority of patients with ERBB2-expressing breast cancer developed ERBB2-specific type 1 T cells following administration of the 100 μg ERBB2 ICD plasmid-based vaccine, as determined by the research. Consequently, phase II trials of the randomized study are currently underway to evaluate the vaccine's efficacy and safety in greater detail (Disis et al. 2023). The study provides important insights into the safety and efficacy of plasmid DNA vaccines in breast cancer treatment.

Also, studies are being done to improve how DNA vaccines to treat breast cancer are made and given. Scientists are looking into new methods like nanoparticle-based transport systems, viral vectors, and adjuvants to improve these vaccines and make people more immune to them.

Finding the best antigen targets for DNA vaccines is a significant step that needs more study. Researchers are trying to find specific tumor-associated antigens that can trigger a strong immune reaction against breast cancer cells while having fewer effects than are not intended. This means that the molecular and genetic features of different types of breast cancer must be carefully studied so that vaccinations can be tailored to each patient.

The tumor microenvironment is another critical thing to consider when making DNA medicines. The area around tumors is often immunosuppressive, which makes it harder for the immune system to find and kill cancer cells. Overcoming immunosuppression is an essential goal, and researchers are looking into many different ways to do so, such as mixing DNA vaccines with immune checkpoint inhibitors or other immunomodulatory drugs to boost the immune response against breast cancer.

Also, before DNA vaccines can be used in people, they must be carefully tested for safety and effectiveness in human tests. DNA vaccine immunogenicity, dose, and possible side effects are being tested in phase I and II clinical studies on people with breast cancer. These trials will tell us much about how DNA vaccines could treat diseases and help guide future studies.

To sum up, DNA vaccines show promise as a possible treatment for breast cancer, but they need more research and growth to be sure. For DNA vaccines to reach their full potential in cancer treatment, there needs to be progress in vaccine design, antigen selection, and ways to study the microenvironment of tumors. With more scientific and clinical study, DNA vaccines could become an essential part of the arsenal of breast cancer drugs, giving patients new ways to get better.

Whole-cell vaccines hold potential as an immunotherapy approach for patients with advanced or metastatic breast cancer. These vaccines use tumor cells to stimulate the patient's immune system against cancer cells, allowing for more tailored treatment and better outcomes. To accomplish this goal, the cancer cells are first altered to be more evident to the patient's immune system, then reintroduced into the patient. By using modified breast cancer cells as a vaccine, the immune system can be stimulated to produce T cells that can recognize and attack breast cancer cells.

Because whole tumor cells already have all possible antigens, there is no need to determine which antigen is the most effective to target in a given kind of cancer when utilizing them as the basis for a vaccine. This is a significant advantage of employing whole tumor cells. Whole-cell vaccines have the added benefit of being able to target numerous tumor antigens at the same time, eliminating the issue of tumor antigen loss. Also, the reactions of vaccinated lymphocytes and antibodies can be used to find new tumor antigens or figure out how vital a response to a specific tumor antigen is. This can be accomplished by contrasting immune responses obtained before and after vaccination and associating those responses with the patient's prognosis.

Patients are given cells that carry the same tumor antigens present in their tumor when a whole-cell vaccination is developed using autologous tumor cells, as Keenan and Jaffee (2012) point out (Keenan and Jaffee 2012). This is a benefit of using autologous tumor cells to create a whole-cell vaccination. However, this strategy is difficult since it necessitates the removal of tumor cells and the development of a vaccination line that produces the same quantity of cytokine every time. These steps can take a long time and cost money (Simons et al. 1997). In addition, this method is not the most effective method for treating cancer.

On the other hand, allogeneic cell vaccines present an option by employing several cell lines originating from a variety of cancers within the vaccine itself. This approach makes it more likely that antigens made by the vaccination cells will be found in the patient's tumor. Some of these antigens could be important tumor antigens that are overexpressed or changed in many cases of that type of cancer.

When whole-cell vaccinations use allogeneic cells, there is a chance that the HLA on the patient's cells won't match the HLA on the vaccine cells. This could make the immune system react to alien HLA molecules instead of tumor antigens. Studies have shown that anti-HLA reactions can happen, but they don't stop the response to tumor antigens and are linked to how well the whole-cell vaccine works (Nemunaitis et al. 2006; Thomas et al. 1998). Antigen-presenting cells (APCs) derived from the patient's own immune cells are critical in triggering CD8+ T cell responses to tumor antigens present in irradiated vaccination cells. The presence of foreign major histocompatibility complex (MHC) molecules in these cells has the potential to improve cross-priming efficiency. APCs can effectively display tumor antigens to CD8+ T lymphocytes by integrating foreign MHC molecules, triggering a powerful immune response against tumor cells. The presence of foreign MHC molecules facilitates this technique of cross-priming, which contributes to the potential efficacy of immunotherapy in cancer treatment. This is critical in triggering specific immune responses against cancer cells. Ongoing research attempts to expand our understanding of APC-based immunization techniques and optimize them for more effective cancer treatment (Thomas et al. 2004; Thomas et al. 1998).

According to Simons et al. (1997), genetic modifications have been made to whole-cell vaccines to express cytokines, chemokines, or costimulatory molecules. This modification sets up an immunological response in reaction to the tumor cells' irradiation into the patient. Numerous phase I and phase II investigations have provided evidence that this approach is safe for individuals with various types of cancer. Historically, delayed-type hypersensitivity reactions (DTH) to autologous tumor cells have been used as an indicator of successful immune response induction by vaccines. Interestingly, patients receiving gene-modified vaccination cells have shown a correlation between DTH responses and improved survival outcomes, despite tumor cells themselves not triggering DTH responses (Simons et al. 1997). Prior research conducted before clinical trials has demonstrated the superior efficacy of granulocyte-macrophage colony-stimulating factor (GM-CSF) compared to other cytokines such as IL-2, IL-4, IL-6, TNF-α, and IFN-γ when introduced into irradiated melanoma cells (Dranoff et al. 1993). The comparison of its activity with other cytokines led to this finding. Dendritic cells (DCs), macrophages, eosinophils, and T lymphocytes have all been shown to amass in greater numbers at the injection site when GM-CSF is included in a whole tumor cell vaccination (Simons et al. 1997). Following vaccination, necrotic tumors with a substantial infiltration of T lymphocytes and plasma cells were identified upon resection of lesions (Soiffer et al. 1998).

On the other hand, infiltrates were not present in the tumors of non-vaccinated individuals. Several published research studies have described the reasons why GM-CSF is a superior cytokine for generating an effective T cell response targeted at tumors. According to Dranoff et al. (1993), GM-CSF has the ability to attract DCs to the location of irradiated cells, which in turn stimulates the absorption, processing, and presentation of antigens (Dranoff et al. 1993). By cross-priming CD8+ T cells after ingesting whole tumor cells that release GM-CSF, DCs enhance the T cell response (Thomas et al. 2004).

Cancer vaccines are a significant focus of research in immunotherapy due to their specific targeting and low toxicity. In addition, various immunogens can induce or enhance antitumor responses in vaccinated patients. The idea of an allogeneic tumor cell vaccination is based on the finding that most T cell priming results from host antigen-presenting cells ingesting tumor cells, which then transmit tumor antigens to T cells. This method offers several advantages over the autologous tumor cell vaccination method regarding practical applications. A summary of the limited experience with allogeneic whole-cell vaccinations in breast cancer patients was provided in a study conducted by Dols et al. (2003). In the study conducted by Dols et al. (2003), the immunological effects of allogeneic tumor cell vaccinations were examined in vitro using peripheral blood mononuclear cells from women diagnosed with stage IV breast carcinoma. The phase I study utilized a human leukocyte antigen-A2 matched, CD80-modified, allogeneic, whole-cell vaccination. The results of the study demonstrated favorable immunological and clinical outcomes in stage IV patients, highlighting the potential effectiveness of allogeneic tumor cell vaccinations in this patient population (Dols et al. 2003).

Emens et al. (2009) conducted a study to investigate whether the use of an allogeneic, GM-CSF-secreting tumor vaccine for HER2-positive breast cancer could be enhanced by the administration of cyclophosphamide (CY) and doxorubicin (DOX). In the course of the research, there were a total of 28 patients with advanced breast cancer who were given three injections of the tumor vaccine on a 3-monthly basis, in addition to a boost administered between 6 and 8 months following study admission. To discover the best chemotherapy doses that optimize HER2-specific immunity, patients were treated with a 3×3 factorial design with escalating doses of cyclophosphamide (CY) and doxorubicin (DOX). The primary objectives were to assess the safety and determine the chemotherapy doses that enhance the immune response to the vaccine. The results of the study show that giving an allogeneic, HER2-positive tumor vaccine that secretes granulocyte-macrophage colony-stimulating factor (GM-CSF) as a monotherapy or in combination with cyclophosphamide and doxorubicin is not only well tolerated but also induces the development of HER2-specific immune responses in patients with metastatic breast cancer. This shows that such therapeutic approaches can potentially improve immune-mediated targeting of HER2-positive tumors in advanced breast cancer patients. Furthermore, the study revealed that low-dose cyclophosphamide (CY) has a narrow therapeutic window, and it is recommended that the optimal dose should not exceed 200 mg/m^2. This finding suggests that careful consideration should be given to the dosage of CY to ensure its effectiveness while minimizing potential side effects in the treatment of metastatic breast cancer. Because of how the trial was designed, researchers could determine which drug combinations were most beneficial to patients. It is necessary to do additional research to investigate the effect of chemotherapy on immunity established by the vaccination (Emens et al. 2009).

For the objective of the study, Chen et al. (2014) administered an allogeneic GM-CSF-secreting tumor vaccine to twenty HER2-positive breast cancer patients who were simultaneously undergoing weekly trastuzumab treatment. The research demonstrated a clinical benefit rate of 55% after 6 months, with a progression-free survival (PFS) median of 7 months and an overall survival (OS) median of 42 months. There may be an association between the presence of HER2-specific delayed-type hypersensitivity (DTH) reactions and prolonged progression-free survival and overall survival. These DTH responses were found in seven individuals. According to the study's findings, combining the vaccination with cyclophosphamide and trastuzumab is likely safe and could result in therapeutic advantages; however, additional research is required to optimize therapy combinations and doses to achieve the highest possible level of efficacy (Chen et al. 2014).

In conclusion, the use of whole-cell vaccinations holds promise as a potential therapeutic approach for breast cancer treatment. These vaccinations effectively stimulate the immune system to target and attack cancer cells, potentially leading to improved clinical outcomes. Studies have demonstrated the safety and efficacy of whole-cell vaccines in treating patients with HER2-positive breast cancer. However, further research is necessary to optimize the dosage, timing, and combination therapies associated with these vaccines, aiming to maximize their effectiveness. Overall, whole-cell vaccines have the potential to be a valuable immunotherapeutic treatment option for breast cancer, and continued investigation in this field is crucial.

Nanoparticle-based vaccines are a promising new approach to breast cancer immunotherapy. These vaccines use tiny particles, typically ranging from 10 to 200 nm in size, to deliver vaccine components to immune cells in the body. In addition, the particles can be engineered to have various properties, such as specific targeting or controlled release of the vaccine components.

In a study conducted by Zupani and colleagues (2014), the potential of PEGylated-Poly-lactide-co-glycolic acid-based (PLGA-PEG) nanoparticles (NPs) as advanced delivery vehicles for cancer immunotherapy was investigated. Ovalbumin (OVA) was employed as a model antigen during the synthesis of the NPs, with the addition of glycol chitosan and block co-polymer pluronic F127 to enhance their effectiveness. The results showed that the OVA-loaded PLGA-PEG NPs exhibited a narrow size distribution, measuring an average size of 167 nm, a low polydispersity index (PdI) of 0.167, and a zeta potential close to neutrality. These characteristics indicate their potential suitability as particulate cancer vaccines. The study also revealed an encapsulation efficiency (EE) of 57.5% and a loading capacity (LC) of 29 μg/mg. Interestingly, the incorporation of pluronic F127 into the PLGA-PEG NPs slightly improved their Z-average and zeta potential, but resulted in lower EE and LC values (Zupančič et al. 2014).

Nanoparticles have gained attention in cancer treatment for their potential to deliver therapeutic agents effectively. Despite the highly enhanced penetration and retention (EPR) impact, systemic injection demonstrated that only a tiny percentage of these particles, around 1%, accumulate in the tumor microenvironment (TME). Hence, new delivery systems and strategies are necessary to achieve a practical impact. In this context, plant-virus-based nanoparticles, also known as PVNPs, have emerged as potentially useful building blocks because of their ability to transport therapeutic compounds and serve as an in situ vaccination method. PVNPs that have several functions can load, protect, and govern the targeted release of cargo, all while naturally exhibiting immunomodulatory properties. In addition, PVNPs can facilitate the demise of immunogenic cells and change the TME. Against breast cancer tumors, PVNP-based medication delivery and in situ vaccination have been shown to have a synergistic action, according to research conducted before clinical trials. It is also possible to combine these methods with several additional therapeutic approaches in order to develop an all-encompassing therapy system. The possibility exists for the development of a unique in situ cancer vaccine by combining chemotherapy with PVNP in situ immunization. This would provide patients with breast cancer with improved results and prospective treatment alternatives (Shahgolzari et al. 2022).

Interactions between cellular and non-cellular compartments are present in the breast tumor microenvironment (TME) (Li and Burgess 2020; Mittal et al. 2018; Noh et al. 2021) just as they are in the TMEs of other solid tumors (Li and Burgess 2020). This microenvironment of the tumor is extremely important in the development and evolution of breast cancer. It also plays a significant part in predicting how a patient will respond to treatment, and it serves as a barrier that inhibits conventional methods from effectively treating tumors. Therefore, plant-virus-based nanoparticles (PVNPs) have emerged as a promising platform for delivering therapeutic agents and in situ vaccination due to their ability to target tumor barriers and fine-tune immunological responses (Li and Burgess 2020).

PVNPs have inherent properties that contribute to their capacity to successfully target tumors. Nanoparticulate traits such as composition, size, surface properties, payload, targetability via nano-engineering, and immune-stimulatory potential are among these. These properties boost PVNPs' tumor-targeting ability, making them a promising strategy for cancer therapy (Wen et al. 2018; Shahgolzari 2020). In addition to their tumor-targeting properties, PVNPs have demonstrated efficacy in inhibiting the growth of cancer cells in animal models. This effectiveness can be attributed to several factors, including the structural differences between tumor-associated blood vessels and normal blood vessels, as well as the overexpression of biomarkers on tumor cells compared to normal cells. These unique characteristics of tumor biology can enhance the effectiveness of PVNPs in combating cancer, offering potential therapeutic benefits (Hanahan and Weinberg 2011).

PVNPs find utility in various approaches for treating breast cancer, including both non-targeted (passive) and targeted (active) delivery methods. In passive targeted delivery, PVNPs take advantage of the phenomenon of extravasation and accumulation in tumor tissues, leading to higher concentrations at the tumor site. On the other hand, active targeted delivery involves conjugating PVNPs with specific targeting molecules that are overexpressed on tumor cells. These tumor-targeted immunotherapies can be combined with other treatment modalities. Successful non-targeted delivery relies on the extravasation process and achieving a high concentration of PVNPs within tumors to maximize therapeutic efficacy (Li and Burgess 2020).

Nanoparticle-based vaccines show promise as a potential treatment option for breast cancer. However, more research is needed to fully understand their potential benefits and limitations. Clinical trials are ongoing to determine their safety and effectiveness in humans.

Preclinical research has shown promising results for breast cancer vaccines, and ongoing clinical trials are currently evaluating their safety and effectiveness. The primary objective of these vaccines is to enhance the immune system's ability to recognize and eliminate cancer cells within the body. They can be tailored to target specific cancer cells or antigens and may be used alone or with other treatments such as chemotherapy or immunotherapy.

However, several questions remain regarding the efficacy and limitations of breast cancer vaccines. Optimal dosing and administration schedules are still being investigated, along with determining the potential synergies with other therapies. It is important to note that not all patients may respond equally to breast cancer vaccines, and there may be associated side effects that require further evaluation.

Further research is essential to gain a better understanding of these factors and to identify the patients who are most likely to benefit from breast cancer vaccines. By addressing these concerns, we can enhance the efficacy and implementation of breast cancer treatment approach.

5 Conclusion

In recent years, immunotherapy has emerged as a promising treatment option for breast cancer, harnessing the power of the immune system to target and eliminate cancer cells. Various techniques, including immune checkpoint inhibitors, cancer vaccines, adoptive cell transfer, and oncolytic viruses, have been developed and evaluated in preclinical and clinical studies. Encouraging results from clinical trials have demonstrated significant benefits for some patients with advanced breast cancer.

However, there are still notable challenges that need to be addressed in the field of breast cancer immunotherapy. These include identifying reliable biomarkers to predict patient response, overcoming resistance mechanisms that can limit the effectiveness of immunotherapy, and exploring combination therapies that can enhance treatment outcomes. Despite these challenges, immunotherapy represents a promising new frontier in the field of cancer treatment, particularly for breast cancer.

Continued research and development in breast cancer immunotherapy hold the potential to improve outcomes and offer new avenues for patients. With a deeper understanding of immune response patterns and advancements in treatment strategies, it is hoped that immunotherapy may play an increasingly important role in the control of breast cancer in the coming years.

References

Adams S, Schmid P, Rugo HS, Winer EP, Loirat D, Awada A, Cescon DW, Iwata H, Campone M, Nanda R, Hui R, Curigliano G, Toppmeyer D, O'Shaughnessy J, Loi S, Paluch-Shimon S, Card D, Zhao J, Karantza V, Cortes J (2017) Phase 2 study of pembrolizumab (pembro) monotherapy for previously treated metastatic triple-negative breast cancer (mTNBC): KEYNOTE-086 cohort A. J Clin Oncol 35:1008. https://doi.org/10.1200/JCO.2017.35.15_suppl.1008

Brahmer JR, Tykodi SS, Chow LQM, Hwu W-J, Topalian SL, Hwu P, Drake CG, Camacho LH, Kauh J, Odunsi K, Pitot HC, Hamid O, Bhatia S, Martins R, Eaton K, Chen S, Salay TM, Alaparthy S, Grosso JF et al (2012) Safety and activity of anti-PD-L1 antibody in patients with advanced cancer. N Engl J Med 366:2455–2465. https://doi.org/10.1056/NEJMoa1200694

Chen G, Gupta R, Petrik S, Laiko M, Leatherman JM, Asquith JM, Daphtary MM, Garrett-Mayer E, Davidson NE, Hirt K, Berg M, Uram JN, Dauses T, Fetting J, Duus EM, Atay-Rosenthal S, Ye X, Wolff AC, Stearns V et al (2014) A feasibility study of cyclophosphamide, trastuzumab, and an allogeneic GM-CSF-secreting breast tumor vaccine for HER2+ metastatic breast cancer. Cancer Immunol Res 2:949–961. https://doi.org/10.1158/2326-6066.CIR-14-0058

Cohen EP (2000) Cancer therapy with DNA-based vaccines. Immunol Lett 74:59–65. https://doi.org/10.1016/S0165-2478(00)00250-9

Corti C, Venetis K, Sajjadi E, Zattoni L, Curigliano G, Fusco N (2022) CAR-T cell therapy for triple-negative breast cancer and other solid tumors: preclinical and clinical progress. Expert Opin Investig Drugs 31:593–605. https://doi.org/10.1080/13543784.2022.2054326

Dees S, Ganesan R, Singh S, Grewal IS (2020) Emerging CAR-T cell therapy for the treatment of triple-negative breast cancer. Mol Cancer Ther 19:2409–2421. https://doi.org/10.1158/1535-7163.MCT-20-0385

Disis MLN, Guthrie KA, Liu Y, Coveler AL, Higgins DM, Childs JS, Dang Y, Salazar LG (2023) Safety and outcomes of a plasmid DNA vaccine encoding the ERBB2 intracellular domain in patients with advanced-stage ERBB2-positive breast cancer: a phase 1 nonrandomized clinical trial. JAMA Oncol 9:71–78. https://doi.org/10.1001/jamaoncol.2022.5143

Dols A, Meijer SL, Smith JW 2nd, Fox BA, Urba WJ (2003) Allogeneic breast cancer cell vaccines. Clin Breast Cancer 3(Suppl 4):S173–S180. https://doi.org/10.3816/cbc.2003.s.008

Dranoff G, Jaffee E, Lazenby A, Golumbek P, Levitsky H, Brose K, Jackson V, Hamada H, Pardoll D, Mulligan RC (1993) Vaccination with irradiated tumor cells engineered to secrete murine granulocyte-macrophage colony-stimulating factor stimulates potent, specific, and long-lasting anti-tumor immunity. Proc Natl Acad Sci U S A 90:3539–3543. https://doi.org/10.1073/pnas.90.8.3539

Emens LA, Asquith JM, Leatherman JM, Kobrin BJ, Petrik SVP, Laiko M, Levi J, Daphtary MM, Biedrzycki BA, Wolff AC, Stearns V, Disis ML, Ye X, Piantadosi S, Fetting JH, Davidson NE, Jaffee EM (2009) Timed sequential treatment with cyclophosphamide, doxorubicin, and an allogeneic granulocyte-macrophage colony-stimulating factor-secreting breast tumor vaccine: a chemotherapy dose-ranging factorial study of safety and immune activation. J Clin Oncol 27(35):5911–5918

Hanahan D, Weinberg RA (2011) Hallmarks of cancer: the next generation. Cell 144:646–674. https://doi.org/10.1016/j.cell.2011.02.013

Kagihara JA, Andress M, Diamond JR (2020) Nab-paclitaxel and atezolizumab for the treatment of PD-L1-positive, metastatic triple-negative breast cancer: review and future directions. Expert Rev Precis Med Drug Dev 5:59–65. https://doi.org/10.1080/23808993.2020.1730694

Keenan BP, Jaffee EM (2012) Whole cell vaccines – past progress and future strategies. Semin Oncol 39:276–286. https://doi.org/10.1053/j.seminoncol.2012.02.007

Larkin J, Chiarion-Sileni V, Gonzalez R, Grob JJ, Cowey CL, Lao CD, Schadendorf D, Dummer R, Smylie M, Rutkowski P, Ferrucci PF, Hill A, Wagstaff J, Carlino MS, Haanen JB, Maio M, Marquez-Rodas I, McArthur GA, Ascierto PA et al (2015) Combined nivolumab and ipilimumab or monotherapy in untreated melanoma. N Engl J Med 373:23–34. https://doi.org/10.1056/NEJMoa1504030

Li J, Burgess DJ (2020) Nanomedicine-based drug delivery towards tumor biological and immunological microenvironment. Acta Pharm Sin B 10:2110–2124. https://doi.org/10.1016/j.apsb.2020.05.008

Lopes A, Vandermeulen G, Préat V (2019) Cancer DNA vaccines: current preclinical and clinical developments and future perspectives. J Exp Clin Cancer Res 38:146. https://doi.org/10.1186/s13046-019-1154-7

Maalej KM, Merhi M, Inchakalody VP, Mestiri S, Alam M, Maccalli C, Cherif H, Uddin S, Steinhoff M, Marincola FM, Dermime S (2023) CAR-cell therapy in the era of solid tumor treatment: current challenges and emerging therapeutic advances. Mol Cancer 22:20. https://doi.org/10.1186/s12943-023-01723-z

Mittal S, Brown NJ, Holen I (2018) The breast tumor microenvironment: role in cancer development, progression and response to therapy. Expert Rev Mol Diagn 18:227–243. https://doi.org/10.1080/14737159.2018.1439382

Mizutani N, Luo Y, Mizutani M, Reisfeld RA, Xiang R (2004) DNA vaccines suppress angiogenesis and protect against growth of breast cancer metastases. Breast Dis 20:81–91. https://doi.org/10.3233/bd-2004-20110

Nanda R, Chow LQM, Dees EC, Berger R, Gupta S, Geva R, Pusztai L, Pathiraja K, Aktan G, Cheng JD, Karantza V, Buisseret L (2016) Pembrolizumab in patients with advanced triple-negative breast cancer: Phase Ib KEYNOTE-012 study. J Clin Oncol Off J Am Soc Clin Oncol 34:2460–2467. https://doi.org/10.1200/JCO.2015.64.8931

Nasiri F, Kazemi M, Mirarefin SMJ, Mahboubi Kancha M, Ahmadi Najafabadi M, Salem F, Dashti Shokoohi S, Evazi Bakhshi S, Safarzadeh Kozani P, Safarzadeh Kozani P (2022) CAR-T cell therapy in triple-negative breast cancer: hunting the invisible devil. Front Immunol 13:1018786. https://doi.org/10.3389/fimmu.2022.1018786

NCT00729664 (2015) Multiple ascending dose (MDX1105-01) (Anti-PDL1). https://clinicaltrials.gov/ct2/show/NCT00729664

NCT01570036 (2013) Combination immunotherapy with herceptin and the HER2 vaccine NeuVax. https://clinicaltrials.gov/ct2/show/NCT01570036

NCT01730118 (2013) Ad/HER2/Neu dendritic cell cancer vaccine testing. https://clinicaltrials.gov/ct2/show/NCT01730118

NCT01837602 (2018) cMet CAR RNA T cells targeting breast cancer. https://clinicaltrials.gov/ct2/show/NCT01837602

NCT02349724 (2014) A clinical research of CAR T cells targeting CEA positive cancer. https://clinicaltrials.gov/ct2/show/NCT02349724

NCT02414269 (2015) Malignant pleural disease treated with autologous T cells genetically engineered to target the cancer-cell surface antigen mesothelin. https://clinicaltrials.gov/ct2/show/NCT02414269

NCT02425891 (2022) A study of atezolizumab in combination with nab-paclitaxel compared with placebo with nab-paclitaxel for participants with previously untreated metastatic triple-negative breast cancer (IMpassion130). https://clinicaltrials.gov/ct2/show/NCT02425891

NCT02447003 (2020) Study of pembrolizumab (MK-3475) monotherapy for metastatic triple-negative breast cancer (MK-3475-086/KEYNOTE-086). https://clinicaltrials.gov/ct2/show/NCT02447003

NCT02580747 (2018) Treatment of relapsed and/or chemotherapy refractory advanced malignancies by CART-meso. https://clinicaltrials.gov/ct2/show/NCT02580747

NCT02587689 (2015) Phase I/II study of anti-mucin1 (MUC1) CAR T cells for patients with MUC1+ advanced refractory solid tumor https://clinicaltrials.gov/ct2/show/NCT02587689

NCT02706392 (2021) Genetically modified T-cell therapy in treating patients with advanced ROR1+ malignancies. https://clinicaltrials.gov/ct2/show/NCT02706392

NCT02792114 (2023) T-cell therapy for advanced breast cancer. https://clinicaltrials.gov/ct2/show/NCT02792114

NCT02915445 (2016) EpCAM CAR-T for treatment of advanced solid tumors. https://clinicaltrials.gov/ct2/show/NCT02915445

NCT03060356 (2020) Autologous T cells expressing MET scFv CAR (RNA CART-cMET). https://clinicaltrials.gov/ct2/show/NCT03060356

NCT03635632 (2019) C7R-GD2.CART cells for patients with relapsed or refractory neuroblastoma and other GD2 positive cancers (GAIL-N). https://clinicaltrials.gov/ct2/show/NCT03635632

NCT03682744 (2018) CAR-T intraperitoneal infusions for CEA-expressing adenocarcinoma peritoneal metastases or malignant ascites (IPC). https://www.clinicaltrials.gov/ct2/show/NCT03682744

NCT03696030 (2018) HER2-CAR T cells in treating patients with recurrent brain or leptomeningeal metastases https://clinicaltrials.gov/ct2/show/NCT03696030

NCT03740256 (2023) Binary oncolytic adenovirus in combination with HER2-specific autologous CAR VST, advanced HER2 positive solid tumors (VISTA). https://clinicaltrials.gov/ct2/show/NCT03740256

NCT04020575 (2020) Autologous huMNC2-CAR44 T cells for breast cancer targeting cleaved form of MUC1 (MUC1*). https://clinicaltrials.gov/ct2/show/NCT04020575

NCT04025216 (2019) A study of CART-TnMUC1 in patients with TnMUC1-positive advanced cancers. https://clinicaltrials.gov/ct2/show/NCT04025216

NCT04107142 (2019) Haplo/allogeneic NKG2DL-targeting chimeric antigen receptor-grafted γδ T cells for relapsed or refractory solid tumour. https://clinicaltrials.gov/ct2/show/NCT04107142

NCT04348643 (2020) Safety and efficacy of CEA-targeted CAR-T therapy for relapsed/refractory CEA+ cancer. https://www.clinicaltrials.gov/ct2/show/NCT04348643

NCT04427449 (2020) 4SCAR-CD44v6 T cell therapy targeting cancer. https://clinicaltrials.gov/ct2/show/NCT04427449

NCT04430595 (2020) Multi-4SCAR-T therapy targeting breast cancer. https://clinicaltrials.gov/ct2/show/NCT04430595

NCT04511871 (2023) A phase I trial of CCT303-406 in patients with relapsed or refractory HER2 positive solid tumors. https://clinicaltrials.gov/ct2/show/NCT04511871

NCT04650451 (2020) Safety and activity study of HER2-targeted dual switch CAR-T cells (BPX-603) in subjects with HER2-positive solid tumors. https://clinicaltrials.gov/ct2/show/record/NCT04650451

Nemunaitis J, Dillman RO, Schwarzenberger PO, Senzer N, Cunningham C, Cutler J, Tong A, Kumar P, Pappen B, Hamilton C, DeVol E, Maples PB, Liu L, Chamberlin T, Shawler DL, Fakhrai H (2006) Phase II study of belagenpumatucel-L, a transforming growth factor beta-2 antisense gene-modified allogeneic tumor cell vaccine in non-small-cell lung cancer. J Clin Oncol Off J Am Soc Clin Oncol 24:4721–4730. https://doi.org/10.1200/JCO.2005.05.5335

Nicolás-Morales ML, Luisa-Sanjuan A, Gutiérrez-Torres M, Vences-Velázquez A, Ortuño-Pineda C, Espinoza-Rojo M, Navarro-Tito N, Cortés-Sarabia K (2022) Peptide-based vaccines in clinical phases and new potential therapeutic targets as a new approach for breast cancer: a review. Vaccine 10. https://doi.org/10.3390/vaccines10081249

Noh M-G, Kim SS, Kim YJ, Jung T-Y, Jung S, Rhee J-H, Lee J-H, Lee J-S, Cho J-H, Moon K-S, Park H, Lee K-H (2021) Evolution of the tumor microenvironment toward immune-suppressive seclusion during brain metastasis of breast cancer: implications for targeted therapy. Cancers (Basel) 13. https://doi.org/10.3390/cancers13194895

O'Donnell JS, Teng MWL, Smyth MJ (2019) Cancer immunoediting and resistance to T cell-based immunotherapy. Nat Rev Clin Oncol 16:151–167. https://doi.org/10.1038/s41571-018-0142-8

Pardoll DM (2012) The blockade of immune checkpoints in cancer immunotherapy. Nat Rev Cancer 12:252–264. https://doi.org/10.1038/nrc3239

Peres LP, da Luz FAC, Pultz BA, Brígido PC, de Araújo RA, Goulart LR, Silva MJB (2015) Peptide vaccines in breast cancer: the immunological basis for clinical response. Biotechnol Adv 33:1868–1877. https://doi.org/10.1016/j.biotechadv.2015.10.013

Schmid P, Adams S, Rugo HS, Schneeweiss A, Barrios CH, Iwata H, Diéras V, Hegg R, Im S-A, Shaw Wright G, Henschel V, Molinero L, Chui SY, Funke R, Husain A, Winer EP, Loi S, Emens LA (2018) Atezolizumab and nab-paclitaxel in advanced triple-negative breast cancer. N Engl J Med 379:2108–2121. https://doi.org/10.1056/NEJMoa1809615

Schmid P, Cortes J, Dent R, Pusztai L, McArthur H, Kümmel S, Bergh J, Denkert C, Park YH, Hui R, Harbeck N, Takahashi M, Untch M, Fasching PA, Cardoso F, Andersen J, Patt D, Danso M, Ferreira M et al (2022) Event-free survival with pembrolizumab in early triple-negative breast cancer. N Engl J Med 386:556–567. https://doi.org/10.1056/NEJMoa2112651

Schreiber RD, Old LJ, Smyth MJ (2011) Cancer immunoediting: integrating immunity's roles in cancer suppression and promotion. Science 331:1565–1570. https://doi.org/10.1126/science.1203486

Shahgolzari M, Dianat-Moghadam H, Yavari A, Fiering SN, Hefferon K (2022) Multifunctional plant virus nanoparticles for targeting breast cancer tumors. Vaccine 10. https://doi.org/10.3390/vaccines10091431

Shumway NM, Ibrahim N, Ponniah S, Peoples GE, Murray JL (2009) Therapeutic breast cancer vaccines: a new strategy for early-stage disease. BioDrugs 23:277–287. https://doi.org/10.2165/11313490-000000000-00000

Simons JW, Jaffee EM, Weber CE, Levitsky HI, Nelson WG, Carducci MA, Lazenby AJ, Cohen LK, Finn CC, Clift SM, Hauda KM, Beck LA, Leiferman KM, Owens AHJ, Piantadosi S, Dranoff G, Mulligan RC, Pardoll DM, Marshall FF (1997) Bioactivity of autologous irradiated renal cell carcinoma vaccines generated by ex vivo granulocyte-macrophage colony-stimulating factor gene transfer. Cancer Res 57:1537–1546

Soiffer R, Lynch T, Mihm M, Jung K, Rhuda C, Schmollinger JC, Hodi FS, Liebster L, Lam P, Mentzer S, Singer S, Tanabe KK, Cosimi AB, Duda R, Sober A, Bhan A, Daley J, Neuberg D, Parry G et al (1998) Vaccination with irradiated autologous melanoma cells engineered to secrete human granulocyte-macrophage colony-stimulating factor generates potent antitumor immunity in patients with metastatic melanoma. Proc Natl Acad Sci U S A 95:13141–13146. https://doi.org/10.1073/pnas.95.22.13141

Thomas MC, Greten TF, Pardoll DM, Jaffee EM (1998) Enhanced tumor protection by granulocyte-macrophage colony-stimulating factor expression at the site of an allogeneic vaccine. Hum Gene Ther 9:835–843. https://doi.org/10.1089/hum.1998.9.6-835

Thomas AM, Santarsiero LM, Lutz ER, Armstrong TD, Chen Y-C, Huang L-Q, Laheru DA, Goggins M, Hruban RH, Jaffee EM (2004) Mesothelin-specific CD8(+) T cell responses provide evidence of in vivo cross-priming by antigen-presenting cells in vaccinated pancreatic cancer patients. J Exp Med 200:297–306. https://doi.org/10.1084/jem.20031435
WHO (2022) Fact sheets: cancer
WHO (2023) Global breast cancer initiative implementation framework: assessing, strengthening and scaling-up of services for the early detection and management of breast cancer
Wolchok JD, Kluger H, Callahan MK, Postow MA, Rizvi NA, Lesokhin AM, Segal NH, Ariyan CE, Gordon R-A, Reed K, Burke MM, Caldwell A, Kronenberg SA, Agunwamba BU, Zhang X, Lowy I, Inzunza HD, Feely W, Horak CE et al (2013) Nivolumab plus ipilimumab in advanced melanoma. N Engl J Med 369:122–133. https://doi.org/10.1056/NEJMoa1302369
Yang B, Jeang J, Yang A, Wu TC, Hung C-F (2014) DNA vaccine for cancer immunotherapy. Human Vaccine Immunother 10:3153–3164. https://doi.org/10.4161/21645515.2014.980686
Yang Y-H, Liu J-W, Lu C, Wei J-F (2022) CAR-T cell therapy for breast cancer: from basic research to clinical application. Int J Biol Sci 18:2609–2626. https://doi.org/10.7150/ijbs.70120
You Z, Zhou W, Weng J, Feng H, Liang P, Li Y, Shi F (2021) Application of HER2 peptide vaccines in patients with breast cancer: a systematic review and meta-analysis. Cancer Cell Int 21:489. https://doi.org/10.1186/s12935-021-02187-1
Zupančič E, Silva J, Videira MA, Moreira JN, Florindo HF (2014) Development of a novel nanoparticle-based therapeutic vaccine for breast cancer immunotherapy. Proc Vaccinol 8: 62–67. https://doi.org/10.1016/j.provac.2014.07.011

Molecular Genetics of Breast Cancer: From Mechanistic Insights to the Therapeutic Approaches

Mohammad Muzaffar Mir, Ahmed Yahia Al Ameer, Shahzada Khalid Sohail, Syeda Fatima Rizvi, Jaber Alfaifi, Saba Muzaffar Mir, and Masoud Ishag Elkhalifa Adam

Abstract

Breast cancer is the second most common malignancy in US women, with 3.8 million affected, and 300,590 new cases were expected in 2023. Survival rates have improved due to advancements in diagnostic and treatment modalities, with noninvasive breast cancer showing 90% and 84% survival rates at 5 and 10 years, respectively. Risk factors for breast cancer include genetic alterations, lifestyle factors, and hormonal influences. Classification of breast cancers into intrinsic subtypes has improved early detection accuracy. Genetic testing plays a crucial role in risk assessment and prevention, with mutations in genes like *BRCA1* and *BRCA2* increasing cancer risks. High- and moderate-penetrance gene alterations

M. M. Mir (✉)
Department of Clinical Biochemistry, College of Medicine, University of Bisha, Bisha, Kingdom of Saudi Arabia
e-mail: mmmir@ub.edu.sa

A. Y. Al Ameer
Department of Surgery (Breast and Endocrine Surgery), College of Medicine, University of Bisha, Bisha, Kingdom of Saudi Arabia

S. K. Sohail · S. F. Rizvi
Department of Pathology, College of Medicine, University of Bisha, Bisha, Kingdom of Saudi Arabia

J. Alfaifi
Department of Child Health, College of Medicine, University of Bisha, Bisha, Kingdom of Saudi Arabia

S. M. Mir
Department of Microbiology, Government Medical College, Srinagar, J and K, India

M. I. E. Adam
Department of Internal Medicine, College of Medicine, University of Bisha, Bisha, Kingdom of Saudi Arabia

© The Author(s), under exclusive license to Springer Nature Switzerland AG 2024
Interdisciplinary Cancer Research, https://doi.org/10.1007/16833_2024_247
Published online: 29 May 2024

in *TP53, PTEN, CDH1, ATM, CHEK2, PALB2, RAD51C, RAD51D, NF1, a*nd *MSH6* cause 20–25% breast cancer risk. Population-based genetic studies have enhanced risk detection and understanding of breast cancer subtypes. Functional genomics and polygenic risk scores are essential for identifying target genes and predicting risk. Current treatments for breast cancer include chemotherapy, targeted therapy, and immunotherapies, with ongoing research focusing on improving long-term benefits for patients.

Keywords

Breast cancer · Genetic alterations · Immunotherapy · Polygenic risk scores · Risk factors · Targeted therapy

1 Introduction

Breast cancer (BC) is a complex condition influenced by multiple factors and has become a significant financial and social burden worldwide. Globally, the incidence of female breast cancer has now exceeded that of lung cancer, making it the most often diagnosed form of cancer. Breast cancer is the second most common type of cancer among women in the United States, after skin cancer. It represents 33% of newly diagnosed female malignancies each year. The United States was projected to have 300,590 new breast cancer cases in 2023 with 297,790 females and 2800 males having the disease (Siegel et al. 2022). Presently, the United States has a population of over 3.8 million women who are currently living with or have survived breast cancer. The total number of deaths from breast cancer were 43,700 in 2023, out of which males are 530 and females 43,170 (Siegel et al. 2022). Globally, female breast cancer ranks as the sixth most prevalent cause of mortality. With over 2.3 million new cases and 685,000 deaths in 2020, breast cancer is the most commonly diagnosed cancer worldwide (Arnold et al. 2022). Breast cancer incidence rates vary widely across regions, with rates ranging from <40 per 100,000 females in some Asian and African countries to over 80 per 100,000 in Australia/New Zealand, Northern America, and parts of Europe. By 2040, population growth and aging are expected to lead to over three million new breast cancer cases and one million deaths annually.

The advances in diagnosis and treatment have significantly improved the survival rates for those diagnosed with breast cancer. According to the American Cancer Society, the current 5-year and 10-year overall survival rates for noninvasive breast cancer are about 90% and 84%, respectively (American Cancer Society: Breast Cancer Survival Rates. Available at: https://www.cancer.org/cancer/breast-cancer/understanding-a-breast-cancer-diagnosis/breast-cancer-survival-rates.html). This percentage has consistently increased over the past few decades due to early identification through mammography, breakthroughs in treatment choices such as targeted treatments and immunotherapies and increasing disease knowledge and understanding. Besides, the 5-year survival rates can vary depending on the stage of the cancer, its features, and the patient's age and condition. Access to high-quality healthcare and treatment adherence also affects survival rates. Increasing breast cancer survival rates show that the fight against this illness is progressing.

2 Risk Factors of Breast Cancer

Various risk factors as delineated by the majority of epidemiological studies are summarized in Fig. 1. Genetic alterations are significant contributors to the BC risk to varying extents and will be discussed in more detail in coming sections. Late marriage, first childbirth, and menopause are significant risk factors for BC development. Women who married at age 30 or older had about 7.0% risk compared to those who married at a younger age (~20 years) with a 1.44% risk. Estrogen genotoxicity, lack of breast tissue differentiation, and increased exposure to

Fig. 1 Schematic representation of risk factors of breast cancer

non-estrogenic mutagens are consequences of late marriage and motherhood (Dey et al. 2009). Delayed menopause prolongs estrogen exposure (Surakasula et al. 2014; Dall and Britt 2017) and is also considered a risk factor for BC. A meta-analysis found that early pregnancy and extended breast feeding decreases estrogen receptor (ER)-positive and ER-negative breast cancer risk (Fortner et al. 2019; Anstey et al. 2017). Another breast cancer risk factor is an imbalanced diet. Non-vegetarian and high-animal-fat diets are associated with increased breast cancer risk (Chang et al. 2017; Khodarahmi and Azadbakht 2014). Another substantial risk factor for breast cancer in postmenopausal women and poor disease outcome in all ages is obesity (high waist-to-hip ratio) (Picon-Ruiz et al. 2017; Guo et al. 2018). Research indicates that postmenopausal women with increased BMI $\geq +5.0$ and abdomen circumference ≥ 90 cm are at higher risk for breast cancer (Suzuki et al. 2017). Studies show that alcohol use may increase breast cancer risk by 46% involving DNA methylation and other events (Starek-Świechowicz et al. 2023; Zhou et al. 2022). Higher doses of alcohol are metabolized to acetaldehyde in breast tissue by alcohol dehydrogenase. Acetaldehyde, the metabolic product of alcohol, downregulates BRCA. Accumulated acetaldehyde also disrupts the antioxidant defense system, DNA synthesis, and repair mechanism (Liu et al. 2015; Dumitrescu and Shields 2005). Long-term use of hormonal contraception and hormone replacement therapy also increases breast cancer risk (Satish et al. 2023; Torres-de la Roche et al. 2023).

3 Grading Staging and Molecular Classification of Breast Cancer

3.1 Grading

The classification of breast cancer has undergone significant changes in the past century. In 1925, Greenough introduced the initial histological grading system for breast cancer, which relied on seven histologic characteristics (Giuliano et al. 2018; Van Dooijeweert et al. 2022). In 1957, Bloom and Richardson enhanced and streamlined the grading system and established a correlation between the grading system and survival rates (Giuliano et al. 2018; Meyer et al. 2005). Three histological characteristics of breast cancer, namely tubule development, pleomorphism, and hyperchromatosis/mitosis, were assigned a score ranging from 1 to 3. Elston and Ellis in 1991 revised Bloom and Richardson grading system and used semiquantitative criteria to improve inter-observer variability. This modification of Scarff-Bloom-Richardson grading system is also known as the Nottingham grading system (NGS) and is the most widely used grading system worldwide (Giuliano et al. 2018; Łukasiewicz et al. 2021; Van Dooijeweert et al. 2022).

Tubular differentiation is evaluated based on the proportion of tubule/acinar development. Tumors with more than 75% tubule formation are assigned a score of 1, tumors with tubule formation between 10% and 75% are assigned a score of

2, and tumors with less than 10% tubule formation are assigned a score of 3. Grade score does not include hyperchromatic figures, and the mitotic count is conducted in the most proliferative and typically solid region of the tumor. A score of 1–3 is assigned to indicate the degree of mitotic increase within a specific microscopic field area. In NGS, the mitotic count is the primary determinant of the grade score, with nuclear pleomorphism and tubule formation being secondary factors (Van Dooijeweert et al. 2022). Nuclear pleomorphism is assessed in the least differentiated region of the tumor. A score of 1–3 is given based on the deviation in nuclear size and shape compared to that of a normal cell. The final score (3–9) is determined by adding together the three scores. The final score of 3–5 corresponds to grade I tumors, which are well differentiated. Scores of 6–7 correspond to grade II tumors, which are moderately differentiated, and scores of 8–9 correspond to grade III tumors, which are poorly differentiated (Van Dooijeweert et al. 2022).

NGS functions as a stand-alone prognostic indicator. It has been incorporated in Nottingham prognostic index (NPI) along with tumor size and lymph node status to determine the chances of survival in operable patients. With the advancement of screening practices, small-sized and lymph node–negative tumors are being diagnosed more often. This further increases the clinical importance of histologic grade. Histologic grade is a component of clinical guidelines for breast cancer. The latest updates from the American Joint Committee on Cancer (AJCC) and Union for International Cancer Control (UICC) include the grading in the prognostic stage grouping of patients (Sawaki et al. 2019; Ginter et al. 2021).

3.2 Staging

The American Joint Comittee on Cancer-Tumor size, Nodal status, Metastasis (AJCC-TNM) staging system categorizes breast cancer into five stages (0–IV) based on the clinical and pathological characteristics of the tumor. Tumor size is represented by the variable T, regional lymph node status is represented by the variable N, and distant metastases are represented by the variable M (Tsang and Tse 2020). This staging system is of prognostic significance and reflects the size of the tumor. However, it is not possible to correctly predict the response to treatment and clinical results. The diagnostic and therapeutic approach to breast cancer has been enhanced due to advancements in our understanding of molecular pathways and gene expression profiling.

The AJCC has integrated biomarkers into the staging systems for prostate and thyroid malignancies. The eighth edition of the AJCC-TNM staging system (2017) for breast cancer utilized a similar technique, using prognostically significant biomarkers to establish the prognostic stage group (Giuliano et al. 2018; Łukasiewicz et al. 2021). The histologic grade, estrogen receptor (ER), progesterone receptor (PR), human epidermal growth factor receptor 2 (HER2) status, and multigene assay were incorporated into the anatomic TNM staging system. This exhibits superior prognostic and predictive efficacy compared to traditional staging systems. Before any treatment, clinical prognostic stage groups are assigned, while pathologic stage groups are assigned after resection. The results of biomarkers before treatment and the anatomical information after surgery (including T and N status) determine the final pathologic prognostic stage group, which is the most precise indicator of the outcome. The pathologic prognostic staging is performed prior to any neoadjuvant therapy. If a patient undergoes neoadjuvant therapy prior to tumor removal, the remaining tumor size and lymph node status will be indicated by ypT and ypN, respectively, and will serve as indicators of the treatment's effectiveness.

The AJCC eighth edition incorporated the commercially accessible 21-gene assay, Oncotype DX, into the staging system (Giuliano et al. 2018). This assay is particularly beneficial for tumors that are small in size (<5 cm), node negative, and test positive for ER/PR while testing negative for HER2. An allocation of a recurrence score (RS) is determined based on this assay. Research indicates that among individuals in the low-risk group (with a risk score of less than 11 RS), 98.6% of cases treated alone with endocrine therapy experience a disease-free survival period of 6.9 years. Hence, systemic chemotherapy is unnecessary for this particular cohort. The addition of prognostic factors to the anatomic stage may upgrade or downgrade a tumor to give more accurate prognostic information. A patient diagnosed with an anatomically staged IA triple-negative tumor will be reclassified as prognostically staged IIA (upstaged). Likewise, if a patient has an anatomic stage IIIA, the prognosis stage will be IB if the tumor is grade 1, ER positive, and HER2 positive (downstaged). The AJCC-TNM staging system is more intricate than classic anatomic staging. However, a user-friendly smartphone application called TNM8 Calculator has been developed to assess the conventional TNM and prognosis stage of breast cancer. The UICC has not made it compulsory to employ biomarkers for the staging of breast cancer. Instead, they have adopted the TNM staging system and incorporated a prognostic factor grid. This is a result of the constraints in the availability and evaluation of biomarkers across various countries.

3.3 Histologic Classification

Over 20 histologic types of invasive breast cancer have been established based on the microscopic appearance of the tumor (Tsang and Tse 2020). Various criteria are employed to categorize a tumor, including the type of tumor cells (such as tumors with apocrine differentiation), the presence of extracellular secretion (mucinous carcinoma), the architectural pattern (papillary and medullary carcinoma), and the immunohistochemical profile (carcinoma with neuroendocrine features). The common type, comprising approximately 80% of all breast cancer cases, is invasive ductal carcinoma, specifically classified as not otherwise specified (IDC-NOS). Additional distinct subtypes include invasive lobular (10%), papillary, micropapillary, mucinous, cribriform, tubular, medullary, metaplastic, and apocrine carcinomas (Tsang and Tse 2020; Sohail et al. 2020).

3.4 Molecular Classification

Perou et al. in 2000 introduced the notion of molecular classification of breast cancer. This was based on gene expression profiling (GEP) and classified all types of breast carcinoma, independent of histologic classification, into four classes, namely luminal, HER2-overexpression, basal-like breast cancers (BLBC), and normal-like tumor. Later, normal-like type was considered an artifact and removed from classification. Further studies divided the luminal type into luminal A and luminal B. At present, there are four distinct molecular categories of breast cancer determined by mRNA GEP: luminal A, luminal B, HER2 enriched, and basal-like. These subtypes are referred to as intrinsic subtypes and are regarded as the most significant predictive factor for the fate of early breast cancer, in addition to tumor size and nodal status (Łukasiewicz et al. 2021; Tsang and Tse 2020; Johnson et al. 2021). These molecular subtypes exhibit distinct variations in terms of occurrence, radiological characteristics, disease advancement, treatment response, and prognosis. The determination of appropriate treatment options for emerging targeted treatments is contingent upon the intrinsic subtype of breast cancer.

The GEP approach is both costly and time-intensive, and its availability is limited. Immunohistochemistry (IHC) is commonly employed in the histological assessment of breast cancer to detect the presence of ER, PR, and HER2. These markers are autonomous indicators that can predict and forecast the effectiveness of hormonal and anti-HER2 therapy (Tsang and Tse 2020; Elston and Ellis 1991). IHC data are being utilized as a substitute for the molecular categorization of breast cancer. Nevertheless, a significant discrepancy of up to 30% has been observed in the data obtained from IHC and GEP subtyping. Certain scenarios necessitate precise intrinsic subtyping using GEP. Oncotype Dx and Mammaprint are commercially available and verified for risk evaluation by delivering a numerical score. They are used to guide treatment decisions. For example, an early-stage ER+, PR+, and HER2- tumor with a low recurrence score will be treated by endocrine therapy alone, whereas the same tumor with a high recurrence score will need adjuvant systemic chemotherapy. The recent edition of AJCC manual has classified ER+, PR+, HER2- as luminal subtype, and HER2+ (regardless of hormonal expression) as HER2 subtype.

3.4.1 Luminal A Subtype

Luminal A subtype accounts for around 40% to 50% of all cases of invasive breast cancer, and tumors are characterized as ER positive, PR positive, and HER2 negative, with a low proliferative index (Ki 67 $< 20\%$). Usually, the tumors in this particular subtype are of a low grade and exhibit the most favorable prognosis. The mammography reveals irregular lumps without any accompanying calcifications. These tumors are best treated with endocrine therapy (Yin et al. 2020; Ginter et al. 2021).

3.4.2 Luminal B Subtype

Luminal B tumors are characterized as HER2 negative, with varying levels of ER/PR expression and a higher proliferative index (Ki67+ $\geq 20\%$). In contrast to luminal A, these tumors exhibit aggressive behavior and are characterized by a higher grade and a bad prognosis. On mammography, they also show irregular masses without associated calcifications but have more frequent involvement of axillary lymph nodes at the time of presentation. They often require chemotherapy in addition to endocrine therapy (Yin et al. 2020; Tsang and Tse 2020).

3.4.3 HER2-Enriched Subtype

HER2-enriched breast cancer subtype accounts for 15% of cases, characterized by the presence of HER2 positive results on immunohistochemistry or HER2 overexpression on in situ hybridization. The intrinsic subtype of GEP and IHC overlap with that of the luminal B type. They are aggressive tumors, present in higher stage, and have an overall poor prognosis with a high relapse rate. Mammography commonly shows a mass with pleomorphic calcifications or as calcifications alone. The response to anti-HER2 therapy (Herceptin) with chemotherapy has markedly improved the outcome (Yin et al. 2020; Tsang and Tse 2020; Meyer et al. 2005).

3.4.4 Triple-Negative Breast Cancer (TNBC)

TNBC refers to cancers that lack the expression of ER, PR, and HER2. Basal-like cancer is a subtype of TNBC and represents about 15% of invasive breast cancers. Basal-like tumors are triple negative but show IHC expression of basal cytokeratin (CK5/6) and proliferation markers. About 80% of TNBCs are basal-like cancers; hence, these terms are used interchangeably in the literature. TNBC has the following subtypes on GEP: basal-like, mesenchymal, mesenchymal stem-like, immunomodulatory, luminal androgen receptor, and unspecified. TNBC mostly occurs in premenopausal and African American women and is commonly associated with BRCA mutation. TNBCs are the most aggressive among all molecular subtypes of breast cancer. Their size, grade, proliferation index, and rate of recurrence are higher than other intrinsic subtypes. They frequently relapse within 5–10 years after diagnosis. Due to rapid growth and larger size, they mostly present with palpable lumps. Endocrine or targeted therapy cannot be used in these tumors, and the only treatment option available is adjuvant and neoadjuvant chemotherapy. The high proliferation index and lack of ER expression make them good candidates for chemotherapy, and studies have shown a high response rate, yet the prognosis is poor. This phenomenon is called triple-negative paradox and is currently an area of active research. In summary, the technical difficulties and cost-intensive GEP make it difficult to implement in all the settings. Although imperfect, IHC surrogate is widely used for molecular classification. Oncologic guidelines for therapeutic decisions are based on IHC results. Molecular subtypes with their most common IHC expression are as follows (Tsang and Tse 2020):

- Luminal A (ER+, PR+, HER2−, and Ki67 $<$20%)
- Luminal B (ER+, PR+ $<$20%, HER2−, and/or Ki67 $\geq$20%)
- HER2 overexpression (ER−, PR−, and HER2+)
- Triple negative (ER−, PR−, and HER2−)

Wherever available, molecular classification should be used as an adjunct to histopathologic classification.

4 Genetic Alterations in Breast Cancer

The role of genetics in BC has been the subject of extensive research over the past few decades that has led to significant advances in our understanding of the disease. Genetic factors play a key role in determining an individual's risk of developing BC (Torabi Dalivandan et al. 2021). Understanding the genetic basis of BC has important implications for both the prevention and the treatment of the disease (Torabi Dalivandan et al. 2021; Roberts et al. 2023). Genetic testing can help in risk stratification and targeted screening and laying out elaborate preventive modalities (Evans et al. 2022). Ultimately, a better understanding of the genetic factors involved in BC can lead to the development of more effective and personalized treatment options. Different types of genes including oncogenes, tumor suppressor genes, and DNA repair genes have been implicated in BC to varying degrees. Depending upon their BC susceptibility, these genes are categorized into high, moderate, intermediate, and low risk genes as depicted in Fig. 1 and are discussed as follows:

4.1 High-Risk BC Genes

BRCA1 and *BRCA2* (BReast-CAncer susceptibility genes 1 and 2) are human genes that produce tumor suppressor proteins and are involved in the repair process of damaged DNA and prevent cell growth from becoming uncontrollable. Mutations in these genes can lead to an increased risk of developing BC and other types of cancer. Specifically, mutations in *BRCA1* and *BRCA2* are associated with hereditary breast and ovarian cancer syndrome, and females carrying *BRCA1* and *BRCA2* pathogenic variants have an approximately 70% risk of having breast cancer by age 80. *BRCA1* and *BRCA2* variants in males lead to higher risks of BC susceptibility of ~1–5% and 5–10%, respectively, as compared to general male population (Filippini and Vega 2013; Torabi Dalivandan et al. 2021).

BRCA1 and BRCA2 proteins are crucial for DNA damage repair, supporting chromosomal structure integrity by repairing double strand breaks (DSBs). They activate cell cycle checkpoints, repair DNA by homologous recombination (HR) and apoptosis, and modify chromatin. The wild-type allele's loss of heterozygosity (LOH) with inherited pathogenic mutations in *BRCA1* or *BRCA2* is predicted to cause gene loss of function (LOF) and genomic instability during breast cancer development. More than 2,200 and 2,600 pathogenic variants of *BRCA1* or *BRCA2* genes, respectively, have been detected, and mostly the variations are found throughout each gene's coding sequence that lead to truncated proteins and LOF. Despite being substantial genetic risk factors for breast cancer, pathogenic mutations in *BRCA1* and *BRCA2* only contribute to ~20% of familial risk. Carriers of *BRCA1* mutations are more likely to lack ER, PR, and HER2-neu receptors and have a somatic *P53* mutation. In addition to BC, *BRCA1* and *BRCA2* gene mutation carriers are at higher risk for colon, prostate, pancreatic, melanoma, and gastric cancers. The most common mutations include minor insertions, deletions, nonsense mutations, missense mutations, premature transcription termination, and splicing issues. Frameshifts result from deletion and insertion mutations. Splicing site mutations cause nonfunctional proteins. Most breast cancer–causing mutations in *BRCA1* and *BRCA2* genes create truncated protein products due to nonsense, frameshift, and splicing alterations (Muranen et al. 2023).

TP53, popularly known as the "guardian of genome", is located on 17p13 and has been implicated in a variety of cancers including BC. The genetic variations in *TP53* contribute to the lifetime risk of having BC to an extent of 56–90% (Sidransky et al. 1992; Torabi Dalivandan et al. 2021). The pathogenic variants of *TP53* have compromised capacity in regulating cell cycle arrest and apoptosis. Most *p53* alterations occur in the DNA-binding domain and result from missense point mutations that hinder its ability to activate transcription. These mutations exhibit a strong correlation with specific tumor subtypes and are correlated with a worse likelihood of survival in patients with BC. The incidence of *p53* mutation is the least common in the luminal-like subtype, with a frequency of 26%. Among the luminal subtypes, luminal-A exhibits a lower mutation frequency of 17% compared to luminal-B, which has a mutation frequency of 41%. Basal-like tumors exhibit the highest occurrence of *p53* mutation, with a frequency of 88%. In comparison, approximately 53% of HER2-amplified tumors contain *p53* mutations (Kaur et al. 2018). A significantly poorer prognosis is observed in cases when there is a presence of *p53* mutation and *HER2* overexpression, as compared to other subtypes. Research has demonstrated that inflammatory BC exhibits a greater prevalence of *p53* mutations (50%) compared to non-inflammatory BC (20–30%). A recent study revealed that the presence of *p53* mutations is linked to reduced overall survival in metastatic BC, regardless of the subtype (Marvalim et al. 2023). Patients with advanced-stage and high-grade tumors exhibit a greater prevalence of *p53* mutations. Furthermore, these genetic alterations have been linked to subtypes of BC that exhibit aggressive characteristics, such as TNBC or HER2-amplified BC (Marvalim et al. 2023).

PTEN, also known as "phosphatase and tensin homolog", suppresses tumors and negatively regulates the *PI3K*/Akt pathway and contributes to about 40–60% of absolute lifetime risk of BC. Loss of *PTEN* activity is found in primary and metastatic malignancies like BC (Pradella et al. 2014; Sajjadi et al. 2021; Huber-Keener 2022). Tumor growth, progression, and therapeutic resistance are linked to this deficit. *PTEN* status has been studied for prognostic and predictive purposes in BC. Patients with HER2+ breast tumors who respond poorly to trastuzumab have altered *PTEN* and Akt isoforms. *PTEN* expression may also be used to assess BC mismatch repair. This may help identify immune-checkpoint blockage candidates, including hormone receptor-positive tumors. Recent research shows that *PTEN* loss is related to a worse prognosis but does not cause trastuzumab resistance in HER2+

BC patients. A comprehensive review of 27 studies involving 10,231 BC cases suggests that loss of *PTEN* may indicate aggressive BC. However, *PTEN* status has not been linked to patient outcomes in later clinical and translational studies. The level of reliability of *PTEN* testing in clinical practice for BC patients is still uncertain.

CDH1, also known as E-cadherin gene 1, located on 16q22, is the sole germline abnormality linked to hereditary diffuse gastric and lobular breast cancers (Corso et al. 2020; Torabi Dalivandan et al. 2021). Its product is a calcium-dependent cell-to-cell adhesion molecule and a tumor suppressor protein. The *CDH1* aberrations contribute to about 60% of the absolute lifetime risk of BC. Through intercellular adhesion complexes, the E-cadherin protein is essential for forming and sustaining polarized and differentiated epithelia. Deregulation of this invasion suppressor molecule is common in advanced epithelial carcinomas.

In BC, deletion or dysregulation of E-cadherin can affect tumor infiltration and metastasis by disrupting the cadherin-catenin complex, leading to decreased cell adhesion and increased cell movement. Somatic *CDH1* inactivation is linked to aggressive breast cancer, including lymphovascular invasion and metastases in axillary lymph nodes (Corso et al. 2020; Torabi Dalivandan et al. 2021).

4.2 Moderate-Risk BC Genes

ATM *(Ataxia-Telangiectasia Mutated)* The *ATM* gene, located on chromosome 11q23, encodes a 350-KDa protein. It functions as an onco-suppressor and is a member of the superfamily of phosphatidylinositol 3-kinase–related protein kinases. When double-strand breaks in DNA activate the ATM protein, it adds phosphate groups to a gene that codes for a protein kinase, which plays a crucial part in the DNA damage response. Homozygous germline pathogenic mutations in the *ATM* gene are linked to ataxia-telangiectasia, while heterozygous mutations increase the likelihood of developing breast cancer. The genetic alterations in the *ATM* gene confer 25–40% lifetime risk of having BC (Renwick et al. 2006; Torabi Dalivandan et al. 2021). The risk of breast cancer is two- to fivefold higher for obligatory heterozygous ATM mutation carriers, according to recent epidemiologic research. Radiation-treated BC patients with mutant ATM variations are susceptible to second malignancy sooner than those without ATM mutations.

CHEK2 (checkpoint kinase2), located on chromosome 22 (22q12.1), is a tumor suppressor gene responsible for encoding the protein CHK2, which is a serine-threonine kinase. *CHK2* plays a role in the process of DNA repair, as well as in causing cell cycle arrest or death when DNA damage occurs. Alterations to the *CHEK2* gene have been associated with various types of malignancies, including BC, and contribute to about 27–40% lifetime risk of BC (Torabi Dalivandan et al. 2021; Yadav et al. 2023). A significant observation is that the removal of a solitary DNA nucleotide at position 1100 in exon 10 (1100delC) results in an inactive form of the CHK2 protein, which is shortened at the kinase domain. Dysfunction of the CHK2 protein results in uncontrolled cell proliferation, DNA damage accumulation, and formation of tumors. Studies have shown that women who have harmful genetic variations in the *CHEK2* gene are more likely to develop premenopausal breast cancers that are ER+ and HER2−. These types of tumors are resistant to treatments comprising endocrine monotherapies.

PALB2 (partner and localizer of *BRCA2*), located on the chromosome 16p12.2, is a tumor suppressor gene that plays a key role in the homologous recombination (HR) pathway as mediator between *BRCA1* and the *BRCA2/RAD51* complex. It is also responsible for *BRCA2* nuclear localization and DNA damage repair. The loss of both *PALB2* alleles causes the activation of the nonhomologous end joining, with consequent genomic instability. Pathogenic variants in *PALB2* are associated with an increased risk of breast and pancreatic cancers. It is estimated that 0.6–3% of patients with BC harbor a pathogenic variant in *PALB2*, and approximately 1% of patients with TNBC carry such a mutation. The overall lifetime risk of having BC with mutations in *PALB2* gene ranges between 25% and 40% (Torabi Dalivandan et al. 2021; Yadav et al. 2023).

BARD1 gene, also referred to as *BRCA1*-associated RING domain protein 1, is located on chromosome 2 and is encoded by the sequence 2q34-q35. Its protein product is composed of 777 amino acids and consists of an N-terminal RING-finger domain, three ankyrin repeat domains, and two tandem *BRCA1* C-terminal domains. The *BARD1* protein structure has similarities to that of *BRCA1*, although it diverges from that of *BRCA2*. The N-terminal RING finger domains of *BARD1* and *BRCA1* possess the capacity to interact and form a heterodimer, leading to the establishment of a stable complex. The full-length *BARD1* protein has tumor suppressor characteristics, regardless of whether it operates as a heterodimer or through *BRCA1*-independent mechanisms. However, the unconventional forms of *BARD1* splice variants exhibit carcinogenic characteristics. The two main isoforms associated with the progression of BC are *BARD1β* and *BARD1δ* (Torabi Dalivandan et al. 2021; Graffeo et al. 2022). The mutated *BARD1* genes have the potential to elevate the probability of developing breast cancer by around 20% in families lacking the *BRCA1* and *BRCA2* gene abnormalities (Torabi Dalivandan et al. 2021). The *BARD1-BRCA1* heterodimer serves as a vital E3 ubiquitin ligase in various cellular regulation functions. Ubiquitin primarily functions to enable the binding of ubiquitin molecules to different proteins, thus marking them for subsequent degradation. The *BARD1-BRCA1* heterodimer is essential in the DNA damage response mechanism as it promotes apoptosis through the stabilization of *p53*.

RAD51C (*RAD51* paralog C) gene is situated on chromosome 17q23 and has a role in two distinct subcomplexes: one including *XRCC2*, *RAD51B*, and *RAD51D* and the other involving *XRCC3*. *RAD51C* participates in DNA repair and plays many roles in the DNA damage response and the preservation of genomic integrity. *RAD51C* plays a crucial role in the repair of DNA by homologous recombination. Multiple potentially harmful *RAD51C* mutations have been detected in breast and ovarian cancer families that do not have the *BRCA1* and *BRCA2* gene mutations. The majority of women who possess a defective *RAD51C* gene face a roughly 20% probability of acquiring breast cancer over their lifespan (Torabi Dalivandan et al.

2021; Graffeo et al. 2022). Notably, *RAD51C* is situated in the chromosomal region 17q23, which undergoes amplification in primary breast cancer tumors. The MCF-7 breast cancer cell line harbors a fusion gene called *RAD51C-ATXN7*, which is composed of *RAD51C* exons 1–7 and *ATXN7* exons 6–13. This fusion gene provides additional evidence of its involvement in breast cancer.

RAD51D (*RAD51* paralog D) gene, situated on chromosome 17 at position 17q12, plays a crucial role in DNA repair and is specifically engaged in the homologous recombination process. BC patients who test negative for *BRCA1/2* mutations may possess harmful *RAD51D* germline mutations, which increase their likelihood of acquiring breast cancer by around 20% throughout their lifetime. *RAD51D* is predominantly a gene that confers a modest level of susceptibility to BC. Mouse cells lacking *RAD51D* display significant genomic instabilities, including aneuploidy, chromosomal segments, deletions, rearrangements, and a spontaneous elevation in random mutagenesis, which ultimately leads to the development of tumors. Recent research indicates that individuals with *RAD51D* mutations experience worse survival rates.

NF1, also known as neurofibromatosis1 is a tumor suppressor gene and is located in chromosome 17q11. Its protein product, neurofibromin, plays a crucial role in various cell signaling pathways, including *Ras/MAPK*, *Akt/mTOR*, *ROCK/LIMK/* cofilin, and cAMP/PKA pathways. Neurofibromin is responsible for regulating essential cellular processes such as proliferation and migration. Germline mutations in the *NF1* tumor suppressor gene, whether inherited or occurring spontaneously, are responsible for neurofibromatosis type 1. This autosomal dominant disorder is the most prevalent condition in humans that predisposes individuals to develop tumors. *NF1* has been linked to the development of contralateral breast cancer as well as the initial occurrence of primary breast cancer. This gene poses a moderate risk and increases the lifetime risk of breast cancer by up to 25% (Torabi Dalivandan et al. 2021; Evans et al. 2022).

Women diagnosed with NF1 who develop breast cancer have a significantly lower chance of long-term survival compared to those without NF1. The 20-year survival rate for NF1 patients is only 37%, as compared to 84% of non-*NF1* BC patients of similar ages.

MSH6 (MutS homolog 6) gene, located on chromosome 2p16.3, codes for a DNA mismatch repair protein belonging to the MutS family. The protein product has the role of identifying mismatched nucleotides and getting them ready for repair. Various types of mutations, such as missense mutations, nonsense mutations, silent mutations, frameshift insertions and deletions, and in-frame deletions, are commonly seen in numerous malignancies, including BC. The pathogenic variations of this have been documented in approximately 1.5% of breast cancers. *MSH6* genetic mutations elevate the lifetime risk of BC approximately by 20% (moderate risk) (Torabi Dalivandan et al. 2021; Nassar et al. 2022).

4.3 Low-Risk BC Genes

Disease-causing mutations in high- and moderate-penetrance genes account for 20–25% of breast cancer susceptibility (Fig. 2). The unaccounted-for heritability is likely caused by a mix of rare and common genetic risk variants with varying population impacts (Torabi Dalivandan et al. 2021). Most prevalent low-risk variants with a minor allele frequency (MAF) >0.05% have a risk of less than 1.5-fold. Population-based genetic association studies have become increasingly useful in detecting risk variations during the past decade. The techniques of these investigations divide them into three main categories. Candidate gene association studies, multistage genome-wide association studies (GWAS), large-scale GWAS, and meta-analyses are used. The latest breast cancer GWAS used the Illumina OncoArray BeadChip technology to genotype 122,977 cases and 105,974 controls of European ancestry utilizing 570,000 SNVs. This study also included 14,068 breast cancer cases as well as 13,104 control specimens of East Asian ancestry.

After combining their findings with the iCOGS and GWAS datasets, Michailidou et al. in 2017 found 65 new breast cancer–risk loci in their thorough investigation. Noncoding DNA areas around these loci often contained regulatory elements and transcription factor binding sites (TFBSs). The comprehensive GWAS have indicated that common low penetrance risk alleles account for 18% of breast cancer genetic influence (Mavaddat et al. 2019), as shown in Fig. 2. However, thousands of common risk alleles may remain undiscovered. New risk alleles can be found using several essential methods.

GWAS-identified low-risk alleles face challenges; 60–80% of breast cancer–associated mutations occur in non-protein-coding DNA. These non-protein coding

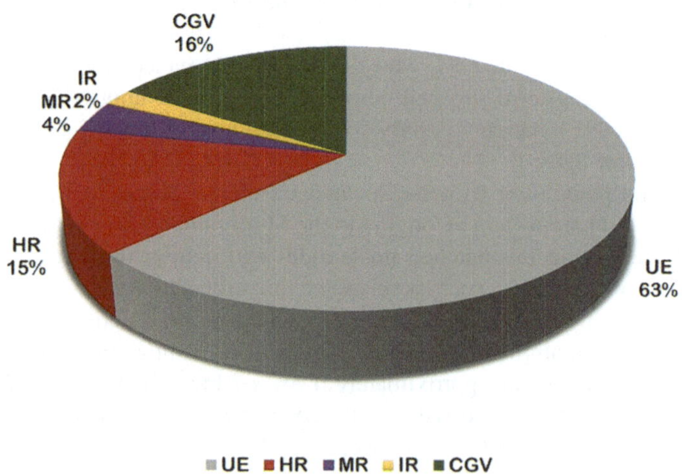

Fig. 2 Schematic illustration of breast cancer genes with proportional susceptibility risk. *UE* unexplained inheritance, *HR* high-risk genes (*BRCA1/2*, *TP53*, *PTEN*, and CDH1), *MR* moderate risk genes (*ATM*, *CHEK2*, *PALB2*, *BARD1*, *RAD51C*, *RAD51D*, *NF1*, *MSH6*), *IR* intermediate risk genes

sections, or noncoding DNA, have no effective genetic code. GWAS risk alleles have low individual risks; therefore, each mutation should have few functional effects. Few functional validations of noncoding risk alleles have shown how difficult it is to understand their biological impact. Noncoding risk variations can affect regulatory elements and TFBSs, which affect target susceptibility genes and gene expression networks. The complexity of tissue-specific transcriptional regulation makes interpreting these noncoding changes difficult. These tasks involve finding the genetic variants that cause illness, finding the genes at risk in these genetic locations, understanding how they contribute to disease development, and determining how these risk variants affect gene expression. As discovery and function diverge, functional genomics (FG) is essential. The main goal is to understand the target gene and gain mechanistic insights beyond risk variations. FG finds causative alterations and genes using bioinformatics, transcriptomics, and proteomics. It also uses molecular and cell biology to prove causality and identify biochemical pathways (Gasperskaja and Kučinskas 2017).

4.4 Genetic Association of BC Subtypes

A lot of recent research has helped in establishing a link between breast cancer subtypes and both high- and moderate-risk BC genes. *BRCA1* mutations have been associated with TNBC, while *TP53* germline alterations may be more common in *HER2*-positive BCs (Wilson et al. 2010; Melhem-Bertrandt et al. 2012). The large number of breast cancer cases in Breast Cancer Association Consortium's recently released targeted sequencing analysis revealed subtype-specific associations among all cancer risk genes. *ATM* and *CHEK2* genes were more strongly linked to ER^{+} BC, high odds ratios. ER-negative breast cancer had stronger correlations for *BARD1*, *BRCA1*, *BRCA2*, *PALB2*, *RAD51C*, and *RAD51D* (Roberts et al. 2023). TNBC is more strongly correlated with *BARD1*, *BRCA1*, and *BRCA2* than other ER-negative breast cancers. *CHEK2* and *CDH1* are connected to ER-positive breast cancer, while *BRCA1* is linked to ER-negative breast cancer (Roberts et al. 2023). However, the new CARRIERS study shows breast cancer subtype–specific hazards. The risk of ER-negative breast cancer and TNBC is linked to *BARD1*, *RAD51C*, and *RAD51D*. It is also reported that *ATM*, *CDH1*, and *CHEK2* pathogenic mutations increase ER-positive breast cancer risk (Roberts et al. 2023). Fine-mapping the breast cancer risk loci identified by GWAS showed that out of 196 strong risk signals, 66 (34%) were associated with ER-positive breast cancer (*CCND1*, *CHEK2*, and *FGFR2*), 29 (15%) with ER-negative disease (*BRCA2*, *CREBBP*, and *ESR1*), and the remaining 51% with similar risks for both (Fachal et al. 2020).

Commonly discovered low-risk alleles utilizing GWAS face challenges. Most breast cancer–associated mutations (60–80%) are in nonprotein-coding DNA. Non-protein-coding regions lack a trustworthy genetic code to identify their function. GWAS risk alleles have a lower individual risk, implying that any mutation will have little functional effects. Few functional validations of noncoding risk alleles have shown how difficult it is to understand their biological impact. Noncoding risk variations affect regulatory elements and TFBSs which affect target susceptibility genes and gene expression networks. The complexity of tissue-specific transcriptional regulation makes interpreting these noncoding changes difficult. The contemporary challenges include connecting these variants causally to disease development, characterizing the gene(s) vulnerable at-risk regions, and understanding how these variants alter gene expression through regulatory mechanisms. These challenges will be circumvented by using the approach of "functional genomics" that uses a combination of strategies including bioinformatics, transcriptomics, and proteomics. It will enrich the mechanistic insights beyond the risk variant detection (Gasperskaja and Kučinskas 2017).

4.5 Polygenic Risk Scoring in BC Prediction

Polygenic risk scores (PRS) are crucial to breast cancer risk prediction and could improve screening and prevention. PRS accounts for around 30% of breast cancer heredity by combining GWAS risk from single nucleotide polymorphisms (SNPs). PRS's more personalized risk assessment helps identify women at higher risk of breast cancer and enables tiered screening and prevention. PRS have improved breast cancer risk prediction and will improve personalization.

Well-validated BC PRS panels include SNP18, SNP77, SNP143, and SNP313. SNP313 from Breast Cancer Association Consortium studies predicts breast cancer risk with an OR of 1.61 (95% CI 1.57–1.65) in white European populations (Roberts et al. 2023). PRS panels are important for BC subtyping. In 2019, Mavaddat et al. used the SNP77 and SNP313 PRS in 11,428 BC patients and 18,323 controls and found that both were predictive for ER-positive and ER-negative breast cancer as well as overall BC (Mavaddat et al. 2019). As expected, SNP313 discriminated better than SNP77. Subtype-specific disease odds ratios for ER-positive breast cancer and overall breast cancer were greater than for ER-negative tumors in both the 77-SNP and 313-SNP PRS (Roberts et al. 2023).

While PRS has shown promise for predicting breast cancer risk, it has yet to be widely applied in clinical settings (Yanes et al. 2020). Adding PRS to existing risk models that involve clinical and lifestyle variables and mammographic density can enhance their therapeutic utility (Sud et al. 2021). Although PRS are new in breast cancer risk prediction, it is unclear how to effectively communicate a personalized risk score to patients and if they prefer this information over regular screening. Three large-scale studies, PROCAS (Predicting the Risk of Cancer at Screening), WISDOM (Women Informed to Screen Depending on Measures of Risk), and MyPeBS (My Personalized Breast Screening), examine the feasibility and implementation of breast cancer PRS (Harkness et al. 2020; Eklund et al. 2019; Roux et al. 2022; Roberts et al. 2023).

To deploy PRS in clinical settings, it is important to evaluate the individual impact on patients. A breast cancer PRS can help women at higher risk make informed decisions about preventative interventions, including medication, bilateral mastectomies, and lifestyle changes (Shah 2021).

However, there is a distinction between relative risk and absolute risk, which helps individuals understand their specific outcomes. It is crucial for clinicians to use PRS responsibly and convey it to patients to avoid unneeded fear and radiation from genetic risk assessments and mammographic screening. Several obstacles must be overcome before PRS can be applied population wide. A major issue is the personalization of PRS in non-White European women as they have currently overpredicted susceptibility risks (Roberts et al. 2023).

Current Therapeutic Approaches for BC

Following a confirmed diagnosis of BC, the progression of the disease is evaluated by using standard clinical diagnostic modalities. This helps to determine whether preoperative (neoadjuvant) systemic therapy is required. The metastatic confirmed tumor is an incurable disease. It is treated by palliative chemotherapy unless there is an indication of local surgical control (Moo et al. 2018; Waks et al. 2019). An essential part of clinical evaluation of non-metastatic BC is to identify the criteria for irresectability that require neoadjuvant chemotherapy. Among these are ipsilateral arm lymphedema, inflammatory carcinoma, tumor fixation to the bony chest wall (ribs, sternum), fixed/matted axillary lymphadenopathy, extensive skin involvement with ulceration or satellite skin nodules, and involvement of the axilla's neurovascular structures. In these situations, administration of neoadjuvant chemotherapy shrinks the tumor and makes it resectable in 80% of cases (Hortobagyi and Buzdar 1988). In those patients who present with resectable disease, the use of neoadjuvant chemotherapy helps to shrink the tumor and makes breast conservative surgery possible when the mastectomy would otherwise be needed. It has a role in reducing the need for axillary lymph node dissection in resectable BC (Pilewskie et al. 2016). Resection of the tumor is the first step in care for most patients who present with stage I and II BC; patients can choose between mastectomy and breast conservation. After surgery for early breast cancer, chemotherapy is administered to reach cancer cells which follow the circulation to distant organs, and radiation therapy is administered to decrease the local recurrence. The response of BC is variable to different treatment options. This response variability necessitates looking for reasons that appear after the exploration of the molecular structure of cancer cells. As a result, treatment modalities are directed to genomic or specific molecular structures to kill the cancer cells.

Systemic Therapies

Systemic therapies include the treatments to kill breast cancer cells after local control has been achieved. It could be used to make local control possible or less aggressive.

Chemotherapy

Chemotherapy is an antitumor medication that stops cancer growth either by stopping them from dividing or by killing them. The patient obtains this type of therapy as cycles followed by a recovery period before receiving another dose. The sensitive subtypes of chemotherapy are basal-like subtypes TNBC, HER2 positive, and

Table 1 Summary of selected chemotherapeutics with brief mechanisms of action

Chemotherapy agent	Classification	Mode of action
Anthracycline (doxorubicin, epirubicin)	Antitumor antibiotic	DNA damage by targeting topoisomerase II and chromatin damage
Mitoxantrone		Topoisomerase inhibitors
Taxanes (docetaxel, paclitaxel)		Antimitotic inhibit protein formation
Vinca alkaloids (viblastine)		
Cisplatin and carboplatin	Alkylating agent	DNA damage
5-fluorouracil, capecitabine, and gemcitabine	Antimetabolite	Act as a substitute for the normal building of DNA and RNA and destruct them

high-risk luminal subtypes. It may be used in ER-positive cases only when there is an urgent risk to life as a means of administering disease management because it induces ovarian failure (Fedele et al. 2017).

Chemotherapy regimens based on anthracycline reduce breast cancer mortality by about one third. Combining multiple drugs is more effective. Patients at risk of cardiac complications may benefit from a non-anthracycline and taxane regimen. Multiparameter molecular tests can identify patients requiring chemotherapy avoidance. There are several regimens of chemotherapy used in BC which are summarized in Table 1.

Hormonal Therapy

Hormonal therapy is used to restore the hormonal balance in postmenopausal women or block the estrogen receptors in premenopausal women. The ER-positive tumors are luminal A and luminal B. Luminal A is sensitive to hormonal therapy and is considered a slowly growing tumor. This type is not sensitive to cytotoxic chemotherapy. Luminal B is further subdivided into HER2 positive and HER2 negative. The luminal B HER2 positive is less sensitive to chemotherapy than the other HER2-positive types. Chemotherapy sensitivity is highly dependent on the proliferation rate. There are several hormonal therapy agents with different modes of action. Selective estrogen receptor blockers (Tamoxifen and Fulvestrant) are recomended for premenopausal patients who are still producing estrogen from their ovaries. Aromatase inhibitors for postmenopausal BC cases decrease the estrogen production from adipose tissue and adrenal gland. Luteinizing hormone–releasing hormone analog (LHRHa) decreases the production of estrogen and progesterone and can be used in combination with tamoxifen in premenopausal settings in advanced or resistant cases (Tan and Wolff 2007). Hormonal therapy usually used as adjuvant therapy decreases the recurrence rate and increases the overall survival. Target therapy can be used in combination with hormonal therapy to augment its action and alleviate the resistance.

Targeted Therapy

The targeted therapies (TT) in BC involve the use of medicines/agents which are targeted against the specific proteins that are involved in the causation, progression, and longevity of BC.

TT is commonly used in combination with other modalities such as chemotherapy and hormonal therapy and is less toxic to healthy tissues while having high efficacy due to the use of nanotechnology approaches. The nanoparticle-delivery system increases the contact time and elimination time of TT with the action site on the cancer cells. This technology decreases the resistance of cancer cells to target therapy. Nanoparticles, polymeric micelles, dendrimers, carbon nanotubes, liposomes, and quantum dots are examples of nanocarriers used in TT for breast cancer (Rizvi and Saleh 2018). Combining trastuzumab with chemotherapy reduces recurrence risk but requires less than one-year prescription due to cardiotoxicity.

Trastuzumab resistance can be reversed with drugs like lapatinib, a small-molecule tyrosine kinase inhibitor, and pertuzumab, a second-generation monoclonal antibody. For TNBC, chemotherapy with or without monoclonal antibody bevacizumab (a vascular endothelial growth factor (VEGF) blocker) is the only accepted systemic treatment option, but it has higher recurrence and metastasis rates compared to other tumor types. The different TT agents with their indication and modes of action are summarized in Table 2 (Costa et al. 2020). The schematic modes of action of selected TT agents are represented in Fig. 3.

Immunotherapy

Immunotherapy is a recent advancement in breast cancer treatment. Recent years have seen a sharp increase in the clinical experience of immunotherapy for the treatment of breast cancer (Hu and McArthur 2018). The wide clinical research landscape may expand immune checkpoint inhibitors (ICI) and other immunotherapies in BC beyond TNBC. Clinical research on immunotherapies for all BC subtypes is increasing. ICI with chemotherapy in TNBC will be FDA approved, giving real-world data to evaluate this therapeutic strategy in underrepresented landmark clinical trial populations like Black patients. Comprehensive translational research and biomarkers will prevent "add-on designs" that add a novel immune drug to a clinically validated modality without patient-specific approaches.

The first biomarker studies in immunotherapy trials for TNBC demonstrate disease heterogeneity and the necessity to identify the tumor microenvironment (TME) to adjust treatment (Debien et al. 2023). Retrospective clinical trials have evaluated tumor-infiltrating lymphocyte numbers, tertiary lymphoid architecture, and immune gene signatures for prognostic value. Programmed death-ligand 1 (PD-L1) immunohistochemistry (IHC) expression is the most widely used biomarker, but not sufficient, as it only appears to have predictive value in metastatic TNBC. The availability of many monoclonal antibodies and scoring methodologies and the poor inter-observer agreement of PD-L1 scoring make its clinical use controversial. Blood-based biomarker research may make liquid biopsies a noninvasive treatment response prediction and monitoring alternative to tissue biopsies.

Table 2 A summary of different classes of targeted therapy agents with their indications and modes of action

Agents for targeted therapy	Class	Mode of action	Indication
Trastuzumab (Herceptin)	Monoclonal antibody	Inhibits ligand-independent HER2 signaling	Early and metastatic HER2+ BC
Pertuzumab	Monoclonal antibody	Inhibits ligand-dependent HER2 and HER3 by blocking their dimerization. Induces ADCC	Metastatic HER2+ breast cancer which is not already treated with chemotherapy or In combination with trastuzumab and docetaxel for inoperable HER2+ BC
Trastuzumab emtansine	Antibody-drug conjugate	The conjugated emtansine is a microtubule inhibitor and is released inside the tumor cell and induces mitotic arrest and apoptosis	Metastatic HER2+ breast cancer has already been treated with trastuzumab and taxane
Trastuzumab deruxtecan	Antibody-drug conjugate	The conjugate Deruxtecan is released into the cell and it is a topoisomerase I inhibitor	Clinical testing is ongoing
Lapatinib	Tyrosine kinase inhibitor	Inhibits tyrosine kinase of HER2 and EGF receptors	Metastatic HER2+ breast cancer is used in combination with capecitabine, trastuzumab, or an aromatase inhibitor
Bevacizumab	Monoclonal antibody	A vascular endothelial growth factor (VEGF) blocker	In combination with chemotherapy in triple-negative breast cancer
Margetuximab	Fc-modified monoclonal antibody	Binds to the same epitope as trastuzamab	Phase 3 clinical testing ongoing
Everolimus	Immunosuppressive macrolide	Mammalian target of rapamycin (mTOR) inhibitor	Postmenopause with HER2 breast cancer
Olaparib or talazoparib	Poly(ADP-ribose) polymerase	(PARP) inhibitors	BRCA gene–mutated breast cancer
Alpelisib	Phosphatidylinositol 3-kinase (PI3K) inhibitor	(PI3K) inhibitor	Used in combination with fulvestrant in estrogen receptor positive

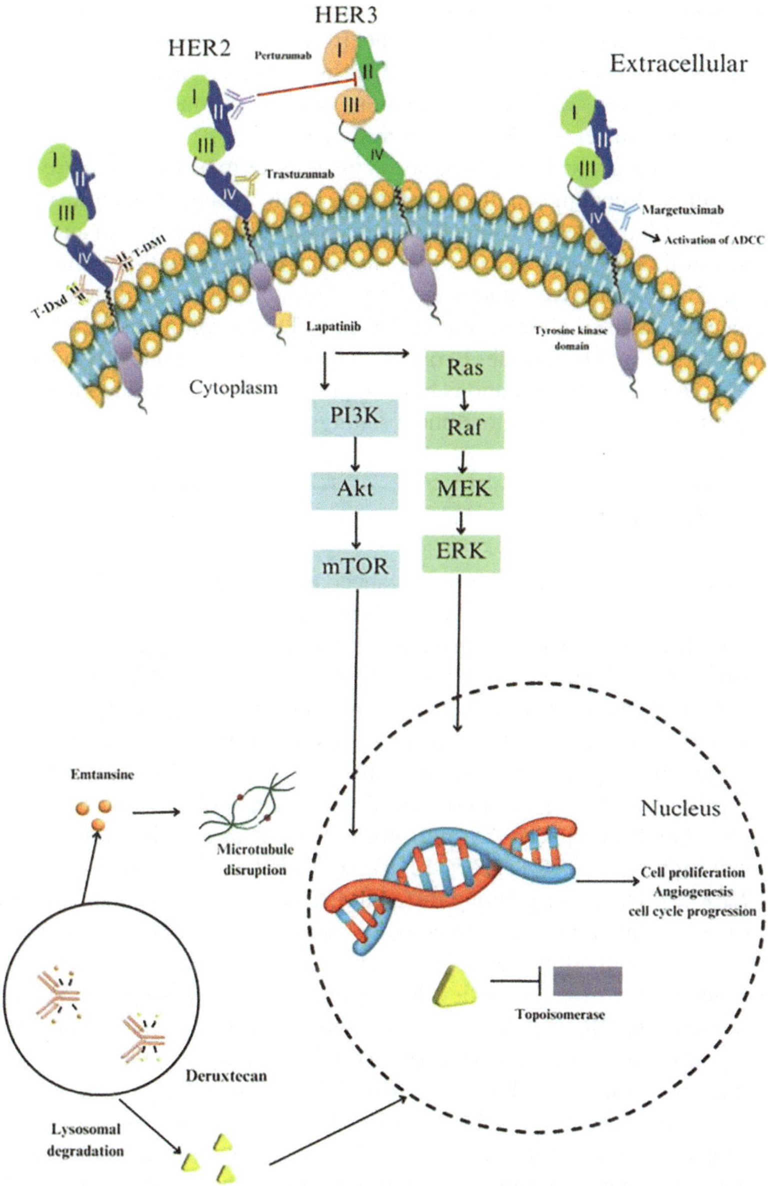

Fig. 3 Schematic representation of mode of action and interaction of target therapy: Trastuzumab (yellow) binds to the ECD IV of HER2, thereby inhibiting the HER2 signaling pathway leading to cell cycle arrest. Pertuzumab (purple) binds to the ECD II and inhibits dimer formation. Margetuximab, which binds trastuzumab to an altered Fc-g domain that is involved in antibody-dependent cellular cytotoxicity (ADCC); Lapatinib (orange) directly prevents the PI3K pathway from being activated by binding to the intracellular tyrosine kinase domain of HER2; T-DM1: trastuzumab emtansine (orange circle) releases the emtansine moiety after the antibody-drug conjugate (ADC) is phagocytosed by lysosomes; T-Dxd: trastuzumab deruxtecan (yellow triangle), another ADC combining trastuzumab and deruxtecan, a potent topoisomerase I inhibitor

Multidisciplinary teamwork is needed to address novel immune-modulating medicines, as immune-related adverse events from immunotherapy are often severe. ICI responses differ from chemotherapeutic drug responses; hence, immunotherapy development requires response evaluation, which is one of the formidable challenges, and is currently under active research.

Future clinical trials will need to focus on the long-term benefits of these therapies for BC patients. Neoadjuvant immunotherapy trials without chemotherapy (e.g., NCT04427293) have been designed due to the good outcomes in patients with high tumor-infiltrating lymphocytes and early-stage TNBC without chemotherapy (Park et al. 2019). Improved understanding of tumor-immune interactions and the TME, especially with single-cell studies and spatial transcriptomics, may reveal new therapeutic targets and techniques to overcome resistance for non-responders (Aldea et al. 2021). New drug combinations have improved BC patient outcomes, and they will be gradually introduced into clinical settings.

5 Conclusion

Breast cancer is the predominant form of cancer among women globally. The genetic makeup of women has a role in determining their susceptibility to developing breast cancer. Pathologic hereditary DNA mutations can significantly elevate the likelihood of developing breast cancer and account for disease risk in vulnerable families. In addition, BC can also be caused by several reasons, including lifestyle choices and hormonal variables. Nevertheless, survival rates for noninvasive breast cancer also have considerably improved as a result of breakthroughs in diagnoses and therapy. Using genetic markers, breast tumors are divided into four separate groups for the purpose of classification. GWAS and other research approaches revealed altered alleles with high (e.g., *BRCA1*, *BRCA2*, *TP53*, *PTEN,* etc.) to moderate risks (e.g., *ATM*, *CHEK2*, *PALB2*, etc.) among different BC patients. To accurately predict the possibility of getting a disease, polygenic risk scores are an indispensable tool.

Targeting the mutant proteins involved in breast cancer has emerged as an effective therapeutic strategy for developing novel drugs. Among the several treatment approaches for BC are chemotherapy, hormone therapy, and targeted therapy. Dual-targeted therapy using trastuzumab and pertuzumab has emerged as an effective therapeutic strategy in HER2-positive breast cancer. Everolimus, a mTOR inhibitor, and abemaciclib, a CDK4/6 inhibitor, have been shown to effectively reverse resistance to targeted agents and endocrine agents. Additionally, the combination of trastuzumab, lapatinib, and hormonal therapy has been confirmed as an effective therapeutic strategy for breast cancer with hormonal receptor positivity and increased HER2 protein expression in the luminal B/HER2-enriched subtype. Currently clinical trials are underway to evaluate the extended advantages that these therapies can offer to patients with different stages of breast cancer.

Acknowledgments The authors are thankful to the Deanship of Graduate Studies and Scientific Research at the University of Bisha for supporting this work through the Fast-Track Research Support Program. The authors express their gratitude to Yahia Mognea for his artistic contribution in drawing Fig. 3.

Ethical Approval Not applicable.

References

Aldea M, Andre F, Marabelle A, Dogan S, Barlesi F, Soria JC (2021) Overcoming resistance to tumor-targeted and immune-targeted therapies. Cancer Discov 11(4):874–899. https://doi.org/10.1158/2159-8290.CD-20-1638

American Cancer Society. Breast Cancer Survival Rates. Available at: https://www.cancer.org/cancer/breast-cancer/understanding-a-breast-cancer-diagnosis/breast-cancer-survival-rates.html. Accessed on 7 December 2023

Anstey EH, Shoemaker ML, Barrera CM, O'Neil ME, Verma AB, Holman DM (2017) Breastfeeding and breast cancer risk reduction: implications for black mothers. Am J Prev Med 53(3S1):S40–S46. https://doi.org/10.1016/j.amepre.2017.04.024

Arnold M, Morgan E, Rumgay H et al (2022) Current and future burden of breast cancer: global statistics for 2020 and 2040. Breast 66:15–23. https://doi.org/10.1016/j.breast.2022.08.010

Chang YJ, Hou YC, Chen LJ et al (2017) Is vegetarian diet associated with a lower risk of breast cancer in Taiwanese women? BMC Public Health 17(1):800. Published 2017 Oct 10. https://doi.org/10.1186/s12889-017-4819-1

Corso G, Figueiredo J, De Angelis SP et al (2020) E-cadherin deregulation in breast cancer. J Cell Mol Med 24(11):5930–5936. https://doi.org/10.1111/jcmm.15140

Costa B, Amorim I, Gärtner F, Vale N (2020) Understanding breast cancer: from conventional therapies to repurposed drugs. Eur J Pharm Sci 151:105401. https://doi.org/10.1016/j.ejps.2020.105401

Dall GV, Britt KL (2017) Estrogen effects on the mammary gland in early and late life and breast cancer risk. Front Oncol 7:110. Published 2017 May 26. https://doi.org/10.3389/fonc.2017.00110

Debien V, De Caluwé A, Wang X et al (2023) Immunotherapy in breast cancer: an overview of current strategies and perspectives. NPJ Breast Cancer 9(1):7. Published 2023 Feb 13. https://doi.org/10.1038/s41523-023-00508-3

Dey S, Boffetta P, Mathews A, Brennan P, Soliman A, Mathew A (2009) Risk factors according to estrogen receptor status of breast cancer patients in Trivandrum, South India. Int J Cancer 125(7):1663–1670. https://doi.org/10.1002/ijc.24460

Dumitrescu RG, Shields PG (2005) The etiology of alcohol-induced breast cancer. Alcohol 35(3):213–225. https://doi.org/10.1016/j.alcohol.2005.04.005

Eklund M, Broglio K, Yau C, Connor JT, Stover Fiscalini A, Esserman LJ (2019) The WISDOM personalized breast cancer screening trial: simulation study to assess potential bias and analytic approaches. JNCI Cancer Spectr 2(4):pky067. Published 2019 Jan 8. https://doi.org/10.1093/jncics/pky067

Elston CW, Ellis IO (1991) Pathological prognostic factors in breast cancer. I. The value of histological grade in breast cancer: experience from a large study with long-term follow-up. Histopathology 19(5):403–410. https://doi.org/10.1111/j.1365-2559.1991.tb00229.x

Evans DGR, van Veen EM, Harkness EF et al (2022) Breast cancer risk stratification in women of screening age: incremental effects of adding mammographic density, polygenic risk, and a gene panel. Genet Med 24(7):1485–1494. https://doi.org/10.1016/j.gim.2022.03.009

Fachal L, Aschard H, Beesley J et al (2020) Fine-mapping of 150 breast cancer risk regions identifies 191 likely target genes. Nat Genet 52(1):56–73. https://doi.org/10.1038/s41588-019-0537-1

Fedele M, Cerchia L, Chiappetta G (2017) The epithelial-to-mesenchymal transition in breast cancer: focus on basal-like carcinomas. Cancers (Basel) 9(10):134. Published 2017 Sept 30. https://doi.org/10.3390/cancers9100134

Filippini SE, Vega A (2013) Breast cancer genes: beyond BRCA1 and BRCA2. Front Biosci (Landmark Ed) 18(4):1358–1372. Published 2013 June 1. https://doi.org/10.2741/4185

Fortner RT, Sisti J, Chai B et al (2019) Parity, breastfeeding, and breast cancer risk by hormone receptor status and molecular phenotype: results from the nurses' health studies. Breast Cancer Res 21(1):40. Published 2019 Mar 12. https://doi.org/10.1186/s13058-019-1119-y

Gasperskaja E, Kučinskas V (2017) The most common technologies and tools for functional genome analysis. Acta Med Litu 24(1):1–11. https://doi.org/10.6001/actamedica.v24i1.3457

Ginter PS, Idress R, Alfonso TMD, Fineberg S, Jaffer S, Sattar AK et al (2021) Histologic grading of breast carcinoma: a multi-institution study of interobserver variation using virtual microscopy. Mod Pathol 34(4):701–709. https://doi.org/10.1038/s41379-020-00698-2. Epub 2020 Oct 19

Giuliano AE, Edge SB, Hortobagyi GN (2018) Eighth edition of the AJCC cancer staging manual: breast cancer. Ann Surg Oncol 25:1783–1785. https://doi.org/10.1245/s10434-018-6486-6

Graffeo R, Rana HQ, Conforti F et al (2022) Moderate penetrance genes complicate genetic testing for breast cancer diagnosis: ATM, CHEK2, BARD1 and RAD51D. Breast 65:32–40. https://doi.org/10.1016/j.breast.2022.06.003

Guo W, Key TJ, Reeves GK (2018) Adiposity and breast cancer risk in postmenopausal women: results from the UK biobank prospective cohort. Int J Cancer 143(5):1037–1046. https://doi.org/10.1002/ijc.31394

Harkness EF, Astley SM, Evans DG (2020) Risk-based breast cancer screening strategies in women. Best Pract Res Clin Obstet Gynaecol 65:3–17. https://doi.org/10.1016/j.bpobgyn.2019.11.005

Hortobagyi GN, Buzdar AU (1988) Management of locally advanced breast cancer. Am J Clin Oncol 11(5):597–601. https://doi.org/10.1097/00000421-198810000-00018

Hu ZI, McArthur HL (2018) Immunotherapy in breast cancer: the new frontier. Curr Breast Cancer Rep 10(2):35–40. https://doi.org/10.1007/s12609-018-0274-y

Huber-Keener KJ (2022) Cancer genetics and breast cancer. Best Pract Res Clin Obstet Gynaecol 82:3–11. https://doi.org/10.1016/j.bpobgyn.2022.01.007

Johnson KS, Conant EF, Soo MS (2021) Molecular subtypes of breast cancer: a review for breast radiologists. J Breast Imaging 3:12–24. https://doi.org/10.1093/jbi/wbaa110

Kaur RP, Vasudeva K, Kumar R, Munshi A (2018) Role of p53 gene in breast cancer: focus on mutation spectrum and therapeutic strategies. Curr Pharm Des 24(30):3566–3575. https://doi.org/10.2174/1381612824666180926095709

Khodarahmi M, Azadbakht L (2014) The association between different kinds of fat intake and breast cancer risk in women. Int J Prev Med 5(1):6–15

Liu Y, Nguyen N, Colditz GA (2015) Links between alcohol consumption and breast cancer: a look at the evidence. Womens Health (Lond) 11(1):65–77. https://doi.org/10.2217/whe.14.62

Łukasiewicz S, Czeczelewski M, Forma A, Baj J, Sitarz R, Stanisławek A (2021) Breast cancer-epidemiology, risk factors, classification, prognostic markers, and current treatment strategies-an updated review. Cancers (Basel) 13(17):4287. Published 2021 Aug 25. https://doi.org/10.3390/cancers13174287

Marvalim C, Datta A, Lee SC (2023) Role of p53 in breast cancer progression: an insight into p53 targeted therapy. Theranostics 13(4):1421–1442. Published 2023 Feb 27. https://doi.org/10.7150/thno.81847

Mavaddat N, Michailidou K, Dennis J et al (2019) Polygenic risk scores for prediction of breast cancer and breast cancer subtypes. Am J Hum Genet 104(1):21–34. https://doi.org/10.1016/j.ajhg.2018.11.002

Melhem-Bertrandt A, Bojadzieva J, Ready KJ et al (2012) Early onset HER2-positive breast cancer is associated with germline TP53 mutations [published correction appears in Cancer. 2012 May 1;118(9):2561]. Cancer 118(4):908–913. https://doi.org/10.1002/cncr.26377

Meyer JS, Alvarez C, Milikowski C, Olson L, Russo I, Russo J et al (2005) Breast carcinoma malignancy grading by Bloom–Richardson system *vs* proliferation index: reproducibility of grade and advantages of proliferation index. Mod Pathol 2005(18):1067–1078. https://doi.org/10.1038/modpathol.3800388

Michailidou K, Lindström S, Dennis J et al (2017) Association analysis identifies 65 new breast cancer risk loci. Nature 551(7678):92–94. https://doi.org/10.1038/nature24284

Moo TA, Sanford R, Dang C, Morrow M (2018) Overview of breast cancer therapy. PET Clin 13(3):339–354. https://doi.org/10.1016/j.cpet.2018.02.006

Muranen TA, Morra A, Khan S et al (2023) PREDICT validity for prognosis of breast cancer patients with pathogenic BRCA1/2 variants. NPJ Breast Cancer 9(1):37. Published 2023 May 12. https://doi.org/10.1038/s41523-023-00546-x

Nassar A, Zekri AN, Kamel MM et al (2022) Frequency of pathogenic germline mutations in early and late onset familial breast cancer patients using multi-gene panel sequencing: an Egyptian study. Genes (Basel) 14(1):106. Published 2022 Dec 29. https://doi.org/10.3390/genes14010106

Park JH, Jonas SF, Bataillon G et al (2019) Prognostic value of tumor-infiltrating lymphocytes in patients with early-stage triple-negative breast cancers (TNBC) who did not receive adjuvant chemotherapy. Ann Oncol 30(12):1941–1949. https://doi.org/10.1093/annonc/mdz395

Perou C, Sørlie T, Eisen M et al (2000) Molecular portraits of human breast tumours. Nature 406: 747–752. https://doi.org/10.1038/35021093

Picon-Ruiz M, Morata-Tarifa C, Valle-Goffin JJ, Friedman ER, Slingerland JM (2017) Obesity and adverse breast cancer risk and outcome: mechanistic insights and strategies for intervention. CA Cancer J Clin 67(5):378–397. https://doi.org/10.3322/caac.21405

Pilewskie M, Jochelson M, Gooch JC, Patil S, Stempel M, Morrow M (2016) Is preoperative axillary imaging beneficial in identifying clinically node-negative patients requiring axillary lymph node dissection? J Am Coll Surg 222(2):138–145. https://doi.org/10.1016/j.jamcollsurg.2015.11.013

Pradella LM, Evangelisti C, Ligorio C et al (2014) A novel deleterious PTEN mutation in a patient with early-onset bilateral breast cancer. BMC Cancer 14:70. Published 2014 Feb 6. https://doi.org/10.1186/1471-2407-14-70

Renwick A, Thompson D, Seal S et al (2006) ATM mutations that cause ataxia-telangiectasia are breast cancer susceptibility alleles. Nat Genet 38(8):873–875. https://doi.org/10.1038/ng1837

Rizvi SAA, Saleh AM (2018) Applications of nanoparticle systems in drug delivery technology. Saudi Pharm J 26(1):64–70. https://doi.org/10.1016/j.jsps.2017.10.012

Roberts E, Howell S, Evans DG (2023) Polygenic risk scores and breast cancer risk prediction. Breast 67:71–77. https://doi.org/10.1016/j.breast.2023.01.003

Roux A, Cholerton R, Sicsic J et al (2022) Study protocol comparing the ethical, psychological and socio-economic impact of personalised breast cancer screening to that of standard screening in the "My Personal Breast Screening" (MyPeBS) randomised clinical trial. BMC Cancer 22(1): 507. Published 2022 May 6. https://doi.org/10.1186/s12885-022-09484-6

Sajjadi E, Venetis K, Piciotti R et al (2021) Combined analysis of PTEN, HER2, and hormone receptors status: remodeling breast cancer risk profiling. BMC Cancer 21(1):1152. Published 2021 Oct 28. https://doi.org/10.1186/s12885-021-08889-z

Satish S, Moore JF, Littlefield JM, Bishop IJ, Rojas KE et al (2023) Breast Cancer (Dove Med Press) 15:227–235. Published 2023 Mar 22. https://doi.org/10.2147/BCTT.S390664

Sawaki M, Shien T, Iwata H (2019) TNM classification of malignant tumors (Breast Cancer Study Group). Jpn J Clin Oncol 49:228–231. https://doi.org/10.1093/jjco/hyy182

Shah PD (2021) Polygenic risk scores for breast cancer-can they deliver on the promise of precision medicine? JAMA Netw Open 4(8):e2119333. Published 2021 Aug 2. https://doi.org/10.1001/jamanetworkopen.2021.19333

Sidransky D, Tokino T, Helzlsouer K et al (1992) Inherited p53 gene mutations in breast cancer. Cancer Res 52(10):2984–2986

Siegel RL, Miller KD, Fuchs HE, Jemal A (2022) Cancer statistics, 2022. CA Cancer J Clin 72(1): 7–33. https://doi.org/10.3322/caac.21708

Sohail SK, Sarfraz R, Imran M, Kamran M, Qamar S (2020) Estrogen and progesterone receptor expression in breast carcinoma and its association with clinicopathological variables among the Pakistani population. Cureus 12(8):e9751. https://doi.org/10.7759/cureus.9751

Starek-Świechowicz B, Budziszewska B, Starek A (2023) Alcohol and breast cancer. Pharmacol Rep 75(1):69–84. https://doi.org/10.1007/s43440-022-00426-4

Sud A, Turnbull C, Houlston R (2021) Will polygenic risk scores for cancer ever be clinically useful? NPJ Precis Oncol 5(1):40. Published 2021 May 21. https://doi.org/10.1038/s41698-021-00176-1

Surakasula A, Nagarjunapu GC, Raghavaiah KV (2014) A comparative study of pre- and post-menopausal breast cancer: risk factors, presentation, characteristics and management. J Res Pharm Pract 3(1):12–18. https://doi.org/10.4103/2279-042X.132704

Suzuki Y, Tsunoda H, Kimura T, Yamauchi H (2017) BMI change and abdominal circumference are risk factors for breast cancer, even in Asian women. Breast Cancer Res Treat 166(3): 919–925. https://doi.org/10.1007/s10549-017-4481-4

Tan SH, Wolff AC (2007) Luteinizing hormone-releasing hormone agonists in premenopausal hormone receptor-positive breast cancer. Clin Breast Cancer 7(6):455–464. https://doi.org/10.3816/CBC.2007.n.002

Torabi Dalivandan S, Plummer J, Gayther SA (2021) Risks and function of breast cancer susceptibility alleles. Cancers (Basel) 13(16):3953. Published 2021 Aug 5. https://doi.org/10.3390/cancers13163953

Torres-de la Roche LA, Acevedo-Mesa A, Lizarazo IL et al (2023) Hormonal contraception and the risk of breast cancer in women of reproductive age: a meta-analysis. Cancers (Basel) 15(23): 5624. Published 2023 Nov 28. https://doi.org/10.3390/cancers15235624

Tsang JYS, Tse GM (2020) Molecular classification of breast cancer. Adv Anat Pathol 27(1): 27–35. https://doi.org/10.1097/PAP.0000000000000232

Van Dooijeweert C, van Diest PJ, Ellis IO (2022) Grading of invasive breast carcinoma: the way forward. Virchows Arch 480(1):33–43. https://doi.org/10.1007/s00428-021-03141-2

Waks AG, Stover DG, Guerriero JL, et al. (2019) The immune microenvironment in hormone receptor-positive breast cancer before and after preoperative chemotherapy. Clin Cancer Res 25(15):4644–4655. https://doi.org/10.1158/1078-0432.CCR-19-0173

Wilson JR, Bateman AC, Hanson H et al (2010) A novel HER2-positive breast cancer phenotype arising from germline TP53 mutations [published correction appears in J Med Genet. 2011 Mar;48(3):216]. J Med Genet 47(11):771–774. https://doi.org/10.1136/jmg.2010.078113

Yadav S, Boddicker NJ, Na J et al (2023) Contralateral breast cancer risk among carriers of germline pathogenic variants in *ATM*, *BRCA1*, *BRCA2*, *CHEK2*, and *PALB2*. J Clin Oncol 41(9):1703–1713. https://doi.org/10.1200/JCO.22.01239

Yanes T, McInerney-Leo AM, Law MH, Cummings S (2020) The emerging field of polygenic risk scores and perspective for use in clinical care. Hum Mol Genet 29(R2):R165–R176. https://doi.org/10.1093/hmg/ddaa136

Yin L, Duan JJ, Bian XW, Yu S (2020) Triple-negative breast cancer molecular subtyping and treatment progress. Breast Cancer Res 22:61. https://doi.org/10.1186/s13058-020-01296-5

Zhou X, Yu L, Wang L et al (2022) Alcohol consumption, blood DNA methylation and breast cancer: a Mendelian randomisation study. Eur J Epidemiol 37(7):701–712. https://doi.org/10.1007/s10654-022-00886-1

Interleukins in the Development and Progression of Breast Cancer

Md. Abdul Aziz, Sarah Jafrin, Mohammad Safiqul Islam, and Yearul Kabir

Abstract

Breast cancer is one of the most frequently diagnosed cancers globally, with the highest mortality in women. Despite remarkable advances in treatment approaches, breast cancer still holds the top position in cancer-associated deaths in women. Over the last few decades, researchers have been trying to reveal the exact etiology of this malignancy, for example, revealing molecular mechanisms behind this. Inflammation is a critical factor leading to breast cancer development and progression. Many inflammatory cytokines or biomarkers, especially interleukins (ILs), are extensively studied due to their prominent role in inflammation leading to breast tumors' initiation, promotion, and progression. Exposure of breast epithelial cells to these cytokines promotes excessive proliferation, enhanced survival, malignant transformation, and cancer development. Previous studies reported that some ILs are associated with acute inflammatory processes, including IL-8 and IL-12, whereas some with chronic inflammatory responses include IL-2, IL-3, IL-6, IL-10, IL-12, and others. Moreover, chronic inflammation leads to different signaling mechanisms and transcription factors and alters both the patient's innate and adaptive immune responses. This chapter will discuss the major ILs and their role in the development and progression of breast cancer. Besides, the authors will try to provide an insight into the mechanisms in both the tumor cell and tumor microenvironment that result in breast cancer progression. They anticipate the data described in this chapter will

Md. A. Aziz · S. Jafrin · M. S. Islam
Department of Pharmacy, Faculty of Science, Noakhali Science and Technology University, Noakhali, Bangladesh

Y. Kabir (✉)
Department of Biochemistry and Molecular Biology, University of Dhaka, Dhaka, Bangladesh
e-mail: ykabir@yahoo.com

© The Author(s), under exclusive license to Springer Nature Switzerland AG 2022
Interdisciplinary Cancer Research, https://doi.org/10.1007/16833_2022_24
Published online: 21 October 2022

play an important role in proposing newer therapeutic approaches to treat breast cancer.

Keywords

Breast cancer · Cytokines · Inflammation · Interleukins

1 Introduction

Breast cancer is a major health problem in women all around the world. The latest update on worldwide cancer shows that breast cancer is the most frequently diagnosed cancer in women, with 2.3 million new cases (11.7%) and the primary cause of all-inclusive cancer death with 685,000 deaths (Sung et al. 2021). Epidemiological studies have suggested many risk factors that can exacerbate normal cell growth, such as previous family history, environmental carcinogens, oral contraceptive pills, poor dietary habits, and socioeconomic conditions (Aziz et al. 2022; Momenimovahed and Salehiniya 2019; Coughlin 2019). However, the pathogenesis and etiology of the disease continue to confound the healthcare professionals, which are crucial for disease prognosis and early diagnosis of this cancer (Shabnaz et al. 2016). For the past few decades, the crucial role of inflammation in different diseases, especially cancer, has gained the attention of scientists. The role of cytokines, especially interleukins, has been evaluated in the development and progression of breast tumors. Several lines of evidence revealed the molecular mechanisms of interleukins related to the initiation, migration, and advancement of this cancer (Fasoulakis et al. 2018). This chapter will discuss the major interleukins and their role in breast cancer development and progression. Besides, the authors will try to provide an insight into the mechanisms in both the tumor cell and tumor microenvironment that result in breast cancer progression.

2 Interleukins and Breast Cancer

Cytokines are natural biomolecules, usually glycoproteins or polypeptides, with a molecular weight of less than 30 kilodaltons (kDa). Cytokines have a key role in cellular differentiation, cell growth, and inflammatory or anti-inflammatory response. They strongly bind to the high-affinity surface receptors in target cells and trigger a variety of intracellular signaling cascades, resulting in gene transcription alteration. Therefore, cytokines lead to cell proliferation and differentiation that attenuate the functions of specific cells (Berraondo et al. 2019). Interleukins (ILs) belong to the cytokines superfamily that has been widely investigated for their complex immunological activities in different cells. The primary functions of ILs include modulation of cell adhesion, migration, maturation, and proliferation associated with the inflammatory response (Dmitrieva et al. 2016). Sources, targets, and functions of different ILs in breast cancer are presented in Table 1. ILs are linked

Table 1 Sources, targets, and functions of different interleukins in breast cancer

Interleukins	Sources	Targets	Function in breast cancer
IL-1	B cells; epithelial cells; monocytes; neutrophils	B cells; endothelial cells; T cells	Proinflammatory cytokine; induction of IL-1α and IL-6 expression; induction of Th17 cells
IL-2	Th1 cells; NK cells	B cells; T cells; non-lymphoid cells	Immunoregulatory cytokine; activates NK cells and monocytes; main growth factor for B and T cells
IL-4	Basophils; mast cells; Th2 cell	B cells; T cells	Anti-inflammatory effect; TNF and IL-6 production by monocytes; inhibition of the Th17 pathway
IL-6	Fibroblasts; macrophages; T cells	B cells; hepatocytes	Angiogenesis; promotes epithelial-to-mesenchymal transition; induction of Th17 cells
IL-7	Fibroblasts; stromal cells; thymus	B cells; T cells	Induces Type 1 immune response; increases CD8+ specific cytotoxicity
IL-8	Endothelial cells; fibroblasts; macrophages; T cells	Endothelial cells; granulocytes	Immunomodulatory activities; neutrophil chemoattractant and activator
IL-10	B cells; monocytes; T cells	Macrophages	Promotes angiogenesis and immunosuppressive action
IL-11	Fibroblasts; monocytes; stromal cells of bone marrow	Hepatocytes	Hematopoietic growth factor; proliferation, growth, and metastasis
IL-13	Basophils; CD8+ T cells; Eosinophils; mast cells; Th2 cells	B cells; endothelial cells; macrophages	Inhibits proinflammatory cytokines; invasiveness
IL-17	NK cells; Th2 cells; Th17 cells	Fibroblasts; macrophages	Proinflammatory cytokine; modifies the tumor microenvironment
IL-18	Monocytes	B cells; endothelial cells; macrophages	Proinflammatory cytokine; induces IFN-γ and NK cells
IL-19	B cells; monocytes	Lymphocytes; macrophages	Increases growth, proliferation, and cancer progression
IL-20	B cells; monocytes; T cells	Macrophages	Proinflammatory cytokine; angiogenesis and metastasis
IL-21	Lymphocytes; Th17 cells	B cells; macrophages; T cells	Increases proliferation, metastasis, and invasion
IL-22	B cells; T cells; monocytes	Neutrophils; NK cells	Promotes inflammation; cellular proliferation
IL-23	Monocytes; M1 dendritic cells; Th1 cells	Granulocytes; NK cells; T cells	Promotes inflammation; Th17 expansion and stabilization
IL-32	Keratinocytes	Fibroblasts; monocytes; T cells	Proinflammatory cytokine; angiogenesis and endothelial cell activities

(continued)

Table 1 (continued)

Interleukins	Sources	Targets	Function in breast cancer
IL-33	Mast cells; Th2 cells	Eosinophils; macrophages; Neutrophils	Decreases apoptosis of myeloid-derived suppressor cells; induces angiogenesis
IL-35	B cells; T cells	Macrophages; T cells	Immunosuppressive cytokine; downregulating differentiation and development of Th17 cells

[a]*TNF* tumor necrosis factor, *Th cell* T helper cell, *NK cell* natural killer cell, *IFN* interferon, *IL* interleukin

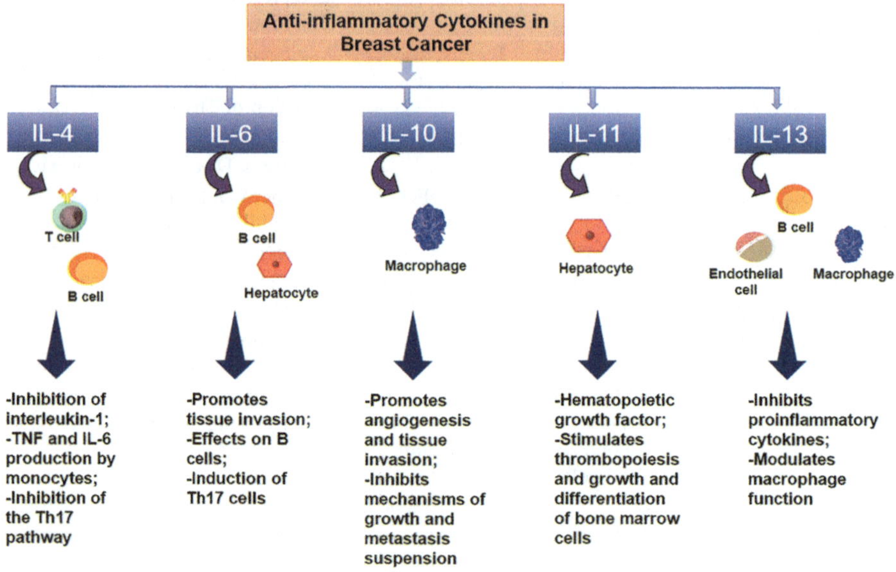

Fig. 1 Role of anti-inflammatory cytokines in breast cancer

with both acute and chronic inflammation through binding and stimulating particular receptors on the cell membrane following the activation of a specific signaling mechanism. About 38 distinct ILs have been discovered till now, each of which binds to a distinct receptor, possessing a particular source, structure, and function. Most of the ILs, including anti-inflammatory (Fig. 1) and proinflammatory (Fig. 2), demonstrated their role in the initiation, promotion, and progression and metastasis of breast malignancy (Fasoulakis et al. 2018).

2.1 Interleukin-1

Interleukin-1 (IL-1) is a critical proinflammatory cytokine having a molecular weight of 17–20 kDa. There are three major representatives of IL-1, including two activators, namely, IL-1α and IL-1β, and one inhibitor, IL-1 receptor antagonist

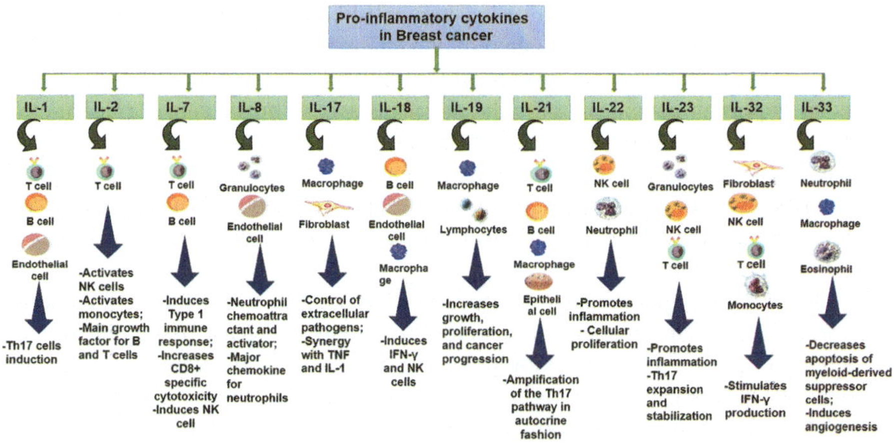

Fig. 2 Role of proinflammatory cytokines in breast cancer

(IL-1ra), all of which are associated with the initiation and progression of inflammatory responses. Through binding with the IL-1 receptor (IL-1R), IL-1α and IL-1β activate downstream signaling molecules following the expression of inflammatory genes. IL-1 is characterized by different biological activities (Baker et al. 2019). Increased stimulation of IL-1 has been observed in various cancers. For example, an autocrine mechanism was proposed between IL-1β and nuclear factor-κB (NF-κB). In metastatic breast carcinoma cell lines, it was reported that the secretion of IL-1β and other related factors affect mesenchymal stem cells (MSCs). In combination with chemokines, MSCs predispose the risk of metastatic-invasive breast cancer (Escobar et al. 2015). Breast tumor-dependent IL-1α induces thymic stromal lymphopoietin (TSLP) expression responsible for breast tumor metastasis. Kuan and Ziegler (2018) recently reported new IL-1α-TSLP-based crosstalk between breast tumor tissues and tumor-infiltrating myeloid cells in metastatic breast cancer. Another recent study demonstrated the significant role of IL-1α while evaluating the link between human epidermal growth factor receptor 2 (HER-2) expression, inflammation, and the spread of cancer stem cells (CSCs). The study explicated that IL-1α and IL-6 expression was caused by a positive feed-forward activation loop induced by the HER2 expression in breast cancer cells, which thus induces signal transducer and activator of transcription 3 (STAT3) and NF-κB responsible for production and maintenance of CSCs (Liu et al. 2018).

Activation of NLR family pyrin domain containing 3 (NLRP3) inflammasome and release of IL-1β cause the macrophage-induced formation of new lymphatic vessels known as lymphangiogenesis. In breast cancer patients, inflammasome expression in macrophages leads to lymph node invasion, metastasis, and survival (Weichand et al. 2017). Again, IL-1β triggers the secretion of IL-6 in mammary carcinoma cells, relevant interleukins, and other growth factors like transforming growth factor-β (TGF-β), epidermal growth factor (EGF), and tumor necrosis factor-α (TNF-α) that aggravate the influence of IL-1β on IL-6 gene expression.

Besides, in mammary carcinoma cells, stem-cell-like phenotypes and epithelial-to-mesenchymal transition were altered via treatment with anti-IL-1β/anti-IL-6 antibody (Oh et al. 2016). An investigation in 2016 delineated that IL-1 expression plays an important part in the growth and metastasis of breast cancer (Holen et al. 2016). Before that, Filippi et al. (2015) reported that IL-1 advances breast cancer metastasis by chemokine (C-X-C motif) receptor (CXCR) 1 upregulation. A previous study showed that different agonists (IL-1, leptin), as well as antagonists (IL-1ra, adiponectin) in breast adipocytes, affect cellular proliferation, invasion, and migration (Perrier et al. 2009).

Multiple investigations have demonstrated that both autocrine and paracrine activation of IL-1 and associated receptors in malignant breast tissue enhance the possibility of pro-tumorigenic cytokines (e.g., IL-8) expression, which subsequently triggers cancer-associated events, including angiogenesis, invasion, and proliferation. While IL-1β has been correlated with promoting the invasiveness in breast cancer, IL-1α has been found to promote the metastatic event through triggering pro-metastatic genes expression in both stromal and breast cancer cells (Méndez-García et al. 2019).

2.2 Interleukin-2

Interleukin-2 (IL-2) is a 15.5 kDa cytokine found in the cell membrane, generally expressed in normal body tissues, intestinal epithelium, and endothelial cells. It is a lymphocytotrophic polypeptide that triggers the cytolytic functions of the natural killer (NK) cells by playing a substantial part in the differentiation of B cells, T cells, and non-lymphoid cells. IL-2 receptor (IL-2R) consists of three common subunits (α, β, γ) and is produced with the help of CD4+ T helper (Th) lymphocytes. This cytokine was first linked to mammary carcinoma progression in patients with a 4.7% relapse rate (normal IL-2 plasma level). In comparison, 33.3% elevated relapse rate was reported in patients (with low IL-2 levels) after 10–12 months of follow-up (Fasoulakis et al. 2018). It is also observed that IL-2 and its three receptor subunits' expression levels are more elevated in breast-infiltrating cancer than in situ surgical specimens (Garcia-Tuñón et al. 2004). Hu et al. (2013) found that variants in the IL-2 were linked with enhanced breast carcinoma risk, and IL-2 might be applied as a prognostic biomarker for breast cancer. Another study explicated that IL-2 could trigger the induction and subsequent growth of breast malignancy in patients with HER-2 negative and overexpressing status due to a substantially reduced percentage of IL-2 (Muraro et al. 2011).

2.3 Interleukin-4

T lymphocytes and macrophages primarily synthesize Interleukin-4 (IL-4). Among the cytokines, IL-4 regulates the maturation of T helper 2 (Th2) cells and has been correlated with the pathogenesis of a variety of cancer. An elevated IL-4 plasma concentration is closely correlated with the risk of acute myeloid leukemia, breast cancer, colon cancer, melanoma, prostate cancer, head and neck squamous carcinoma, renal cell carcinoma, and small cell lung cancer (Chu et al. 2020). Multiple studies reported the impact of IL-4 in breast cancer progression and metastasis as well as apoptotic resistance (Koenig et al. 2016). The IL-4 receptor (IL-4R) is mandatory for IL-4 functions in tumors, and IL-4R is prominently overexpressed in breast tumor cells (Balasubramanian et al. 2006). To examine this, a study used an IL-4Rα antagonist in mammary gland tumors that blocked IL-4 and inhibited mammary carcinoma growth, proliferation, and invasion through downregulation of the mitogen-activated protein kinase (MAPK) signaling mechanism (Gaggianesi et al. 2017). Line of evidence also showed that IL-4 and IL-4/IL-4R pathways induce breast cancer metastasis and macrophage-dependent inflammatory breast cancer (Venmar and Fingleton 2014; Bankaitis and Fingleton 2015). IL-4 also helps to differentiate tumor-associated macrophages (TAM), which activates the EGF receptor (EGFR) in the presence of colony stimulating factor-1 in epithelial breast tumor cells. This event promotes the metastasis of breast cancer cells. Again, IL-4 binds to its surface receptor, IL-4Rα, which stimulates mammary carcinoma colonization to distant metastatic areas, for example, in lungs. This anomaly occurs through the activation of multiple signaling pathways, including Akt, Erk 1/2, and mTOR pathways (Méndez-García et al. 2019).

2.4 Interleukin-6

Interleukin-6 (IL-6) is a frequently synthesized interleukin family member in the tumor microenvironment. Tumor-related fibroblasts, granulocytes, macrophages, and tumors are all major sources of IL-6. It precisely acts on tumor cells for stimulating STAT3-targeted gene expression, which further encodes proteins that mediate cancer cell proliferation and survival, advancing various events, including angiogenesis, immunosuppression, invasion, and metastasis. Additionally, IL-6 promotes other proinflammatory cytokine production by employing different tumor microenvironment immune cells. Notably, the transcriptional factor NF-κB is a prominent IL-6 expression enhancer, and overactive NF-κB is generally reported in multiple human cancers (Sanguinetti et al. 2015). The rectified concentration of IL-6 was detected in patients suffering from a range of several cancers, particularly breast, cervical, and colorectal cancer. Moreover, the level of IL-6 is raised by radiotherapy and surgery in patients with recurrent tumors (Hu et al. 2021; Taher et al. 2018). In addition, IL-6 stimulates the synthesis of the signaling molecule vascular endothelial growth factor (VEGF). IL-6 controls the VEGF level in platelets and promotes the multiplication of vascular endothelial cells. A study observed that the culture of human mammary adenocarcinoma cell line (MCF-7) with IL-6 and IL-6 soluble receptors regulates the growth of these cell lines (Jiang et al. 2011).

Again, an upregulated serum concentration of IL-6 in breast cancer patients is usually correlated with a lower survival rate and poor prognosis. Studies explained that breast carcinoma cell sensitivity to IL-6 greatly relies on estrogen receptors

(ER) and progesterone receptors (PR) expression. For instance, in breast cancer patients, ER-positive cells mainly produce the sIL-6R, and ER-negative cells primarily produce the mIL-6R. Furthermore, it was reported that ER-negative cells were repressed by IL-6 following an autocrine pathway under normal physiologic conditions. IL-6 also enhances circulatory and tumor site-specific estrogen concentration via activation of estrogen-releasing enzymes, such as 17 beta-hydroxysteroid dehydrogenase, estrone sulfatase, and aromatase. Estrogen sulfate acts as a reservoir of estrogen because it stays in circulation for a prolonged time, even longer than estrogen. Conversely, estrone sulfatase is reported to be overexpressed in breast cancer tissues. IL-6 also acts as one of the main activators of STAT3 transcriptional factor, which is found to be overexpressed in greater than 50% of malignant breast tissues (Masjedi et al. 2018).

Moreover, Lin et al. (2015) reported that upregulated IL-6 expression is linked with a poor prognosis. IL-6 also promotes its mRNA and notch-3-based overexpression of the carbonic anhydrase IX gene, which triggers the hypoxia-resistant/invasive characteristics in breast cancer cell lines. A study performed by Hou et al. (2018) revealed that IL-6 is enhanced in more than 80% of malignant breast tissue with negative estrogen compared to that of the adjoined healthy tissues. The study revealed that IL-6 provokes the migration of tumor cells through the Hippo kinase pathway activation, which phosphorylates and inactivates the YES-associated protein, a transcriptional factor responsible for breast cancer cell migration and proliferation (Hou et al. 2018). Furthermore, IL-6 was found to be responsible for chemoresistance in breast cancer. It induces resistance to tamoxifen and trastuzumab during the treatment of breast cancer (Dittmer and Leyh 2015).

2.5 Interleukin-7

Interleukin-7 (IL-7) is a ~25 kDa hematopoietic growth factor generally secreted by the stromal cells, the thymus-dependent proliferation of B cells in combination with hepatocyte growth factors (HGFs) located in the marrow and T-cell progenitors in the thymus gland. IL-7 binds to the cell membrane receptor (IL-7R), which comprises the IL-7Rα chain and the γc chain positioned on chromosome 5p13. It is commonly found on vascular endothelial cells, and the anomalous IL-7 and IL-7Rα variant expressions have distinctive activities in mammary cancer based on their molecular subclass (Vitiello et al. 2018). A recent study reported that IL-7-expressing carcinoma-related fibroblasts (CAFs) increased the growth of breast carcinoma and rendered crucial niches to maintain the stemness of breast carcinoma. At the same time, CXCL-12 was demonstrated to be a prominent niche element in CAFs (Boesch et al. 2018). Previously, IL-7 was demonstrated to show profound activity during the development of mammary cancer, stimulating the survival and progression of carcinoma tissues in culture media. The study proposed that IL-7 enhanced breast cancer cell growth via a wortmannin-sensitive mechanism in vitro (Al-Rawi et al. 2004).

2.6 Interleukin-8

Interleukin-8 (IL-8) is a major chemokine with a molecular weight of 8 kDa produced by macrophages, fibroblasts, and vascular endothelial cells. The main action of IL-8 can be observed in the two members of the immune system: T-lymphocytes and neutrophils. It executes a potential chemotactic effect on the mentioned cells and enhances the receptor-binding properties of CD11b/CD18 (leukocyte adhesion receptor). IL-8 exerts (positively or negatively) antiviral and immunomodulatory activities and shows antiproliferative characteristics by interfering with cellular inflammation and migration. As a representative of the chemokine family, IL-8 shows a tendency to be overexpressed in the human neoplastic cell lines. Excessive production and expression of IL-8 have been detected in both normal and various neoplastic cell lines of the breast, ovaries, prostate, thyroid, and many others (Fasoulakis et al. 2018).

A plethora of cytokines get engaged in both healthy and cancerous breast tissues. Green et al. (1997) investigated several mRNA transcripts to screen cytokines and detected several cytokines, including IL-a, IL-1β, IL-2 to IL-8, interferon-gamma (IFN-γ), and TNF-β. Their thorough analysis reported an enhanced level of IL-8 in comparison to the other cytokines. In a different investigation, 20 healthy and 73 cancerous mammary tissues were inspected for the mRNA profiling of 13 cytokines. Two types of neoplasms (MCF-7 and MDA-435) were investigated for metastatic potency. The MDA-435 cell line showed a greater extent of metastasis (almost 50% of cases) by aggressively invading the anatomical sites of distant areas; on the contrary, the MCF-7 cell line did not undergo metastasis. Among the two neoplastic cell lines, the MDA-435 cell type is the maximum IL-8 secreting cell line compared to the MCF-7 cell line. The serum concentration of IL-8 was detected to be the highest among all the 13 cytokines under observation (De Larco et al. 2001). It was reported that ER-human malignant breast tumors and cell lines possess greater IL-8 levels than cell lines with ER+ status. Also, increased IL-8 serum levels were observed in breast cancer patients with HER2+ status and lymph node metastasis (Méndez-García et al. 2019).

IL-8 has a potential neoplastic index as it presents both autocrine and paracrine characteristics that can enhance tumor prognosis actively. Both ER+ and ER- breast carcinoma subjects were observed to possess a high expression of IL-8 that ultimately excites the invasiveness and metastasis phenomena of breast cancer. Endothelial cells, fibroblasts, and tumor cells secrete IL-8 during infectious clinical conditions, thus showing its action in inflammatory clinical conditions. The IL-8RA and the IL-8RB are two receptors possessed by IL-8, and the noticeable higher IL-8 expression implicates the invasive nature of breast carcinoma cell types. The elevated serum IL-8 signifies the initiation and propagation of breast cancer cells (Todorović-Raković and Milovanović 2013). IL-8 promotes the invasiveness of ER- mammary carcinoma cells in the later stages. A previous in vitro observation validated this phenomenon and showed that IL-8 did not enhance tumor cell proliferation, which indicates that IL-8 can alter the tumor microenvironment. The study also reported that IL-8 could directly increase several events like angiogenesis, proliferation, survival, and matrix metalloproteinase expression in endothelial cells (Li et al. 2003).

2.7 Interleukin-10

Interleukin-10 (IL-10) is a monomeric interleukin with an approximate molecular weight of 18 kDa. IL-10 gene is pleiotropic, that means the generation of multiple effects from a single entity or gene. IL-10 is generated by T helper cells (Th0, Th1, and Th2), regulatory T cells, activated monocytes, mast cells, and cytotoxic T cells. This cytokine can restrict the synthesis of specific cytokines and be primarily known as the human cytokine synthesis inhibitory factor for this particular nature. IL-10 is a prominent anti-inflammatory cytokine that has been comprehensively evaluated for its powerful action in the immune system (Ekmekcioglu et al. 2008). IL-10 can be used as a potential biomarker for detecting cancer prognosis and disease-free state in various breast cancer subtypes such as ER+, PR+, non-basal, and non-triple-negative breast malignancy (Ahmad et al. 2017). The mammary tissues of patients showed elevated IL-10 mRNA expression in many former studies indicating a potential influence of IL-10 in mammary carcinogenesis progression. This multifunctional cytokine was found to be correlated with the carcinogenesis, progression, metastasis of the neoplastic cells, and development of breast cancer due to its antiangiogenic and immunosuppressive actions (Joimel et al. 2010; Chavey et al. 2007; Lyon et al. 2008). Anti-inflammatory IL-10 was shown to advance the development of the cellular microenvironment that is correlated with the promotion of tumor growth and the inhibition of anticancer immune activity.

A study by Kozlowski and colleagues showed that breast cancer patients had higher levels of serum IL-10 compared to the healthy subjects. They have also demonstrated that IL-10 was overexpressed in ER- patients in contrast to ER+ patients (Kozlowski et al. 2003). A recent study with breast cancer subjects reported that atypical hyperplasia patients had an increased level of serum IL-10 than healthy subjects, suggesting its significant role in breast cancer promotion. Moreover, stage II and stage III ductal carcinoma patients had a greater concentration of IL-10 compared to stage I carcinoma patients (Lv et al. 2018). It is an extremely pleiotropic interleukin that has a prominent function in both the initiation and the development of breast cancer. Greater expression of IL-10 was observed in individuals with breast cancer compared to healthy volunteers. An updated study suggested that IL-10 can be a potential risk factor for the development and progression of breast cancer (Li et al. 2020). Another study described the inhibitory effect of IL-10 on the migration of mammary carcinoma cell lines. The study showed that patients with malignant breast tumors having greater levels of IL-10 present a better disease-free survival rate and breast tumor-associated survival during the primary stages of invasive breast tumor (Ahmad et al. 2017).

2.8 Interleukin-11

Closely related to IL-6, interleukin-11 (IL-11) is a protein (65–85 kDa) coding cytokine that takes part in immune regulation and plays a role as a hematopoietic growth factor. IL-11 is produced from stromal cells of the bone marrow and fibroblast. IL-11 only possesses a single transmembrane receptor, IL-11Rα (IL-11 receptor-alpha), that initiates rapid growth and spreading of neoplastic cells. Stimulation of breast cancer cells rapid proliferation, growth, and metastasis into distant organs are found to be initiated by IL-11. Breast cancer patients with poor prognosis and survival rates showed significantly higher mRNA expression of IL-11. As such, a high histological grade of IL-11 can be an indicator of worsening conditions in breast cancer patients (Johnstone et al. 2015). IL-11 is generally correlated with tumorigenic events, such as angiogenesis, metastasis, proliferation, chemoresistance, radioresistance, and antiapoptotic nature. It stimulates tumor progression through activation of the JAK-STAT3 signaling mechanism. It is produced from both tumor cells and tumor-associated cells, such as fibroblasts and myeloid cells. The overexpression of IL-11 has been found to be linked with several cancer types, including both epithelial cancers and hematopoietic cancers. Analysis of IL-11 in malignant breast tissues revealed that this cytokine is secreted from the breast cancer tissues and associated with breast cancer-dependent osteolytic pathway. Furthermore, overexpressed IL-11 is linked with enhanced bone metastasis (Maroni et al. 2021).

A previous study demonstrated the key role of hypoxic conditions in inducing IL-11 in metastasis of mammary cancer. The authors revealed that the invasiveness and motility of breast cancer cells in the presence of hypoxia are stimulated by this cytokine (Hanavadi et al. 2006). Multiple studies explicated the role of IL-11 in mammary carcinoma-associated bone metastasis in human. They showed that breast tumor patients with bone metastasis have an elevated level of osteoclasts and serum IL-11 levels in comparison to non-metastatic patients. These events collectively reduce the survival rate of breast cancer patients (Maroni et al. 2021). Multiple genetic factors responsible for breast cancer development are reported to be correlated with IL-11 and IL-11Rα overexpression. A previous study validated the link between the expression of IL-11 and IL-11Rα with Ki-67, HER-2, ER, and PR in breast cancer patients. They also identified that the co-expression of both IL-11 and IL-11Rα was closely related to that of the angiogenic property and microvessel density in breast cancer neoplasms (Cui et al. 2016). Furthermore, there is a positive link between IL-11Rα expression in cancer tissues and bone metastasis (Irawan et al. 2016).

2.9 Interleukin-13

Interleukin-13 (IL-13) is homologous to IL-4 with a molecular weight of approximately 12 kDa. IL-13 is mainly secreted by T helper cells (Th2) and other subsets of T helper cells, including CD8+ T cells, mast cells, eosinophils, and basophils. Due to its homologous nature, IL-13 shares many biological actions with IL-4. It functions on epithelial cells, endothelial cells, mononuclear phagocytic cells, and B cells to initiate biological actions along with IL-4, although the respondent cells react lesser to IL-13 compared to IL-4. The overexpression of IL-13 mRNA was reported in breast cancer neoplasms (Lai et al. 2009). Several recent studies thoroughly evaluated the susceptibility of IL-13 in mammary carcinogenesis. Breast cancer neoplasms show high infiltration with CD4+ cells. This T-helper cell subset secretes IL-13 and IFN-γ. The breast tumor cells appoint dendritic cells to enhance CD4+ cell secretion, which helps to boost tumor growth. An inverse correlation between the expression level of IL-13 and the expression of ER and PR has been discovered in a study suggesting a possible activity of IL-13 in the invasiveness of ER-breast tumors. IL-13 is also investigated to be a strong prognostic indicator for breast cancer. Some studies also approached to find out the impact of IL-13 by inhibiting its secretion (Chavey et al. 2007).

2.10 Interleukin-17

Interleukin-17 (IL-17) is another proinflammatory cytokine that is secreted from mainly two types of cells: CD8 Tc17 and CD4 Th17. Among these cell types, T helper cell-17 (Th17) generally handles the IL-17 production. The IL-17 cytokine family comprises six representatives: IL-17A, IL-17B, IL-17C, IL-17D, IL-17E, and IL-17F. IL-17 shows its actions on different cells and stimulates prostaglandins, chemokines, and nitric oxide. Due to its potent protumor nature, IL-17 cytokines tend to change the tumor cell microenvironment which can exaggerate tumor cells' invasiveness and metastatic rate. This protumor effect can be directly observed in cancer cells or maybe in a parallel way that affects the anticancer response of cancer patients (Chen et al. 2013a).

Previous studies demonstrated that IL-17 is linked with the growth and angiogenesis of cancer tissues in patients with breast, cervical, colon, ovary, trachea, colorectal, and hepatocellular carcinoma. IL-17 has a predominant effect on tumor initiation and development via a wide range of signaling mechanisms, such as COX-2, MAPK, STAT3, and Src/PI3K/Akt/NF-κB pathways. It plays a critical function in angiogenesis, metastatic, and tumorigenesis. IL-17 is also associated with the enhancement of tumor severity and poor prognosis (Murugaiyan and Saha 2009). In 2013, Chen et al. (2013a) also reported that the negative status of ER/PR, higher histopathological grade, and triple-negative subclass of breast cancers showed an increased level of IL-17-secreting cells. The elevated level of cytokines indicated a potential correlation between IL-17 and breast cancer. Overexpression of IL-17-generating cells in the mammary gland of breast cancer patients can modify the tumor microenvironment, resulting in poor prognosis, delayed recovery, and low survival rate (Welte and Zhang 2015). A recent study with breast cancer subjects reported that ductal carcinoma and atypical hyperplasia patients had an increased level of serum IL-17A than healthy subjects, suggesting its significant role in breast cancer promotion. Moreover, stage III ductal carcinoma patients had a greater concentration of IL-17A compared to stage I and stage II carcinoma patients (Lv et al. 2018).

2.11 Interleukin-18

Interleukin-18 (IL-18) is a cytokine (24 kDa) that influences the generation of IFN-γ. IL-18 was first discovered in1989 and also designated as an IFN-γ-triggering factor for this specific characteristic. IL-18 enhances the IFN-γ synthesis from anti-CD3-induced T helper cell-1 (Th1) when IL-12 is present. Micromolecular analysis for the isolation and characterization to identify the molecular structure of IL-18 revealed that it has a similar structure to IL-1. IL-18 is primarily produced as a pro-peptide (inactive) like IL-1β that undergoes further activation via proteolytic cleavage in the NLRP3 inflammasome by caspase-1. The final mature, bioactive product is an 18 kDa molecule. The cytokine receptor, IL-18R, possesses one inducible component, IL-18Rα, and one subsequently expressed component, IL-18Rβ. IL-18Rα induces the IL-1 receptor-related protein. IL-18Rβ is a highly expressed protein-like element related to the IL-1R accessory protein-like. It shows biological actions by activating a pathway identical to that of IL-1. The IL-18 binding protein is a naturally occurring high-affinity protein and can suppress the activity of IL-18 by binding with IL-18. As a result, IL-18 fails to occupy its receptor, IL-18R. IL-18 plays a vital role in both innate and adaptive immune systems and is largely expressed in several cell types (Baker et al. 2019).

Past studies have revealed the susceptibility of IL-18 to cancer development by identifying its pro-carcinogenic and anti-carcinogenic actions on tumor cells. The elevated serum concentration and overexpression of IL-18 have been observed in tumor cells of cancer patients, which indicated the positive impact of IL-18 on tumor growth and development. Breast cancer patients having enhanced levels of IL-18 showed poor prognostic and low survival rates. IL-18 might play a pivotal function in cancer cell invasion, migration, angiogenesis, and metastasis (Baker et al. 2019). Exogenous IL-18 was also reported to suppress the claudin-12 protein expression (a tight junction protein), which stimulates the migration and invasion of breast carcinoma cells into distant parts of the body (Yang et al. 2015).

2.12 Interleukin-19

Interleukin-19 (IL-19) is an immunosuppressive cytokine that originates from monocytes secretion and is a congener of the IL-10 cytokine family. The IL-19 works mainly on fibronectin (FN), a stromal-binding protein that enhances migration and metastasis of breast carcinoma. IL-19 helps to assemble and stimulate the expression of FN, which increases the proliferation of breast cancer cells and the rate of metastasis. Multiple combinations of actions, such as cell-cell interactions or cell-matrix overproduction of cytokines, can upgrade metastasis and the invasion process. The inhibition of FN has been targeted as a treatment approach for cancer.

Antibody inhibitors and peptides that inhibit FN actions effectively prevent metastasis and provide a new path for cancer research. Many cellular factors are accountable for upregulating IL-19, such as the macrophage colony-stimulating factor, granulocyte, lipopolysaccharide, TNF-α, and IL-6. IL-19 also controls genetic profiles of TGF-β, cytokines (IL-1β and IL-6), matrix metalloproteinase (MMP2, MMP9), and CXCR4 (Hsing et al. 2012; Chen et al. 2013b).

IL-19 is closely associated with the overall pathogenesis of breast carcinoma and is responsible for the autocrine action in these cells. Thus, it is considered a vital prognostic indicator in multiple cancer types along with breast carcinoma. Breast cancer patients with an elevated level of IL-19 showed poor clinical outcomes. IL-19 promotes cancer progression directly by enhancing proliferation and migration of the tumor cells or indirectly by creating an efficient microenvironment for the growth and development of tumors (Hsing et al. 2012; Chen et al. 2013b). IL-19 directly induces the expression of cytokines and chemokines that highly affect cell proliferation, tumor progression, and migration in breast cancer. The advancement of the disease stages, metastasis, and higher mitotic rates are correlated with cytokine expression levels. IL-19 expression levels can detect and differentiate between metastasis-free survival and worse disease-specific survival conditions. Investigations also assumed IL-19 functions generally as a localized influencer in the tumor microenvironment via stimulating breast tumor cells (Hsing et al. 2012).

2.13 Interleukin-20

Interleukin-20 (IL-20) is a pleiotropic proinflammatory cytokine from the family of IL-10 cytokine, which acts through modulation of the signal transduction by two common heterodimeric receptor cascades IL-20R1/IL-20R2 and IL-22R1/IL-20R2. It is a key angiogenic regulator in breast cancer. A recent study in breast cancer cells demonstrated that IL-20 triggers the growth and migration of malignant cells and increases the secretion of regulatory molecules like cathepsin G, cathepsin K, MMP9, and MMP12 by serving as an upstream modulator in these cells. As these are responsible for metastatic activities, the authors concluded that IL-20 might lead to breast tumor metastasis (Wang et al. 2018). IL-20 has a pivotal role in breast cancer formation, and it is overexpressed in tissues of breast tumor bone metastasis. It is linked with higher tumor stage and poor clinical outcomes (Hsu et al. 2012).

2.14 Interleukin-21

Interleukin-21 (IL-21) is located at the chromosome locus 4q26-q27 and encoded by the IL-21 gene. IL-21 has a four-helix-bundle cytokine sequence domain consisting of 131 amino acid. It is a potent immune modulator and structurally identical to IL-2 and IL-15. IL-21 is synthesized from the secretion of CD4+ T cells and NK cells. This pleiotropic IL-21 participates in diverse immunological and biological activities by activating its possessed receptor, IL-21R. In a recent experiment, the level of

IL-21 expression was determined in triple-negative breast carcinoma cell lines (MDA-231). The study revealed that IL-21 holds the ability to increase proliferation, metastasis, and invasion of the MDA-231 cell line and can intensify the invasive nature of breast tumor cells (Wang et al. 2015).

2.15 Interleukin-22

Interleukin-22 (IL-22) is secreted by CD4+ T cells and has recently been investigated for detecting its cancer susceptibility. This cytokine of interest has powerful tumor-promoting properties. Activation of NLRP3 inflammasome and release of IL-1β from both T cells and myeloid cells induce the generation and secretion of IL-22 in the human body. Anakinra, a potent antagonist of the IL-1R, was administered in a murine mammary carcinoma model. The antagonist was found to restrict the generation of IL-22 and control tumor expansion in the cancer model (Oh et al. 2016). Breast cancer was found to progress with an enhanced level of IL-22 and infiltration of innate lymphoid cell group 3. A recent study reported that IL-22 triggered the proliferation of mammary carcinoma cell lines (including MDA-MB-231 cells, ER+ MCF7 cells, and triple-negative 4 T1 cells) in a STAT3-dependent manner (Zhang et al. 2020). The exposure of IL-22 in triple-negative breast cancer cells activates JAK-STAT3/MAPKs/AKT signaling mechanisms. Studies reported that IL-22 has been linked with in vitro migration and transformation of malignant breast tumor cells (Wang et al. 2018).

2.16 Interleukin-23

Interleukin-23 (IL-23) is a heterodimeric cytokine from the IL-6 cytokine superfamily and has a close structural similarity to IL-12. IL-23 has become a crucial therapeutic option for discovering efficient treatments of several autoimmune and chronic immune-inflammatory diseases because of its specific nature of causing cellular differentiation of local CD4+ T cells into other cells like Th17. An investigation of patients with breast cancer revealed that their overall survival was inversely correlated with an enhanced IL-23 level (Gangemi et al. 2012). Moreover, an increased level of IL-23/IL-23R gene expression was observed in those cancer cells. The correlation between IL-23/IL-23R expression levels and tumor size, development, stage, invasion, and metastasis was also reported (Sheng et al. 2018).

2.17 Interleukin-32

Interleukin (IL-32) is a proinflammatory novel cytokine encoded by the IL-32 gene. IL-32 generally originates from activated T-cell isolation. It has multiple roles in infections, autoimmune diseases, and neoplasia. IL-32 can influence the tumor environment by initiating inflammation and can also block the multiplication of the virus. It plays a crucial function in angiogenesis and endothelial cell activities (Sloot et al. 2018). According to a previous report, the expression of IL-32 was found to be related to the increased rate of invasion and expansion in breast tumors. It showed an influencing impact on the development and survival of neoplastic breast tissues. IL-32 also blocked the apoptotic characteristics and showed high proliferative potential on breast carcinoma cells (Wang et al. 2014).

2.18 Interleukin-33

Interleukin-33 (IL-33) is one of the members of the IL-1 cytokine superfamily. First, it was discovered to be a nuclear factor, and later it was identified as a ligand acting on the ST2 receptor (ST2R). The ST2R has two subtypes: the soluble subtype (sST2) and the full-length membrane type (ST-2 L). This sST2 receptor maintains proper regulation and an alteration of IL-33 identical to that of IL-1R. The most powerful IL-33 expression is observed in non-hematopoietic cells, such as fibroblasts, endothelial cells, keratinocytes, epithelial cells, smooth muscle cells, and fibrocytes. Among the hematopoietic cells, ST2 is potentially expressed on the Th2 cells surface, mast cells, and ILC2s. The IL-33/ST2 complex signaling axis is closely linked to wound healing-related activities, such as fibrosis, angiogenesis, matrix components synthesis, and immune system regulation. These pathways are critically involved in tumor pathogenesis. Several research groups have recently evaluated the correlation of the IL-33/ST2 complex with tumor pathogenesis, implicating that IL-33 plays a central role in tumor initiation, expansion, invasion, and metastasis in different malignancies, such as breast, colorectal, lung, and gastric cancer. Moreover, IL-33 acts as pro-carcinogenic and anti-carcinogenic agents, similar to other members of the IL-1 superfamily (Baker et al. 2019). The comparison of the IL-33 expression level between breast cancer tissues and normal breast tissues revealed overexpression of IL-33 and HER-2 status in cancerous tissues (Liu et al. 2014). In stage IV breast cancer samples, significantly high serum levels of IL-33 and IL-33/IL-12 ratio were observed compared to other cancer stages and controls. An imbalance in Th cell responses (Th1/Th2) was also detected in breast cancer, proving the contribution of IL-33 to breast cancer pathogenesis (Jafarzadeh et al. 2015).

2.19 Interleukin-35

Interleukin 35 (IL-35) is a novel inhibitory cytokine from the IL-12 family. It corresponds to a heterodimer of two subunits, namely, Epstein–Barr virus-induced gene 3 (EBI3) and p35. IL-35 is primarily secreted by B and T cells and contributes to the progression of autoimmune disorders. Recent investigations have shown that this novel interleukin is overexpressed in different tumor cells, including breast, colon, liver, and pancreatic cancer cells (Hao et al. 2018). IL-35 is an immunosuppressive cytokine that potentially inhibits the proliferation of T cells, converts conventional T cells into induced regulatory T cells, and downregulates the differentiation and development of Th17 cells (Collison et al. 2010). A cutting-edge study by Hao et al.

(2018) illustrated that breast tumor cells could express and discharge IL-35 cytokine that can suppress the proliferation of conventional T cells and further conversion of regulatory T cells. This breast cancer cell-derived IL-35 induces breast cancer development and is associated with a poor prognosis. Moreover, the enhanced IL-35 expression level in malignant breast cells microenvironment was significantly associated with a higher aggressive tumor phenotype, such as ER- and PR-expression and poor differentiation (Hao et al. 2018).

3 Conclusion

Interleukins are the most vital members of the immune system for maintaining the body's homeostatic balance during any hostile pathogenic conditions. They play a crucial role in the initiation of the healing process and propagation of pathologic mechanisms. Interleukins directly or indirectly affect pathogenic pathways of the disease like angiogenesis, tumor growth, survival rate, prognosis, and metastatic profile. The function of interleukins in breast cancer pathology has been evaluated for years. Several approaches were taken to establish the actual contribution of interleukins in breast cancer pathology during the past decades. After years of research, the possible association of interleukins with breast cancer initiation, proliferation, and metastasis is justified. Besides, clinicians have been utilizing interleukins for the treatment of breast cancer as these are responsible for the regulation of immunosurveillance. Clinical trials have also been conducted using cytokines for their critical role in breast cancer. Although several valuable pieces of information highlighted the influence of interleukins on breast cancer pathogenesis, some unclear evidence on some specific interleukins needs to be clarified through further additional research.

Acknowledgments None.

Compliance with Ethical Standards The authors declare that there is no conflict of interest.

Glossary

Angiogenesis:	The process of new blood vessel formation.
Cytokines:	Small protein molecules required for maintaining cellular communication.
Anti-inflammatory Cytokines:	Cytokines that mainly act by suppressing proinflammatory cytokines.
Interleukins:	Cytokines that are mainly produced from leukocytes and associated with immune responses.
Invasion:	The process of direct extension and penetration of tumor cells to the surrounding tissues.
Metastasis:	The process of spreading tumor cells to distant organs.
Proinflammatory Cytokines:	Cytokines that are responsible for mediating inflammation.
Proliferation:	The process of increasing the number of tumor cells.

References

Ahmad N, Ammar A, Storr SJ, Green AR, Rakha E, Ellis IO, Martin SG (2017) IL-6 and IL-10 are associated with good prognosis in early stage invasive breast cancer patients. Cancer Immunol Immunother 67:537–549. https://doi.org/10.1007/s00262-017-2106-8

Al-Rawi MA, Rmali K, Mansel RE, Jiang WG (2004) Interleukin 7 induces the growth of breast cancer cells through a wortmannin-sensitive pathway. Br J Surg 91:61–68. https://doi.org/10.1002/bjs.4449

Aziz MA, Akter T, Sarwar MS, Islam MS (2022) The first combined meta-analytic approach for elucidating the relationship of circulating resistin levels and RETN gene polymorphisms with colorectal and breast cancer. Egypt J Med Hum Genet 23:27. https://doi.org/10.1186/s43042-022-00240-w

Baker KJ, Houston A, Brint E (2019) IL-1 family members in cancer; two sides to every story. Front Immunol 10:1197. https://doi.org/10.3389/fimmu.2019.01197

Balasubramanian SP, Azmy IA, Higham SE, Wilson AG, Cross SS, Cox A, Brown NJ, Reed MW (2006) Interleukin gene polymorphisms and breast cancer: a case control study and systematic literature review. BMC Cancer 6:188. https://doi.org/10.1186/1471-2407-6-188

Bankaitis KV, Fingleton B (2015) Targeting IL4/IL4R for the treatment of epithelial cancer metastasis. Clin Exp Metastasis 32:847–856. https://doi.org/10.1007/s10585-015-9747-9

Berraondo P, Sanmamed MF, Ochoa MC, Etxeberria I, Aznar MA, Pérez-Gracia JL, Rodríguez-Ruiz ME, Ponz-Sarvise M, Castañón E, Melero I (2019) Cytokines in clinical cancer immunotherapy. Br J Cancer 120:6–15. https://doi.org/10.1038/s41416-018-0328-y

Boesch M, Onder L, Cheng HW, Novkovic M, Mörbe U, Sopper S, Gastl G, Jochum W, Ruhstaller T, Knauer M, Ludewig B (2018) Interleukin 7-expressing fibroblasts promote breast cancer growth through sustenance of tumor cell stemness. Onco Targets Ther 7:e1414129. https://doi.org/10.1080/2162402X.2017.1414129

Chavey C, Bibeau F, Gourgou-Bourgade S, Burlinchon S, Boissière F, Laune D, Roques S, Lazennec G (2007) Oestrogen receptor negative breast cancers exhibit high cytokine content. Breast Cancer Res 9:15. https://doi.org/10.1186/bcr1648

Chen WC, Lai YH, Chen HY, Guo HR, Su IJ, Chen HHW (2013a) Interleukin-17-producing cell infiltration in the breast cancer tumour microenvironment is a poor prognostic factor. Histopathology 63:225–233. https://doi.org/10.1111/his.12156

Chen YY, Li CF, Yeh CH, Chang MS, Hsing CH (2013b) Interleukin-19 in breast cancer. Clin Dev Immunol 2013:29432. https://doi.org/10.1155/2013/294320

Chu CN, Wang YC, Chang WS, Wang ZH, Liu LC, Wang SC, Lin CC, Liu TY, Chang JG, Tsai CW, Yu CC, Bau DT (2020) Association of interleukin-4 polymorphisms with breast cancer in Taiwan. In Vivo 34:1111–1116. https://doi.org/10.21873/invivo.11882

Collison LW, Chaturvedi V, Henderson AL, Giacomin PR, Guy C, Bankoti J, Finkelstein D, Forbes K, Workman CJ, Brown SA, Rehg JE, Jones ML, Ni HT, Artis D, Turk MJ, Vignali DA (2010) IL-35-mediated induction of a potent regulatory T cell population. Nat Immunol 11:1093–10101. https://doi.org/10.1038/ni.1952

Coughlin SS (2019) Epidemiology of breast cancer in women. Adv Exp Med Biol 1152:9–29. https://doi.org/10.1007/978-3-030-20301-6_2

Cui M, Zhang Q, Qiu X, Jin F (2016) The role of IL-11 and IL-Ra in angiogenesis of breast cancer. Int J Clin Exp Pathol 9:11682–11687

De Larco JE, Wuertz BR, Rosner KA, Erickson SA, Gamache DE, Manivel JC, Furcht LT (2001) A potential role for interleukin-8 in the metastatic phenotype of breast carcinoma cells. Am J Pathol 158:639–646. https://doi.org/10.1016/S0002-9440(10)64005-9

Dittmer J, Leyh B (2015) The impact of tumor stroma on drug response in breast cancer. Semin Cancer Biol 31:3–15. https://doi.org/10.1016/j.semcancer.2014.05.006

Dmitrieva OS, Shilovskiy IP, Khaitov MR, Grivennikov SI (2016) Interleukins 1 and 6 as main mediators of inflammation and cancer. Biochemist 81:80–90. https://doi.org/10.1134/S0006297916020024

Ekmekcioglu S, Kurzrock R, Grimm EA (2008) Hematopoietic growth factors and cytokines. In: The molecular basis of cancer. Elsevier, Amsterdam, pp 605–619

Escobar P, Bouclier C, Serret J, Bièche I, Brigitte M, Caicedo A, Sanchez E, Vacher S, Vignais ML, Bourin P, Geneviève D, Molina F, Jorgensen C, Lazennec G (2015) IL-1β produced by aggressive breast cancer cells is one of the factors that dictate their interactions with mesenchymal stem cells through chemokine production. Oncotarget 6:9–12. https://doi.org/10.18632/oncotarget.4732

Fasoulakis Z, Kolios G, Papamanolis V, Kontomanolis EN (2018) Interleukins associated with breast cancer. Cureus 10:e3549. https://doi.org/10.7759/cureus.3549

Filippi I, Carraro F, Naldini A (2015) Interleukin-1β affects MDAMB231 breast cancer cell migration under hypoxia: role of HIF-1α and NFκB transcription factors. Mediat Inflamm 2015:789414. https://doi.org/10.1155/2015/789414

Gaggianesi M, Turdo A, Chinnici A, Lipari E, Apuzzo T, Benfante A, Sperduti I, Di Franco S, Meraviglia S, Lo Presti E, Dieli F, Caputo V, Militello G, Vieni S, Stassi G, Todaro M (2017) IL4 primes the dynamics of breast cancer progression via DUSP4 inhibition. Cancer Res 77:3268–3279. https://doi.org/10.1158/0008-5472

Gangemi S, Minciullo P, Adamo B, Franchina T, Ricciardi GR, Ferraro M, Briguglio R, Toscano G, Saitta S, Adamo V (2012) Clinical significance of circulating interleukin-23 as a prognostic factor in breast cancer patients. J Cell Biochem 113:2122–2125. https://doi.org/10.1002/jcb.24083

García-Tuñón I, Ricote M, Ruiz A, Fraile B, Paniagua R, Royuela M (2004) Interleukin-2 and its receptor complex (alpha, beta and gamma chains) in in situ and infiltrative human breast cancer: an immunohistochemical comparative study. Breast Cancer Res 6:1–7. https://doi.org/10.1186/bcr730

Green AR, Green VL, White MC, Speirs V (1997) Expression of cytokine messenger RNA in normal and neoplastic human breast tissue: identification of interleukin-8 as a potential regulatory factor in breast tumours. Int J Cancer 72:937–941. https://doi.org/10.1002/(SICI)1097-0215(19970917)72:6%3C937::AID-IJC3%3E3.0.CO;2-Q

Hanavadi S, Martin TA, Watkins G, Mansel RE, Jiang WG (2006) Expression of interleukin 11 and its receptor and their prognostic value in human breast cancer. Ann Surg Oncol 13:802–808. https://doi.org/10.1245/ASO.2006.05.028

Hao S, Chen X, Wang F, Shao Q, Liu J, Zhao H, Yuan C, Ren H, Mao H (2018) Breast cancer cell-derived IL-35 promotes tumor progression via induction of IL-35-producing induced regulatory T cells. Carcinogenesis 39:1488–1496. https://doi.org/10.1093/carcin/bgy136

Holen I, Lefley DV, Francis SE, Rennicks S, Bradbury S, Coleman RE, Ottewell P (2016) IL-1 drives breast cancer growth and bone metastasis in vivo. Oncotarget 7(46):75571–75584. https://doi.org/10.18632/oncotarget.12289

Hou L, Xie S, Li G, Xiong B, Gao Y, Zhao X, Hu J, Deng S, Jiang J (2018) IL-6 triggers the migration and invasion of oestrogen receptor-negative breast cancer cells via regulation of hippo pathways. Basic Clin Pharmacol Toxicol 123:549–557. https://doi.org/10.1111/bcpt.13042

Hsing CH, Cheng HC, Hsu YH, Chan CH, Yeh CH, Li CF, Chang MS (2012) Upregulated IL-19 in breast cancer promotes tumor progression and affects clinical outcome. Clin Cancer Res 18:713–725. https://doi.org/10.1158/1078-0432.CCR-11-1532

Hsu YH, Hsing CH, Li CF, Chan CH, Chang MC, Yan JJ, Chang MS (2012) Anti-IL-20 monoclonal antibody suppresses breast cancer progression and bone osteolysis in murine models. J Immunol 188:1981–1991. https://doi.org/10.4049/jimmunol.1102843

Hu XB, Ouyang LZ, Tang LL (2013) Interleukin-2 gene polymorphisms and prognosis of breast cancer. Genet Test Mol Biomarkers 17:453–457. https://doi.org/10.1089/gtmb.2012.0494

Hu F, Song D, Yan Y, Huang C, Shen C, Lan J, Chen Y, Liu A, Wu Q, Sun L, Xu F, Hu F, Chen L, Luo X, Feng Y, Huang S, Hu J, Wang G (2021) IL-6 regulates autophagy and chemotherapy resistance by promoting BECN1 phosphorylation. Nat Commun 12:3651. https://doi.org/10.1038/s41467-021-23923-1

Irawan C, Atmakusumah D, Siregar NC, Tean TB, Kong LW, Kiat OC, Abdulmuthalib A, Harahap A, Mansyur M (2016) Expression of biomarkers CXCR4, IL11-RA, TFF1, MLF1P in advanced breast cancer patients with bone metastatic: a diagnostic study. Acta Med Indones 48:261–268

Jafarzadeh A, Minaee K, Farsinejad AR, Nemati M, Khosravimashizi A, Daneshvar H, Mohammadi MM, Sheikhi A, Ghaderi A (2015) Evaluation of the circulating levels of IL-12 and IL-33 in patients with breast cancer: influences of the tumor stages and cytokine gene polymorphisms. Iran J Basic Med Sci 18:1189–1198

Jiang XP, Yang DC, Elliott RL, Head JF (2011) Down-regulation of expression of interleukin-6 and its receptor results in growth inhibition of MCF-7 breast cancer cells. Anticancer Res 31:2899–2906

Johnstone CN, Chand A, Putoczki TL, Ernst M (2015) Emerging roles for IL-11 signaling in cancer development and progression: focus on breast cancer. Cytokine Growth Factor Rev 26:489–498. https://doi.org/10.1016/j.cytogfr.2015.07.015

Joimel U, Gest C, Soria J, Pritchard LL, Alexandre J, Laurent M, Blot E, Cazin L, Vannier JP, Varin R, Li H, Soria C (2010) Stimulation of angiogenesis resulting from cooperation between macrophages and MDA-MB-231 breast cancer cells: proposed molecular mechanism and effect of tetrathiomolybdate. BMC Cancer 10:375. https://doi.org/10.1186/1471-2407-10-375

König A, Vilsmaier T, Rack B, Friese K, Janni W, Jeschke U, Andergassen U, Trapp E, Jückstock J, Jäger B, Alunni-Fabbroni M, Friedl T, Weissenbacher T (2016) SUCCESS STUDY GROUP. Determination of interleukin-4,-5,-6,-8 and-13 in serum of patients with breast cancer before treatment and its correlation to circulating tumor cells. Anticancer Res 36:3123–3130

Kozłowski L, Zakrzewska I, Tokajuk P, Wojtukiewicz MZ (2003) Concentration of interleukin-6 (IL-6), interleukin-8 (IL-8) and interleukin-10 (IL-10) in blood serum of breast cancer patients. RoczAkad Med Bialymst 48:82–84

Kuan EL, Ziegler SF (2018) A tumor-myeloid cell axis, mediated via the cytokines IL-1alpha and TSLP, promotes the progression of breast cancer. Nat Immunol 19:366–374. https://doi.org/10.1038/s41590-018-0066-6

Lai EW, Joshi BH, Martiniova L, Dogra R, Fujisawa T, Leland P, de Krijger RR, Lubensky IA, Elkahloun AG, Morris JC, Puri RK, Pacak K (2009) Overexpression of interleukin-13 receptor-α2 in neuroendocrine malignant pheochromocytoma: a novel target for receptor directed anti-cancer therapy. J Clin Endocrinol Metab 94:2952–2957. https://doi.org/10.1210/jc.2009-0309

Li A, Dubey S, Varney ML, Dave BJ, Singh RK (2003) IL-8 directly enhanced endothelial cell survival, proliferation, and matrix metalloproteinases production and regulated angiogenesis. J Immunol 170:3369–3376. https://doi.org/10.4049/jimmunol.170.6.3369

Li M, Yue C, Zuo X, Jin G, Wang G, Guo H, Wu F, Huang S, Zhao X (2020) The effect of interleukin 10 polymorphisms on breast cancer susceptibility in Han women in Shaanxi Province. PLoS One 15:e0232174. https://doi.org/10.1371/journal.pone.0232174

Lin S, Gan Z, Han K, Yao Y, Min D (2015) Interleukin-6 as a prognostic marker for breast cancer: a meta-analysis. Tumori 101:535–541. https://doi.org/10.5301/tj.5000357

Liu J, Shen JX, Hu JL, Huang WH, Zhang GJ (2014) Significance of interleukin-33 and its related cytokines in patients with breast cancers. Front Immunol 5:141. https://doi.org/10.3389/fimmu.2014.00141

Liu S, Lee JS, Jie C, Park MH, Iwakura Y, Patel Y, Soni M, Reisman D, Chen H (2018) HER2 overexpression triggers an il1alpha proinflammatory circuit to drive tumorigenesis and promote chemotherapy resistance. Cancer Res 78:2040–2051. https://doi.org/10.1158/0008-5472.CAN-17-2761

Lv Z, Liu M, Shen J, Xiang D, Ma Y, Ji Y (2018) Association of serum interleukin-10, interleukin-17A and transforming growth factor-α levels with human benign and malignant breast diseases. Exp Ther Med 15:5475–5480. https://doi.org/10.3892/etm.2018.6109

Lyon DE, McCain NL, Walter J, Schubert C (2008) Cytokine comparisons between women with breast cancer and women with a negative breast biopsy. Nurs Res 57:51–58. https://doi.org/10.1097/01.NNR.0000280655.58266.6c

Maroni P, Bendinelli P, Ferraretto A, Lombardi G (2021) Interleukin 11 (IL-11): role(s) in breast cancer bone metastases. Biomedicine 9:659. https://doi.org/10.3390/biomedicines9060659

Masjedi A, Hashemi V, Hojjat-Farsangi M, Ghalamfarsa G, Azizi G, Yousefi M, Jadidi-Niaragh F (2018) The significant role of interleukin-6 and its signaling pathway in the immunopathogenesis and treatment of breast cancer. Biomed Pharmacother 108:1415–1424. https://doi.org/10.1016/j.biopha.2018.09.177

Méndez-García LA, Nava-Castro KE, Ochoa-Mercado TL, Palacios-Arreola MI, Ruiz-Manzano RA, Segovia-Mendoza M, Solleiro-Villavicencio H, Cázarez-Martínez C, Morales-Montor J (2019) Breast cancer metastasis: are cytokines important players during its development and progression? J Interf Cytokine Res 39:39–55. https://doi.org/10.1089/jir.2018.0024

Momenimovahed Z, Salehiniya H (2019) Epidemiological characteristics of and risk factors for breast cancer in the world. Breast Cancer 11:151–164. https://doi.org/10.2147/BCTT.S176070

Muraro E, Martorelli D, Turchet E, Miolo G, Scalone S, Comaro E, Talamini R, Mastorci K, Lombardi D, Perin T, Carbone A, Veronesi A, Crivellari D, Dolcetti R (2011) A different immunologic profile characterizes patients with HER-2-overexpressing and HER-2-negative locally advanced breast cancer: implications for immune-based therapies. Breast Cancer Res 13:117. https://doi.org/10.1186/bcr3060

Murugaiyan G, Saha B (2009) Protumor vs antitumor functions of IL-17. J Immunol 183:4169–4175. https://doi.org/10.4049/jimmunol.0901017

Oh K, Lee OY, Park Y, Seo MW, Lee DS (2016) IL-1beta induces IL-6 production and increases invasiveness and estrogen-independent growth in a TG2-dependent manner in human breast cancer cells. BMC Cancer 16:724. https://doi.org/10.1186/s12885-016-2746-7

Perrier S, Caldefie-Chézet F, Vasson MP (2009) IL-1 family in breast cancer: potential interplay with leptin and other adipocytokines. FEBS Lett 583:259–265. https://doi.org/10.1016/j.febslet.2008.12.030

Sanguinetti A, Santini D, Bonafè M, Taffurelli M, Avenia N (2015) Interleukin-6 and pro inflammatory status in the breast tumor microenvironment. World J Surg Onc 13:129. https://doi.org/10.1186/s12957-015-0529-2

Shabnaz S, Ahmed MU, Islam MS, Islam MR, Al-Mamun MM, Islam MS, Hasnat A (2016) Breast cancer risk in relation to TP53 codon 72 and CDH1 gene polymorphisms in the Bangladeshi women. Tumour Biol 37:7229–7237. https://doi.org/10.1007/s13277-015-4612-7

Sheng S, Zhang J, Ai J, Hao X, Luan R (2018) Aberrant expression of IL-23/IL-23R in patients with breast cancer and its clinical significance. Mol Med Rep 17:4639–4644. https://doi.org/10.3892/mmr.2018.8427

Sloot YJE, Smit JW, Joosten LAB, Netea-Maier RT (2018) Insights into the role of IL-32 in cancer. Semin Immunol 38:24–32. https://doi.org/10.1016/j.smim.2018.03.004

Sung H, Ferlay J, Siegel RL, Laversanne M, Soerjomataram I, Jemal A, Bray F (2021) Global cancer statistics 2020: GLOBOCAN estimates of incidence and mortality worldwide for 36 cancers in 185 countries. CA Cancer J Clin 71:209–249. https://doi.org/10.3322/caac.21660

Taher MY, Davies DM, Maher J (2018) The role of the interleukin (IL)-6/IL-6 receptor axis in cancer. Biochem Soc Trans 46:1449–1462. https://doi.org/10.1042/BST20180136

Todorović-Raković N, Milovanović J (2013) Interleukin-8 in breast cancer progression. J Interf Cytokine Res 33:563–570. https://doi.org/10.1089/jir.2013.0023

Venmar KT, Fingleton B (2014) Lessons from immunology: IL4R directly promotes mammary tumor metastasis. Onco Targets Ther 3:955373. https://doi.org/10.4161/21624011.2014.955373

Vitiello GAF, Losi Guembarovski R, Amarante MK, Ceribelli JR, Carmelo ECB, Watanabe MAE (2018) Interleukin 7 receptor alpha Thr244Ile genetic polymorphism is associated with susceptibility and prognostic markers in breast cancer subgroups. Cytokine 103:121–126. https://doi.org/10.1016/j.cyto.2017.09.019

Wang S, Chen F, Tang L (2014) IL 32 promotes breast cancer cell growth and invasiveness. Oncol Lett 9:305–307. https://doi.org/10.3892/ol.2014.2641

Wang LN, Cui YX, Ruge F, Jiang WG (2015) Interleukin 21 and its receptor play a role in proliferation, migration and invasion of breast cancer cells. Cancer Genomics Proteomics 12:211–221

Wang S, Yao Y, Yao M, Fu P, Wang W (2018) Interleukin-22 promotes triple negative breast cancer cells migration and paclitaxel resistance through JAK-STAT3/MAPKs/AKT signaling pathways. Biochem Biophys Res Commun 503:1605–1609. https://doi.org/10.1016/j.bbrc.2018.07.088

Weichand B, Popp R, Dziumbla S, Mora J, Strack E, Elwakeel E, Frank AC, Scholich K, Pierre S, Syed SN, Olesch C, Ringleb J, Ören B, Döring C, Savai R, Jung M, von Knethen A, Levkau B, Fleming I, Weigert A, Brüne B (2017) S1PR1 on tumor-associated macrophages promotes lymphangiogenesis and metastasis via NLRP3/IL-1beta. J Exp Med 214:2695–2713. https://doi.org/10.1084/jem.20160392

Welte T, Zhang XHF (2015) Interleukin-17 could promote breast cancer progression at several stages of the disease. Mediat Inflamm 2015:804347. https://doi.org/10.1155/2015/804347

Yang Y, Cheon S, Jung MK, Song SB, Kim D, Kim HJ, Park H, Bang SI, Cho D (2015) Interleukin-18 enhances breast cancer cell migration via down-regulation of claudin-12 and induction of the p38 MAPK pathway. Biochem Biophys Res Commun 459:379–386. https://doi.org/10.1016/j.bbrc.2015.02.108

Zhang Y, Liu C, Gao J, Shao S, Cui Y, Yin S, Pan B (2020) IL-22 promotes tumor growth of breast cancer cells in mice. Aging (Albany NY) 12:13354–13364. https://doi.org/10.18632/aging.103439

Shaping of the Immune Landscape by Chemokine Receptors that Impacts the Clinical Outcome in Triple-Negative Breast Cancer

Dharmindra Dulal, Andrew R. Boring, David Terrero, Amit K. Tiwari, and Dayanidhi Raman

Abstract

Chronic inflammation triggers a cytokine or chemokine storm in the tissue microenvironment. The unresolved, persistent inflammatory microenvironment has a tendency to initiate tumors or likely to favor the colonization of disseminated tumor cells (DTCs) into the inflamed tissue. Chemokines and their receptors present in the tumor cells and its microenvironment can orchestrate the progression of primary breast cancer and seed circulating tumor cells in distant organs such as the lungs. Chemokines can facilitate the infiltration of immune cells into the tumor and are often pro-tumor in nature. As the tumor-immune microenvironment (TIME) is infiltrated with myeloid cells and lymphocytes along with the acellular stroma, it can form a supportive bionetwork that could promote primary tumor growth and prepare and establish a premetastatic niche at distant sites for colonization of circulating tumor cells from the blood and lymphatic system. Also, TIME can facilitate colonization and growth of DTCs at the predilection sites such as the

Authors Dharmindra Dulal, Andrew R. Boring, David Terrero and Dayanidhi Raman have equally contributed to this chapter.

D. Dulal · A. R. Boring · D. Raman (✉)
Department of Cell & Cancer Biology, College of Medicine & Life Sciences, University of Toledo Health Science Campus, Toledo, OH, USA
e-mail: dayanidhi.raman@utoledo.edu

D. Terrero
Department of Pharmacology & Experimental Therapeutics, College of Pharmacy and Pharmaceutical Sciences, University of Toledo, Toledo, OH, USA

A. K. Tiwari
Department of Cell & Cancer Biology, College of Medicine & Life Sciences, University of Toledo Health Science Campus, Toledo, OH, USA

Department of Pharmacology & Experimental Therapeutics, College of Pharmacy and Pharmaceutical Sciences, University of Toledo, Toledo, OH, USA

© The Author(s), under exclusive license to Springer Nature Switzerland AG 2022

Interdisciplinary Cancer Research, https://doi.org/10.1007/16833_2022_20
Published online: 29 September 2022

lungs, bones, and the brain. Antagonizing the chemokine network can alter the TIME, allowing one to control the colonization and establishment of the metastatic seed. The overall goal is to improve the quality of life and prolong the lives of patients with metastatic TNBC.

Keywords

B lymphocytes · Chemokine GPCRs · Chemokines · Dendritic cells · Immune cells · M2 macrophages · N2 neutrophils · T lymphocytes

Abbreviations

APC	Antigen-presenting cells
ATRA	All trans retinoic acid
BCSCs	Breast cancer stem-like cells
BMDM	Bone marrow derived macrophages
CHI3L1	Chitinase3-like protein
COX2	Cyclooxygenase 2
CTGF	Connective tissue growth factor
CTIME	Circulating tumor immune microenvironment
DC	Dendritic cells
DTC	Disseminated tumor cells
ECM	Extracellular matrix
EGF	Epidermal growth factor
EMT	Epithelial to mesenchymal transition
ER	Estrogen receptor
ERK	Extracellular signal regulated kinase
ET1	Endothelin 1
FDA	Food and drug administration
Gpr132	G-protein coupled receptor 132
HDAC	Histone deacetylase
ICI	Immune checkpoint inhibition
IFN-γ	Interferon-γ
iNOS	inducible nitric oxide synthase
LAG-3	lymphocyte activation gene 3
LOX	Lysyl oxidase
LSD1	Lysine-specific demethylase 1
MAM	Metastasis associated macrophages
M-CSF	Macrophage colony stimulating factor
MDSC	Myeloid-derived suppressor cells
MMP	Matrix metalloproteinases
mTNBC	Metastatic triple-negative breast cancer
NACT	Neoadjuvant chemotherapy
NET	Neutrophil extracellular trap

NFATc1	Nuclear factor of activated T cells c1
NK Cells	Natural killer cells
PD-1	Programmed death receptor 1
PD-L1	Programmed death ligand 1
PGE2	Prostaglandin E2
PI3K	Phosphatidylinositol-3-kinase
pMHC	peptide-Major Histocompatibility Complex
RANTES	Regulated upon Activation, Normal T cell Expressed and Presumably Secreted
ROS	Reactive oxygen species
SOX4	SRY-related HMG box transcription factor 4
SPARC	Secreted protein acidic and rich in cysteine
STAT3	Signal transducers and activators of transcription 3
TAM	Tumor-associated macrophages
TAN	Tumor-associated neutrophils
TCR	T-cell receptor
TGF-β	Transforming growth factor β
TIGIT	T lymphocyte immunoreceptor with Ig and ITIM domains
TILs	Tumor-infiltrating lymphocytes
TIM-3	T lymphocyte immunoglobulin mucin 3
TIME	Tumor immune microenvironment
TLR	Toll-like receptor
TME	Tumor microenvironment
TMEM	Tumor microenvironment of metastasis
TNBC	Triple-negative breast cancer
TNF-α	Tumor necrosis factor α
TRAIL	TNF-related apoptosis-inducing ligand
Tregs	Regulatory T cells
VCAM-1	Vascular cellular adhesion molecule 1
VEGF	Vascular endothelial growth factor
VISTA	V-domain Ig suppressor of T cell activation

1 Introduction

Persistent irritants or inciting factors in the tissue induce production and release of a cytokine/chemokine storm. The battery of chemotactic cytokines (chemokines) that are released would promote local inflammation (Germano et al. 2008). Inflammation is one of the hallmarks of cancer (Hanahan and Weinberg 2011; Hanahan and Coussens 2012). If the acute inflammation is not resolved, it leads to a chronic inflammatory state. This can be a predisposing factor for initiation of preneoplastic foci which subsequently converts to a neoplasm with the lingering inflammation (Allavena et al. 2008; Mantovani et al. 2008a, b; Mantovani and Pierotti 2008). Such neoplasms are in constant reciprocal communication and interaction with its surrounding cellular and acellular stroma, termed as the tumor microenvironment (TME). There is ample evidence to suggest that the autocrine signaling in the breast cancer cells and paracrine signaling between the tumor and stromal cells can immensely support tumor cell survival, growth, and metastasis (Knowles and Harris 2007; Sica et al. 2008; Noy and Pollard 2014; Zurlo et al. 2022).

The stromal cells from the triple-negative breast cancer (TNBC) are comprised of carcinoma activated fibroblasts (CAFs); M2 macrophages; N2 neutrophils; dendritic cells (DCs); $CD4^{+}$-, $CD8^{+}$-, and regulatory T lymphocytes, B lymphocytes, and plasmablasts; endothelial and lymphendothelial cells; myeloid-derived tumor suppressor cells (MDSCs); and adipocytes (Raman et al. 2007; Lyes et al. 2019; Plava et al. 2019) (Fig. 1). The chemokines and their receptors govern the pro-tumorigenic and pro-metastatic effect of stromal immune cells and thus shape the immune landscape or the tumor immune microenvironment (TIME) that facilitates/orchestrates several events such as (1) primary tumor growth, (2) preparation of the premetastatic niche (metastatic predilection sites), (3) dissemination of tumor

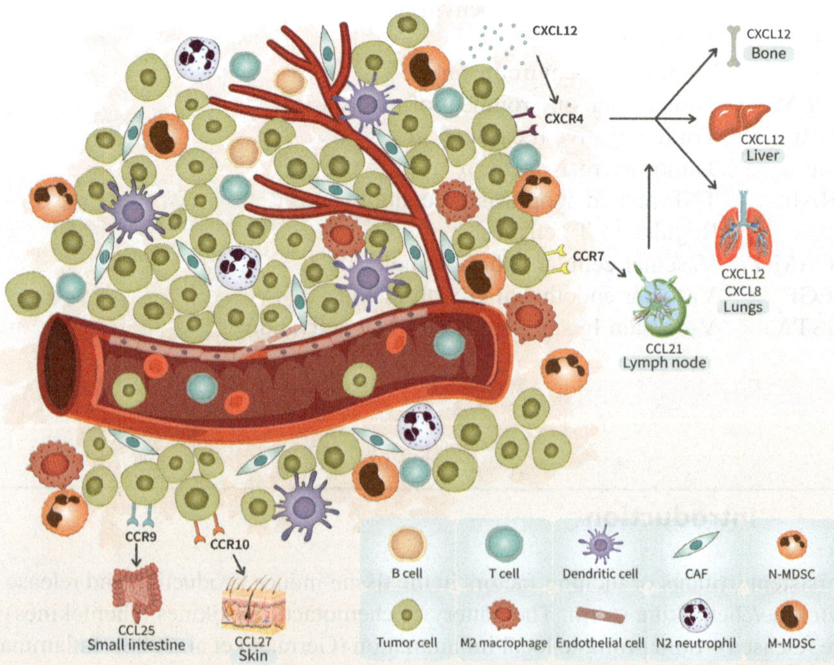

Fig. 1 A cartoon depicting a slice of the tumor with a variety of immune cells surrounding the tumor cells in the tumor immune microenvironment. This schema illuminates a hot tumor where there is heavy infiltration of immune cells that are pro- or anti-tumor in nature

cells into circulation through lympho-vascular invasion (intravasation), (4) survival of the circulating tumor cell clusters, (5) extravasation from local capillaries, and (6) colonization of the metastatic seed at the distant sites (Nagarsheth et al. 2017). Depending on the spatiotemporal context, only some subsets of these immune cells along with specific chemokines and matrix metalloproteinases (MMPs) in the TME orchestrate primary tumor growth, enable chemoresistance, and facilitate and maintain metastases (Hanahan and Coussens 2012). Collectively, such TIME events contribute to distant dissemination of the tumor cells as polyclonal clusters in the blood stream. Overall, we will discuss the impact of the chemokines and their receptors on the TIME and their collective contribution to immune cell trafficking which is a dynamic continuum that shapes the primary tumorigenesis, development of drug resistance, and metastasis. In particular, the pro-tumorigenic and pro-metastatic role of macrophages, neutrophils, myeloid-derived tumor suppressor cells, and $CD8^+$ T lymphocytes will be described in terms of their impact on clinical outcome in mTNBC (Oner et al. 2019).

2 Macrophages

Macrophages have the ability to eliminate or restrict the tumor growth when they encounter the tumor cells in its very early stages. Classically activated, or M1 polarized, macrophages are an important part of the innate immune system that is tumoricidal and can eliminate small tumor cell clusters at an incipient stage. M1 macrophages secrete tumor necrosis factor alpha (TNF-α) which can kill tumor cells extracellularly through the inducible nitric oxide (iNOS) system (Huang et al. 2021). M1 macrophages also secrete pro-inflammatory cytokines such as interleukin-12 (IL-12), CXC-chemokine ligand 10 (CXCL10), and interferon-γ (IFN-γ) (Huang et al. 2021) as well depending on the spatiotemporal context.

In physiological homeostatic situations, alternatively activated macrophages or M2 macrophages serve as an important mediator in wound healing. They facilitate wound healing through secretion of cytokines, such as interleukin-4 (IL-4), interleukin-10 (IL-10), and interleukin-13 (IL-13). They restore blood flow to hypoxic regions of the wound through stimulation of angiogenesis by secreting vascular endothelial growth factors (VEGFs) (Huang et al. 2021). Another wound healing function of M2 macrophages is the secretion of extracellular matrix (ECM) remodeling factors such as matrix metalloproteinases, transforming growth factor β (TGF-β), and arginase 1 (Huang et al. 2021).

The cytokines in the TME such as TGF-β and interleukin-6 (IL-6) can subvert the anti-tumor immunity through conversion of M1 to M2 macrophages. These are also called tumor-associated macrophages (TAMs) and are pro-tumorigenic in nature. TAMs are an important aspect of the TIME of TNBC, and they impact the primary tumor progression and metastasis through production of several growth factors and cytokines/chemokines that activate the tumor cells and induce vascularization in a paracrine manner. Additionally, these secreted factors can amplify the tumor-associated inflammation (Knowles and Harris 2007; Noy and Pollard 2014; Mantovani and Allavena 2015; Belgiovine et al. 2016).

In TNBC, the high levels of TAMs are correlated with poor prognosis as they hinder the therapeutic responses (Kryczek et al. 2006; De Palma and Lewis 2013; Galluzzi et al. 2014; Zhang et al. 2018). At the primary tumor site, breast cancer cells proliferate they secrete macrophage colony stimulating factor (M-CSF), which increases the number of circulating monocytes (monocytosis). The increase in the intratumor levels of secreted M-CSF induces extravasation of the circulating monocytes and facilitates monocyte trafficking into the TME (Lu et al. 2019). Such monocytes become polarized to M2 phenotype due to the paracrine influence from cytokines IL-4 and IL-10 already present in the TME, and this influence arises from preexisting M2 macrophages (Sousa et al. 2015). This forms a vicious self-feeding loop generating more M2 macrophages in TNBC. There is also a positive feedback loop involving cancer cells and M2 macrophages within the TME. Both cancer cells and TAMs secrete TNF-α, which in turn induces TAMs to secrete CC chemokine ligand 8 (CCL8) (Cassetta et al. 2019). In response to CCL8, the cancer cells secrete M-CSF, which stimulates production of more monocytes which increases the number of circulating monocytes. The chemokine CCL8 is one of the ligands for CC chemokine receptor 2 (CCR2) present on the circulating monocytes and facilitates infiltration of monocytes into tumors which eventually polarizes into tumor-resident TAMs (Cassetta et al. 2019). CCR2 is the target of many cytokines produced by cancer cells including CC chemokine ligand 2 (CCL2) (Brummer et al. 2020). CCL2 has also been shown to enhance the proliferation of endothelial cells which promotes neo-angiogenesis (Brummer et al. 2020). Together, CCL2 and M-CSF are the two of the main chemotactic factors driving recruitment of circulating monocytes to tumors (Huang et al. 2021). The tumor-infiltrated monocytes after conversion to tumor-resident M2 macrophages continue to express CCR2 and are one of the key chemokine receptors in their repertoire for their functionality.

2.1 Macrophages and PD-L1

TAMs have a strong immunosuppressive activity in the TME by displaying the integral membrane protein, programmed death ligand (PD-L1), on their plasma membrane. The presence of PD-L1 on the tumor cells and antigen-presenting cells (APC) can alter the response to immunotherapy (Herbst et al. 2014; Tumeh et al. 2014; Zhu et al. 2014; Jing et al. 2020). PD-L1 ligand binds to its cognate receptor, programmed cell death protein-1 (PD-1), present on the plasma membrane of $CD8^+$ T lymphocytes in the TME. Persistent activation of the PD-1 receptor by PD-L1 leads to an inhibitory signaling in $CD8^+$ T lymphocytes. This results in a reduction in their proliferative capacity and inhibits their cytotoxic activity as well due to lower output of several cytokines including IFN-γ (Mantovani et al. 2005; Kryczek et al. 2006). Increase in cell surface expression of PD-L1 on the tumor cells can be achieved via activation of different pathways involving immunosuppressive cytokines such as IL-10 and TGF-β (Hussain et al. 2021; Lim et al. 2021). However, the overexpression of the inhibitory ligand PD-L1 on the plasma membrane of

TAMs has shown to be a critical step in the deactivation of the $CD8^+$ T lymphocyte arm of the anti-tumor immunity (Hartley et al. 2017). TAMs are one of the preponderant immune cell types in the TIME of TNBC (Niu et al. 2016).

Human PD-1 (CD279) receptor is a transmembrane glycoprotein encoded by the *PDCD1* gene (Jiang et al. 2019). PD-1 receptor is expressed on the cell surface of B lymphocytes, monocytes, dendritic cells (DCs), activated $CD8^+$ T lymphocytes, regulatory T lymphocytes (Tregs), and natural killer cells (NK cells) (Jiang et al. 2019). The physiological role of PD-1 receptor is to maintain self-tolerance and prevent or decrease the development of immunopathological conditions or autoimmune disorders. During lingering inflammatory conditions such as chronic infections and cancer, however, PD-1 is overexpressed on $CD8^+$ T lymphocytes. This leads to an exhausted, anergic, and immunosuppressive phenotype of $CD8^+$ T lymphocytes (Pauken and Wherry 2015).

PD-L1 (CD274) is also a transmembrane glycoprotein ligand with its gene located in chromosome 9 (Jiang et al. 2019). PD-L1 is expressed on tumor cells and antigen-presenting cells (APCs) especially macrophages in TNBC. Although many cancers execute the activation of the PD-L1/PD-1 axis for immuno-evasion, our understanding of this axis is enhanced by the extensive research on its role in TNBC. As monocytes infiltrate into the TME and get polarized into TAMs, the tumor cells secrete versican, an ECM proteoglycan (Hartley et al. 2017). Versican along with other ECM protein forms a stable scaffolding network for "landing" of immune cells into the TME following their extravasation. Versican from the ECM binds to TAMs and mediates the secretion of TNF-α in a toll-like receptor 2 (TLR-2)-dependent manner (Hartley et al. 2017). The secreted TNF-α in the TME upregulates the expression of PD-L1 on TAMs (Hartley et al. 2017). This self-feeding cycle leads to a positive feedback, and TAMs are capable of elevating their PD-L1 levels via an autocrine signaling mechanism culminating in immunosuppression.

One might ask how does the overexpression of PD-L1 on TAMs enable the tumor to escape an immune response? The answer lies in the interaction between tumor-infiltrating lymphocytes (TILs) and macrophages. As TILs infiltrate into TME, especially $CD8^+$ T lymphocytes, they interact with TAMs to get educated about neoantigens arising out of the tumor cells in order to mount an efficacious immune response (Hartley et al. 2018). However, during this interaction process, the higher levels of PD-L1 on the surface of macrophages bind to PD-1 receptor on $CD8^+$ T lymphocytes. Activation of PD-L1/PD-1 axis leads to the initiation of an inhibitory signaling cascade inside of $CD8^+$ T lymphocytes that suppresses their growth and anti-tumor functionality (Hartley et al. 2018). The intricate molecular mechanisms involving the suppression of cytotoxic activity of $CD8^+$ T lymphocytes via PD-L1/PD-1 axis are not fully understood yet. Nonetheless, it has been demonstrated that the elevated levels of PD-1 expression on $CD8^+$ T lymphocytes inhibit the spreading of $CD8^+$ T lymphocytes on the peptide-major histocompatibility complex (pMHC) surface on macrophages. This is due to an induction of negative cooperativity between T-cell receptors (TCR)-pMHC-CD8 complex assembly due to inefficient binding (Li et al. 2021). This reduced interaction between $CD8^+$ T lymphocytes and displayed neoantigens indicates a disruption of the neoantigen recognition process (Li et al. 2021). Furthermore, PD-L1/PD-1 interaction also suppresses TCR-CD8-dependent intracellular Ca^{2+} signaling in $CD8^+$ T lymphocytes resulting in reduced downstream signaling events and eventually an ineffective cytotoxic capability by $CD8^+$ T lymphocytes (Li et al. 2021).

2.2 Macrophages in Metastasis

Tumor hypoxia plays a role in intravasation of the tumor cells (Donato et al. 2020). The hypoxic TME is abundant with lactate accumulation from anaerobic glycolysis. Binding of lactate to its cognate receptor, G-protein coupled receptor 132 (Gpr132), on macrophages further polarizes them towards M2 phenotype (Laoui et al. 2014). The secretion of VEGF-A ligand by M2 macrophages has been shown to increase local vascular permeability (Harney et al. 2015), and thus, the activation of Gpr132 receptor by lactate promotes metastasis (Chen et al. 2017a, b).

Macrophages play a vital role in tumor progression and metastasis as they promote local invasion, neovascularization, vascular permeability, intravasation, survival in bloodstream as tumor cell clusters along with macrophages and platelets, and extravasation of the tumor cells out of the local capillary network into the metastatic site (Binnewies et al. 2018). Invasion of the local environment of the primary tumor by tumor cells that have undergone epithelial-mesenchymal transition (EMT) is the first basic step of tumor metastasis. EMT is enabled by epigenetic repression of E-cadherins resulting in loss of intrinsic cell polarity and disruption of the cell-cell junctions (Lamouille et al. 2014). TAMs facilitate EMT of tumor cells by secreting interleukins, such as interleukin-1β (IL-1β), interleukin-18 (IL-18), TNF-α, as well as ECM-degrading enzymes, such as cathepsins, matrix metalloproteinases (MMP7, MMP2, and MMP9), serine proteases, and importantly the "CXC-chemokine Ligand 1" (CXCL1) that facilitates cell migration (Kessenbrock et al. 2010). Furthermore, the studies on breast cancer have demonstrated that the production of chitinase 3-like protein (CHI3L1) by TAMs increases the expression of MMPs through interactions with interleukin-13 receptor α2 (IL-13Rα2). This triggers the activation of mitogen-activated protein kinase (MAPK) signaling pathways (Chen et al. 2017a, b). Recent evidence also suggests that chemokine CXCL1 plays a vital role in breast cancer EMT and invasion (Wang et al. 2018). CXCL1 can activate the chemokine receptors CXCR1 and CXCR2 present on neutrophils, MDSCs, and breast cancer stem-like cells (BCSCs) and bulk tumor cells. Production of CXCL1 by TAMs increases the level of SRY-related HMG box transcription factor 4 (SOX4), which represses the expression of E-cadherin while increasing the expression of pro-invasive EMT proteins such as β-catenin, vimentin, and N-cadherin (Wang et al. 2018). Once the tumor degrades ECM with the help of MMPs and other proteases, tumor cells migrate through the acellular stroma/ECM fibers that terminally converge on the vasculature (Barker et al. 2005). This migration on collagen fibers is enabled by "secreted protein acidic and rich in cysteine" (SPARC) (Sangaletti et al. 2019) produced by TAMs that facilitate the interactions between the integrins on tumor cells and ECM components, such as fibronectin and vitronectin in the TME, generating a traction force allowing tumor cells to migrate along the matrix (haptotaxis) towards the vasculature network (Barker et al. 2005).

Intravasation of breast tumor cells from the primary sites is assisted by the production of VEGF-A by TAMs which increases the vascular permeability allowing tumor cells to gain entry into vasculature (Harney et al. 2015). Once in the circulation, tumor cells form a tripartite arrangement of TAMs, tumor cells, and endothelial cells called the "tumor microenvironment of metastasis" (TMEM) or circulatory tumor immune microenvironment (CTIME) (Lin et al. 2019). TAMs in the TMEM produce epidermal growth factor (EGF) to promote tumor cell survival and migration, while the tumor cells produce growth factor M-CSF to promote survival and replication of TAMs. This reciprocal paracrine activation of tumor cells and TAMs create a positive feedback loop (Lin et al. 2019). The TAMs in the TMEM prevent triggering of an immune response by activating the phosphatidylinositol-3-kinase (PI3K)/Akt survival signaling cascade by engaging vascular cell adhesion molecule 1 (VCAM-1) via α4 integrins (Lin et al. 2019). The active PI3K cascade protects the tumor cells from pro-apoptotic cytokines, such as TNF-related apoptosis-inducing ligand (TRAIL), secreted by other immune cells (Lin et al. 2019). Considering all these factors, TAMs play an important role in intravasation as well as the protection of tumor cell clusters during dissemination through the bloodstream, allowing them to efficiently metastasize to distant sites.

In order for a successful metastasis, the circulating TMEM must find a secondary location/tissue for colonization, called premetastatic niche, which is predetermined and primed by metastasis-associated macrophages (MAMs), neutrophils, and MDSCs (Condamine et al. 2015a, b; Høye and Erler 2016). It has been demonstrated that the primary breast tumor secretes "matrisome," such as connective tissue growth factor (CTGF), fibronectin, and periostin among others, to recruit bone marrow-derived macrophages (BMDMs) and transform them to MAMs that prepares the premetastatic niche (Høye and Erler 2016). Furthermore, primary tumor cell-derived exosomes, vesicular carriers for intercellular activity, colocalize and interact with MAMs in premetastatic niche via TLR-8 and induce the expression of TNF-α and IL-6, increasing the infiltration of more macrophages (Hoshino et al. 2015). Along with recruitment of BMDMs, signals from primary tumors can also promote the formation of premetastatic niche by impacting tissue resident macrophages. For example, lysyl oxidase (LOX) secreted by breast cancer cells increases the activity of "nuclear factor of activated T cells c1" (NFATc1) transcription factor, a key regulator of osteoclastogenesis (Cox et al. 2015). This increase will skew the homeostasis of bone towards osteoclastogenesis, forming a lytic lesion in the bone that is favorable for growth of metastatic cancer cells (Cox et al. 2015).

During extravasation, the activation of the CC chemokine ligand 3/CC chemokine receptor 1 (CCL3/CCR1) axis in MAMs allows the association of MAMs with tumor cells through their overexpressed VCAM-1 protein via an α4-integrin-dependent mechanism (Kitamura et al. 2015). This leads to pro-survival, proliferative, and colonizing characteristics, promoting the secondary tumor formation at the predilection sites (Chen et al. 2011). Additionally, macrophage-derived factors, such as VEGF and endothelin-1 (ET1), promote metastatic colonization (Quintero-Fabián et al. 2019). VEGF acts on its receptor VEGFR1 that is expressed in macrophages and promotes the secretion of matrix metalloproteinase 9 (MMP9). This enhances the vascular permeability contributing to tumor progression and chemokine CXCL12 which further recruits BMDM and other tumor cells that express the chemokine receptor CXCR4 (Quintero-Fabián et al. 2019). The presence of macrophages in premetastatic niche provides a roadmap for the metastasis of tumor cells. Therefore, reprogramming of these MAMs could be a big step towards the prevention and treatment of TNBC.

2.3 Macrophages as a Therapy Target

One of the most abundant immune cell types in TME of the TNBC are TAMs, and their role in promoting the ability of tumor cells to replicate, evade immune response, and metastasize cannot be understated. TAMs are double-edged swords, they can impair standard of cancer therapy and promote metastasis, but they can also enhance the therapy and produce desirable responses. For example, treatment of breast cancer using doxorubicin and paclitaxel leads to accumulation of perivascular TAMs that secrete VEGF and promote angiogenesis as well as high rate of relapse after chemotherapy (Hughes et al. 2015). Conversely, the neoadjuvant chemotherapeutic agent, cyclophosphamide, can induce production of CCL4, CXCL8, and TNF-α by tumor cells, increasing macrophage infiltration and phagocytosis in the TME (Pallasch et al. 2014). Therefore, it is important to consider the mechanism of specific chemotherapeutic agents while targeting the TAMs.

Initially, many studies focused on strategies that deplete or inactivate macrophages by targeting M-CSF and its receptor M-CSFR and CCR2 signaling ability (Mantovani et al. 2017; Cassetta and Pollard 2018). However these strategies led to systemic inactivation/depletion of macrophages rendering the immune system vulnerable to external insults (Mantovani et al. 2017, Cassetta and Pollard 2018). Recently, an approach of reprogramming or re-educating TAMs from pro-tumoral to anti-tumoral type has been considered as a superior therapeutic approach (Gatti-Mays et al. 2019).

There are various ways to reprogram macrophages; here we discuss some of the widely accepted and understood approaches. One of the approaches is to target macrophage polarization signaling pathways. This can be achieved via several mechanisms including histone deacetylase inhibitors (HDACs), phosphoinositide 3-kinase γ (PI3Kγ) inhibitors, and TLR agonists. The HDAC inhibitor, TMP195, is an epigenetic modulator that has demonstrated to increase the production of CCL1 and decrease CCL2 levels, promoting apoptosis and chemotaxis (Guerriero et al. 2017). PI3Kγ is a lipid kinase in immune cells that acts as a molecular switch to increase immunosuppressive activity in macrophages via downstream signaling events (Kaneda et al. 2016). PI3Kγ inhibitors, such as SF1126, have shown to increase the expression of major histocompatibility complex-II (MHC-II) antigen and induce secretion of proinflammatory cytokines and reduce the release of immunosuppressive cytokines (Kaneda et al. 2016). TLRs are pattern recognition receptors and are part of innate immunity that has shown to polarize macrophages towards proinflammatory phenotypes (M1 type) (Rodell et al. 2018). Hence, employment of TLR agonists can also be beneficial in converting macrophages from M2 to M1 phenotype.

Macrophages can also be reprogrammed by targeting their phagocytic and immunosuppressive activities. Antiphagocytic signals secreted by tumor cells via CD47, CD24, and β2-microglobulin (β2-M) bind inhibitory receptors such as macrophage signal regulatory protein 1α (SIRP1α), macrophage sialic acid-binding Ig-like lectin (SIGLEC)-10, and macrophage inhibitory receptor LILRBI, respectively (Barkal et al. 2018). These interactions suppress the ability of the macrophages to phagocytose tumor cells and have shown to be a good target for immunotherapeutic agents, such as Hu5F9-G4 and CD24Fc (Barkal et al. 2019). Moreover, macrophage immunosuppressive activities can be exploited by inhibiting the PD-L1/PD1 axis. PD1/PD-L1 inhibitors such as Pembrolizumab, Atezolizumab, and Nivolumab have been approved by the food and drug administration (FDA) and have shown to have clinical efficacy along with neoadjuvant chemotherapy (Jiang et al. 2019). All of these studies enhance our understanding of role of macrophages in cancer and how we can effectively harness the macrophage biology in TME to reprogram them in order to eliminate the TNBC.

3 Neutrophils

Neutrophils are the first line of defense in the body especially against bacterial infections. The circulating leukocytes are comprised of neutrophils 50–70% and 10–25% in human and mice, respectively. They play a key role in the activation and regulation of innate immunity. They also serve downstream of T helper 17 (T_H17) lymphocytes in the adaptive immunity arm. In cancer biology, tumor-associated neutrophils (TANs) or N2 neutrophils are now considered as a vital part of the TME. The cytokine and chemokine profile in the TME is such that the tumors retain N2 neutrophils longer than its lifetime of 6–8 h (Borregaard 2010; Lawrence et al. 2018). Hence, the TANs are aged neutrophils that express CXCR4 waiting to ingress back into the bone marrow for destruction as part of its homeostasis but remain associated with cancer cells longer than its stipulated time outside of bone marrow. This is due to the cytokines in the TME that retain them. Morphologically, the nuclei of such aged neutrophils are hypersegmented (5–6 lobes) than the normal circulating neutrophils (3–4 lobes) (Casanova-Acebes et al. 2013; Zhang et al. 2015; Adrover et al. 2019). Physiologically, neutrophils are endowed with high levels of two key chemokine receptors CXCR1 and CXCR2 when they egress out of the bone marrow and remain in circulation. Conversely, they upregulate the expression levels of their cell surface CXCR4 in order to ingress into the bone marrow and get destroyed. In TNBC, this last step is blocked as they are retained in the TME with high CXCR4 on the cell surface.

Neutrophils can infiltrate into solid tumors at differing levels, and it is a dynamic continuum and not static. So, the level of TANs at the time of biopsy and treatment is likely to differ in the immune cell compartment of the TME reflecting their dynamic nature. In human tumors, neutrophils are identified by the marker CD66b (and Ly6G in mice). TANs are potent suppressors of anti-tumor immunity. Such an infiltration in the tumor along with a high neutrophil-to-lymphocytes ratio (NLR) in circulation is associated with poor prognosis (Gabrilovich et al. 2012; Coffelt et al. 2016; Shaul and Fridlender 2019). TANs participate as a pro-tumor factor by promoting angiogenesis, remodeling of ECM, immunosuppression, and in the case of metastasis by preparing the premetastatic niche (Jaillon et al. 2020). Metabolic switch in the neutrophils due to deprivation of glucose in the TME led to an increased oxidation of fatty acids, elevated levels of reactive oxygen species (ROS), and subsequent reduction in cytotoxic activity of $CD8^+$ T lymphocytes in a murine mammary carcinoma model (Rice et al. 2018). The high levels of ROS can compromise $CD8^+$ T cell activity through nitric oxide (NO) production via inducible of iNOS in another mouse model (Coffelt et al. 2015). Subversion of $CD8^+$ T lymphocyte-mediated antitumor function is also facilitated by "Arginase 1" enzymatic activity by the TANs which reduces L-arginine in the TME (Fridlender et al. 2009; Sagiv et al. 2015). Tumor cells that express ligands such as CXCL8 for the chemokine receptors CXCR1 and CXCR2 on the cell surface of neutrophils can promote the production of neutrophil extracellular traps (NET) called NETosis. The NET protects the tumor cells from cytotoxic activity of $CD8^+$ T lymphocytes (Teijeira et al. 2020).

3.1 Neutrophil Targeting and Reprogramming

A clear understanding of the molecular underpinnings of the intricate role of neutrophils in TIME of TNBC will provide a mechanistic basis for developing novel and efficacious therapeutic approaches. The heterogeneity in neutrophil phenotypes and the observed plasticity pose a challenge to target neutrophils in a precise fashion. The chemokine receptors CXCR1 and CXCR2 that are expressed on the plasma membrane of the neutrophils can be targeted. Alternatively, the ligand CXCL8 is abundant in most of the solid tumors and is being evaluated using an antibody-based approach targeting CXCL8 (BMS-986253; NCT03400332). These are vital for their recruitment into TIME. Resident aged neutrophils in breast TIME can be targeted with newer small molecule inhibitors against the chemokine receptor CXCR4 such as Balixafortide from Polyphor in combination with generic anti-inflammatory agents (Karpova et al. 2017; Pernas et al. 2018; Zhang et al. 2019; Miao et al. 2020).

Myeloid-Derived Suppressor Cells

Myeloid-derived suppressor cells (MDSCs) are crucial to creating an immunosuppressive TIME (Tcyganov et al. 2018; Veglia et al. 2018; Zhou et al. 2018). They can promote metastasis by assisting in the formation of premetastatic niche, by increasing angiogenesis and promoting invasion of tumor cells (Condamine et al. 2015a, b). This is well studied in preclinical mouse models as well as in human patients. Human predisposing traits such as aging and obesity reportedly produce a proinflammatory cytokines in tissue microenvironment that eventually may lead to elevated levels of MDSCs in circulation. MDSCs in the TIME may be a heterogeneous population consisting of cells of either monocytic (M-MDSC) or neutrophilic (or granulocytic or polymorphonuclear; N- or G- or PMN-MDSC) lineage. The markers for N-MDSCs in mice and humans are $CD11b^{+}/Ly6G^{+}/CD84^{+}$ and $CD15^{+}/CD66b^{+}/CD14^{-}/LOX1^{+}$ (or $CD15^{+}/CD66b^{+}/CD14^{-}/CD84^{+}$), respectively (Veglia et al. 2021). The markers for M-MDSCs in mice and humans are $CD11b^{+}/Ly6G^{-}/Ly6C^{hi}/CD84^{+}$ and $CD14^{+}/CD66b^{-}/CXCR1^{+}$ (or $CD14^{+}/CD66b^{-}/CD84^{-}$), respectively (Veglia et al. 2021). The development of these two MDSC populations occurs in two partially overlapping phases. In phase I, the expansion and conditioning of granulocytic and monocytic precursors occur in bone marrow and spleen. The second phase involves pathological activation into MDSCs in peripheral tissues (Condamine et al. 2015a, b). Functionally, N- and M-MDSCs share some features and also have some dissimilar features. Biochemical features such as elevated levels of signal transducer and activator of transcription 3 (STAT3), ER stress induction, expression of "book-marking" chemokines such as S100A8 and S100A9 which facilitates in the preparation of premetastatic niche, and expression of arginase 1 are shared between them. N-MDSCs mediate immunosuppression through production of ROS, peroxynitrite, and prostaglandin E2. Conversely, M-MDSCs employ NO and cytokines such as IL-15 and TGF-β to mediate immunosuppression (Gabrilovich 2017). These are the functional features that are differentially expressed or executed by N- and M-MDSCs.

3.2 Role of MDSCs in the Formation of Premetastatic Niche

Mature neutrophils and N-MDSCs express chemokine receptors CXCR2 and CXCR4 which facilitate their recruitment into the premetastatic niche or predilection sites such as the lungs. They can prime the site for facile colonization of the tumor cells by promoting angiogenesis, through remodeling of the extracellular matrix and by immunosuppression (Veglia et al. 2021). The MDSCs at the premetastatic niche also secrete the book-marking chemokines such as S100A8 and S100A9 which flags the site for colonization by the tumor cells. Additionally, they also produce matrix metalloproteinases such as MMP8 and MMP9 which prepares the sites for facile colonization (Veglia et al. 2021). All of these processes enable survival, proliferation, and colonization of the tumor cells. The tumor cell clusters in circulation are often found clumped with neutrophils, macrophages, platelets, and MDSCs, and such heterogeneous cell clusters can easily get engrafted and stabilized (Veglia et al. 2021) at the metastatic sites.

3.3 Targeting of MDSCs

The recruitment into the premetastatic niche may be targeted with anti-CXCR2- and anti-CXCR4-based therapies with either small molecule inhibitors (Highfill et al. 2014; Karpova et al. 2017; Pernas et al. 2018) or neutralizing therapeutic antibodies. It is interesting that blockade of CXCR2 signaling and migration of MDSCs with small molecule inhibitor enhanced the efficacy of anti-PD-1 therapy in established tumors (Highfill et al. 2014). The ligands for the CXCR2 mainly CXCL1, CXCL2, and CXCL8 may be neutralized with antibodies. The MDSCs can also be reprogrammed or differentiated into mature monocytes, dendritic cells, and neutrophils by treating them with all-*trans* retinoic acid (ATRA) (Lee et al. 2012; Long et al. 2016; Peng et al. 2021). ATRA in fact induced death in neutrophils after conversion from N-MDSCs and differentiated M-MDSCs into monocytes. Targeting of systemic PGE2 with cyclooxygenase 2 (COX2) inhibitors curtailed the expansion of MDSCs in a mouse model. Fatty acid transport protein 2 (FATP2) is high in MDSCs residing in the tumors, and so targeting FATP2 will target MDSCs in the tumor and cut their PGE2 levels and hence their clonal expansion (Adeshakin et al. 2021). Inhibition of Aurora A mitotic kinase hampered MDSCs, and this led to an efficacious anti-PD-L1 therapy (Yin et al. 2019). Interestingly, a sex disparity has been observed in terms of MDSC functionality, severity of cancer in terms of survival, and treatment response in glioblastoma multiforme (Bayik et al. 2020). It remains to be seen if such as disparity will be observed in more lethal male breast cancer and female breast cancer in terms of morbidity, severity (mortality), and treatment outcome.

4 T Lymphocytes

The subtypes of T lymphocytes that infiltrate the TME of TNBC are comprised of $CD8^+$, $CD4^+$, and regulatory T cells. The chemokine "Regulated upon Activation, Normal T Cell Expressed and Presumably Secreted" (RANTES) or CCL5 mediates the infiltration of $CD8^+$ T lymphocytes into the TME. Usually, the TNBC tumors are highly immunosuppressive or have "immunologically cold" TME; they lack the chemokine CCL5. Recently, inhibition of the epigenetic molecule, lysine-specific demethylase 1 (LSD1) in the tumor cells, has been shown to induce the secretion of CCL5 into the TME (Sheng et al. 2018). This approach may be considered as a two-pronged dagger that elicits anti-tumor response through (1) pharmacological inhibition of LSD1 in the tumor cells and (2) recruitment of $CD8^+$ T lymphocytes into the TME due to induced secretion of CCL5. Additionally, T helper 1 (T_H1) and natural killer (NK) cells express IFN-γ which produces strong anti-tumor response. The IFN-γ also produces the release of CCL5 from the tumor cells by acting on its IFN-γ receptor. The regulatory T lymphocytes (Tregs) express the chemokine receptor CXCR4, and they can employ this receptor to infiltrate into the TNBC nodules provided the TME expresses its ligand CXCL12. CXCL12 is secreted by the carcinoma-associated fibroblasts (CAFs) (Orimo et al. 2005; Orimo and Weinberg 2006). Hence, T_{regs} and plasmacytoid dendritic cells can migrate into the tumor through CXCL12-CXCR4 axis. They promote expansion of MDSCs in the TME.

These cells are involved in vascularization (Kryczek et al. 2005), primary tumor progression, and metastasis (Muller et al. 2001). Another way to recruit $CD8^+$ T lymphocytes, T_H1, and NK cells into the tumor is through the chemokines CXCL9 and CXCL10 in the TME as these cells possess their cognate chemokine receptor CXCR3.

Once in the TME, the infiltrated $CD8^+$ T lymphocytes that are specific for tumor-associated neoantigens may exert an anti-tumor response through secretion of its effector molecules such as granzyme B and perforin (Chinnaiyan et al. 1996). Additionally, the exhausted $CD8^+$ T lymphocytes due to overstimulation of PD-1 receptor on them by PD-L1 in the TME form a basis for employment of immune checkpoint inhibitors (ICI). This is to reinvigorate these exhausted $CD8^+$ T lymphocytes back into mounting an efficacious anti-tumor immune response (Schmid et al. 2020; Braso-Maristany et al. 2021). In 2019, Atezolizumab has been approved along with nab-paclitaxel by the FDA for PD-L1 positive locally unresectable, advanced, or metastatic TNBC (mTNBC) (Schmid et al. 2020).

5 Conclusions

The TNBC microenvironment is enriched with a variety of chemokines that activate their cognate receptors on tumor and immune cells in an autocrine and paracrine manner. The chemokine and its receptor network orchestrate primary tumor progression, prepare premetastatic niches, and facilitate facile dissemination, seeding, survival, and colonization of tumor cells at the metastatic (predilection) sites. Targeting the chemokine-chemokine receptor axes such as CCL21-CCR7, CCL2-CCR2, CXCL1,2 and 8-CXCR2 and CXCL12-CXCR4 and other novel axes may reprogram the TIME in order to reinvigorate the $CD8^+$ T lymphocytes, convert M2 macrophages to M1 phenotype, release TANs (N2 neutrophils) back into circulation to home into bone marrow, and get destroyed. Finally, M-MDSCs can be converted into monocytes or DCs and PMN-MDSCs into neutrophils, through differentiation therapy. Overall, a combination of such reprogramming approaches may facilitate synergism or additivity with chemotherapy involving small molecule inhibitors and immunotherapy targeting novel molecules such as T lymphocyte immunoglobulin mucin 3 (TIM-3), V-domain Ig suppressor of T-cell activation (VISTA) (He et al. 2021), T lymphocyte immunoreceptor with Ig and ITIM domains (TIGIT), and lymphocyte activation gene-3 (LAG3).

Acknowledgments We sincerely thank Sara Virginia Hernandez Mendosa for her skillful and dedicated artwork in Fig. 1. This manuscript has been supported in part by National Institute of Health (NIH)/National Cancer Institute (NCI grant R21CA256462) and University of Toledo startup grants (F110796) (to D.R.) and Susan G. Komen Breast Cancer Foundation (CCR18548498) and Department of Defense – Breast Cancer Research Program Level 2 award (1W81XWH-21-1-0053) (to A.K.T.). The views expressed in this article are those of authors and may not reflect the official policy or position of the National Institute of Health, Department of the Army, Department of Defense or the US Government or Susan. G. Komen Breast Cancer Foundation.

Conflict of Interests No conflict of interest was perceived or disclosed.

Compliance with Ethical Standards The authors declare that there is no conflict of interest.

References

Adeshakin AO, Liu W, Adeshakin FO, Afolabi LO, Zhang M, Zhang G, Wang L, Li Z, Lin L, Cao Q, Yan D, Wan X (2021) Regulation of ROS in myeloid-derived suppressor cells through targeting fatty acid transport protein 2 enhanced anti-PD-L1 tumor immunotherapy. Cell Immunol 362:104286

Adrover JM, Del Fresno C, Crainiciuc G, Cuartero MI, Casanova-Acebes M, Weiss LA, Huerga-Encabo H, Silvestre-Roig C, Rossaint J, Cossio I, Lechuga-Vieco AV, Garcia-Prieto J, Gomez-Parrizas M, Quintana JA, Ballesteros I, Martin-Salamanca S, Aroca-Crevillen A, Chong SZ, Evrard M, Balabanian K, Lopez J, Bidzhekov K, Bachelerie F, Abad-Santos F, Munoz-Calleja-C, Zarbock A, Soehnlein O, Weber C, Ng LG, Lopez-Rodriguez C, Sancho D, Moro MA, Ibanez B, Hidalgo A (2019) A neutrophil timer coordinates immune defense and vascular protection. Immunity 50(390–402):e310

Allavena P, Garlanda C, Borrello MG, Sica A, Mantovani A (2008) Pathways connecting inflammation and cancer. Curr Opin Genet Dev 18:3–10

Barkal AA, Weiskopf K, Kao KS, Gordon SR, Rosental B, Yiu YY, George BM, Markovic M, Ring NG, Tsai JM, McKenna KM, Ho PY, Cheng RZ, Chen JY, Barkal LJ, Ring AM, Weissman IL, Maute RL (2018) Engagement of MHC class I by the inhibitory receptor LILRB1 suppresses macrophages and is a target of cancer immunotherapy. Nat Immunol 19:76–84

Barkal AA, Brewer RE, Markovic M, Kowarsky M, Barkal SA, Zaro BW, Krishnan V, Hatakeyama J, Dorigo O, Barkal LJ, Weissman IL (2019) CD24 signalling through macrophage Siglec-10 is a target for cancer immunotherapy. Nature 572:392–396

Barker TH, Baneyx G, Cardó-Vila M, Workman GA, Weaver M, Menon PM, Dedhar S, Rempel SA, Arap W, Pasqualini R, Vogel V, Sage EH (2005) SPARC regulates extracellular matrix organization through its modulation of integrin-linked kinase activity. J Biol Chem 280:36483–36493

Bayik D, Zhou Y, Park C, Hong C, Vail D, Silver DJ, Lauko A, Roversi G, Watson DC, Lo A, Alban TJ, McGraw M, Sorensen M, Grabowski MM, Otvos B, Vogelbaum MA, Horbinski C, Kristensen BW, Khalil AM, Hwang TH, Ahluwalia MS, Cheng F, Lathia JD (2020) Myeloid-derived suppressor cell subsets drive glioblastoma growth in a sex-specific manner. Cancer Discov 10:1210–1225

Belgiovine C, D'Incalci M, Allavena P, Frapolli R (2016) Tumor-associated macrophages and anti-tumor therapies: complex links. Cell Mol Life Sci 73:2411–2424

Binnewies M, Roberts EW, Kersten K, Chan V, Fearon DF, Merad M, Coussens LM, Gabrilovich DI, Ostrand-Rosenberg S, Hedrick CC, Vonderheide RH, Pittet MJ, Jain RK, Zou W, Howcroft TK, Woodhouse EC, Weinberg RA, Krummel MF (2018) Understanding the tumor immune microenvironment (TIME) for effective therapy. Nat Med 24:541–550

Borregaard N (2010) Neutrophils, from marrow to microbes. Immunity 33:657–670

Braso-Maristany F, Sanso M, Chic N, Martinez D, Gonzalez-Farre B, Sanfeliu E, Ghiglione L, Carcelero E, Garcia-Corbacho J, Sanchez M, Soy D, Jares P, Peg V, Saura C, Munoz M, Prat A, Vivancos A (2021) Case report: a case study documenting the activity of Atezolizumab in a PD-L1-negative triple-negative breast cancer. Front Oncol 11:710596

Brummer G, Fang W, Smart C, Zinda B, Alissa N, Berkland C, Miller D, Cheng N (2020) CCR2 signaling in breast carcinoma cells promotes tumor growth and invasion by promoting CCL2 and suppressing CD154 effects on the angiogenic and immune microenvironments. Oncogene 39:2275–2289

Casanova-Acebes M, Pitaval C, Weiss LA, Nombela-Arrieta C, Chevre R, González NA, Kunisaki Y, Zhang D, van Rooijen N, Silberstein LE, Weber C, Nagasawa T, Frenette PS, Castrillo A, Hidalgo A (2013) Rhythmic modulation of the hematopoietic niche through neutrophil clearance. Cell 153:1025–1035

Cassetta L, Pollard JW (2018) Targeting macrophages: therapeutic approaches in cancer. Nat Rev Drug Discov 17:887–904

Cassetta L, Fragkogianni S, Sims AH, Swierczak A, Forrester LM, Zhang H, Soong DYH, Cotechini T, Anur P, Lin EY, Fidanza A, Lopez-Yrigoyen M, Millar MR, Urman A, Ai Z, Spellman PT, Hwang ES, Dixon JM, Wiechmann L, Coussens LM, Smith HO, Pollard JW (2019) Human tumor-associated macrophage and monocyte transcriptional landscapes reveal cancer-specific reprogramming, biomarkers, and therapeutic targets. Cancer Cell 35:588–602 e510

Chen Q, Zhang XH, Massagué J (2011) Macrophage binding to receptor VCAM-1 transmits survival signals in breast cancer cells that invade the lungs. Cancer Cell 20:538–549

Chen Y, Zhang S, Wang Q, Zhang X (2017a) Tumor-recruited M2 macrophages promote gastric and breast cancer metastasis via M2 macrophage-secreted CHI3L1 protein. J Hematol Oncol 10:36

Chen P, Zuo H, Xiong H, Kolar MJ, Chu Q, Saghatelian A, Siegwart DJ, Wan Y (2017b) Gpr132 sensing of lactate mediates tumor-macrophage interplay to promote breast cancer metastasis. Proc Natl Acad Sci U S A 114:580–585

Chinnaiyan AM, Hanna WL, Orth K, Duan H, Poirier GG, Froelich CJ, Dixit VM (1996) Cytotoxic T-cell-derived granzyme B activates the apoptotic protease ICE-LAP3. Curr Biol 6:897–899

Coffelt SB, Kersten K, Doornebal CW, Weiden J, Vrijland K, Hau CS, Verstegen NJM, Ciampricotti M, Hawinkels L, Jonkers J, de Visser KE (2015) IL-17-producing gammadelta T cells and neutrophils conspire to promote breast cancer metastasis. Nature 522:345–348

Coffelt SB, Wellenstein MD, de Visser KE (2016) Neutrophils in cancer: neutral no more. Nat Rev Cancer 16:431–446

Condamine T, Mastio J, Gabrilovich DI (2015a) Transcriptional regulation of myeloid-derived suppressor cells. J Leukoc Biol 98:913–922

Condamine T, Ramachandran I, Youn JI, Gabrilovich DI (2015b) Regulation of tumor metastasis by myeloid-derived suppressor cells. Annu Rev Med 66:97–110

Cox TR, Rumney RMH, Schoof EM, Perryman L, Høye AM, Agrawal A, Bird D, Latif NA, Forrest H, Evans HR, Huggins ID, Lang G, Linding R, Gartland A, Erler JT (2015) The hypoxic cancer secretome induces pre-metastatic bone lesions through lysyl oxidase. Nature 522:106–110

De Palma M, Lewis CE (2013) Macrophage regulation of tumor responses to anticancer therapies. Cancer Cell 23:277–286

Donato C, Kunz L, Castro-Giner F, Paasinen-Sohns A, Strittmatter K, Szczerba BM, Scherrer R, Di Maggio N, Heusermann W, Biehlmaier O, Beisel C, Vetter M, Rochlitz C, Weber WP, Banfi A, Schroeder T, Aceto N (2020) Hypoxia triggers the Intravasation of clustered circulating tumor cells. Cell Rep 32:108105

Fridlender ZG, Sun J, Kim S, Kapoor V, Cheng G, Ling L, Worthen GS, Albelda SM (2009) Polarization of tumor-associated neutrophil phenotype by TGF-beta: "N1" versus "N2" TAN. Cancer Cell 16:183–194

Gabrilovich DI (2017) Myeloid-derived suppressor cells. Cancer Immunol Res 5:3–8

Gabrilovich DI, Ostrand-Rosenberg S, Bronte V (2012) Coordinated regulation of myeloid cells by tumours. Nat Rev Immunol 12:253–268

Galluzzi L, Vacchelli E, Bravo-San Pedro JM, Buque A, Senovilla L, Baracco EE, Bloy N, Castoldi F, Abastado JP, Agostinis P, Apte RN, Aranda F, Ayyoub M, Beckhove P, Blay JY, Bracci L, Caignard A, Castelli C, Cavallo F, Celis E, Cerundolo V, Clayton A, Colombo MP, Coussens L, Dhodapkar MV, Eggermont AM, Fearon DT, Fridman WH, Fucikova J, Gabrilovich DI, Galon J, Garg A, Ghiringhelli F, Giaccone G, Gilboa E, Gnjatic S, Hoos A, Hosmalin A, Jager D, Kalinski P, Karre K, Kepp O, Kiessling R, Kirkwood JM, Klein E, Knuth A, Lewis CE, Liblau R, Lotze MT, Lugli E, Mach JP, Mattei F, Mavilio D, Melero I, Melief CJ, Mittendorf EA, Moretta L, Odunsi A, Okada H, Palucka AK, Peter ME, Pienta KJ,

Porgador A, Prendergast GC, Rabinovich GA, Restifo NP, Rizvi N, Sautes-Fridman C, Schreiber H, Seliger B, Shiku H, Silva-Santos B, Smyth MJ, Speiser DE, Spisek R, Srivastava PK, Talmadge JE, Tartour E, Van Der Burg SH, Van Den Eynde BJ, Vile R, Wagner H, Weber JS, Whiteside TL, Wolchok JD, Zitvogel L, Zou W, Kroemer G (2014) Classification of current anticancer immunotherapies. Oncotarget 5:12472–12508

Gatti-Mays ME, Balko JM, Gameiro SR, Bear HD, Prabhakaran S, Fukui J, Disis ML, Nanda R, Gulley JL, Kalinsky K, Abdul Sater H, Sparano JA, Cescon D, Page DB, McArthur H, Adams S, Mittendorf EA (2019) If we build it they will come: targeting the immune response to breast cancer. NPJ Breast Cancer 5:37

Germano G, Allavena P, Mantovani A (2008) Cytokines as a key component of cancer-related inflammation. Cytokine 43:374–379

Guerriero JL, Sotayo A, Ponichtera HE, Castrillon JA, Pourzia AL, Schad S, Johnson SF, Carrasco RD, Lazo S, Bronson RT, Davis SP, Lobera M, Nolan MA, Letai A (2017) Class IIa HDAC inhibition reduces breast tumours and metastases through anti-tumour macrophages. Nature 543:428–432

Hanahan D, Coussens LM (2012) Accessories to the crime: functions of cells recruited to the tumor microenvironment. Cancer Cell 21:309–322

Hanahan D, Weinberg RA (2011) Hallmarks of cancer: the next generation. Cell 144:646–674

Harney AS, Arwert EN, Entenberg D, Wang Y, Guo P, Qian BZ, Oktay MH, Pollard JW, Jones JG, Condeelis JS (2015) Real-time imaging reveals local, transient vascular permeability, and tumor cell Intravasation stimulated by TIE2hi macrophage-derived VEGFA. Cancer Discov 5:932–943

Hartley G, Regan D, Guth A, Dow S (2017) Regulation of PD-L1 expression on murine tumor-associated monocytes and macrophages by locally produced TNF-α. Cancer Immunol Immunother 66:523–535

Hartley GP, Chow L, Ammons DT, Wheat WH, Dow SW (2018) Programmed cell death ligand 1 (PD-L1) signaling regulates macrophage proliferation and activation. Cancer Immunol Res 6:1260–1273

He HX, Gao Y, Fu JC, Zhou QH, Wang XX, Bai B, Li PF, Huang C, Rong QX, Ping LQ, He YX, Mao JY, Chen X, Huang HQ (2021) VISTA and PD-L1 synergistically predict poor prognosis in patients with extranodal natural killer/T-cell lymphoma. Onco Targets Ther 10:1907059

Herbst RS, Soria JC, Kowanetz M, Fine GD, Hamid O, Gordon MS, Sosman JA, McDermott DF, Powderly JD, Gettinger SN, Kohrt HE, Horn L, Lawrence DP, Rost S, Leabman M, Xiao Y, Mokatrin A, Koeppen H, Hegde PS, Mellman I, Chen DS, Hodi FS (2014) Predictive correlates of response to the anti-PD-L1 antibody MPDL3280A in cancer patients. Nature 515:563–567

Highfill SL, Cui Y, Giles AJ, Smith JP, Zhang H, Morse E, Kaplan RN, Mackall CL (2014) Disruption of CXCR2-mediated MDSC tumor trafficking enhances anti-PD1 efficacy. Sci Transl Med 6:237ra267

Hoshino A, Costa-Silva B, Shen TL, Rodrigues G, Hashimoto A, Tesic Mark M, Molina H, Kohsaka S, Di Giannatale A, Ceder S, Singh S, Williams C, Soplop N, Uryu K, Pharmer L, King T, Bojmar L, Davies AE, Ararso Y, Zhang T, Zhang H, Hernandez J, Weiss JM, Dumont-Cole VD, Kramer K, Wexler LH, Narendran A, Schwartz GK, Healey JH, Sandstrom P, Labori KJ, Kure EH, Grandgenett PM, Hollingsworth MA, de Sousa M, Kaur S, Jain M, Mallya K, Batra SK, Jarnagin WR, Brady MS, Fodstad O, Muller V, Pantel K, Minn AJ, Bissell MJ, Garcia BA, Kang Y, Rajasekhar VK, Ghajar CM, Matei I, Peinado H, Bromberg J, Lyden D (2015) Tumour exosome integrins determine organotropic metastasis. Nature 527:329–335

Høye AM, Erler JT (2016) Structural ECM components in the premetastatic and metastatic niche. Am J Physiol Cell Physiol 310:C955–C967

Huang YK, Busuttil RA, Boussioutas A (2021) The role of innate immune cells in tumor invasion and metastasis. Cancers (Basel) 13:5885

Hughes R, Qian BZ, Rowan C, Muthana M, Keklikoglou I, Olson OC, Tazzyman S, Danson S, Addison C, Clemons M, Gonzalez-Angulo AM, Joyce JA, De Palma M, Pollard JW, Lewis CE (2015) Perivascular M2 macrophages stimulate tumor relapse after chemotherapy. Cancer Res 75:3479–3491

Hussain SM, Kansal RG, Alvarez MA, Hollingsworth TJ, Elahi A, Miranda-Carboni G, Hendrick LE, Pingili AK, Albritton LM, Dickson PV, Deneve JL, Yakoub D, Hayes DN, Kurosu M, Shibata D, Makowski L, Glazer ES (2021) Role of TGF-beta in pancreatic ductal adenocarcinoma progression and PD-L1 expression. Cell Oncol (Dordr) 44:673–687

Jaillon S, Ponzetta A, Di Mitri D, Santoni A, Bonecchi R, Mantovani A (2020) Neutrophil diversity and plasticity in tumour progression and therapy. Nat Rev Cancer 20:485–503

Jiang Y, Chen M, Nie H, Yuan Y (2019) PD-1 and PD-L1 in cancer immunotherapy: clinical implications and future considerations. Hum Vaccin Immunother 15:1111–1122

Jing W, Guo X, Wang G, Bi Y, Han L, Zhu Q, Qiu C, Tanaka M, Zhao Y (2020) Breast cancer cells promote CD169(+) macrophage-associated immunosuppression through JAK2-mediated PD-L1 upregulation on macrophages. Int Immunopharmacol 78:106012

Kaneda MM, Messer KS, Ralainirina N, Li H, Leem CJ, Gorjestani S, Woo G, Nguyen AV, Figueiredo CC, Foubert P, Schmid MC, Pink M, Winkler DG, Rausch M, Palombella VJ, Kutok J, McGovern K, Frazer KA, Wu X, Karin M, Sasik R, Cohen EE, Varner JA (2016) PI3Kγ is a molecular switch that controls immune suppression. Nature 539:437–442

Karpova D, Brauninger S, Wiercinska E, Kramer A, Stock B, Graff J, Martin H, Wach A, Escot C, Douglas G, Romagnoli B, Chevalier E, Dembowski K, Hooftman L, Bonig H (2017) Mobilization of hematopoietic stem cells with the novel CXCR4 antagonist POL6326 (balixafortide) in healthy volunteers-results of a dose escalation trial. J Transl Med 15:2

Kessenbrock K, Plaks V, Werb Z (2010) Matrix metalloproteinases: regulators of the tumor microenvironment. Cell 141:52–67

Kitamura T, Qian BZ, Soong D, Cassetta L, Noy R, Sugano G, Kato Y, Li J, Pollard JW (2015) CCL2-induced chemokine cascade promotes breast cancer metastasis by enhancing retention of metastasis-associated macrophages. J Exp Med 212:1043–1059

Knowles HJ, Harris AL (2007) Macrophages and the hypoxic tumour microenvironment. Front Biosci 12:4298–4314

Kryczek I, Lange A, Mottram P, Alvarez X, Cheng P, Hogan M, Moons L, Wei S, Zou L, Machelon V, Emilie D, Terrassa M, Lackner A, Curiel TJ, Carmeliet P, Zou W (2005) CXCL12 and vascular endothelial growth factor synergistically induce neoangiogenesis in human ovarian cancers. Cancer Res 65:465–472

Kryczek I, Zou L, Rodriguez P, Zhu G, Wei S, Mottram P, Brumlik M, Cheng P, Curiel T, Myers L, Lackner A, Alvarez X, Ochoa A, Chen L, Zou W (2006) B7-H4 expression identifies a novel suppressive macrophage population in human ovarian carcinoma. J Exp Med 203:871–881

Lamouille S, Xu J, Derynck R (2014) Molecular mechanisms of epithelial-mesenchymal transition. Nat Rev Mol Cell Biol 15:178–196

Laoui D, Van Overmeire E, Di Conza G, Aldeni C, Keirsse J, Morias Y, Movahedi K, Houbracken I, Schouppe E, Elkrim Y, Karroum O, Jordan B, Carmeliet P, Gysemans C, De Baetselier P, Mazzone M, Van Ginderachter JA (2014) Tumor hypoxia does not drive differentiation of tumor-associated macrophages but rather fine-tunes the M2-like macrophage population. Cancer Res 74:24–30

Lawrence SM, Corriden R, Nizet V (2018) The ontogeny of a neutrophil: mechanisms of Granulopoiesis and homeostasis. Microbiol Mol Biol Rev 82:e00057-17

Lee JM, Seo JH, Kim YJ, Kim YS, Ko HJ, Kang CY (2012) The restoration of myeloid-derived suppressor cells as functional antigen-presenting cells by NKT cell help and all-trans-retinoic acid treatment. Int J Cancer 131:741–751

Li K, Yuan Z, Lyu J, Ahn E, Davis SJ, Ahmed R, Zhu C (2021) PD-1 suppresses TCR-CD8 cooperativity during T-cell antigen recognition. Nat Commun 12:2746

Lim YW, Coles GL, Sandhu SK, Johnson DS, Adler AS, Stone EL (2021) Single-cell transcriptomics reveals the effect of PD-L1/TGF-beta blockade on the tumor microenvironment. BMC Biol 19:107

Lin Y, Xu J, Lan H (2019) Tumor-associated macrophages in tumor metastasis: biological roles and clinical therapeutic applications. J Hematol Oncol 12:76

Long AH, Highfill SL, Cui Y, Smith JP, Walker AJ, Ramakrishna S, El-Etriby R, Galli S, Tsokos MG, Orentas RJ, Mackall CL (2016) Reduction of MDSCs with all-trans retinoic acid improves CAR therapy efficacy for sarcomas. Cancer Immunol Res 4:869–880

Lu X, Yang R, Zhang L, Xi Y, Zhao J, Wang F, Zhang H, Li Z (2019) Macrophage Colony-stimulating factor mediates the recruitment of macrophages in triple negative breast cancer. Int J Biol Sci 15:2859–2871

Lyes MA, Payne S, Ferrell P, Pizzo SV, Hollenbeck ST, Bachelder RE (2019) Adipose stem cell crosstalk with chemo-residual breast cancer cells: implications for tumor recurrence. Breast Cancer Res Treat 174:413–422

Mantovani A, Allavena P (2015) The interaction of anticancer therapies with tumor-associated macrophages. J Exp Med 212:435–445

Mantovani A, Pierotti MA (2008) Cancer and inflammation: a complex relationship. Cancer Lett 267:180–181

Mantovani A, Sica A, Locati M (2005) Macrophage polarization comes of age. Immunity 23:344–346

Mantovani A, Allavena P, Sica A, Balkwill F (2008a) Cancer-related inflammation. Nature 454: 436–444

Mantovani A, Marchesi F, Portal C, Allavena P, Sica A (2008b) Linking inflammation reactions to cancer: novel targets for therapeutic strategies. Adv Exp Med Biol 610:112–127

Mantovani A, Marchesi F, Malesci A, Laghi L, Allavena P (2017) Tumour-associated macrophages as treatment targets in oncology. Nat Rev Clin Oncol 14:399–416

Miao M, De Clercq E, Li G (2020) Clinical significance of chemokine receptor antagonists. Expert Opin Drug Metab Toxicol 16:11–30

Muller A, Homey B, Soto H, Ge N, Catron D, Buchanan ME, McClanahan T, Murphy E, Yuan W, Wagner SN, Barrera JL, Mohar A, Verastegui E, Zlotnik A (2001) Involvement of chemokine receptors in breast cancer metastasis. Nature 410:50–56

Nagarsheth N, Wicha MS, Zou W (2017) Chemokines in the cancer microenvironment and their relevance in cancer immunotherapy. Nat Rev Immunol 17:559–572

Niu M, Valdes S, Naguib YW, Hursting SD, Cui Z (2016) Tumor-associated macrophage-mediated targeted therapy of triple-negative breast cancer. Mol Pharm 13:1833–1842

Noy R, Pollard JW (2014) Tumor-associated macrophages: from mechanisms to therapy. Immunity 41:49–61

Oner G, Altintas S, Canturk Z, Tjalma W, Verhoeven Y, Van Berckelaer C, Berneman Z, Peeters M, Pauwels P, van Dam PA (2019) Triple-negative breast cancer-role of immunology: a systemic review. Breast J 26:995–999

Orimo A, Weinberg RA (2006) Stromal fibroblasts in cancer: a novel tumor-promoting cell type. Cell Cycle 5:1597–1601

Orimo A, Gupta PB, Sgroi DC, Arenzana-Seisdedos F, Delaunay T, Naeem R, Carey VJ, Richardson AL, Weinberg RA (2005) Stromal fibroblasts present in invasive human breast carcinomas promote tumor growth and angiogenesis through elevated SDF-1/CXCL12 secretion. Cell 121:335–348

Pallasch CP, Leskov I, Braun CJ, Vorholt D, Drake A, Soto-Feliciano YM, Bent EH, Schwamb J, Iliopoulou B, Kutsch N, van Rooijen N, Frenzel LP, Wendtner CM, Heukamp L, Kreuzer KA, Hallek M, Chen J, Hemann MT (2014) Sensitizing protective tumor microenvironments to antibody-mediated therapy. Cell 156:590–602

Pauken KE, Wherry EJ (2015) Overcoming T cell exhaustion in infection and cancer. Trends Immunol 36:265–276

Peng SY, Chen L, Deng RH, Li H, Liu XH, Zheng DW, Wu CC, Liu CJ, Sun ZJ, Zhang XZ (2021) Non-depleting reformation of immunosuppressive myeloid cells to broaden the application of anti-PD therapy. Nanoscale 13:4420–4431

Pernas S, Martin M, Kaufman PA, Gil-Martin M, Gomez Pardo P, Lopez-Tarruella S, Manso L, Ciruelos E, Perez-Fidalgo JA, Hernando C, Ademuyiwa FO, Weilbaecher K, Mayer I, Pluard TJ, Martinez Garcia M, Vahdat L, Perez-Garcia J, Wach A, Barker D, Fung S, Romagnoli B, Cortes J (2018) Balixafortide plus eribulin in HER2-negative metastatic breast cancer: a phase 1, single-arm, dose-escalation trial. Lancet Oncol 19:812–824

Plava J, Cihova M, Burikova M, Matuskova M, Kucerova L, Miklikova S (2019) Recent advances in understanding tumor stroma-mediated chemoresistance in breast cancer. Mol Cancer 18:67

Quintero-Fabián S, Arreola R, Becerril-Villanueva E, Torres-Romero JC, Arana-Argáez V, Lara-Riegos J, Ramírez-Camacho MA, Alvarez-Sánchez ME (2019) Role of matrix metalloproteinases in angiogenesis and cancer. Front Oncol 9:1370

Raman D, Baugher PJ, Thu YM, Richmond A (2007) Role of chemokines in tumor growth. Cancer Lett 256:137–165

Rice CM, Davies LC, Subleski JJ, Maio N, Gonzalez-Cotto M, Andrews C, Patel NL, Palmieri EM, Weiss JM, Lee JM, Annunziata CM, Rouault TA, Durum SK, McVicar DW (2018) Tumour-elicited neutrophils engage mitochondrial metabolism to circumvent nutrient limitations and maintain immune suppression. Nat Commun 9:5099

Rodell CB, Arlauckas SP, Cuccarese MF, Garris CS, Li R, Ahmed MS, Kohler RH, Pittet MJ, Weissleder R (2018) TLR7/8-agonist-loaded nanoparticles promote the polarization of tumour-associated macrophages to enhance cancer immunotherapy. Nat Biomed Eng 2:578–588

Sagiv JY, Michaeli J, Assi S, Mishalian I, Kisos H, Levy L, Damti P, Lumbroso D, Polyansky L, Sionov RV, Ariel A, Hovav AH, Henke E, Fridlender ZG, Granot Z (2015) Phenotypic diversity and plasticity in circulating neutrophil subpopulations in cancer. Cell Rep 10:562–573

Sangaletti S, Talarico G, Chiodoni C, Cappetti B, Botti L, Portararo P, Gulino A, Consonni FM, Sica A, Randon G, Di Nicola M, Tripodo C, Colombo MP (2019) SPARC is a new myeloid-derived suppressor cell marker licensing suppressive activities. Front Immunol 10:1369

Schmid P, Rugo HS, Adams S, Schneeweiss A, Barrios CH, Iwata H, Dieras V, Henschel V, Molinero L, Chui SY, Maiya V, Husain A, Winer EP, Loi S, Emens LA, Investigators IM (2020) Atezolizumab plus nab-paclitaxel as first-line treatment for unresectable, locally advanced or metastatic triple-negative breast cancer (IMpassion130): updated efficacy results from a randomised, double-blind, placebo-controlled, phase 3 trial. Lancet Oncol 21:44–59

Shaul ME, Fridlender ZG (2019) Tumour-associated neutrophils in patients with cancer. Nat Rev Clin Oncol 16:601–620

Sheng W, LaFleur MW, Nguyen TH, Chen S, Chakravarthy A, Conway JR, Li Y, Chen H, Yang H, Hsu PH, Van Allen EM, Freeman GJ, De Carvalho DD, He HH, Sharpe AH, Shi Y (2018) LSD1 ablation stimulates anti-tumor immunity and enables checkpoint blockade. Cell 174:549–563 e519

Sica A, Allavena P, Mantovani A (2008) Cancer related inflammation: the macrophage connection. Cancer Lett 267:204–215

Sousa S, Brion R, Lintunen M, Kronqvist P, Sandholm J, Mönkkönen J, Kellokumpu-Lehtinen PL, Lauttia S, Tynninen O, Joensuu H, Heymann D, Määttä JA (2015) Human breast cancer cells educate macrophages toward the M2 activation status. Breast Cancer Res 17:101

Tcyganov E, Mastio J, Chen E, Gabrilovich DI (2018) Plasticity of myeloid-derived suppressor cells in cancer. Curr Opin Immunol 51:76–82

Teijeira A, Garasa S, Gato M, Alfaro C, Migueliz I, Cirella A, de Andrea C, Ochoa MC, Otano I, Etxeberria I, Andueza MP, Nieto CP, Resano L, Azpilikueta A, Allegretti M, de Pizzol M, Ponz-Sarvise M, Rouzaut A, Sanmamed MF, Schalper K, Carleton M, Mellado M, Rodriguez-Ruiz ME, Berraondo P, Perez-Gracia JL, Melero I (2020) CXCR1 and CXCR2 chemokine receptor agonists produced by tumors induce neutrophil extracellular traps that interfere with immune cytotoxicity. Immunity 52(856–871):e858

Tumeh PC, Harview CL, Yearley JH, Shintaku IP, Taylor EJ, Robert L, Chmielowski B, Spasic M, Henry G, Ciobanu V, West AN, Carmona M, Kivork C, Seja E, Cherry G, Gutierrez AJ, Grogan TR, Mateus C, Tomasic G, Glaspy JA, Emerson RO, Robins H, Pierce RH, Elashoff DA, Robert C, Ribas A (2014) PD-1 blockade induces responses by inhibiting adaptive immune resistance. Nature 515:568–571

Veglia F, Perego M, Gabrilovich D (2018) Myeloid-derived suppressor cells coming of age. Nat Immunol 19:108–119

Veglia F, Sanseviero E, Gabrilovich DI (2021) Myeloid-derived suppressor cells in the era of increasing myeloid cell diversity. Nat Rev Immunol 21:485–498

Wang N, Liu W, Zheng Y, Wang S, Yang B, Li M, Song J, Zhang F, Zhang X, Wang Q, Wang Z (2018) CXCL1 derived from tumor-associated macrophages promotes breast cancer metastasis via activating NF-κB/SOX4 signaling. Cell Death Dis 9:880

Yin T, Zhao ZB, Guo J, Wang T, Yang JB, Wang C, Long J, Ma S, Huang Q, Zhang K, Ma X, Liu C, Liu S, Lian ZX, Yang Z (2019) Aurora A inhibition eliminates myeloid cell-mediated immunosuppression and enhances the efficacy of anti-PD-L1 therapy in breast cancer. Cancer Res 79:3431–3444

Zhang D, Chen G, Manwani D, Mortha A, Xu C, Faith JJ, Burk RD, Kunisaki Y, Jang JE, Scheiermann C, Merad M, Frenette PS (2015) Neutrophil ageing is regulated by the microbiome. Nature 525:528–532

Zhang WJ, Wang XH, Gao ST, Chen C, Xu XY, Sun Q, Zhou ZH, Wu GZ, Yu Q, Xu G, Yao YZ, Guan WX (2018) Tumor-associated macrophages correlate with phenomenon of epithelial-mesenchymal transition and contribute to poor prognosis in triple-negative breast cancer patients. J Surg Res 222:93–101

Zhang J, Pang Y, Xie T, Zhu L (2019) CXCR4 antagonism in combination with IDO1 inhibition weakens immune suppression and inhibits tumor growth in mouse breast cancer bone metastases. Onco Targets Ther 12:4985–4992

Zhou J, Nefedova Y, Lei A, Gabrilovich D (2018) Neutrophils and PMN-MDSC: their biological role and interaction with stromal cells. Semin Immunol 35:19–28

Zhu Y, Knolhoff BL, Meyer MA, Nywening TM, West BL, Luo J, Wang-Gillam A, Goedegebuure SP, Linehan DC, DeNardo DG (2014) CSF1/CSF1R blockade reprograms tumor-infiltrating macrophages and improves response to T-cell checkpoint immunotherapy in pancreatic cancer models. Cancer Res 74:5057–5069

Zurlo IV, Schino M, Strippoli A, Calegari MA, Cocomazzi A, Cassano A, Pozzo C, Di Salvatore M, Ricci R, Barone C, Bria E, Tortora G, Larocca LM, Basso M, Martini M (2022) Predictive value of NLR, TILs (CD4+/CD8+) and PD-L1 expression for prognosis and response to preoperative chemotherapy in gastric cancer. Cancer Immunol Immunother 71:45–55

The Prognostic Impact of Tumor-Infiltrated Lymphocytes (TILs) in Breast Cancer

Caterina Fattorini, Federica Petrelli, Grazia Gallippi, and Alessandro Ginori

Abstract

The immune compartment of tumor microenvironment is formed by cells of innate immune system, such as neutrophils and tumor-associated macrophages and cells of the adaptive immune system, namely, tumor-infiltrating lymphocytes (TILs). The role of TILs in invasive breast cancer has been studied since several decades. Scientists first noted the presence of a brisk lymphocytic infiltrate in the peritumoral stroma of the so-called "medullary carcinoma" and suggested that its better prognosis should be related to this type of infiltrate as a sort of "host antitumoral response." It is been a long time since then, and many subsequent studies examined the association between tumor lymphocytic infiltrates and prognosis in other histological types of breast cancer, not without contradicting results. A systematic review was conducted using studies sourced from the PubMed, Embase, and Cochrane literature databases. The relevant publications discussing TILs in breast cancer are analyzed. The guidelines for histopathological assessment of TILs are discussed and summarized. TILs play a prognostic role in breast cancer, especially in triple-negative and HER-2-positive tumors. TIL assessment is mandatory in every breast cancer subtype and also in the neoadjuvant chemotherapy setting. It seems to have also therapeutic implications in conventional chemotherapy and chemotherapy/immunotherapy regimens.

The authors "Caterina Fattorini, Federica Petrelli" have contributed equally to this chapter.

C. Fattorini · F. Petrelli · G. Gallippi · A. Ginori (✉)
USL Northwest Tuscany, "Monterosso" Polyspecialistic Center, Pathology Unit, Carrara, Italy
e-mail: alessandro.ginori@uslnordovest.toscana.it

© The Author(s), under exclusive license to Springer Nature Switzerland AG 2022
Interdisciplinary Cancer Research, https://doi.org/10.1007/16833_2022_33
Published online: 28 October 2022

Keywords

Breast cancer · Checkpoint inhibitors · Guidelines · Histopathological assessment · PD-L1 · Prognosis · Prognostic impact · TILs · Tumor microenvironment

1 Introduction

According to the GLOBOCAN 2020 report, breast cancer (BC) is the most common tumor worldwide excluding nonmelanoma skin cancer, with 2.26 million new cases diagnosed in 2020 (11.7% of all tumors) (GLOBOCAN – Global Cancer Observatory 2020). Incidence per 100,000 inhabitants varies from 26.2 in South–Central Asia to 95.5 in Australia and New Zealand, with an average of 47.8. In recent years, the incidence rate for this cancer has risen in very low and low human development index countries (Bray et al. 2018). Breast carcinoma is also an important cause of death, ranking fifth among cancers with 684,996 deaths in 2020 and being the first cause of death for cancer in women (WHO Classification of Tumours 2019). The most important risk factors are gender (less than 1% of BC cases occur in men), age (the incidence reaches its peak around menopause, then decreases or remains stable), and socioeconomic status. Women with higher socioeconomic status have a higher risk of developing BC, and this is mainly due to lifestyle factors such as a high-fat diet, poor physical activity, and older age at first childbirth. On the other hand, these women have better access to care and prevention programs, leading to better overall and disease-free survival (DFS). Estrogens also are well-known carcinogenic factors in the mammary gland. They act through estrogen receptor (ER) binding, inducing terminal duct lobular unit proliferation. Moreover, previous proliferative atypical breast lesions, such as atypical ductal hyperplasia and ductal carcinoma in situ (DCIS), increase the risk by 4–5× and 10×, respectively, compared with the general population. Finally, hereditary factors play an important role. A family history of BC is considered one of the major risk factors. BRCA1 and BRCA2 are two high-penetrance genes that greatly increase the risk of developing BC. Seventy-two percent of BRCA1 and 69% of BRCA2 germline mutation carriers develop the disease before the age of 80. The 5-year survival rate has seen a marked improvement over the last decades. This is mainly due to early diagnosis, effective screening programs, and novel therapeutic protocols. Prognostic stratification of patients is mainly based on a correct staging following the TNM staging system (Amin et al. 2017) and evaluation of histological grade (the Nottingham combined histological grade, modified by Scoff–Bloom–Richardson grading system), lymph-vascular invasion, proliferation rate, ER, and progesterone receptor (PR) expression and HER2-neu gene status. Finally, tumor microenvironment (TME) appears to be an important host factor that is able to influence patients' prognosis and response to certain therapies. TME is the complex background where the tumor settles and grows, and it consists of all cellular and noncellular components that have close interaction with tumor cells. The cellular component of TME is composed mainly of endothelial cells and pericytes that form blood vessels, immune cells, and stromal cells such as fibroblasts, whereas noncellular component consists of extracellular matrix (ECM) and signaling molecules such as cytokines (Baghban et al. 2020). The immune compartment of TME is formed by cells of innate immune system, such as neutrophils and tumor-associated macrophages (TAMs), and cells of the adaptive immune system, namely, tumor-infiltrating lymphocytes (TILs). The role of TILs in invasive BC has been studied since several decades. Scientists first noted the presence of a brisk lymphocytic infiltrate in the peritumoral stroma of the so-called "medullary carcinoma" and suggested that its better prognosis should be related to this type of infiltrate as a sort of "host antitumoral response" (Moore and Foote 1949). It has been a long time since then, and many subsequent studies examined the association between tumor lymphocytic infiltrates and prognosis in other histological types of BC, not without contradicting results.

2 TILs in BC

As mentioned earlier, TME consists of all cellular and noncellular components that have close interaction with tumor cells (Baghban et al. 2020). The close connection between these two compartments enables the tumor to develop and finally acquire a progressive metastatic phenotype. The interaction of tumor cells with immune compartment is fundamental to the so-called "immunoediting," that is, the ability of the tumor to escape the immune system. ECM and cytokines make up the noncellular compartment of TME, whereas the cellular compartment consists of endothelial cells, pericytes, stromal cells, and immune cells. Both cells of the adaptive and innate immune system concur to form the immune compartment of TME, whose composition and intensity greatly vary among different types of neoplasia. The main components of the adaptive immune system are B and T lymphocytes, the so-called TILs. Most BCs show lymphocytes around and within the tumor, so TILs always have been thoroughly studied. Starting from the 1940s to the 1950s, scientists first studied peritumoral lymphocytes infiltrates. For example, Moore and coworkers noted the presence of a brisk immune infiltrate in a category of BC with relatively better prognosis, the so-called "medullary-type" BC (Moore and Foote 1949). A few decades later, Horst and Horny began to systematically study the composition of BC immune infiltrates with the aid of immunohistochemistry (Horny and Horst 1986; Horst and Horny 1988). Later, many scientists have studied the prognostic and predictive value of TILs in BC. Location, density, and composition of infiltrates were taken into consideration. Stromal and intratumoral TILs must be distinguished: stromal TILs are defined as lymphocytes dispersed in the stroma and not in direct contact with tumor cells, while intratumoral TILs are lymphocytes present in the tumor area without intervening stroma and directly interacting with tumor cells (Salgado et al. 2015a). BCs with brisk lymphocytes infiltrates are called "lymphocytes predominant breast cancers" (LPBCs) and are defined as tumors with

50% of cancer area occupied by TILs (Burugu et al. 2017). Immunohistochemical studies allowed analyzing the different subpopulations of the immune infiltrates that were constituted by CD8+ lymphocytes, CD4+ lymphocytes, and B lymphocytes. CD8+ cytotoxic lymphocytes concur to eliminate cancer cells by releasing perforin and granzyme after recognition of specific epitopes on major histocompatibility complex (MHC) class I molecules. Some studies have highlighted the greater presence of CD8+ cells in triple-negative BCs (TNBCs) and hormone-negative/HER2+ BC. Those BC types are associated with increased level of genomic instability, and the subsequent production of neoantigens recognized by CD8+ cells explained this phenomenon (Burugu et al. 2017). Moreover, many studies evaluating the possible prognostic role of CD8+ cells reported that high level of CD8+ cells in TILs is associated with better clinical outcome. For example, in 2011 Mahmoud studied the antitumoral activity of CD8+ cells in BC and found that their presence had a favorable effect on patients' survival (Mahmoud et al. 2011a, b). In 2012, Liu and coworkers found that CD8+ cells were an independent prognostic factor associated with better survival in basal-like BC but not in the other subtypes (Liu et al. 2012). Another study associated high number of CD8+ cells with reduction in relative risk of death in hormone-positive BC (ER+/HER2- and ER+/HER2+ BC) (Ali et al. 2014). CD4+ lymphocytes are composed of Th1, Th2, regulatory T cells (Tregs), and follicular T helper cell (Tfh) subtypes. Th1 cells activate CD8+ cytotoxic cells via IFNγ. Higher number of Th1 cells are related to favorable prognosis and better DFS in many tumor types (Gu-Trantien et al. 2013). Th2 cells secrete many types of cytokines, but their role in BC is still less clear. Treg cells are identified by the nuclear immunohistochemical marker forkhead box protein (FOXP) 3 and normally limit excessive immune responses that could cause damage to normal tissue. Their role in BC is somewhat controversial as some scientists found that they do not have a clear prognostic role in BC and that a higher number of Treg cells were not an independent prognostic factor (Mahmoud et al. 2011a, b). West and coworkers found instead that FOXP3 cells positively correlate with CD8+ cells and are associated with favorable prognosis. Patients with hormone-negative tumors seemed to have prolonged recurrence-free survival (RFS) (West et al. 2013). Also, Lee et al. in 2013 identified Treg cells as an independent prognostic factor for overall survival (OS) and progression-free survival (PFS), especially in TNBC (Lee et al. 2013). Higher FOXP3 cells were also associated with improved survival. Tfh cells function as mediators of B-cells in germinal centers and were linked to better prognosis especially in HER2+ BCs (Gu-Trantien et al. 2013). Finally, also the presence of B lymphocytes in BC has been studied. They have multiple functions as antibody production, antigen presentation to CD4+ cells, and antitumor activity modulation. There were only a few studies in the literature trying to characterize their significance in BC (Mahmoud et al. 2012; Mohammed et al. 2013). In 2012, Mahmoud found that higher B-cell infiltrates are associated with higher tumor grade, estrogen, and progesterone receptor negativity and basal phenotype (Mahmoud et al. 2012). The cited results are summarized in Table 1.

Table 1 Prognostic value of lymphocytic subpopulations

	Study, year	Number of patients	Time interval	BC phenotype	TIL evaluation method	Outcome parameter	Results
CD8+ T cytotoxic lymphocytes	Mahmoud, 2011	1902	1987–1998	TN, ER/PR+, HER2+	IHC for CD8	BCSS	Higher total CD8+ count independently associated with longer BCSS (HR 0.55, $P = 0.001$)
					Intratumoral CD8+, adjacent stromal CD8+, distant stromal CD8+ and total CD8+ count		Distant stromal CD8+ count conferred an independent good prognostic value (HR, 0.50; $P < 0.001$)
	Liu, 2012	3992	1986–1992	Intrinsic subtypes: luminal A, luminal B, HER2+/ER-, TN (core basal, 5NP)	IHC for CD8	BCSS	High intratumoral CD8+ associated with superior BCSS in the whole cohort
					Intratumoral CD8+, stromal CD8+ count		On multivariate analysis, the favorable prognostic effect of high CD8+ was significant only in the core basal intrinsic subgroup (HR 0.35, $P < 0.001$)
	Ali, 2014	12,439	1986–1998	TNBC, ER/PR+, HER2+	IHC for CD8 Intratumoral CD8+, stromal CD8+ count	BCSS, relative risk of death from BC	Intratumoral and stromal CD8+ were independently associated with a reduced relative risk of death from breast cancer in ER+ and ER- BC

(continued)

Table 1 (continued)

	Study, year	Number of patients	Time interval	BC phenotype	TIL evaluation method	Outcome parameter	Results
CD4+ Th1/Th2 /Tfh/ Treg lymphocytes	Gu-Trantien, 2013	794	Not available	TN, ER/PR+, HER2+	Flow cytometry and gene expression analysis	DFS and pCR after neoadjuvant chemotherapy	Extensively infiltrated tumors have a higher frequency of TLS associated with detectable CXCL13-producing CD4+ Tfh cells
					Assessing of CD4+ subpopulations gene signatures		Tfh signature was prognostic, demonstrating a linear association with survival ($P = 0.0036$)
	Lee, 2013	175	2000–2009	TN, ER/PR+, HER2+	IHC for CD3, CD8, and FOXP3	pCR after neoadjuvant chemotherapy	LI was a significant independent predictor of pCR (OR, 1.26; $P = 0.024$)
					Peritumoral lymphocytic infiltrates (LI)		Patients who achieved pCR showed significantly higher LI, CD3, CD8 and FOXP3 percentages
	West, 2013	255	1988–2000	ER-	IHC for FOXP3 on TMA	RFS, BCSS	Higher 10-year RFS rates for FOXP3-high patients compared to low-FOXP3 patients ($P = 0.0002$)

B lymphocytes	Mahmoud, 2012	1902	1987–1998	TN, ER/PR+, HER2+	IHC for CD20 on TMA Intratumoral and stromal compartment	BCSS and DFI	The total number of CD20+ cells, irrespective of location, was associated with significantly better BCSS ($P = 0.037$) and longer DFI ($P = 0.001$)
	Mohammed, 2013	388	1995–1998	TN, ER/PR+, HER2+	IHC for CD4, CD8, CD20, and CD138 on TMA	BCSS	On univariate analysis, CD138+ B lymphocytic infiltrate ($P < 0.001$) was associated with poorer cancer-specific survival. On multivariate analysis, CD138+ B lymphocytic infiltrate ($P < 0.001$) was independently associated with cancer survival

BC breast cancer, *BCSS* Breast cancer-specific survival, *DFI* Disease-free interval, *DFS* disease-free survival, *ER* estrogen receptor, *HER2* human epidermal growth factor receptor 2, *HR* hazard ratio, *IHC* immunohistochemistry, *OR* odds ratio, *P* P–value, *pCR* pathological complete response, *PR* progesterone receptor, *RFS* recurrence-free survival, *TLS* tertiary lymphoid structures, *TMA* tissue microarray, *TN* triple–negative, *5NP* five negative phenotype

3 TILs and Tumor Subtypes

TILs' prognostic and predictive role in BC seems to vary according to different tumor subtypes, and many interesting randomized trials have been brought forward for this purpose (Table 2).

3.1 TILs and TNBCs

TNBCs are well known for having the highest peri- and intratumoral immune infiltrate (Fig. 1). Higher levels of genomic instability and clonal heterogeneity associated with higher mutational load of cancer cells in comparison with luminal subtypes account for the greater stimulation of host immune response against tumor cells in TNBC (Solinas et al. 2017a, b). The phase III BIG-02-98 trial analyzed the prognostic role of TILs in node-positive TNBC patients. Scientists concluded that there was a certain reduction in risk of recurrence and deaths every 10% of stromal and intratumoral TIL level increase. Moreover, they noted that LPBC phenotype was associated with better DFS and OS (Loi et al. 2013). Similar results were noted also in other randomized trials, such as the International Cancer Study Group Trial 22-00 (Krishnamurti et al. 2017) and the Eastern cooperative Oncology Group Trial (Adams et al. 2014), demonstrating that each 10% increase in stromal TILs was significantly associated with better OS and DFS. In the FinHER trial, Loi and coworkers also found that every 10% increase in TILs was associated with reduction in death and distant recurrence percentage but no significant association was noted between stromal TILs and OS (Loi et al. 2014). In their meta-analysis, Ibrahim et al. noted that higher levels of TILs in TNBC were associated with lower risk of recurrence, distal recurrence, death, and better OS (Ibrahim et al. 2014). Leon Ferre and coworkers studied early-stage TNBC and concluded that lower TILs levels were associated with worse OS and DFS in both treated and untreated patients (Leon-Ferre et al. 2018). Then, Loi in 2019 studied the role of TILs in early-stage TNBC. She found that, after adjuvant chemotherapy, for each 10% increase of TILs there was an increase in DFS and OS especially for node-negative patients, resulting in better survival and confirming the strong prognostic role of TILs in early TNBC. According to the importance of integrating biomarkers into the TNM staging emphasized in the eighth edition of the American Joint Committee on Cancer (AJCC) Staging System and based on the same cohort of patients of the study of 2019, Loi has also recently noted that TIL scores are seen to strongly up- and downstage traditional pathological staging. She suggested that TILs should be taken into account in clinical and research studies in addition to stage because patients with stage II TNBC and high TILs had a better outcome than patients with stage I and low TILs regardless of the histological grade (Loi et al. 2022).

Table 2 Prognostic value of TILs in BC different subtypes

	Study, year	Number of patients	Time interval	Chemotherapy	BC phenotype	TIL evaluation method	Outcome parameter	Results
TNBC	Loi, 2013 (BIG 02-98 trial)	2887	1998–2001	Adjuvant, comparing doxorubicin-only CHT versus doxorubicin + docetaxel CHT	TN, ER/PR+, HER2+	Histopathological analysis of TILs on hematoxylin–eosin slides Intratumoral and stromal TILs	DFS, OS	Each 10% increase in intratumoral and stromal TILs was associated with 17% and 15% reduced risk of relapse (adjusted $P = 0.1$ and $P = 0.025$), respectively, and 27% and 17% reduced risk of death in TNBC regardless of CHT type
	Adams, 2014 (ECOG 2197, ECOG 2199 trials)	ECOG 2197: 776 ECOG 2199: 5052	ECOG 2197: 1998–2000 ECOG 2199: 1999–2002	Adjuvant ECOG 2197: doxorubicin + cyclophosphamide/ docetaxel ECOG 2199: doxorubicin + cyclophosphamide and taxane	TNBC (ECOG 2197: 250 patients/ ECOG 2199: 926 patients)	Histopathological analysis of TILs on hematoxylin–eosin slides Intratumoral (iTILs) and stromal TILs (sTILs)	DFS, OS, DRFI	Higher TILs were associated with better prognosis: for every 10% increase in sTILs, a 14% reduction of risk of recurrence or death ($P = 0.02$), 18% reduction of risk of distant recurrence ($P = 0.04$), and 19% reduction of risk of death ($P = 0.01$) were observed

(continued)

Table 2 (continued)

Study, year	Number of patients	Time interval	Chemotherapy	BC phenotype	TIL evaluation method	Outcome parameter	Results
Loi, 2014 (FinHER trial)	1010 of which 778 HER2- and 232 HER2+	2000–2003	Adjuvant	TN, ER/PR+, HER2+	Histopathological analysis of TILs on hematoxylin–eosin slides	DDFS	Each 10% increase in TILs was significantly associated with decreased DDFS in TNBC
			CHT+/– trastuzumab		Intratumoral (iTILs) and stromal TILs (sTILs)		
Krishnamurti, 2017	344	2004–2013	–/+	TN, ER/PR+, HER2+	Histopathological analysis of TILs on hematoxylin–eosin slides	OS, DFS	Peripheral TILs were significantly associated with better OS (HR: 0.95; $P = 0.0354$) and DFS (HR: 0.95; $P = 0.0314$) in univariate and multivariate analysis
					Stromal TILs (sTILs)		
Leon Ferre, 2018	605	1985–2012	–	TNBC	Histopathological analysis of TILs on hematoxylin–eosin slides	IDFS, OS	On multivariate analysis, lower TILs and absence of adjuvant chemotherapy were associated with worse IDFS and OS
					Intratumoral and stromal TILs		

	Loi, 2019	2148	Not available	Adjuvant	Early stage TNBC	Histopathological analysis of TILs on Hematoxylin-Eosin slides	DFS, OS	sTILs has a strong prognostic role in early-stage TNBC
				Anthracycline-based chemotherapy with or without taxane		Stromal TILs		Each 10% increment in sTILs corresponded to an HR of 0.87 for iFDS, 0.83 for D-DFS, and 0.84 for OS
HER2+ BC	Denkert, 2010 (GeparDuo and GeparTrio trials)	GeparDuo: 993	GeparDuo: 1999–2001	Neoadjuvant	TN, ER/PR+, HER2+	Histopathological analysis of TILs on hematoxylin–eosin slides	pCR	The percentage of iTILs was a significant independent parameter for pathologic complete response (pCR)
		GeparTrio: 2090	GeparTrio: 2002–2005	Anthracycline/ taxane-based		Intratumoral and stromal TILs		
	Loi, 2013 (BIG 02-98 trial)	2887	1998–2001	Adjuvant, comparing doxorubicin-only CHT versus doxorubicin + docetaxel CHT	TN, ER/PR+, HER2+	Histopathological analysis of TILs on hematoxylin–eosin slides Intratumoral and stromal TILs	DFS, OS	In HER2+ BC, there was a significant interaction between increasing stromal TILs and benefit with anthracycline-only CHT (DFS, interaction $P = 0.042$; OS, $P = 0.018$)

(continued)

Table 2 (continued)

	Study, year	Number of patients	Time interval	Chemotherapy	BC phenotype	TIL evaluation method	Outcome parameter	Results
	Loi, 2014 (FinHER trial)	1010 of which 778 HER2- and 232 HER2+	2000–2003	Adjuvant	TN, ER/PR+, HER2+	Histopathological analysis of TILs on Hematoxylin-Eosin slides	DDFS	For HER2+ BC each 10% increase in TILs was significantly associated with DDFS in patients randomized to the trastuzumab arm ($P = 0.025$).
				CHT+/− trastuzumab		Intratumoral (iTILs) and stromal TILs (sTILs)		
	Dieci, 2015	823	1989–1995	Adjuvant	TN, ER/PR+, HER2+	Histopathological analysis of TILs on hematoxylin–eosin slides	OS	For HER2+ BC, both iTILs and sTILs were strong prognostic factors for OS (HR 0.85, $P = 0.003$; HR 0.89, $P = 0.005$ for iTILs and sTILs, respectively
				Patients were randomized to receive 5-fluorouracil, doxorubicin, or epirubicin and cyclophosphamide, or no chemotherapy		Intratumoral (iTILs) and stromal TILs (sTILs)		
	Bianchini, 2015 (NeoSphere trial)	417	2007–2009	Neoadjuvant docetaxel with trastuzumab, pertuzumab, or both, or monoclonal antibodies alone	TN, ER/PR+, HER2+	Histopathological analysis of TILs on hematoxylin–eosin slides	pCR	Neither iTILs nor sTILs were significantly associated with pCR in HER2+ BC
						Intratumoral (iTILs) and stromal TILs (sTILs)		

	Perez, 2016 (N9831 trial)	3505	2000–2005	Adjuvant	HER2+ BC	Histopathological analysis of TILs on hematoxylin–eosin slides	RFS	sTILs were prognostically associated with RFS in patients treated with CHT alone, but not in patients treated with CHT+ trastuzumab
				CHT alone or CHT plus trastuzumab		Stromal TILs (sTILs)		
	Ignatiadis, 2019 (TRYPHAENA trial)	255	2009–2011	Neoadjuvant	HER2+ BC	Histopathological analysis of TILs on hematoxylin–eosin slides	pCR, EFS	sTILs % was not associated with pCR (OR for every 10% unit increase in TILs =1.12, $P = 0.17$)
				Trastuzumab/ pertuzumab concurrently or sequentially with an anthracycline-containing regimen or concurrently with an anthracycline-free regimen		Stromal TILs		At a median follow-up of 4.7 years, for every increase in baseline TILs of 10%, there was a 25% reduction in the hazard for an EFS event (OR = 0.75, $P = 0.05$)
	Angelico, 2021	47	Not available	Neoadjuvant/ adjuvant CHT plus trastuzumab	HER2+ BC	Histopathological analysis of TILs on hematoxylin–eosin slides	pCR	There was a positive correlation between TILs level and response to both adjuvant and neoadjuvant treatments in HER-2+ BC
						Stromal TILs		

(continued)

Table 2 (continued)

	Study, year	Number of patients	Time interval	Chemotherapy	BC phenotype	TIL evaluation method	Outcome parameter	Results
Hormone +/HER2- BC	Loi, 2013 (BIG 02-98 trial)	2887	1998–2001	Adjuvant, comparing doxorubicin-only CHT versus doxorubicin + docetaxel CHT	TN, ER/PR+, HER2+	Histopathological analysis of TILs on hematoxylin–eosin slides	DFS, OS	There was no significant prognostic association in the ER + HER2- population
						Intratumoral and stromal TILs		
	Loi, 2014 (FinHER trial)	1010 of which 778 HER2- and 232 HER2+	2000–2003	Adjuvant	TN, ER/PR+, HER2+	Histopathological analysis of TILs on hematoxylin–eosin slides	DDFS	There was no significant prognostic association in the ER + HER2- population
				CHT+/– trastuzumab		Intratumoral and stromal TILs		
	Dieci, 2015	823	1989–1995	Adjuvant	TN, ER/PR+, HER2+	Histopathological analysis of TILs on hematoxylin–eosin slides	OS	There was no significant prognostic association in the ER + HER2- population
				Patients were randomized to receive 5-fluorouracil, doxorubicin, or epirubicin and cyclophosphamide, or no chemotherapy		Intratumoral (iTILs) and stromal TILs (sTILs)		

	Montagna, 2017 (VEX trial)	92	Not available	Metronomic chemotherapy (cyclophosphamide/ capecitabine/ vinorelbine) in metastatic patients	TN, ER/PR+, HER2+	Histopathological analysis of TILs on hematoxylin–eosin slides Stromal TILs	TTP	For ER+ BC, each 10% increase in TILs strongly predicted a worse TTP (HR: 1.27, $p = 0.008$)

BC breast cancer, *CHT* chemotherapy, *CMF* cyclophosphamide, methotrexate, fluorouracil, *DDFS* distant disease-free survival, *DFS* disease-free survival, *DRFI* distant recurrence-free interval, *EFS* event-free survival, *HER2* human epidermal growth factor receptor 2, *N* lymph nodes status, *OS* overall survival, *pCR* pathological complete response, *RFS* recurrence-free survival, *TNBC* triple-negative breast cancer, *TTP* time to progression

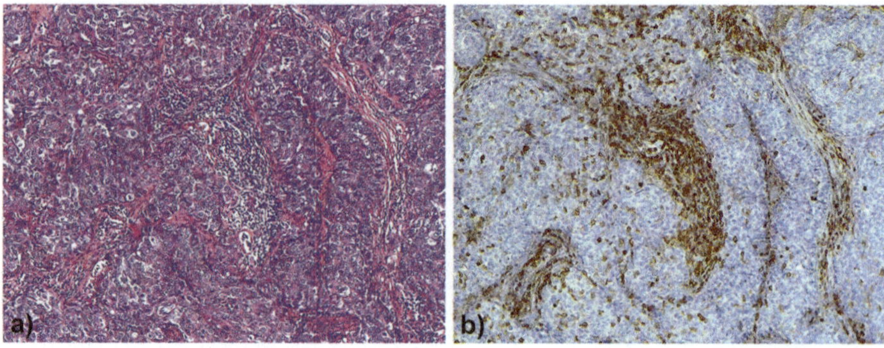

Fig. 1 Poorly differentiated invasive breast carcinoma no special type with heavy peritumoral lymphocytic infiltrate (**a**, hematoxylin and eosin stain, 100×; **b**, CD45 immunohistochemistry stain, 100×)

3.2 TILs and HER2+ BCs

HER2+ BCs also display higher level of immune infiltration compared to hormone-positive/HER2- tumors. The reason behind this characteristic is the higher mutational load that, as for TNBCs, leads to the presence of higher number of neoantigens and to a massive stimulation of immune response. The average number of stromal TILs fluctuates between 15% and 20%, while intratumoral TILs range between 3% and 10% (Solinas et al. 2017a, b). These percentages are higher among HER2+/hormone- cancers compared to HER2+/hormone+ ones. The exact reason for this disparity is still not well known, but most scientists believe that estrogen and progesterone receptor status should influence TIL levels. Scientists have studied the role of TILs in HER2+ tumors in both adjuvant and neoadjuvant settings. Regarding the adjuvant setting, in 2015 Dieci and coworkers found that, in patients treated with anthracyclines but not with trastuzumab, stromal and intratumoral TILs had prognostic effect on OS but cannot be used as a parameter of response to anthracyclines (Dieci et al. 2015). In the aforementioned Big 02-98 Trial, scientists noted that TILs predicted benefit in terms of OS and DFS in patients treated with anthracyclines alone compared to patients treated with a combination of anthracyclines and taxane. They also concluded that intratumoral TILs, especially CD8+ cells, were independent predictors of improved outcome (Loi et al. 2013). In the FinHER trial, patients were treated with trastuzumab alone or in combination with chemotherapy. Patients treated with trastuzumab and chemotherapy with high stromal TILs tended to have lower percentages of distant recurrence compared with patients with lower levels of immune infiltrates (Loi et al. 2014). However, in 2016 Perez and coworkers published conflicting results as they demonstrated that patients who benefited most from treatment with trastuzumab had no LPBC (Perez et al. 2016). Regarding the role of TILs in neoadjuvant setting, one of the first studies analyzing the role of TILs in HER2+ tumors after neoadjuvant chemotherapy (NACT) was the GeparTrio study, carried out by Denkert and coworkers in 2010. They found that patients with LPBC achieved higher pathological complete response (pCR) compared to the group with low or absent TILs (Denkert et al. 2010). Similar results were demonstrated also in the meta-analysis by Solinas in 2017, showing that higher level of TILs was associated with high pCR and that these data were independent of chemotherapy or anti-HER2 agents (Solinas et al. 2017a, b). The Tryphaena trial instead demonstrated that higher TILs and higher pCR were related to dual HER2 blockade (trastuzumab plus pertuzumab) in combination with taxane and anthracyclines/carboplatinum (Ignatiadis et al. 2019). In 2018, Denkert and colleagues analyzed six randomized trials carried out by the German Breast Cancer Group and also concluded that higher levels of TILs were associated with higher complete response rate after chemotherapy and that every 10% increase in TIL value was associated with longer DFS (Denkert et al. 2018). Finally, in 2021, also Angelico and colleagues published the results of a study evaluating TILs in HER2+ BC of patients treated with NACT. They noted a positive correlation between TILs and response rate to therapy. They also concluded that cases with progression of disease displayed lower level of TILs (Angelico et al. 2021). However, the Neosphere trial pointed out opposite results. Scientists divide patients into two arms: one group treated with an anti-HER2 agent (trastuzumab or pertuzumab) alone and the other group treated with anti-HER2 agents in combination with docetaxel. They concluded that higher level of intratumoral or stromal TILs after therapy was not associated with higher pCR and that in the arm treated with both trastuzumab and docetaxel higher level of pCR was achieved in non-LPBC (Bianchini et al. 2015).

3.3 TILs and Hormone+/HER2− BCs

Hormone-positive tumors are characterized by expression of estrogen and/or progesterone receptor (ER/PgR), and they represent the most common tumor subtype. This type of BC is known to be the least immune infiltrated. In fact, the median value of stromal TILs ranges from 7% to 10%, while intratumoral TILs range between 1.5% and 5%. Moreover, the percentage of LPBC in early stages ranges between 2.9% and 15% (Solinas et al. 2017a, b). The exact reason behind the lower immunogenicity of hormone-positive BC is still not yet clear, but some authors hypothesize that the expression of ER is able to promote Th2 response and subsequently decrease MHC class II molecule expression in BC cells, finally resulting in lower density of immune cells (Stanton et al. 2016). Most studies have failed to demonstrate a statistically significant correlation between TILs and survival in hormone-positive BC (Dieci et al. 2015; Loi et al. 2013, 2014). No correlation was found between TIL level and outcome in early stages also in adjuvant setting (Carbognin et al. 2016). Instead, in metastatic disease Montagna and colleagues found that patients treated with chemotherapy and whose tumors had high level of TILs had a shorter time to progression, highlighting an unfavorable prognostic role of TILs (Montagna et al. 2017). Desmetdt and coworkers also found that higher level of TILs was correlated with worse prognosis. Moreover, they compared TIL levels in ductal and lobular subtypes and found that lobular cancers had a lower level of immune infiltrate (Desmedt et al. 2018). In their review, Stanton and colleagues recapitulate all available data on the subject. They concluded that most studies seemed to confirm that patients with hormone+ BC have lower level of TILs and that even patients with LPBC phenotype do not show the same improved survival as HER2+ or TN BC. They also highlighted that some studies indicated that FOXP3+ cells rich in immune infiltrates seemed to predict a worse survival (Stanton et al. 2016). Surely further larger studies are needed to elucidate the exact role of TILs in this BC subtype.

3.4 TILs and Neoadjuvant Chemotherapy

The ability of NACT to stimulate an active immune response against the tumor is very well known (Badr et al. 2020), so many studies evaluated the potential role of TILs in predicting pCR and prognosis of BCs (Table 3). Some authors took into consideration different types of therapies. For example, Denkert in 2010 and Issa-Nummer in 2013 analyzed the use of anthracyclines and taxane and concluded that higher TIL levels were associated with better outcomes (Denkert et al. 2010; Issa-Nummer et al. 2013). In 2012, Yamaguchi came to similar conclusions in patients treated only with anthracyclines-based NACT (Yamaguchi et al. 2012). Other authors evaluated the effect of TIL levels in relation to residual cancer burden (RBC). RBC is an important index that quantifies residual disease after NACT and a predictor of survival (Symmans et al. 2007). In 2019, Luen and colleagues evaluated the possible prognostic value of TILs in TNBC residual disease in predicting patients' survival after NACT. They noted that TIL levels in residual cancer were associated with improved RFS and OS also when including RCB class and especially for RBC class II (Luen et al. 2019). Another study took into consideration both TNBC and HER2+ BC and evaluated the prognostic role of TILs before and after NACT (Ochi et al. 2019). Scientists concluded that in TNBC with residual disease, low pre- and post-TILs were associated with shorter RFS. Another work came also to the similar conclusion that RCB combined with TILs level were statistically significant predictors of BC recurrence after NACT in all BC subtypes (Asano et al. 2017). Contrasting results were obtained by Hamy and coworkers (Hamy et al. 2017). They analyzed a group of HER2+ BC patients treated with NACT alone or in combination with trastuzumab. A decrease in TILs level during chemotherapy was positively associated with response to treatment. Moreover, in patients with residual disease, post-NACT TILs levels were associated with higher RBC and adverse clinical outcome. Finally, other authors focused their work on correlation between TIL subsets and NACT response. One of the first available studies analyzed a group of patients treated with paclitaxel and measured the level of CD3+ lymphocytes in immune infiltrates. Higher CD3+ intratumoral cells were associated with higher response rate to therapy (Demaria et al. 2001). Subsequently, two other works (West et al. 2011; Brown et al. 2014) indicated similar results in patients treated with anthracyclines and anthracyclines plus taxane regimens. CD8+

Table 3 Prognostic value of TILs and chemotherapy

	Study, year	Number of patients	Time interval	Type of chemotherapy	BC phenotype	TIL evaluation method	Outcome parameter	Results
Neoadjuvant setting	Demaria, 2001	25	Not available	Paclitaxel protocol	Not specified	Histopathological analysis of TILs on hematoxylin–eosin slides before and after treatment Intratumoral and stromal TILs	Response to CHT in terms of SD, CPR, CCR/PRD, CCR and apoptotic response	The development of TILs after treatment may be related to the clinical response and to the apoptotic response to neoadjuvant chemotherapy
	Denkert, 2010 (GeparDuo and GeparTrio trials)	GeparDuo: 993 GeparTrio: 2090	GeparDuo: 1999–2001 GeparTrio: 2002–2005	Anthracycline/ taxane-based	TN, ER/PR+, HER2+	Histopathological analysis of TILs on hematoxylin–eosin slides Intratumoral and stromal TILs	pCR	The percentage of iTILs was a significant independent parameter for pathologic complete response (pCR)
	West, 2011 (EORTC 10994/BIG 00-01 trial)	EORTC cohort: 113 MBTB cohort: 255	Not available	Anthracycline based: FEC (fluorouracil, epirubicin, and cyclophosphamide) or TET (docetaxel followed by docetaxel plus epirubicin)	TN, ER/PR+, HER2+	CD3, CD4, CD8, CD20, TIA-1 IHC on TMA	pCR/RD, DFS	In patients with ER- tumors treated with neoadjuvant anthracycline-based CHT, TIL status ($P = 0.001$) was an independent predictor of response

(continued)

Table 3 (continued)

	Study, year	Number of patients	Time interval	Type of chemotherapy	BC phenotype	TIL evaluation method	Outcome parameter	Results
	Yamaguchi, 2012	68	2005–2010	Anthracycline and taxane-based regimen	TN, ER/PR+, HER2+	Histopathological analysis of TILs on Hematoxylin-Eosin slides	pCR	In univariate analysis, high TILs was significantly associated with pCR ($P < 0.0001$)
						Stromal TILs		In multivariate analysis, high TILs was the best independent predictor for pCR (OR 4.7; $P < 0.0001$)
	Issa-Nummer, 2013 (PREDICT substudy of GeparQuinto trial)	313	Not available	Epirubicin and cyclophosphamide followed by taxane	HER2-BC	Histopathological analysis of TILs on Hematoxylin-Eosin slides	pCR	Patients with LPBC had a significantly increased pCR rate compared to non-LPBC patients ($p < 0.001$)
						Intratumoral and stromal TILs		LPBC and sTILs were significantly independent predictors for pCR in multivariate analysis (LPBC: OR 2.7, $p = 0.003$, sTILs: OR 1.2, $p = 0.01$)

	Seo, 2013	153	2004–2010	Anthracycline- or anthracycline/ taxane-based	TN, ER/PR+, HER2+	IHC staining for CD8, CD4, FOXP3, CD44, CD243, ALDH1, and EMT markers	pCR	High levels of CD4, CD8, and FOXP3+ TILs were associated with pCR
						Histopathological analysis of TILs on hematoxylin-eosin slides		CD8+ TILs were found to be an independent predictive factor for pCR
						Intratumoral TILs		
	Brown, 2014	105	2002–2010	Anthracycline and taxane-based regimen	TN, ER/PR+, HER2+	Quantitative immunofluorescence (QIF) assay to measure stromal CD3+, CD8+, and CD20+ lymphocytes	pCR	sTIL count predicted pCR ($p = 0.043$, OR: 4.77)
	Asano, 2016	177	2007–2013	FEC100 followed by paclitaxel	TN, ER/PR+, HER2+	CFR (CD8+/ FOXP3+ TIL ratio)	pCR	pCR rate was significantly higher in the high-CFR group than in the low-CFR group (TNBC: $P = 0{\cdot}022$; HER2BC: $P < 0{\cdot}001$)
				Patients with HER2+ BC also received trastuzumab during paclitaxel treatment				In multivariable analysis, high-CFR status was an independent predictor of a favorable prognosis for TNBC and HER2+ BC

(continued)

Table 3 (continued)

	Study, year	Number of patients	Time interval	Type of chemotherapy	BC phenotype	TIL evaluation method	Outcome parameter	Results
	Hamy, 2017	175	2002–2011	NAC+/− trastuzumab	HER2+ BC	Pre- and post-NAC TILs evaluation	DFS, pCR	After CHT, TIL levels were higher in tumors displaying aggressive features
						Histopathological analysis of TILs on hematoxylin–eosin slides		In the patients with RD, TIL levels >25% at the end of NAC were significantly associated with an adverse outcome (TILs >25%, HR = 7.98, $P = 0.009$) after multivariate analyses
						Stromal TILs		
	Goto, 2018	136	2007–2015	FEC100 followed by paclitaxel	TN, ER/PR+, HER2+	CFR (CD8+/FOXP3+ TIL ratio)	RFS	Patients with a high rate of change in the CFR had significantly better RFS ($p < 0.001$)
				Patients with HER2+ BC also received trastuzumab during paclitaxel treatment				In patients with TNBC and hormone+ BC, the rate of change in the CFR was an

								independent predictor for RFS (HR = 13.021, $p = 0.002$, and HR = 4.377, $p = 0.003$)
	Luen, 2019	375	Not available	Anthracycline/taxane based	TNBC	Histopathological analysis of TILs on hematoxylin–eosin slides in RCB	RFS, OS	Higher RD TILs were significantly associated with improved RFS (HR: 0.86; $P < 0.001$), and improved OS (HR: 0.87; $P < 0.001$) also in multivariate analysis
						Stromal TILs		
						IHC for CD8 and FOXP3, CFR		
	Ochi, 2019	209	2001–2009		TNBC and HER2+ BC		pCR, RFS	In TNBC with RD, low pre-NAT TILs showed significant association with shorter RFS (HR = 3.844, $p = 0.024$) in multivariate analysis
								In HER2+ BC, low pre-NAT TILs were not associated with RFS

(continued)

Table 3 (continued)

	Study, year	Number of patients	Time interval	Type of chemotherapy	BC phenotype	TIL evaluation method	Outcome parameter	Results
Adjuvant setting	Loi, 2013 (BIG 02-98 trial)	2887	1998–2001	Adjuvant, comparing doxorubicin only CHT versus doxorubicin + docetaxel CHT	TN, ER/PR+, HER2+	Histopathological analysis of TILs on hematoxylin–eosin slides	DFS, OS	Each 10% increase in intratumoral and stromal TILs was associated with 17% and 15% reduced risk of relapse (adjusted $P = 0.1$ and $P = 0.025$), respectively, and 27% and 17% reduced risk of death in TNBC regardless of CHT type
						Intratumoral and stromal TILs		In HER2+ BC, there was a significant interaction between increasing stromal TILs and benefit with anthracycline-only CHT (DFS, interaction $P = 0.042$; OS, $P = 0.018$)

	Loi, 2014 (FinHER trial)	1010 of which 778 HER2- and 232 HER2+	2000–2003	CHT+/– trastuzumab	TN, ER/PR+, HER2+	Histopathological analysis of TILs on hematoxylin–eosin slides	DDFS	There was no significant prognostic association in the ER + HER2- population
						Intratumoral and stromal TILs		
	Loi, 2019	2148	Not available	Anthracycline-based chemotherapy with or without taxane	Early stage TNBC	Histopathological analysis of TILs on hematoxylin–eosin slides	DFS, OS	sTILs has a strong prognostic role in early-stage TNBC
						Stromal TILs		Each 10% increment in sTILs corresponded to an HR of 0.87 for iFDS, 0.83 for D-DFS, and 0.84 for OS

BC breast cancer, *CCR* clinical complete response, *CHT* chemotherapy, *CPR* clinical partial response, *DFS* disease-free survival, *EMT* epithelial–mesenchymal transition, *ER* estrogen receptor, *HER2* human epidermal growth factor receptor 2, *IHC* Immunohistochemistry, *LPBC* lymphocytes predominant breast cancer, *OS* overall survival, *PR* progesterone receptor, *PRD* pathological residual disease, *PCR* pathological complete response, *RCB* residual cancer burden, *RD* residual disease, *SD* stable disease, *TMA* tissue microarray, *TNBC* triple-negative breast cancer cells also seem to be independent predictive factors for pCR in anthracyclines or anthracyclines plus taxane-treated BC irrespective of BC subtypes (Seo et al. 2013). In 2016, Asano and colleagues evaluate the usefulness of CD8/FOXP3 lymphocyte ratio (CFR): higher CFRs are related to higher pCR percentage as well as better prognosis (Asano et al. 2017). In 2018, Goto and coworkers found that these results were particularly true for TNBC and HER2+ BC, where CFR was an independent predictor for RFS (Goto et al. 2018).

3.5 TILs and Adjuvant Chemotherapy

In the adjuvant setting, high stromal or intratumoral TILs have been associated with a good prognosis in patients with TNBC in most studies (Criscitiello et al. 2016). For example, in 2011, Loi and coworkers found that for every 10% increase of TILs also prognosis increased. In the FinHER trial, scientists noted that in patients with TNBC every 10% increase in TIL values corresponded to a decrease in the risk of distant recurrence (Loi et al. 2014). Moreover, in 2019, Loi concluded that TILs represented a strong prognostic factor for adjuvant-treated TNBC (Loi et al. 2019). See Table 3 for the summarized results.

3.6 Microenvironment and Metastatic BC

If, on the one hand, many studies have been conducted allowing us to understand the composition and prognostic significance of TILs in primary BC, there is still little knowledge regarding the biology of TILs during metastatic progression and in metastatic lesions (mTILs). Various retrospective studies investigated mTILs in different metastatic anatomical sites and compared them with the corresponding primary tumor. In all these studies over the years, important differences were discovered, and a relevant immune cell depletion in secondary lesions emerged. Ogiya et al. (2016) observed a higher percentage of CD4+ and CD8+ T-cells in primary tumors, suggesting a crucial role of immune escape in tumor progression, while FOXP3+ T-cell levels were very low and stable in primary and secondary sites. Moreover, TIL composition and distribution turned out to be substantially unaffected in metastases compared to the primary tumor, regardless of the anatomical site. The expression of programmed death-ligand (PD-L) 1 and human leukocyte antigen (HLA) class I was variable from primary to metastatic tumors (Ogiya et al. 2016). This different setting suggests that immune cells could partially contribute to a poor therapeutic response in advanced stages of BC (Li et al. 2021). Sobottka et al. (2016), analyzing the immune infiltrate in liver, bone, brain, and soft tissues metastases, revealed significantly lower levels of TILs than in the primary tumor and confirmed that the metastatic lesions retained the infiltrating lymphocyte pattern of that of the primary tumor. In fact, in metastases, a similar arrangement of lymphocyte subtypes with higher percentages of CD4+ T-cells, followed by CD8+ T-cells and CD20+ B-cells, was observed. These results suggest that the primary tumor somehow can give an imprint to immune infiltrate composition rather than local immunity of metastatic sites (Sobottka et al. 2016). In accordance with the previous studies, Zhu et al. (2019) studied immune cell composition in primary BC and matched metastases in brain, ovary, bone, and gastrointestinal tract and pointed out that metastatic lesions had a lower immune cell total score (including all subsets of lymphocytes and dendritic cells) with the only exception of M2-macrophages that were increased, hinting their potential role in BC progression. Local immunity, however, may affect immune cell composition in metastatic sites since higher TIL levels were observed in lung metastases than in bone, liver, or skin lesions (Wein et al. 2017). Particular attention was also paid to CD8-cytotoxic T-cells infiltrate, the main antitumoral lymphocytic component, whose low levels are associated with better outcome and complete response to NACT with trastuzumab than with lapatinib in patients with HER2-positive BC. Lower TIL levels in patients who develop metastatic lesions than in patients with early-stage BC were also observed (Liu et al. 2017). Starting on the premise that metastatic lesions have escaped immune surveillance at the primary tumor site, Szekely et al. (2018) suggested that metastases may have a different immune microenvironment than the corresponding primary tumor. They found that TIL counts, PD-L1 positivity, immune cell metagenes, and expression of immune-oncology (IO) therapeutic targets were significantly lower in metastases. Vice versa, the expression of macrophage markers, protumorigenic toll-like receptor pathway genes, HLA-E, ecto-nuclease CD73/NT5E, and inhibitory complement receptors remained high in metastases, representing potential therapeutic targets. The results of the cited studies are summarized in Table 4.

3.7 Microenvironment and BC In Situ

It is well known that TILs play a pivotal role in disease progression and response to therapy in patients with invasive BC. However, the immune setting of DCIS and how it changes from DCIS to invasive and metastatic cancer is still partially unknown. Consequently, little is known about its potential prognostic significance. Thompson et al. (2016) investigated the composition of TILs in cases of DCIS with known ER, PR, and HER-2 expression, demonstrating different patterns related to different phenotypes but also to patient age. They observed that ER-negative DCIS and DCIS in younger patients presented higher levels of all TIL subsets compared to ER-positive DCIS and DCIS in older patients. Among subtypes of DCIS, T lymphocytes were the predominant cell type, slightly more CD4-positive than CD8-positive T-cells, followed by B lymphocytes and FOXP3+ Tregs. ER-positive DCIS tended to have a high CD8/FOXP3 ratio than ER-negative ones. Moreover, DCIS with associated infiltrating ductal carcinoma component showed higher CD8+ T-cell and B-cell infiltrates than cases with only in situ disease, while recurring DCIS had higher levels of TILs and CD8+T-cells compared to DCIS that did not recur. Similar results were obtained from subsequent studies in which a higher prevalence of TILs was highlighted in HER-2-positive DCIS and, to a lesser

Table 4 Summary of trials that have assessed TILs in metastatic BC

Reference study (year)	Cohort	Subtypes	Metastatic site	Objective	TIL evaluation	End points	Correlation with outcome
Ogiya et al. (2016)	Patients ($n = 25$) with BC diagnosed between 1990 and 2009	HER2+ (ER+/ER−) TN	Skin, brain, lymph node, lung, bone, bone marrow/ liver/muscle	Compare levels of immune infiltrate in PT vs. MT	TILs (%) TILs positivity rate (%): CD4 CD8 FOXP3 PD-L1	Higher TILs in PT: p 0.004 (paired t-test), p 0.036 (HER2+), p 0.06 (TN)	n.a.
Sobottka et al. (2016)	Patients ($n = 87$) diagnosed between 1995 and 2012	HER2+	Brain, bone, liver, soft tissue	Compare levels and distribution of immune infiltrate in PT vs. MT	iCD4+ T-cells iCD8+ T-cells iCD20+ B-cells	Fewer ($p < 0.001$) TILs in MT than TILs in PT	High numbers of imCD8 + TILs associated with a longer DFS ($p < 0.05$)
		(ER+/ER−; PR+/PR−)			sCD4+ T-cells sCD8+ T-cells sCD20+ B-cells	Similar tendency both for the frequency (sTIL > iTIL) and for the composition of the TIL subsets (CD4+ > CD8+ > CD20+) ($p < 0.001$) irrespectively of the anatomical site at which MT had occurred	Luminal A and luminal B tumors displayed a similar long-term better overall survival than HER2+ and TN cases, which showed also a similar follow-up
		TN			imCD4+ T-cells imCD8+ T-cells imCD20+ B-cells mCD4+ T-cells mCD8+ T-cells mCD20+ B-cells	Within the tumor compartments, imTIL more frequent than sTIL and iTIL both within MT and the matched PT ($p < 0.001$)	

Liu et al. (CCTG MA.31 trial) (2017)	Patients ($n = 652$) with HER2+ metastatic BC recruited from 2008 to 2011	HER2+	n.a.	Investigate the role of cytotoxic CD8+ T-cells in the prediction of outcomes in patients with HER2+ metastatic BC randomized to an antibody-based (trastuzumab) vs. a small molecule-based (lapatinib) anti-HER2 therapy, in combination with taxane chemotherapy for 24 weeks	CD8+ sTIL count CD8+ iTIL count	Higher risk for disease progression among patients with low CD8+ sTILs treated with lapatinib compared with trastuzumab (HR, 2.94; 95% CI, 1.40–6.17; $P = 0.003$) than is evident among patients with high numbers of CD8+ sTILs (HR, 1.36; 95% CI, 1.05–1.75; $p = 0.02$) (univariate analysis)	No significant effect of high vs. low levels of immune infiltrates on PFS
					FOXP3+ sTIL count FOXP3+ iTIL count	Data confirmed in the prespecified stratified stepwise multivariate Cox regression analysis (stratified HR, 0.52; 95% CI, 0.28–0.98; interaction, $p = 0.04$)	Predictive effect on risk for progression with lapatinib compared with trastuzumab among patients with low CD8+ sTIL counts and among those with high CD8+ sTIL counts
					PD-1+ sTIL count PD-1+ iTIL count		
					Overall H&E-stained sTIL count		
Szekely et al. (2018)	Cohort 1: 45 patients with full sections of paired MT and PT	ER+/ER−	Liver, chest wall, soft, tissue, ovary, brain, pericardium, lung, stomach	Compare TIL count, PD-L1 expression by ICH and mRNA levels of immune-related genes using Nanostring	TILs count (%)	TIL counts and PD-L1 positivity significantly lower in metastases	n.a.
	Cohort 2: tissue microarrays	HER2+/ HER2−			PD-L1 tumor		

(continued)

Table 4 (continued)

Reference study (year)	Cohort	Subtypes	Metastatic site	Objective	TIL evaluation	End points	Correlation with outcome
	(TMAs) from 55 other patients including			technology *in* PT and MT		Lower median PD-L1 mRNA expression in MT than in PT ($p = 0.033$)	
	42 paired MT and PT	TN			PD-L1 stroma	CD8, T-helper, T-reg, cytotoxic T, dendritic and mastoid cells metagenes and expression of 13 of 29 immuno-oncology therapeutic targets in clinical development including PD1, PD-L1, and CTLA4 significantly lower in MT	
					PD-L1 stromal macrophages	Down regulation of chemoattractant ligand/receptor pairs (CCL19/CCR7, CXCL9/CXCR3, IL15/IL15R), interferon regulated genes (STAT1, IRF-1,-4,-7, IFI-27,-35), granzyme/granulysin, MHC class I and immune proteasome (PSMB-8,-9,-10) expression in MT	
					PD-L1 stromal lymphocytes	Immunotherapy response predictive signatures lower in MT	

					Immune cell gene analysis: genes grouped into 14 immune cell type metagenes (total T, Th1, Treg, total CD8, exhausted CD8, cytotoxic T, B, NK, NKCD56, mastoid cell, CD45, dendritic cell, macrophage, and neutrophil) and 22 immune functions	Preserved high expression of macrophage markers (CD163, CCL2/CCR2, CSF1/CSFR1, CXCR4/CXCL12), protumorigenic toll-like receptor pathway genes (CD14/TLR-1,-2,-4,-5,-6/MyD88), HLA-E, ecto-nuclease CD73/NT5E and inhibitory complement receptors (CD-59,-55,-46)	
Zhu et al. (2019)	Patients ($n = 50$) with metastatic BC	ER+/ER−	Brain, ovary, bone, gastro-intestinal tract	Examination of exome-capture RNA sequencing data in PT and matched MT		Lower total immune score in MT ($p < 0.001$) especially in brain MT	Association between higher sTILs percentage in MT and longer survival post metastasis
		HER2+/HER2−				The most significant decrease ($p < 0.01$) observed in TN subtype	
		TN				Relative levels of M2-like macrophages increased in MT	

BC breast cancer, *HER2* human epidermal growth factor receptor 2, *ER* estrogen receptor, *TN* triple–negative, *PD-L1* programmed cell death ligand 1, *iTIL* intratumoral TILs, *sTIL* stromal TILs, *imTIL* invasive margin TILs, *mTIL* infiltrating metastasis TILs, *PT* primary tumor, *MT* metastasis, *n.a.* not available, *DFS* disease-free survival, *PFS* progression-free survival extent, in triple-negative DCIS (Hendry et al. 2017a, b; Pruneri et al. 2017). Moreover, HER-2-positive and triple-negative DCIS, as microinvasive forms, were found to be strongly associated to higher levels of tertiary lymphoid structures (TLSs) than invasive carcinoma. Pruneri et al. (2017) also observed a higher prevalence of HER-2 positivity in DCIS than that documented in invasive breast carcinoma, suggesting that HER-2 positivity is negatively selected during progression from DCIS to invasive cancer and that its overexpression in in situ disease could be correlated to a powerful stimulation of immunity. More and more evidence upholds the leading role of immune-checkpoint pathways, such as the PD-1 and PD-L1 axis, in the regulation of the immune response against tumor. Thompson et al. (2016) demonstrated that while PD-L1 is expressed by tumoral cells in invasive BC and TILs, in DCIS the expression of PD-L1 is absent in most cases. Instead, triple-negative DCIS and high nuclear-grade DCIS showed high expression of PD-L1 on infiltrating lymphocytes. PD-L1 expression by infiltrating lymphocytes was associated with higher levels of TILs in DCIS, supporting its significance as a marker of host immune response within DCIS. A descriptive study by Campbell et al. (2017) also attempted to understand the possible prognostic value of TME in DCIS, focusing on its composition in high and low nuclear-grade lesions and looking for a potential correlation with the risk for recurrence and metastases. This study discovered significantly higher percentages of TILs, FOXP3+ T-cells, CD68+ macrophages, HLA-DR+ cells, and CD4+ T-cells and B-cells in high nuclear grade DCIS (Fig. 2). This setting is considered to be associated with a suppressive immune microenvironment and poor prognosis. Almost all studies on TME in DCIS have focused on tumor-infiltrating T lymphocytes, but little knowledge is available about the role of infiltrating B lymphocytes in tumor progression. Miligy et al. (2017) explored this topic and pointed out that in DCIS were present more infiltrating B lymphocytes and lymphoid aggregates, especially those stromal, compared to cases of DCIS associated with invasive component. Intratumoral infiltrating B lymphocytes turned out to be very rare within DCIS unlike the periductal stromal

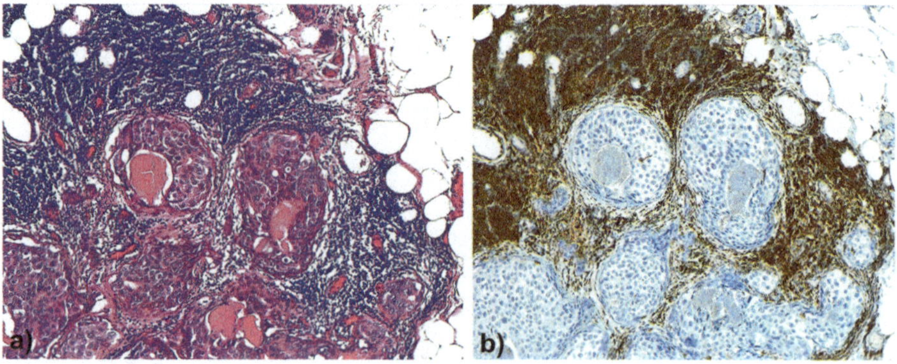

Fig. 2 High-grade in situ ductal breast carcinoma with heavy peritumoral lymphocytic infiltrate (**a**, hematoxylin and eosin stain, 100×; **b**, CD45 immunohistochemistry stain, 100×)

ones, which showed both diffuse and aggregate patterns. This study showed that pure DCIS with higher numbers of peritumoral and paratumoral infiltrating B lymphocytes were significantly associated with higher tumor grade, presence of necrosis, ER and PgR negativity and HER2+ status, and consequently with decreased RFS. The results of the cited studies are summarized in Table 5.

4 Histopathological Assessment of TILs in Invasive BC, In Situ BC, and Post-NACT Residual Disease

A large number of studies conducted over time have demonstrated the clinical relevance of TIL assessment in BC as immunological biomarker, in particular in triple-negative and HER-2-overexpressing phenotypes. There is evidence to support that TILs density has a potential predictive value for response to NACT and, in specific subtypes, a prognostic significance in patients undergoing adjuvant chemotherapy. Accordingly, the quantitative evaluation of TILs in clinical trials and daily histopathological practice can be considered of primary importance as well as the necessity to define accurate guidelines to assist pathologists, clinicians, and researchers in their work (Dieci et al. 2018). For this purpose, the International Immuno-Oncology Biomarker Working Group on BC in 2014 published methodological recommendations for the evaluation of TILs in BC in an attempt to standardize the method of analysis and increase accuracy and repeatability but also to suggest the integration of these standards into future clinical trials, translational research, and diagnostic practice. As the first step, the working group developed a guideline document for TIL evaluation on hematoxylin and eosin (HE)-stained slides of core biopsies and tumor resections. Recommendations are summarized in Table 6 (Salgado et al. 2015a), and they define the various steps and rules to calculate the percentage of stromal TILs in a standardized way. These percentages allow dividing invasive carcinomas into three groups: low (<10%), intermediate (10–40%), and high infiltrate (>40%). After the publication of these international guidelines, two ring studies were conducted in order to collect data on interobserver variability of TIL assessments since the human eye, even if expert, has limitation in quantitative evaluation and highlights difficult areas to optimize the scoring methodology. Further investigations focused on TIL assessments in the post-NACT residual disease as a stratifying marker in clinical trials, but also in the DCIS setting with or without associated invasive component, which is an emerging research topic. Current evidence in TILs on residual disease post-NACT mainly concerns TNBCs. Nowadays, NACT and subsequent surgical resection are recognized as the standard therapeutic line in locally advanced disease and are also applied in early-stage BCs in order to favor more conservative surgical approaches. pCR at surgery after NACT, that is, defined as the absence of residual invasive carcinoma in breast specimens and in axillary lymph nodes, was suggested as a pointer of long-term outcome. However, it has been observed that the prognostic value of pCR is remarkably variable according to the subtype of BC because it is not necessarily associated to a better outcome. The correlation between the nonachievement of pCR

Table 5 Summary of trials that have assessed TILs in DCIS

Reference study (year)	Cohort	Subtypes	Objective	TIL evaluation	End points	Correlation with outcome
Thompson et al. (2016)	Patients ($n = 27$) with DCIS of mean age of 40	Luminal A	Profile the composition of TILs and PD -L1 expression	TILs density: none (score 0), focal (score 1), moderate (score 2), diffuse (score 3)	Lower TILs density scores in ER+ DCIS and DCIS in older patients than ER− DCIS or DCIS in younger patients ($p = 0.28$ and 0.32, respectively)	n.a.
	No patients received neoadjuvant therapy; all patients received adjuvant therapy	Luminal B		ICH: CD3 CD4 CD8 FOXP3 CD20 PD-L1	CD3+ T-cells predominance in all DCIS subtypes (CD4+ > CD8 + T) followed by CD20 + B-cells and FOXP3+ Treg	
		HER2+			ER+ DCIS was more likely to have to have a high CD8/FOXP3 ratio (>4) than ER− DCIS	
		TN			All luminal A, luminal B and HER-2+ DCIS with PD -L1+ showed low PD-L1 TILs expression while totality of TN DCIS	

					had high expression of PD -L1 ($p = 0.0008$)	
Hendry et al. (2017a, b)	Patients ($n = 138$) with pure DCIS without associated invasive BC, some of which had copy number and mutation data	Luminal A	Quantification of TILs and evaluation of PD-L1 expression and correlation with clinicopathological features (nuclear grade, comedo necrosis, ER status, HER2 status, disease recurrence)	TILs count	TIL score and PD-L1 positivity significantly increased in high-grade tumors ($p = 0.002/0.035$), ER- tumors ($p = 0.002/0.02$), HER2-positive tumors ($p < 0.0001/0.048$)	High TILs associated with improved RFS in TN and HER2-positive tumors after chemotherapy ($p = 0.643$)
		Luminal B	Investigate correlations between genomic alterations and immune setting in DCIS compared with existing genomic information (copy number analysis, fraction of the genome altered (FGA) and specific genetic mutations)	ICH: PD-L1	Comedo necrosis positively associated with TILs ($p < 0.0001$) but not with PD-L1	
		HER2+			TILs score significantly higher in cases with TP53 mutation ($p = 0.03$) but not with PIK3CA or GATA3 mutation	
		TN			In the cases with copy number data, both the fraction of the genome altered and the number of telomeric imbalances were	

(continued)

Table 5 (continued)

Reference study (year)	Cohort	Subtypes	Objective	TIL evaluation	End points	Correlation with outcome
					significantly positively correlated with TILs (both $p < 0.001$)	
Pruneri et al. (2017)	Patients ($n = 1448$) who underwent primary BC surgery between 1997 and 2008 and were diagnosed with DCIS	Luminal A	Evaluation of the prevalence and clinical relevance of TILs in DCIS	sTILs count	Clinical factors significantly associated with TILs ($p \leq 0.001$): intrinsic subtype, nuclear grade, necrosis, type of surgery	No association between TILs and the risk of IBEs ($p = 0.767$)
		Luminal B HER2+ TN			HER2+ DCIS more frequently associated with TILs, followed by TN, Luminal B/HER2+ and Luminal A/B subtypes ($p < 0.0001$)	
Campbell et al. (2017)	Patients ($n = 117$) collected between 1994 and 2014 (52 cases of HG-DCIS and 65 cases of nHG-DCIS)	Luminal A	Investigate immune cell population and correlation with clinicopathological parameters in HG-DCIS vs. nHG-DCIS	sTILs count	Higher TILs, FOXP3+ cells, CD68+ macrophages; CD68 + PCNA+ macrophages, HLA-DR+ cells, CD4+ cells and	Three immune cell populations associated with risk of recurrence: CD8 + HLADR+ T-cells, CD8 + HLADR−

					CD20+ cells in HG-DCIS compared to nHG-DCIS (p value n.a.)	T-cells, and CD115+ cells
		Luminal B HER2+ TN		itTILs count ICH: ki-67 CD3 CD4 CD8 FOXP3 CD20 CD68 HLA-DR, DP, DQ S100 A8/A9 MRC1 PCNA CD115	CD4+ T-cells, CD20+ B-cells, and total TILs associated with high-risk features (VNPI, large tumor size, high grade, comedo necrosis, high ki-67, HER2 positivity, and HR negativity). CD8+ cells correlated with high grade, HER2 positivity, and HR negativity, but not with other high-risk features (p value n.a.)	
Miligy et al. (2017)	Patients ($n = 80$) with DCIS (36 pure DCIS and 44 mixed with invasive BC)	Luminal A	Evaluate the prognostic significance of TIL-Bs lymphocytes	TIL-Bs count	Higher levels of TIL-Bs and lymphoid aggregates in pure DCIS than DCIS with invasion	Outcome analysis showed that pure DCIS associated with higher numbers of B lymphocytes had shorter
		Luminal B		ICH: CD19 CD20 CD138	In pure DCIS, higher peri-tumoural and paratumoral	RFS ($p = 0.04$); however, the association was not significant with

(continued)

Table 5 (continued)

Reference study (year)	Cohort	Subtypes	Objective	TIL evaluation	End points	Correlation with outcome
		HER2+			TIL-Bs associated with large tumor size ($p = 0.016$), HR negative ($p = 0.008$), and HER2+ status ($p = 0.010$)	CD138+ plasma cell count ($p = 0.07$)
		TN			In DCIS with invasive component high TIL-Bs associated with variables of poor prognosis including larger size, high grade, lymphovascular invasion, lymph node metastases, ER/PR-negative and HER2+ status	

DCIS ductal carcinoma in situ, *HER2* human epidermal growth factor receptor 2, *TN* triple–negative, *ICH* immunohistochemistry, *PD-L1* programmed cell death ligand 1, *ER* estrogen receptor, *BC* breast cancer, *RFS* recurrence-free survival, *sTILs* stromal TILS, *IBE* ipsilateral breast event, *HG-DCIS* high-grade ductal carcinoma in situ, *nHG-DCIS* non-high-grade ductal carcinoma in situ, *itTILS* intratumoral TILS, *n.a.* not available, *VNPI* Van Nuis prognostic indicator, *HR* hormone receptor, *TIL-Bs* tumor-infiltrating B lymphocytes

Table 6 Recommendations for assessing TILs in BC

Evaluate only TILs in the stromal compartment. Determine the percentage of stromal TILs using as denominator the area of stromal tissue (i.e., area occupied by mononuclear inflammatory cells over total intratumoral stromal area), not the number of stromal cells.
TILs should be evaluated within the borders of the invasive tumor.
Exclude TILs outside of the tumor border and around DCIS and normal lobules; do not include TILs in tumor zones with crush artifacts, necrosis, regressive hyalinization, as well as in the previous core biopsy site.
All mononuclear cells should be scored, excluding polymorphonuclear leukocytes.
One section (4–5 μm, magnification ×200–400) per patient is currently considered to be sufficient.
Full sections are preferred over biopsies whenever possible. Cores can be used in the pretherapeutic neoadjuvant setting; currently no validated methodology has been developed to score TILs after neoadjuvant treatment.
A full assessment of average TILs in the tumor area by the pathologist should be used. Do not focus on hotspots.
The working group's consensus is that TILs may provide more biological relevant information when scored as a continuous variable since this will allow more accurate statistical analyses, which can later be categorized around different thresholds. Pathologist should report scores in as much detail as the pathologist feels comfortable with.
TILs should be assessed as a continuous parameter. The percentage of stromal TILs is a semiquantitative parameter for this assessment, for example, 80% stromal TILs means that 80% of the stromal area shows a dense mononuclear infiltrate. For assessment of percentage values, the dissociated growth pattern of lymphocytes needs to be taken into account. Lymphocytes typically do not form solid cellular aggregates; therefore, the designation "100% stromal TILs" would still allow some empty tissue space between the individual lymphocytes.
No formal recommendation for a clinically relevant TIL threshold(s) can be given at this stage. The consensus was that a valid methodology is currently more important than issues of thresholds for clinical use, which will be determined once a solid methodology is in place. Lymphocyte predominant breast cancer can be used as a descriptive term for tumors that contain "more lymphocytes than tumor cells." However, the thresholds vary between 50% and 60% stromal lymphocytes.

Source: Salgado et al. (2015b)

and poor prognosis is certainly more evident in more aggressive phenotypes, which are triple-negative and non-luminal HER2+ BCs compared to luminal ones. On the other hand, there are cases with no pCR showing long-term survival and, conversely, patients with complete response demonstrating following relapse and unfavorable prognosis. In this setting, pCR is a limited synonym of drug efficacy and therefore the search for biomarkers that can allow a more accurate risk stratification becomes necessary (Dieci et al. 2018). In this setting, TILs assessment' in post-NACT residual disease has been proposed as a potentially useful marker. A guide for TILs' evaluation was developed by the International Immuno-Oncology Biomarker Working Group on BC according to the previous recommendations for TILs assessment in untreated BC. The point that encountered the greatest problems was the definition of the area for TIL evaluation. According to the guidelines adopted for untreated BC, assessment should be performed within the borders of the residual tumor bed, defined as the largest cross-sectional area between residual invasive cells, which should be examined in full in TIL evaluation. In case of pCR, the area of regression defined by radiological/macroscopic/microscopic findings should be assessed. Recommendations for assessing TILs in residual disease after NACT are listed in Table 7. In residual disease, different conditions have been described: high levels or low levels of TILs and changes in TILs from before to after NACT. The presence of high TILs is clearly a favorable condition for the use of immunotherapies since it indicates the existence of an antitumor immunity to maintain or stimulate. Only a few studies investigated the prognostic role of TILs in DCIS and available data do not support an association with recurrence risk, but preliminary studies support an active adaptive immune response in DCIS that could be targeted by immune-based therapies. Then, TILs' evaluation may provide more learning about the immunogenicity level of the lesion. Recommendations for harmonizing TIL evaluation in DCIS, reported in Table 8, are proposed. Even if the main studies on the subject have focused on DCIS, recommendations are applicable also in lobular carcinoma in situ.

5 TILs' Evaluation and Digital Pathology

TILs' quantification and spatial characterization in BC have an increasingly significant prognostic and predictive value since these can predict clinical outcomes, as well as treatment response in the neoadjuvant and adjuvant setting. In current practice, pathologists routinely assess HE-stained tissue sections to classify invasive BC, providing valuable information for clinical and surgical management, selection of therapeutic approaches, survival, and recurrence. As discussed above, international guidelines for TIL assessment in BC recommend the pathologist to select areas of interest of the tumor in the HE-stained sections and outline stromal areas to calculate the percentage of TILs in stromal compartment, expressed from 0% to 100% within total tumor bed, and so define the infiltrate as low, intermediate, or high. These recommendations help to reduce scoring variations, but limitations still remain because this evaluation is intrinsically qualitative and often can show intra- and interobserver variability. The development of new computational TIL assessment (CTA) methods is a promising solution to reduce these limitations and many new approaches to evaluation have been attempted (Sun et al. 2021). To date, evaluation and scoring of TILs on digital whole-slide images (WSIs) have been analyzed principally by domain experts, but this evaluation is subjective and influenced by human judgment. So fully automated image analysis methods would be useful to reduce labor costs and increase precision, accuracy, and repeatability. Some studies have focused on the morphological characteristics of lymphocytes, which have small dark nuclei and scarce cytoplasm and therefore clearly distinguishable from neoplastic epithelial cells and stromal cells. The first studies have built algorithms to discern cell nuclei from cytoplasm and classify the various cellular components into different categories based on nuclear morphology (Lu et al. 2020). Some recent research groups have proposed deep learning-based algorithms to detect tumor regions in BC WSIs and assess TILs' infiltrate. Sun et al. (2021) compared a deep learning-based computational TIL assessment method with traditional visual

Table 7 Recommendations for assessing TILs in residual disease after NACT

TIL assessment in the residual disease setting should be done within the borders of the residual tumor bed, as defined by the presence of the residual tumor cells, in analogy with the definition of the residual tumor bed of the residual cancer burden (RCB) index. In case of pCR, TILs may, for specific research purposes, be assessed in the area of regression as defined by imaging/ macroscopic/microscopic features.
The entire largest cross-sectional area of the residual tumor bed should be used for histological TIL assessment. One section (4–5 μm) per patient can be considered to be sufficient for practical purposes.
If the residual tumor bed is large than 2 cm, more slides need to be assessed, with one slide for each cm of tumor bed as a minimum. For example, if the largest diameter is >5 cm, then at least five representative slides from the largest cross-sectional area should be considered. If the residual tumor bed is only 2 cm, one slide is considered enough. Assessing numerous slides for each case should thus be possible mentioning the number of assessed slides specifically in the study protocol
To assess TIL cellularity, it is helpful to scan across the sections of tumor bed (50–100× magnification) and then estimate the average TIL cellularity from the different microscopic fields and slides (200–400× magnification).
Areas of tumor necrosis may be replaced by aggregates of histiocytes with cholesterol clefts. TILs associated with these necrotic areas should be excluded.
Exclude TILs in tumor zones with crush artifacts.
Exclude TILs closely related to remaining foci of carcinoma in situ or normal lobules within the residual tumor bed.
TILs should be assessed when tumor cells are embedded within aggregates of histiocytes.
TILs can be assessed on core biopsies after a short period of treatment (targeted or chemotherapy), prior to surgery, only when they contain tumor cells. Cores with scarred and inflammatory stroma but with no tumor cells should not be assessed.
TILs should be reported separately for the stromal compartment (% stromal TILs) and the tumor cell compartment (% intraepithelial tumoral TILs). The reasons are that in many tumors the TIL density in the two compartments is different. The denominator used to determine the% stromal TILs is the area of stromal tissue (i.e., area occupied by mononuclear inflammatory cells over total stromal area), not the number of stromal cells (i.e., fraction of total stromal nuclei that represent mononuclear inflammatory cell nuclei in the stroma). Similarly, for intraepithelial tumoral TILs the tumor cell area is the denominator, not the stromal area.
Do not focus on hotspots; all mononuclear cells (including lymphocytes and plasma cells) should be scored, but polymorphonuclear leukocytes are excluded.
TILs may provide more biological relevant information when scored as a continuous variable since this will allow more accurate statistical analyses, which can later be categorized around different thresholds. However, in daily practice most pathologists will rarely report for example 13.5% and will round up to the nearest 5–10%, in this example thus 15%. Pathologist should report their scores in as much detail as he/she feels comfortable with.
TILs should be assessed as a continuous parameter. The percentage of stromal or intratumoral TILs is a semi-quantitative parameter for this assessment, for example, 80% stromal TILs means that 80% of the stromal area shows a dense mononuclear infiltrate. For assessment of percentage values, the dissociated growth pattern of lymphocytes needs to be taken into account. Lymphocytes typically do not form solid cellular aggregates; therefore, the designation "100% stromal TILs" would still allow some empty tissue space between the individual lymphocytes.
Do not include DCIS, normal lobules, necrosis, large fibrous areas, crush artifacts, and regressive hyalinization.

Source: Dieci et al. (2018)

Table 8 Recommendations for assessing TILs in CIS

TILs can be evaluated for each type of CIS, whether it is ductal or lobular (this should be specified in the report and in advance), with the same methodology.
TILs can be evaluated in pure CIS, CIS with a microinvasive carcinoma, and CIS with an invasive carcinoma. The same methodology applies. TILs should not be evaluated in the areas of invasive cancer.
The final score of TILs in CIS score should not contain the score of TILs in the area containing invasive cells.
Evaluation of all sections containing CIS of one patient is recommended. Full-face sections (4–5 μm) are preferred over biopsies whenever possible. Cores or vacuum-assisted biopsies can be used in the neoadjuvant setting. No validated methodology has been developed to score TILs in CIS after neoadjuvant treatment.
These recommendations are for evaluation of TILs on hematoxylin and eosin-stained sections of formalin fixed/paraffin-embedded tissues.
Only stromal TILs should be reported. TILs located within CIS-lesions (between the in situ tumor cells), that is, intraepithelial tumoral TILs, should not be assessed.
The stromal compartment is the area of the specialized stroma surrounding the CIS lesions. If a DCIS is surrounded by a fibrous rim with TIL immediately adjacent to it, these should be assessed as well. If this area of specialized stroma is not readily distinguishable, one can use a more arbitrary stromal area that extends over 2 HPF (×40). Any type of circumferential TIL infiltration should be taken into account (minimal, partial, subtotal, or total).
If many CIS lesions are encountered, TIL evaluation in all lesions should be done and the mean value should be derived. Do not focus on hot spots.
TILs that are in continuity between the invasive lesions and the in situ lesions, without clear distinction whether they are associated to the invasive or to the in situ lesion, should be assessed within the 2HPF area for DCIS-TILs while the remaining TILs are part of the invasive component.
No distinction to be made between TILs surrounding CIS lesions of different differentiation grades.
TILs around normal lobules, wherever they are found, and TILs in tumor zones with crush artifacts, necrosis, regressive hyalinization, as well as in the previous core biopsy site should not be assessed.
The result of TIL assessment should be reported as % of stromal area surface occupied by TILs. The authors recommend reporting the result as a continuous parameter (0–100%). TIL evaluators should report their results in as much detail they feel comfortable with. In determining the percentage of stromal area surface occupied by TILs, the dissociated growth pattern of lymphocytes needs to be considered. Lymphocytes typically do not form solid cellular aggregates; therefore, the designation "100% stromal TILs" would still allow some empty tissue space between the individual lymphocytes.
No formal recommendation for a clinically relevant TIL threshold(s) can be given at this stage. From our point of view, a valid methodology is currently more important than issues of thresholds for clinical use, which will be determined once a solid methodology is in place.

Source: Dieci et al. (2018)

evaluation based on standardized guidelines and showed that the automated TILs' score was consistent with manual TILs' score. However, computed methods still require a manual preselection of the carcinoma boundary within WSIs to reduce errors of regional identification on subsequent quantification. This detail remains a limit for the overall evaluation of the TILs. Lu et al. (2020) proposed a deep learning-based pipeline for detecting TILs at the cellular level from WSIs of BCs in order to generate high-resolution TIL maps on WSIs. The study consisted of selection of field of views on WSIs, lymphocyte detection to build TIL maps on WSIs, and quantification of TILs for clinical assessment. Then, the authors used TIL maps to explore the relationship between these TIL features and different genomics data for different BC subtypes by using strong enrichment of immune response genes and pathways. This method was validated on multiple data sets, and the evaluation of the correlation between the TIL score and gene transcription levels confirmed the validity of the quantification of TILs and its strong association with immunotherapy pathway and genes such as PD-1 and cytotoxic T lymphocyte-associated antigen (CTLA-4). Therefore, quantitative, automated, and reproducible computational TIL assessment on WSIs based on machine learning could be convenient for further standardization and can be potentially useful in the future for assessing or predicting immunotherapy potential. In 2021, Bai et al. (2021) used an open-source software platform (QuPath) to create a machine learning-based BC TIL scoring algorithm based on recognition of specific cell types (tumor cells, TILs, fibroblasts, and "other" cells) in TNBC. The cells were mathematically combined to create variables that assess the proportion of TILs within different cell populations or the density of TILs within different tissue areas. Using one discovery set and four validation sets from three institutions, the authors found that machine-read TIL variables had stratified patients with TNBC into favorable and poor prognosis groups, where higher TILs scores were significantly associated with better OS.

6 Therapeutic Implications of TILs' Assessment

Many studies have demonstrated that response to conventional anticancer agents (chemotherapy, radiotherapy, targeted therapy) may be partially mediated by effects on immunity. As illustrated above, many clinical trials showed that increases in the number of TILs in BC improved patient prognosis and uniformly produced gains in RFS and OS (De Melo Gagliato et al. 2017). TILs are an inexpensive and strong prognostic marker that represents a surrogate for antitumor immune-mediated response. Their assessment at the time of diagnosis may help clinicians to determine prognosis and is informative for predictive purposes (El Bairi et al. 2018). Therefore, the evaluation of TILs becomes increasingly relevant with the emerging role of immunotherapy in BC treatment alongside surgery, radiotherapy, chemotherapy, and targeted therapy. TILs could be prognostic biomarkers in specific BC subtypes, especially in TNBC and HER2-positive BC. A prognostic value of TILs has been demonstrated in TNBC subtype in patients treated with adjuvant anthracycline chemotherapy, with improved survival in cases showing higher percentages of infiltrate. Similar correlation has been observed in HER2-positive BC, although results were not always statistically significant and sometimes contrasting among different trials. On the other hand, at present no prognostic significance has been found in ER-positive/HER2-negative BC subtypes, probably due to their heterogeneity (Luen et al. 2017; De Melo Gagliato et al. 2017; Ravelli et al. 2017; Wein et al. 2017). In the context of NACT, high levels of TILs seem to be predictive of pCR in

HER2-positive BC treated with HER2-targeting agents but also in TNBC, thus favoring a better prognosis (Ravelli et al. 2017; De Melo Gagliato et al. 2017). It was observed that increasing TILs can maximize trastuzumab efficacy in HER2-positive BC (Luen et al. 2017). These observations suggest an important role of TILs in modulating responses to chemotherapy. Moreover, in recent studies it emerged that TILs in metastatic BC, although present at lower levels than in primary disease and more complex for heterogeneity immune setting in different metastatic sites, still maintain prognostic value (Wein et al. 2017). Then, TIL assessment may have a potential utility in the design of clinical trials by identifying primary HER2-positive and TNBC patients with excellent prognoses and appropriately grouping patients into those who may benefit from standard treatments and those who require novel therapeutic strategies. Moreover, it is well known that the dysregulation of immune-checkpoint protein expression is the major factor of immune escape in BC (Zhang et al. 2021). Immunotherapy is aimed precisely at interacting with those immune checkpoints favoring the reactivation of immune system to fight cancer cells and re-establish the antitumoral immune response of the host. The proven efficacy of drugs employed in immunotherapy has revolutionized therapeutic approaches in a variety of cancers in advanced stage. The success of immunotherapy largely depends on the immunogenic background of the tumor. Clearly, given the heterogeneity of BCs, the question whether certain breast tumors might benefit more from immuno-therapeutic treatment and which cancer intrinsic and microenvironmental factors define the likelihood of inducing a potent and durable antitumor immune response arises. TNBCs and HER2-positive BCs have generally higher TIL levels than ER-positive/HER2-negative BCs, suggesting that these phenotypes are more immunogenic (Luen et al. 2017). The main immune checkpoints investigated in BC include CTLA-4, PD-1, PD-L1, lymphocyte activation gene 3 (LAG-3), and T-cell immunoglobulin domain and mucin 3 (TIM-3) (Zhang et al. 2021). Currently, the majority of BC studies focus on inhibition of the PD1/PD-L1 pathway.

6.1 Inhibitors Targeting PD-1/PD-L1 Axis

PD-1 and PD-L1 are co-inhibitory molecules that inhibit T-cell proliferation, differentiation, and activation, leading to immunosuppression and immune escape. In TME, PD-L1 favors TILs' exhaustion weakening immune surveillance and stimulate immune suppression by inducing Treg cell proliferation, while PD-1 mediates immunosuppression by regulating peripheral T-cells in late response. Then, the PD-1/PD-L1 axis blockade enhances antitumor activity. In the category of inhibitors of this axis, pembrolizumab is a selective humanized monoclonal antibody that blocks the interaction between PD-1 and its ligand PD-L1 and activates T lymphocytes. It has been shown that this drug has antitumor efficacy with low toxicity in patients with early and metastatic TNBC. Atezolizumab is a humanized anti-PD-L1 monoclonal immunoglobulin G1 antibody that restores T-cell function and relieves inhibition of the host immune system against tumor cells by binding to PD-L1 expressed on tumor cells and tumor-infiltrating immune cells (Fig. 3). In a

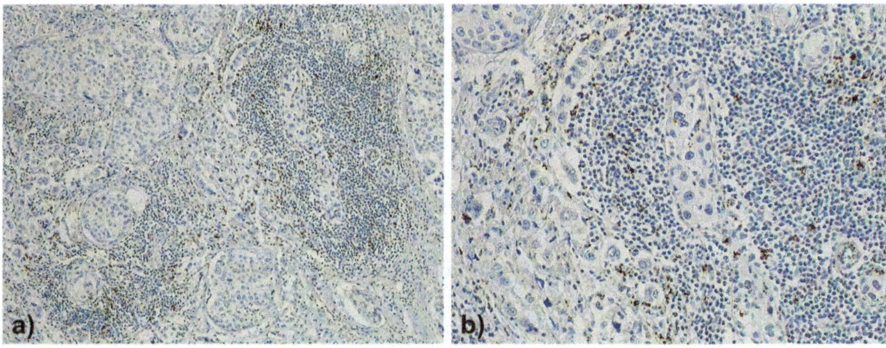

Fig. 3 An example of invasive breast carcinoma with peritumoral immune infiltrate positive to PD-L1 (**a**, PD-L1 SP142 immunohistochemistry stain, 100×; **b**, PD-L1 SP142 immunohistochemistry stain, 200×)

substudy from the phase 3 IMpassion 130 trial, Emens et al. (2021) found that PD-L1 expression in primary or metastatic tumors was linked to PFS and OS benefit with atezolizumab and *nab*-paclitaxel (A + nP) therapy and that intratumoral CD8+ lymphocytes and stromal TILs were associated with PD-L1 expression. Moreover, the authors noted that A + nP was more efficacious in patients with higher tumor immune microenvironment, but clinical benefit was observed only in patients whose tumors expressed PD-L1. Avelumab is a human IgG1 anti-PD-L1 immune-checkpoint inhibitor that can competitively block the binding between PD-1 and PD-L1 promoting tumor infiltration and T lymphocytes to produce cytotoxic molecules, but also mediating Ab-dependent cell cytotoxicity against tumor cells. Combination regimens of PD1/PD-L1 inhibitors with chemotherapy have shown promising results in advanced and early-stage TNBC, suggesting a synergic effect of these treatment lines. In fact, chemotherapy increases tumor cell antigen release, induces the expression of MHC class I molecules, neoantigens, and PD-L1, and promotes dendritic cell activation. Accordingly, it can potentiate immune response following or during immunotherapy. In line with this rationale, evaluation of PD-L1 expression before treatment could be a biomarker to predict efficacy of immunotherapy with PD-1/PD-L1 inhibitors since it was demonstrated that overexpression of PD-L1 is related to significant efficacy of these drugs in many cancers (Thomas et al. 2021). Some studies showed a correlation between higher pretreatment PD-L1 expression and better outcome and revealed a significant improvement of PFS in strong PD-L1-positive metastatic TNBC treated with pembrolizumab in addition to chemotherapy. Moreover, other studies observed that pembrolizumab added to NACT in early and locally advanced stages of HER2-negative BC, including TNBC, favors pCR (Thomas et al. 2021).

6.2 Inhibitors Targeting CTLA-4

CTLA-4 inhibits activated T-cell response by reducing the signal of T-cell receptor (TCR) and CD28 through binding B7 competitively with CD28 or recruiting phosphatase to the intracellular domain of CTLA-4 and by reducing or eliminating CD80 and CD86 expression on antigen-presenting cells (APC), which reduces their availability for CD28 engagement and co-stimulation in T-cell activation. Currently, ipilimumab and tremelimumab are two available antibodies inhibiting CTLA-4, already employed in the treatment of many cancers. Ipilimumab is the first human anti-CTLA-4 antibody that reactivates T-cells against tumor, while tremelimumab is a humanized CTLA-4 monoclonal antibody that highly selectively blocks the interaction of CTLA-4 with B7 and enhances T-cell activation. Although the efficacy of anti-CTLA-4 monotherapy is modest, its combination with radiation therapy, cytotoxic chemotherapy, and other immunotherapies could have a therapeutic potential (Mansour et al. 2017). However, toxicity is the main problem that limits the use of these drugs.

6.3 Inhibitors Targeting LAG-3 and TIM-3

Clinical trials on inhibitors of PD-1/PD-L1 axis and CTLA-4 showed promising results on several cancers but not without revealing side effects and drug resistance. From this fact arises the need to investigate other potential therapeutic targets and agents. Few recent studies have focused on the potential therapeutic role of drugs inhibiting Lag-3 and TIM-3 to modulate immune response. LAG-3 is usually co-expressed with PD-1 on exhausted or dysfunctional T lymphocytes in cancer, and thus the inhibition of both can potentiate immune response. For this reason, current trials investigated the effects of a combined inhibition and observed that it is more effective than single blocking. TIM-3 is expressed on T-cells, Treg lymphocytes, and innate immune cells as a negative regulator for immune response. Only limited studies demonstrated its expression in BC, and at present there is no approved drug targeting TIM-3. Certainly, further studies are needed (Zhang et al. 2021).

7 TILs and PD-L1 Expression as a Potential Predictor of Response to Immunotherapy

Overexpression of PD-L1 is significantly related to the efficacy of PD-1/PD-L1 inhibitors in many cancer types. However, the expression level of PD-L1 has a big difference in the prediction of the efficacy of these drugs in BC since some patients with no expression of PD-L1 may also respond. However, strong PD-L1 positivity is associated with better overall response rates, PFS, and OS in metastatic TNBC patients treated with immune-checkpoint inhibitor monotherapy or in some cases combined with chemotherapy (Thomas et al. 2021). Moreover, high stromal TIL levels are associated with improved response rates in patients with metastatic TNBC treated with pembrolizumab (Zhang et al. 2021). Recently emerging data suggested that TIL quantity could help clinicians to identify patients with BC who benefit most from PD-1/PD-L1 inhibition (Loi et al. 2021). In patients with advanced TNBC and HER-2+ disease, the authors identified a TIL cutoff of 5% or 10%, associated with PD-L1 expression, that define a "immune-enriched" tumor category that seems to have the most clinical relevance in the checkpoint inhibitor scenario. Thus, evaluation of composite biomarkers, such as a combination of TIL and PD-L1, may be useful to select patients most likely to respond to immune-checkpoint inhibitors. However, multiple factors, including tumor mutational burden and composition of immune infiltrate, influence TME and thus responsiveness to immunotherapy (Thomas et al. 2021).

8 Future Perspectives

As described by Zacharakis et al. (2018), an interesting and promising use of TILs is for the treatment of metastatic BC. The authors presented the case of a patient with a chemorefractory HR-positive metastatic BC treated with TILs reactive against mutant version of four proteins (SLC3A2, KIAA0368, CADPS2, and CTSB). Adoptive transfer of these specific TILs associated with IL-2 and checkpoint blockade induced the complete regression of the metastases. This approach represents a potential starting point for immunotherapy in this category of patients and a new way of seeing TILs.

9 Conclusion

TILs play a prognostic role in BC, especially in TNBC and HER-2-positive tumors. Their assessment is mandatory in every BC subtype and also in DCIS and in the NACT setting. TILs seem to have also therapeutic implications in particular due to the even more important role of immunotherapy in BC.

Compliance with Ethical Standards The authors declare that they have no conflict of interest. Neither human participants nor animals have been involved in research studies for the writing of this chapter.

References

Adams S, Gray RJ, Demaria S, Goldstein L, Perez EA, Shulman LN, Martino S, Wang M, Jones VE, Saphner TJ, Wolff AC, Wood WC, Davidson NE, Sledge GW, Sparano JA, Badve SS (2014) Prognostic value of tumor-infiltrating lymphocytes in triple-negative breast cancers from two phase III randomized adjuvant breast cancer trials: ECOG 2197 and ECOG 1199. J Clin Oncol 32(27):2959–2966

Ali HR, Provenzano E, Dawson SJ, Blows FM, Liu B, Shah M, Earl HM, Poole CJ, Hiller L, Dunn JA, Bowden SJ, Twelves C, Bartlett JM, Mahmoud SM, Rakha E, Ellis IO, Liu S, Gao D, Nielsen TO, Pharoah PD, Caldas C (2014) Association between CD8+ T-cell infiltration and breast cancer survival in 12,439 patients. Ann Oncol 25(8):1536–1543

Amin MB, Edge S, Greene F, Byrd DR, Brookland RK, Washington MK, Gershenwald JE, Compton CC, Hess KR, Sullivan DC, Jessup JM, Brierley JD, Gaspar LE, Schilsky RL, Balch CM, Winchester DP, Asare EA, Madera M, Gress DM, Meyer LR (2017) AJCC cancer staging manual, 8th edn. Springer International Publishing, Cham

Angelico G, Broggi G, Caltabiano R, Santoro A, Spadola S, D'Alessandris N, Scaglione G, Valente M, Arciuolo D, Sanchez AM, Franceschini G, Masetti R, Mulè A, Zannoni GF (2021) Histopathological evaluation of tumor-infiltrating lymphocytes (TILs) as predictive biomarker for hormone receptors status, proliferative activity and clinical outcome in her-2 positive breast cancer. Appl Sci 11(15):6788

Asano Y, Kashiwagi S, Goto W, Takada K, Takahashi K, Hatano T, Noda S, Takashima T, Onoda N, Tomita S, Motomura H, Ohsawa M, Hirakawa K, Ohira M (2017) Prediction of survival after neoadjuvant chemotherapy for breast cancer by evaluation of tumor-infiltrating lymphocytes and residual cancer burden. BMC Cancer 17(1):888

Badr NM, Berditchevski F, Shaaban AM (2020) The immune microenvironment in breast carcinoma: predictive and prognostic role in the neoadjuvant setting. Pathobiology 87(2):61–74

Baghban R, Roshangar L, Jahanban-Esfahlan R, Seidi K, Ebrahimi-Kalan A, Jaymand M, Kolahian S, Javaheri T, Zare P (2020) Tumor microenvironment complexity and therapeutic implications at a glance. Cell Commun Signal 18(1):59

Bai Y, Cole K, Martinez-Morilla S, Ahmed FS, Zugazagoitia J, Staaf J, Bosch A, Ehinger A, Nimeus E, Hartman J, Acs B, Rimm DL (2021) An open source, automated tumor infiltrating lymphocyte algorithm for prognosis in triple-negative breast cancer. Clin Cancer Res 27(20): 5557–5565

Bianchini G, Pusztai L, Pienkowski T, Im YH, Bianchi GV, Tseng LM, Liu MC, Lluch A, Galeota E, Magazzù D, de la Haba-Rodríguez J, Oh DY, Poirier B, Pedrini JL, Semiglazov V, Valagussa P, Gianni L (2015) Immune modulation of pathologic complete response after neoadjuvant HER2-directed therapies in the NeoSphere trial. Ann Oncol 26(12):2429–2436

Bray F, Ferlay J, Soerjomataram I, Siegel RL, Torre LA, Jemal A (2018) Global cancer statistics 2018: GLOBOCAN estimates of incidence and mortality worldwide for 36 cancers in 185 countries. CA Cancer J Clin 68(6):394–424

Brown JR, Wimberly H, Lannin DR, Nixon C, Rimm DL, Bossuyt V (2014) Multiplexed quantitative analysis of CD3, CD8, and CD20 predicts response to neoadjuvant chemotherapy in breast cancer. Clin Cancer Res 20(23):5995–6005

Burugu S, Asleh-Aburaya K, Nielsen TO (2017) Immune infiltrates in the breast cancer microenvironment: detection, characterization and clinical implication. Breast Cancer 24(1):3–15

Campbell MJ, Baehner F, O'Meara T, Ojukwu E, Han B, Mukhtar R, Tandon V, Endicott M, Zhu Z, Wong J, Krings G, Au A, Gray JW, Esserman L (2017) Characterizing the immune microenvironment in high-risk ductal carcinoma in situ of the breast. Breast Cancer Res Treat 161(1): 17–28

Carbognin L, Pilotto S, Nortilli R, Brunelli M, Nottegar A, Sperduti I, Giannarelli D, Bria E, Tortora G (2016) Predictive and prognostic role of tumor-infiltrating lymphocytes for early breast cancer according to disease subtypes: sensitivity analysis of randomized trials in adjuvant and neoadjuvant setting. Oncologist 21(3):283–291

Criscitiello C, Esposito A, Trapani D, Curigliano G (2016) Prognostic and predictive value of tumor infiltrating lymphocytes in early breast cancer. Cancer Treat Rev 50:205–207

De Melo Gagliato D, Cortes J, Curigliano G, Loi S, Denkert C, Perez-Garcia J, Holgado E (2017) Tumor-infiltrating lymphocytes in breast cancer and implications for clinical practice. Biochim Biophys Acta Rev Cancer 1868(2):527–537

Demaria S, Volm MD, Shapiro RL, Yee HT, Oratz R, Formenti SC, Muggia F, Symmans WF (2001) Development of tumor-infiltrating lymphocytes in breast cancer after neoadjuvant paclitaxel chemotherapy. Clin Cancer Res 7(10):3025–3030

Denkert C, Loibl S, Noske A, Roller M, Müller BM, Komor M, Budczies J, Darb-Esfahani S, Kronenwett R, Hanusch C, von Törne C, Weichert W, Engels K, Solbach C, Schrader I, Dietel M, von Minckwitz G (2010) Tumor-associated lymphocytes as an independent predictor of response to neoadjuvant chemotherapy in breast cancer. J Clin Oncol 28(1):105–113

Denkert C, von Minckwitz G, Darb-Esfahani S, Lederer B, Heppner BI, Weber KE, Budczies J, Huober J, Klauschen F, Furlanetto J, Schmitt WD, Blohmer JU, Karn T, Pfitzner BM, Kümmel S, Engels K, Schneeweiss A, Hartmann A, Noske A, Fasching PA, Jackisch C, van Mackelenbergh M, Sinn P, Schem C, Hanusch C, Untch M, Loibl S (2018) Tumour-infiltrating lymphocytes and prognosis in different subtypes of breast cancer: a pooled analysis of 3771 patients treated with neoadjuvant therapy. Lancet Oncol 19(1):40–50

Desmedt C, Salgado R, Fornili M, Pruneri G, Van den Eynden G, Zoppoli G, Rothé F, Buisseret L, Garaud S, Willard-Gallo K, Brown D, Bareche Y, Rouas G, Galant C, Bertucci F, Loi S, Viale G, Di Leo A, Green AR, Ellis IO, Rakha EA, Larsimont D, Biganzoli E, Sotiriou C (2018) Immune infiltration in invasive lobular breast cancer. J Natl Cancer Inst 110(7):768–776

Dieci MV, Mathieu MC, Guarneri V, Conte P, Delaloge S, Andre F, Goubar A (2015) Prognostic and predictive value of tumor-infiltrating lymphocytes in two phase III randomized adjuvant breast cancer trials. Ann Oncol 26(8):1698–1704

Dieci MV, Radosevic-Robin N, Fineberg S et al. International Immuno-Oncology Biomarker Working Group on Breast Cancer (2018) Update on tumor-infiltrating lymphocytes (TILs) in breast cancer, including recommendations to assess TILs in residual disease after neoadjuvant therapy and in carcinoma in situ: a report of the International Immuno-Oncology Biomarker Working Group on Breast Cancer. Semin Cancer Biol 52(Pt 2):16–25

El Bairi K, Haynes HR, Blackley E et al (2018) The tale of TILs in breast cancer: a report from the International Immuno-Oncology Biomarker Working Group. NPJ Breast Cancer 7(1):150

Emens LA, Molinero L, Loi S, Rugo HS, Schneeweiss A, Diéras V, Iwata H, Barrios CH, Nechaeva M, Nguyen-Duc A, Chui SY, Husain A, Winer EP, Adams S, Schmid P (2021) Atezolizumab and nab-paclitaxel in advanced triple-negative breast cancer: biomarker evaluation of the IMpassion 130 study. J Natl Cancers Inst 113(8):1005–1016

Global Cancer Observatory (2020) GLOBOCAN. https://gco.iarc.fr/

Goto W, Kashiwagi S, Asano Y, Takada K, Takahashi K, Hatano T, Takashima T, Tomita S, Motomura H, Ohsawa M, Hirakawa K, Ohira M (2018) Predictive value of improvement in the immune tumour microenvironment in patients with breast cancer treated with neoadjuvant chemotherapy. ESMO Open 3(6):e000305

Gu-Trantien C, Loi S, Garaud S, Equeter C, Libin M, de Wind A, Ravoet M, Le Buanec H, Sibille C, Manfouo-Foutsop G, Veys I, Haibe-Kains B, Singhal SK, Michiels S, Rothé F, Salgado R, Duvillier H, Ignatiadis M, Desmedt C, Bron D, Larsimont D, Piccart M, Sotiriou C, Willard-Gallo K (2013) $CD4^+$ follicular helper T cell infiltration predicts breast cancer survival. J Clin Invest 123(7):2873–2892

Hamy AS, Pierga JY, Sabaila A, Laas E, Bonsang-Kitzis H, Laurent C, Vincent-Salomon A, Cottu P, Lerebours F, Rouzier R, Lae M, Reyal F (2017) Stromal lymphocyte infiltration after neoadjuvant chemotherapy is associated with aggressive residual disease and lower disease-free survival in HER2-positive breast cancer. Ann Oncol 28(9):2233–2240

Hendry S, Pang JB, Byrne DJ, Lakhani SR, Cummings MC, Campbell IG, Mann GB, Gorringe KL, Fox SB (2017a) Relationship of the breast ductal carcinoma in situ immune microenvironment with clinicopathological and genetic features. Clin Cancer Res 23(17):5210–5217

Hendry S, Salgado R, Gevaert T et al (2017b) Assessing tumor-infiltrating lymphocytes in solid tumors: a practical review for pathologists and proposal for a standardized method from the International Immunooncology Biomarkers Working Group: part 1: assessing the host immune response, TILs in invasive breast carcinoma and ductal carcinoma in situ, metastatic tumor deposits and areas for further research. Adv Anat Pathol 24(5):235–251

Horny HP, Horst HA (1986) Lymphoreticular infiltrates in invasive ductal breast cancer. A histological and immunohistological study. Virchows Arch 409(2):275–286

Horst HA, Horny HP (1988) Frequency distribution of lymphoreticular infiltrates in invasive carcinoma of the female breast. Cancer Detect Prev 11(3–6):297–301

Ibrahim EM, Al-Foheidi ME, Al-Mansour MM, Kazkaz GA (2014) The prognostic value of tumor-infiltrating lymphocytes in triple-negative breast cancer: a meta-analysis. Breast Cancer Res Treat 148(3):467–476

Ignatiadis M, Van den Eynden G, Roberto S, Fornili M, Bareche Y, Desmedt C, Rothé F, Maetens M, Venet D, Holgado E, McNally V, Kiermaier A, Savage HM, Wilson TR, Cortes J, Schneeweiss A, Willard-Gallo K, Biganzoli E, Sotiriou C (2019) Tumor-infiltrating lymphocytes in patients receiving trastuzumab/pertuzumab-based chemotherapy: a TRYPHAENA substudy. J Natl Cancer Inst 111(1):69–77

Issa-Nummer Y, Darb-Esfahani S, Loibl S, Kunz G, Nekljudova V, Schrader I, Sinn BV, Ulmer HU, Kronenwett R, Just M, Kühn T, Diebold K, Untch M, Holms F, Blohmer JU, Habeck JO, Dietel M, Overkamp F, Krabisch P, von Minckwitz G, Denkert C (2013) Prospective validation of immunological infiltrate for prediction of response to neoadjuvant chemotherapy in HER2-negative breast cancer – a substudy of the neoadjuvant GeparQuinto trial. PLoS One 8(12): e79775

Krishnamurti U, Wetherilt CS, Yang J, Peng L, Li X (2017) Tumor-infiltrating lymphocytes are significantly associated with better overall survival and disease-free survival in triple-negative but not estrogen receptor-positive breast cancers. Hum Pathol 64:7–12

Lee S, Cho EY, Park YH, Ahn JS, Im YH (2013) Prognostic impact of FOXP3 expression in triple-negative breast cancer. Acta Oncol 52(1):73–81

Leon-Ferre RA, Polley MY, Liu H, Gilbert JA, Cafourek V, Hillman DW, Elkhanany A, Akinhanmi M, Lilyquist J, Thomas A, Negron V, Boughey JC, Liu MC, Ingle JN, Kalari KR, Couch FJ, Visscher DW, Goetz MP (2018) Impact of histopathology, tumor-infiltrating lymphocytes, and adjuvant chemotherapy on prognosis of triple-negative breast cancer. Breast Cancer Res Treat 167(1):89–99

Li JJ, Tsang JY, Tse GM (2021) Tumor microenvironment in breast cancer-updates on therapeutic implications and pathologic assessment. Cancers (Basel) 13(16):4233

Liu S, Lachapelle J, Leung S, Gao D, Foulkes WD, Nielsen TO (2012) CD8+ lymphocyte infiltration is an independent favorable prognostic indicator in basal-like breast cancer. Breast Cancer Res 14(2):R48

Liu S, Chen B, Burugu S, Leung S, Gao D, Virk S, Kos Z, Parulekar WR, Shepherd L, Gelmon KA, Nielsen TO (2017) Role of cytotoxic tumor-infiltrating lymphocytes in predicting outcomes in metastatic HER2-positive breast cancer: a secondary analysis of a randomized clinical trial. JAMA Oncol 3(11):e172085

Loi S, Sirtaine N, Piette F, Salgado R, Viale G, Van Eenoo F, Rouas G, Francis P, Crown JP, Hitre E, de Azambuja E, Quinaux E, Di Leo A, Michiels S, Piccart MJ, Sotiriou C (2013) Prognostic and predictive value of tumor-infiltrating lymphocytes in a phase III randomized adjuvant breast cancer trial in node-positive breast cancer comparing the addition of docetaxel to doxorubicin with doxorubicin-based chemotherapy: BIG 02-98. J Clin Oncol 31(7):860–867

Loi S, Michiels S, Salgado R, Sirtaine N, Jose V, Fumagalli D, Kellokumpu-Lehtinen PL, Bono P, Kataja V, Desmedt C, Piccart MJ, Loibl S, Denkert C, Smyth MJ, Joensuu H, Sotiriou C (2014) Tumor infiltrating lymphocytes are prognostic in triple negative breast cancer and predictive for trastuzumab benefit in early breast cancer: results from the FinHER trial. Ann Oncol 25(8): 1544–1550

Loi S, Drubay D, Adams S, Pruneri G, Francis PA, Lacroix-Triki M, Joensuu H, Dieci MV, Badve S, Demaria S, Gray R, Munzone E, Lemonnier J, Sotiriou C, Piccart MJ, Kellokumpu-Lehtinen PL, Vingiani A, Gray K, Andre F, Denkert C, Salgado R, Michiels S (2019) Tumor-infiltrating lymphocytes and prognosis: a pooled individual patient analysis of early-stage triple-negative breast cancers. J Clin Oncol 37(7):559–569

Loi S, Michiels S, Adams S, Loibl S, Budczies J, Denkert C, Salgado R (2021) The journey of tumor-infiltrating lymphocytes as a biomarker in breast cancer: clinical utility in an era of checkpoint inhibition. Ann Oncol 32(10):1236–1244

Loi S, Salgado R, Adams S, Pruneri G, Francis PA, Lacroix-Triki M, Joensuu H, Dieci MV, Badve S, Demaria S, Gray R, Munzone E, Drubay D, Lemonnier J, Sotiriou C, Kellokumpu-Lethinen PL, Vingiani A, Gray K, André F, Denkert C, Piccart M, Roblin E, Michiels S (2022) Tumor infiltrating lymphocyte stratification of prognostic staging of early-stage triple negative breast cancer. NPJ Breast Cancer 8:3

Lu Z, Xu S, Shao W, Wu Y, Zhang J, Han Z, Feng Q, Huang K (2020) Deep-learning-based characterization of tumor-infiltrating lymphocytes in breast cancers from histopathology images and multiomics data. JCO Clin Cancer Inform 4:480–490

Luen SJ, Salgado R, Fox S, Savas P, Eng-Wong J, Clark E, Kiermaier A, Swain SM, Baselga J, Michiels S, Loi S (2017) Tumour-infiltrating lymphocytes in advanced HER2-positive breast cancer treated with pertuzumab or placebo in addition to trastuzumab and docetaxel: a retrospective analysis of the CLEOPATRA study. Lancet Oncol 18(1):52–62

Luen SJ, Salgado R, Dieci MV, Vingiani A, Curigliano G, Gould RE, Castaneda C, D'Alfonso T, Sanchez J, Cheng E, Andreopoulou E, Castillo M, Adams S, Demaria S, Symmans WF, Michiels S, Loi S (2019) Prognostic implications of residual disease tumor-infiltrating lymphocytes and residual cancer burden in triple-negative breast cancer patients after neoadjuvant chemotherapy. Ann Oncol 30(2):236–242

Mahmoud SM, Paish EC, Powe DG, Macmillan RD, Grainge MJ, Lee AH, Ellis IO, Green AR (2011a) Tumor-infiltrating CD8+ lymphocytes predict clinical outcome in breast cancer. J Clin Oncol 29(15):1949–1955

Mahmoud SM, Paish EC, Powe DG, Macmillan RD, Lee AH, Ellis IO, Green AR (2011b) An evaluation of the clinical significance of FOXP3+ infiltrating cells in human breast cancer. Breast Cancer Res Treat 127(1):99–108

Mahmoud SM, Lee AH, Paish EC, Macmillan RD, Ellis IO, Green AR (2012) The prognostic significance of B lymphocytes in invasive carcinoma of the breast. Breast Cancer Res Treat 132(2):545–553

Mansour M, Teo ZL, Luen SJ, Loi S (2017) Advancing immunotherapy in metastatic breast cancer. Curr Treat Options in Oncol 18(6):35

Miligy I, Mohan P, Gaber A, Aleskandarany MA, Nolan CC, Diez-Rodriguez M, Mukherjee A, Chapman C, Ellis IO, Green AR, Rakha EA (2017) Prognostic significance of tumour infiltrating B lymphocytes in breast ductal carcinoma in situ. Histopathology 71(2):258–268

Mohammed ZM, Going JJ, Edwards J, Elsberger B, McMillan DC (2013) The relationship between lymphocyte subsets and clinico-pathological determinants of survival in patients with primary operable invasive ductal breast cancer. Br J Cancer 109(6):1676–1684

Montagna E, Vingiani A, Maisonneuve P, Cancello G, Contaldo F, Pruneri G, Colleoni M (2017) Unfavorable prognostic role of tumor-infiltrating lymphocytes in hormone-receptor positive, HER2 negative metastatic breast cancer treated with metronomic chemotherapy. Breast 34:83–88

Moore OS Jr, Foote FW Jr (1949) The relatively favorable prognosis of medullary carcinoma of the breast. Cancer 2(4):635–642

Ochi T, Bianchini G, Ando M, Nozaki F, Kobayashi D, Criscitiello C, Curigliano G, Iwamoto T, Niikura N, Takei H, Yoshida A, Takei J, Suzuki K, Yamauchi H, Hayashi N (2019) Predictive and prognostic value of stromal tumour-infiltrating lymphocytes before and after neoadjuvant therapy in triple negative and HER2-positive breast cancer. Eur J Cancer 118:41–48

Ogiya R, Niikura N, Kumaki N, Bianchini G, Kitano S, Iwamoto T, Hayashi N, Yokoyama K, Oshitanai R, Terao M, Morioka T, Tsuda B, Okamura T, Saito Y, Suzuki Y, Tokuda Y (2016) Comparison of tumor-infiltrating lymphocytes between primary and metastatic tumors in breast cancer patients. Cancer Sci 107(12):1730–1735

Perez EA, Ballman KV, Tenner KS, Thompson EA, Badve SS, Bailey H, Baehner FL (2016) Association of stromal tumor-infiltrating lymphocytes with recurrence-free survival in the N9831 adjuvant trial in patients with early-stage HER2-positive breast cancer. JAMA Oncol 2(1):56–64

Pruneri G, Lazzeroni M, Bagnardi V, Tiburzio GB, Rotmensz N, DeCensi A, Guerrieri-Gonzaga A, Vingiani A, Curigliano G, Zurrida S, Bassi F, Salgado R, Van den Eynden G, Loi S, Denkert C, Bonanni B, Viale G (2017) The prevalence and clinical relevance of tumor-infiltrating lymphocytes (TILs) in ductal carcinoma in situ of the breast. Ann Oncol 28(2):321–328

Ravelli A, Roviello G, Cretella D, Cavazzoni A, Biondi A, Cappelletti MR, Zanotti L, Ferrero G, Ungari M, Zanconati F, Bottini A, Alfieri R, Petronini PG, Generali D (2017) Tumor-infiltrating lymphocytes and breast cancer: beyond the prognostic and predictive utility. Tumour Biol 39(4):1010428317695023

Salgado R, Denkert C, Campbell C, Savas P, Nuciforo P, Aura C, de Azambuja E, Eidtmann H, Ellis CE, Baselga J, Piccart-Gebhart MJ, Michiels S, Bradbury I, Sotiriou C, Loi S (2015a) Tumor-infiltrating lymphocytes and associations with pathological complete response and event-free survival in HER2-positive early-stage breast cancer treated with Lapatinib and Trastuzumab: a secondary analysis of the NeoALTTO trial. JAMA Oncol 1(4):448–454

Salgado R, Denkert C, Demaria S, Sirtaine N, Klauschen F, Pruneri G, Wienert S, Van den Eynden G, Baehner FL, Penault-Llorca F, Perez EA, Thompson EA, Symmans WF, Richardson AL, Brock J, Criscitiello C, Bailey H, Ignatiadis M, Floris G, Sparano J, Kos Z, Nielsen T, Rimm DL, Allison KH, Reis-Filho JS, Loibl S, Sotiriou C, Viale G, Badve S, Adams S, Willard-Gallo K, Loi S, International TILs Working Group 2014 (2015b) The evaluation of tumor-infiltrating lymphocytes (TILs) in breast cancer: recommendations by an International TILs Working Group 2014. Ann Oncol 26(2):259–271

Seo AN, Lee HJ, Kim EJ, Kim HJ, Jang MH, Lee HE, Kim YJ, Kim JH, Park SY (2013) Tumour-infiltrating CD8+ lymphocytes as an independent predictive factor for pathological complete response to primary systemic therapy in breast cancer. Br J Cancer 109(10):2705–2713

Sobottka B, Pestalozzi B, Fink D, Moch H, Varga Z (2016) Similar lymphocytic infiltration pattern in primary breast cancer and their corresponding distant metastases. Onco Targets Ther 5(6): e1153208

Solinas C, Carbognin L, De Silva P, Criscitiello C, Lambertini M (2017a) Tumor-infiltrating lymphocytes in breast cancer according to tumor subtype: current state of the art. Breast 35: 142–150

Solinas C, Ceppi M, Lambertini M, Scartozzi M, Buisseret L, Garaud S, Fumagalli D, de Azambuja E, Salgado R, Sotiriou C, Willard-Gallo K, Ignatiadis M (2017b) Tumor-infiltrating lymphocytes in patients with HER2-positive breast cancer treated with neoadjuvant chemotherapy plus trastuzumab, lapatinib or their combination: a meta-analysis of randomized controlled trials. Cancer Treat Rev 57:8–15

Stanton SE, Adams S, Disis ML (2016) Variation in the incidence and magnitude of tumor-infiltrating lymphocytes in breast cancer subtypes: a systematic review. JAMA Oncol 2(10): 1354–1360

Sun P, He J, Chao X, Chen K, Xu Y, Huang Q, Yun J, Li M, Luo R, Kuang J, Wang H, Li H, Hui H, Xu S (2021) A computational tumor-infiltrating lymphocyte assessment method comparable with visual reporting guidelines for triple-negative breast cancer. EBioMedicine 70:103492

Symmans WF, Peintinger F, Hatzis C, Rajan R, Kuerer H, Valero V, Assad L, Poniecka A, Hennessy B, Green M, Buzdar AU, Singletary SE, Hortobagyi GN, Pusztai L (2007) Measurement of residual breast cancer burden to predict survival after neoadjuvant chemotherapy. J Clin Oncol 25(28):4414–4422

Szekely B, Bossuyt V, Li X, Wali WB, Patwardhan GA, Frederick C, Silber A, Park T, Harigopal M, Pelekanou V, Zhang M, Yan Q, Rimm DL, Bianchini G, Hatzis C, Pusztai L (2018) Immunological differences between primary and metastatic breast cancer. Ann Oncol 29(11):2232–2239

Thomas R, Al-Khadairi G, Decock J (2021) Immune checkpoint inhibitors in triple negative breast cancer treatment: promising future prospects. Front Oncol 10:600573

Thompson E, Taube JM, Elwood H, Sharma R, Meeker A, Warzecha HN, Argani P, Cimino-Mathews A, Emens LA (2016) The immune microenvironment of breast ductal carcinoma in situ. Mod Pathol 29(3):249–258

Wein L, Savas P, Luen SJ, Virassamy B, Salgado R, Loi S (2017) Clinical validity and utility of tumor-infiltrating lymphocytes in routine clinical practice for breast cancer patients: current and future directions. Front Oncol 7:156

West NR, Milne K, Truong PT, Macpherson N, Nelson BH, Watson PH (2011) Tumor-infiltrating lymphocytes predict response to anthracycline-based chemotherapy in estrogen receptor-negative breast cancer. Breast Cancer Res 13(6):R126

West NR, Kost SE, Martin SD, Milne K, Deleeuw RJ, Nelson BH, Watson PH (2013) Tumour-infiltrating FOXP3(+) lymphocytes are associated with cytotoxic immune responses and good clinical outcome in oestrogen receptor-negative breast cancer. Br J Cancer 108(1):155–162

WHO Classification of Tumours Editorial Board (2019) Breast tumours. WHO classification of tumours, 5th edn. IARC Press, Lyon

Yamaguchi R, Tanaka M, Yano A, Tse GM, Yamaguchi M, Koura K, Kanomata N, Kawaguchi A, Akiba J, Naito Y, Ohshima K, Yano H (2012) Tumor-infiltrating lymphocytes are important pathologic predictors for neoadjuvant chemotherapy in patients with breast cancer. Hum Pathol 43(10):1688–1694

Zacharakis N, Chinnasamy H, Black M, Xu H, Lu YC, Zheng Z, Pasetto A, Langhan M, Shelton T, Prickett T, Gartner J, Jia L, Trebska-McGowan K, Somerville RP, Robbins PF, Rosenberg SA, Goff SL, Feldman SA (2018) Immune recognition of somatic mutations leading to complete durable regression in metastatic breast cancer. Nat Med 24(6):724–730

Zhang W, Kong X, Ai B, Wang Z, Wang X, Wang N, Zheng S, Fang Y, Wang J (2021) Research progresses in immunological checkpoint inhibitors for breast cancer immunotherapy. Front Oncol 11:582664

Zhu L, Narloch JL, Onkar S, Joy M, Broadwater G, Luedke C, Hall A, Kim R, Pogue-Geile K, Sammons S, Nayyar N, Chukwueke U, Brastianos PK, Anders CK, Soloff AC, Vignali DAA, Tseng GC, Emens LA, Lucas PC, Blackwell KL, Oesterreich S, Lee AV (2019) Metastatic breast cancers have reduced immune cell recruitment but harbor increased macrophages relative to their matched primary tumors. J Immunother Cancer 7(1):265

From Genes to Recovery: Precision Medicine and Its Influence on Multidrug Resistant Breast Cancer

Sanchaita Rajkhowa, Safrina Zeenat, Manisha Agarwal, Alaiha Zaheen, Magdi E. A. Zaki, and Subrata Sinha

Abstract

Breast cancer ranks as the most frequently diagnosed and life-threatening form of cancer in women, serving as a leading cause of cancer-related deaths in this population. Among women afflicted with breast cancer, the primary factor contributing to treatment failure and mortality is the resistance to cytotoxic chemotherapy. Individuals unresponsive to chemotherapy often demonstrate resistance to various antineoplastic drugs with different structures and mechanisms of action, indicating a multidrug resistance phenotype in the malignant breast tissue. Recent progress in the field of oncology has led to an enhanced comprehension of the human genome's role in these changes, opening up avenues for the development of targeted therapeutic strategies. Precision medicine, an approach that integrates clinical and diverse data to categorize patients into novel subgroups, plays a pivotal role in these developments. This chapter delves into recent progress in the therapeutic strategies for various solid tumors, emphasizing the significance of precision medicine in addressing breast cancer, especially in cases of multidrug resistance. As we explore the current landscape of precision medicine for breast cancer, it becomes apparent that tailoring treatments based on individual characteristics is transformative. However, this journey is not without its challenges. Therefore, the chapter also highlights the obstacles associated with precision medicine in the context of breast cancer, underscoring the imperative to overcome these hurdles

S. Rajkhowa (✉) · S. Zeenat · M. Agarwal · A. Zaheen
Centre for Biotechnology and Bioinformatics, Dibrugarh University, Dibrugarh, Assam, India
e-mail: s_rajkhowa@dibru.ac.in

M. E. A. Zaki
Department of Chemistry, Imam Mohammad Ibn Saud Islamic University (IMSIU), Riyadh, Saudi Arabia

S. Sinha
Department of Computational Science, Brainware University, Barasat, Kolkata, West Bengal, India

© The Author(s), under exclusive license to Springer Nature Switzerland AG 2024
Interdisciplinary Cancer Research, https://doi.org/10.1007/16833_2024_248
Published online: 25 May 2024

to fully unleash the potential of precision medicine and usher in a new era of more effective and personalized therapies for breast cancer.

Keywords

Breast cancer · Chemotherapy · Multidrug resistance · Precision medicine

1 Introduction

1.1 Overview of Breast Cancer

One of the most commonly identified and life-threatening forms of cancer in women is attributed to breast cancer (BC). It serves as a major contributor to cancer-related deaths in this demographic. The disease originates in the breast tissue. It typically emerges from the inner lining of milk ducts and sometimes from the lobules responsible for milk production. According to statistics provided in 2020, by the World Health Organization (WHO), approximately 2.3 million women were diagnosed with breast cancer, leading to 685,000 fatalities globally. By the end of 2020, approximately 7.8 million women had survived a diagnosis of breast cancer within the preceding five years, solidifying its status as the most widespread cancer globally.

While the majority of breast cancers exhibit a benign nature and are amenable to surgical intervention, a quarter of cases manifest a latent and insidious behavior, characterized by slow growth but early metastasis. Despite the effectiveness of current therapeutic modalities in significantly delaying tumor progression, the inevitability of recurrence contributes to elevated mortality rates. It is crucial to recognize that breast cancer is not a singular ailment but comprises various subtypes, each associated with distinct clinical outcomes. Unraveling this heterogeneity is paramount for the development of targeted interventions in cancer prevention and therapy. Analyzing gene expression profiles reveals the existence of at least five discernible forms of breast cancer, each possessing distinct biological characteristics.

The predominant proportion of breast cancers originate from the epithelial compartment and is categorized as malignant according to the degree of differentiation, spanning from poorly differentiated (high grade) to well differentiated (low grade). This classification reflects the progressive loss of characteristics typical of normal breast cells as the cells undergo malignant transformation. Cancers exhibiting low differentiation levels are associated with a less favorable prognosis. Receptors are found on the surface, in the cytoplasm, and within the nucleus of BC cells, enabling molecular classification through histopathology and simple immunohistochemical techniques. The three most extensively studied receptors are the human epidermal growth factor receptor 2 (HER2, also known as ERBB2), the progesterone receptor (PR), and the estrogen receptor (ER). Clinicians rely on their status to guide specific therapeutic approaches. Cells devoid of expression of these three receptor subtypes are classified as triple negative (TN). Addressing molecular targets in these tumors poses a significant challenge, and currently, no FDA-approved treatments specifically tailored to triple-negative breast cancer (TNBC) are available. This underscores the urgent need for innovative therapeutic approaches targeting alternative pathways in the absence of these receptors.

1.2 What Is Precision Medicine?

Precision medicine, an innovative approach to healthcare, has gained prominence in recent years, propelled by advancements in genomics, data analytics, and personalized treatment strategies. The nomenclature "precision medicine (PM)" has superseded the designation "personalized medicine," originally perceived as interchangeable but later dismissed due to the historical approach of physicians in administering individualized treatments to patients (McGrath and Ghersi 2016). While the personalized approach inherent in the doctor-patient relationship is integral to PM, it is not a novel concept. However, the emergence of novel biomedical information has the potential to augment the understanding beyond observable signs and symptoms. The term "precision medicine" conveys the innovative aspect of this concept. It denotes the amalgamation of a varied array of personalized data, including lifestyle, clinical, and genetic, as well as, additional biomarker information. The comprehensive characterization of numerous tumors has been made possible through the introduction and accessibility of various technologies, including DNA sequencing, mRNA microarrays, and next-generation sequencing (NGS). These tools, along with a range of proteomics assays, contribute to the understanding of mutational profiles and abnormal gene and protein expression in tumors. The process of characterizing tumors can be accomplished through sequencing of the genome sequencing, profiling of the transcriptome, or utilizing integrated multi-omics platforms. Profiling of the genome predominantly concentrates on tumor-causing cells to reveal mutations in the somatic cells (single-nucleotide variants and indels), copy-number variations, or oncogenic fusions. Conversely, profiling of the transcriptome and proteome, encompassing phosphoprotein analysis, provides insights into the activation status of pathways, thus presenting an alternative pathway for the formulation of targeted therapies (Sachdev et al. 2019).

According to its conceptualization, PM seeks to intricately align the distinct genomic alterations of each individual with a therapeutic intervention that is optimally effective. The objective of PM is to tailor medical care and interventions based on the distinctive attributes of each patient, enabling accurate diagnoses, effective treatments, and improved outcomes. This paradigm diverges from the conventional one-size-fits-all model, which formulates treatments and interventions based on the average response observed within a population.

Widely employed conceptualizations of PM encompass the following facets:

1. Priority on the result, particularly tailored therapeutic approaches: PM is defined as "interventions targeted at the distinct needs of individual patients, leveraging genetic, biomarker, phenotypic, or psychosocial attributes that distinguish a given

patient from others presenting similar clinical manifestations" (Larry Jameson and Longo 2015).

2. Prioritization of procedural elements and utilized data is accentuated, with diverse viewpoints highlighting the importance of PM as a framework that amalgamates clinical and varied data for the purpose of classifying patients into novel subgroups. This classification is anticipated to establish a shared basis concerning disease susceptibility and manifestation, potentially enhancing the precision of therapeutic interventions (Robinson 2012).

In the realm of healthcare, PM signifies a fundamental change in approach, transitioning from a population-centric paradigm to a personalized and focused model. It possesses the potential to substantially augment medical methodologies, although its complete integration demands a substantial timeframe. Presently, genetics plays a role in guiding clinical decision-making, with a foreseen augmentation of its contribution. Expedited progress mandates fundamental alterations in data collection, storage, and sharing infrastructure and mechanisms. This transformation aims to establish a perpetually learning healthcare system, fostering seamless integration between clinical care and research. An imperative aspect involves educating patients on the advantages of data sharing (Aronson and Rehm 2015). Building upon this foundation, PM seeks to utilize the accumulated knowledge to redefine the comprehension of diseases, formulate innovative therapeutic interventions, and present auspicious opportunities for forthcoming patient cohorts (Ashley 2016).

1.3 Significance of Precision Medicine in Breast Cancer Management

Recent advancements in the therapeutic approach to various solid tumors exhibit potential by leveraging the innate immune system of the patient to discern and eliminate extant metastatic ailments. While only a restricted portion of BCs adhere to this model, it has established new standards for progress in pharmaceuticals and the swift delivery of effective treatments to patients in the era of PM. BC stands as an instructive exemplar in the PM era, given the protracted utilization and refinement of endocrine therapy and anti-HER2 therapy by researchers and clinicians.

In-depth sequencing inquiries have substantiated the participation of established cancer-associated genes, such as PIK3CA, ERBB2, PTEN, and TP53, in BC. Moreover, these investigations have revealed numerous other genes displaying recurring mutations in BC. Such discoveries highlight the potential of comprehensive genomic profiling to unveil clinically significant mutations prevalent in most BC. This observation suggests the possibility of progressively categorizing BCs into more nuanced subgroups, enabling tailored therapies aligned with the specific mutational profiles of individual tumors. Apart from providing guidance for therapeutic interventions in clinical trials, clinical sequencing has revealed new mechanisms governing both resistance and notable responses to therapy. Meanwhile, specific situations in BC require extra sequencing information, encompassing metastatic disease, uncommon subtypes like inflammatory BC, and sequential samples over time to evaluate tumor evolution and therapeutic resistance.

Accompanying substantial potential, a spectrum of challenges is encountered by PM in BC across various domains, such as encompassing ethical considerations and education of patients (in social domain), entailing the annotation and exploration of fresh mutations (in biological domain), involving the assessment of expenses and the widespread implementation dissemination (in technical domain), and pertaining to data storage and handling (in infrastructure domain). These aspects must be diligently considered as an epoch characterized by genomics-driven PM in the context of BC is ushered by Stover and Wagle (2015).

1.4 List of FDA-Approved Precision Medicine in Breast Cancer

In the field of oncology and malignant hematology, the FDA granted approval to 207 drugs from May 1, 2016, to May 31, 2021, with 39 of them approved for BC. Over the span of 70 years, the FDA has granted approval for a higher number of drugs for indications related to BC as compared to any other type of cancer.

As per the most recent data on FDA drug approvals (Research 2024), covering the period between January 2020 and December 2023, a total of 208 drugs gained approval for oncology and hematologic malignancies, with 13 of them specifically approved for BC.

Figure 1 presents the proportional distribution of anticancer drugs across various categories that have received approval for treating BC. Initially, these medications were authorized for metastatic BC; however, over time, around 30% of these medications obtained the status of supplementary adjuvant (Chaurasia et al. 2023).

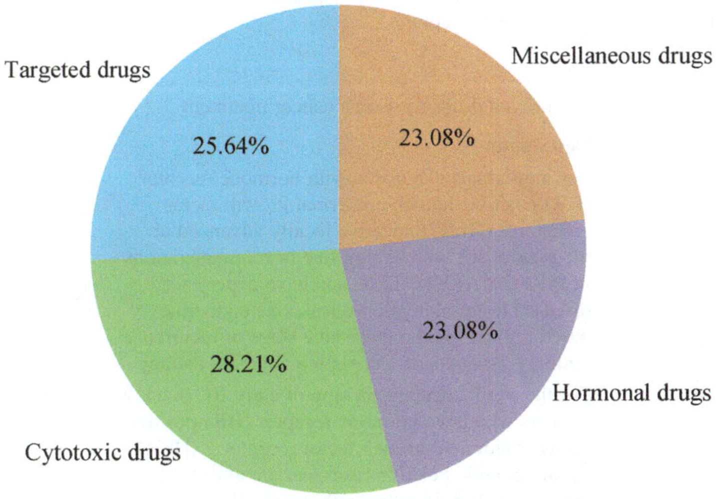

Fig. 1 Proportional distribution of drugs across various categories approved for treating BC

Table 1 provides examples of PMs approved by FDA for use in the therapy of BC. Table 2 includes additional FDA-approved drugs relevant to BC treatment (between January 2020 and December 2023).

Table 1 List of FDA-approved PMs for the treatment of BC

Targeted drugs	Biomarkers	References
Trastuzumab	HER2	Chiavenna et al. (2017), Lu et al. (2020)
Pertuzumab and Trastuzumab emtansine	HER2	Chiavenna et al. (2017), Cobleigh et al. (1999)
Sacituzumab govitecan	Trop-2, HER2	Cinkaya et al. (2016)
Traxtuzumab and pertuzumab	HER2	Musyuni et al. (2022)
Temsirolimus	p-mTOR, PTEN, activation of PI3K/AKT pathway	Thomas (2017)
Piqray "alpelisib"	HR, HER2, and PIK3CA-mutated	Musyuni et al. (2022)
"Tukysa" (tucatinib)	HER2	Musyuni et al. (2022)
Enhertu "fam-trastuzumab deruxtecannxki"	HER2	Musyuni et al. (2022)
"Margenza" (margetuximab-cmkb)	HER2	Musyuni et al. (2022)
"Trodelvy" (sacituzumab govitecan-hziy)	ER, PR, and HER2	Musyuni et al. (2022)

HER2 human epidermal growth factor receptor 2, Trop-2 trophoblast cell-surface antigen 2, p-mTOR phosphorylated mammalian target of rapamycin, PTEN phosphatase and tensin homolog, PI3K phosphoinositide 3-kinase, AKT serine/threonine-specific protein kinase, HR hormone receptor, PIK3CA phosphatidylinositol-4,5-bisphosphate 3-kinase catalytic subunit alpha, ER estrogen receptor, PR progesterone receptor

Table 2 Recently FDA-approved drugs for breast cancer treatment

Drug	Description	Year	Reference
Capivasertib with fulvestrant	For individuals diagnosed with hormone receptor (HR) positive, human epidermal growth factor receptor 2 (HER2) negative locally advanced or metastatic BC, who exhibit one or more alterations in PIK3CA/AKT1/PTEN, and have experienced disease progression after at least one endocrine-based regimen in the metastatic stage or recurrence within 12 months of completing adjuvant therapy	2023	Research (2024)
Abemaciclib with endocrine therapy	For the supplementary therapy of early BC in adult patients who have hormone receptor (HR)-positive, human epidermal growth factor receptor 2 (HER2)-negative, node-positive cases and are at a heightened risk of recurrence	2023	

(continued)

Table 2 (continued)

Drug	Description	Year	Reference
Sacituzumab govitecan-hziy	For individuals with inoperable locally advanced or metastatic BC characterized by hormone receptor (HR) positivity and human epidermal growth factor receptor 2 (HER2) negativity (IHC 0, IHC 1+, or IHC 2+/ISH-), and who have undergone endocrine-based therapy along with a minimum of two other systemic treatments in the metastatic stage	2023	
Elacestrant	For individuals who have reached menopause or are diagnosed with advanced or metastatic BC, specifically with ER-positive, HER2-negative, and ESR1 mutations, and have experienced disease progression after undergoing at least one round of endocrine therapy	2023	
Fam-trastuzumab deruxtecan-nxki	For adult individuals diagnosed with HER2-low (IHC 1+ or IHC 2+/ISH-) BC that is either unresectable or metastatic, and who have undergone previous chemotherapy (CT) in the metastatic stage or experienced a recurrence within six months of completing adjuvant chemotherapy	2022	
Olaparib	To provide additional support in treating high-risk early BC in adults, specifically those with confirmed or suspected deleterious germline BRCA mutations (gBRCAm) and lacking human epidermal growth factor receptor 2 (HER2), who have undergone neoadjuvant chemotherapy (NAC) or adjuvant chemotherapy. Patient eligibility for this treatment is determined using an FDA-approved companion diagnostic for olaparib	2022	
Abemaciclib with endocrine therapy	For the adjuvant therapy of early BC in adults with high recurrence risk, characterized by hormone receptor (HR)-positive, human epidermal growth factor receptor 2 (HER2)-negative, node-positive status, and a Ki-67 score $\geq 20\%$ as per an FDA-approved test; this marks the inaugural approval of a CDK 4/6 inhibitor for adjuvant BC treatment	2021	
Pembrolizumab	For the neoadjuvant treatment of high-risk, early-stage triple-negative breast cancer (TNBC), in conjunction with chemotherapy, followed by its use as a standalone agent for adjuvant treatment post-surgery	2021	
Sacituzumab govitecan	For individuals diagnosed with inoperable locally advanced or metastatic triple-negative breast cancer (mTNBC) and who have undergone two or more previous systemic treatments, with at least one specifically administered for metastatic disease	2021	
Margetuximab	To address metastatic HER2-positive BC in adults, the treatment is indicated for individuals who have undergone two or more prior anti-HER2 regimens, with at least one of them being for metastatic disease	2020	

(continued)

Table 2 (continued)

Drug	Description	Year	Reference
Sacituzumab govitecan-hziy	For individuals diagnosed with metastatic triple-negative BC, who have undergone a minimum of two prior treatments for metastasis in adults	2020	
Tucatinib	Tucatinib, along with trastuzumab and capecitabine, is prescribed for adults dealing with advanced unresectable or metastatic HER2-positive BC. This includes individuals with brain metastases who have undergone one or more prior treatments involving anti-HER2 regimens in the metastatic setting	2020	
Neratinib	Neratinib, when used alongside capecitabine, is indicated for adults with advanced or metastatic HER2-positive BC. This applies to individuals who have undergone two or more previous anti-HER2 treatments in the metastatic setting.	2020	

2 Genomic Landscape of Breast Cancer

2.1 Mutations and Alterations Associated with Breast Cancer

It has been approximated that between five percent to ten percent of instances of cancer in the breast are hereditary, the consequence of aberrant gene transfer from parents to children (*Genetics* n.d.). Casaubon et al. (2017) argue that a considerable portion of familial cases can be attributed to mutations occurring in the autosomal dominant genes, namely BRCA1 and BRCA2. Repairing cellular damage and controlling the proper growth of breast, ovarian, and other cells are the main roles of the BRCA genes. However, the normal operation of these is compromised by hereditary mutation, as they tend to interfere with their functioning. Genetic alterations in the BRCA1/BRCA2 genes may account for as much as ten percent of all BC cases. If a woman carries a mutation in either or both of these genes, her lifetime risk of being diagnosed with BC can reach 72%. Compared to BC in women without these genetic changes, BCs linked to BRCA1 or BRCA2 mutation typically emerge in younger women and affect both breasts more frequently (*Genetics*, n.d.)

Table 3 displays a compilation of genes that may undergo unusual changes and have been associated with a risk of BC and other malignancies in families

2.2 Molecular Subtypes of Breast Cancer and Their Clinical Implications

Breast carcinoma is a heterogeneous disease involving multiple subgroups that have changed in categorization throughout time, both medically and biologically. The predominant and widely acknowledged categorization is immune-histochemical,

Table 3 List of high-risk gene mutations in BC

Gene mutation	Description	References
BRCA1 and ***BRCA2***	The primary reason for hereditary BC is a genetic mutation passed down through families, specifically in the BRCA1 or BRCA2 genes. Normally, these genes play a role in creating proteins that fix damaged DNA. When these genes are mutated, it can cause abnormal cell growth, potentially leading to cancer. Women with a BRCA1 or BRCA2 mutation(s) can face a risk of up to 72% of developing BC in their lifetime. Men with a BRCA2 mutation have a higher likelihood of BC compared to those without, around 8% by the time they reach 80 years old. BRCA1 is associated with triple-negative BC, while young women with BRCA2-related BC are more susceptible to it	*Genetics and Risk of Breast Cancer* (n.d.)
PALB2	PALB2 is the third most common gene linked to a higher risk of BC. PALB2 collaborates with the BRCA2 gene to repair damage and prevent the development of breast cancer. If there is a mutation in the PALB2 gene, the risk of developing BC increases. Around 53% of women with a mutated PALB2 gene are estimated to develop BC by the age of 80. This mutation is connected to a higher risk of triple-negative BC in younger women and male BC. Additionally, PALB2 is associated with an increased risk of ovarian cancer in women, and, in certain families, a risk of pancreatic cancer	*Genetics as a Risk Factor for Breast Cancer* (2024)
PTEN	The PTEN gene plays a role in controlling how cells grow. If there is a problem with the PTEN gene, it can lead to Cowden syndrome, which is a rare condition associated with an increased likelihood of developing noncancerous and cancerous breast tumors. Additionally, individuals with Cowden syndrome may experience growths in the digestive tract, thyroid, uterus, and ovaries. Women with a mutated PTEN gene have an estimated lifetime risk of 25–50% for BC, although some studies suggest a higher risk of 77–85%. On average, people with Cowden syndrome are diagnosed with BC between the ages of 38 and 50	*Genetics as a Risk Factor for Breast Cancer* (2024)
TP53	The TP53 gene provides instructions for producing a protein that hinders tumor growth. Inheriting a defective TP53 gene can result in Li-Fraumeni syndrome, a condition characterized by the early onset of soft tissue cancers. Women with this syndrome face a higher likelihood of developing BC, particularly HER2-positive BC, at a young age. The gene's abnormality heightens the risk of various cancer types, including those affecting children. Women carrying a TP53 mutation have an almost 100% probability of developing any cancer, while the risk in men is about 73%, primarily due to the increased BC risk in women	*Genetics as a Risk Factor for Breast Cancer* (2024)

(continued)

Table 3 (continued)

Gene mutation	Description	References
ATM	The ATM gene plays a role in repairing damaged DNA, which carries genetic information in cells. When an individual inherits two abnormal copies of this gene, it leads to ataxia-telangiectasia, a rare disease impacting brain development. In cases where one abnormal ATM gene is inherited, there is an elevated risk of breast and pancreatic cancer in certain families. This heightened risk is attributed to the abnormal gene hindering the cells' ability to repair damaged DNA. Studies indicate that individuals with ATM mutations face a 33–38% lifetime risk of developing BC by the age of 80	*Genetics as a Risk Factor for Breast Cancer* (2024)
CDH1	The CDH1 gene produces a protein that facilitates cell adhesion for tissue formation in the body. Women with a mutated CDH1 gene face an elevated risk of developing invasive lobular BC, which is the second most prevalent form of BC. The lifetime risk for lobular BC is around 40%. It is important to note that CDH1 does not heighten the risk of ductal BC, which is the more common type of BC	*Genetics as a Risk Factor for Breast Cancer* (2024)
CHEK2	The CHEK2 gene produces a protein that inhibits tumor growth. A mutated CHEK2 gene can potentially double the risk of BC and also elevate the risks of colorectal and prostate cancer. Women with CHEK2 mutations and a family history of BC may face a lifetime risk ranging from 28–37% for developing BC	*Genetics as a Risk Factor for Breast Cancer* (2024)
STK11	The STK11 gene plays a role in controlling cell growth. An abnormal STK11 gene is linked to Peutz-Jeghers syndrome, a rare condition characterized by the development of hamartomatous polyps, primarily in the small intestine, as well as in the stomach and colon. Aside from gastrointestinal cancers, individuals with Peutz-Jeghers syndrome also face elevated risks of BC, lung cancer, and ovarian tumors. For women with this syndrome, the estimated lifetime risk of BC by the age of 70 is around 45%	*Genetics as a Risk Factor for Breast Cancer* (2024)
NF1	A mutation in the NF1 gene leads to a condition known as neurofibromatosis type 1, raising the likelihood of cancers in the central nervous system. Women carrying an NF1 mutation face a somewhat elevated risk of experiencing BC, with an 18% risk over their lifetime. Several studies propose an increased susceptibility to BC, particularly before reaching the age of 50, among women with an NF1 mutation	*Genetics as a Risk Factor for Breast Cancer* (2024)

(continued)

Table 3 (continued)

Gene mutation	Description	References
NBN	The NBN gene regulates the synthesis of a protein known as nibrin, responsible for repairing DNA damage within cells. A defective NBN gene leads to Nijmegen breakage syndrome, a condition associated with stunted growth in infancy and early childhood. Individuals affected by Nijmegen breakage syndrome face an increased susceptibility to various cancers, such as BC and encounter numerous other health challenges. While research is restricted, existing studies indicate that individuals with specific NBN mutations may face a two- to threefold higher lifetime risk of developing BC	*Genetics as a Risk Factor for Breast Cancer* (2024)

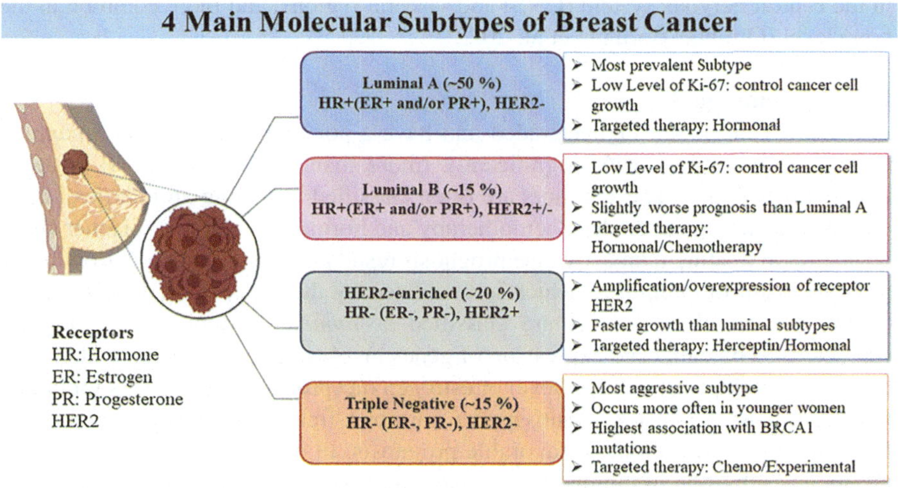

Fig. 2 Molecular subtypes of BC

focusing on hormone receptor expression such as and human epidermal growth factor 2, estrogen, and progesterone. Consequently, four well-recognized subtypes emerge: *HER2-positive*, *triple-negative*, *luminal A*, and *luminal B*. A visual depiction of the many biological types of carcinomas of the breast is presented in Fig. 2. Since ER serves as a crucial diagnostic factor, given that around seventy to seventy-five percent of invasive breast carcinomas exhibit notably elevated ER expression. More than half of ER-positive patients and very seldom those with ER-negative BC have PRs. Because ER affects PR expression control, physiological PR levels serve as a barometer for the health of the ER pathway. Because of their abundant presence in BC cells, both ER and PR are recognized as diagnostic and prognostic biomarkers for BC.

The features of these four subtypes are presented in Table 4.

Luminal Subtype A

Luminal A tumors lack HER2 expression, have ER and PR, and have a small amount of the cell growth indicator Ki-67 (less than 20%). Clinically, they are described as fragile, mild, and associated with the best prognosis—that is, higher rates of survival and fewer rates of recurrence. These cancers react strongly to hormone treatment (tamoxifen or aromatase inhibitors), whereas the effects of chemotherapy are quite small. Therefore, in order to determine which patients in this group would profit from receiving chemotherapy as an adjuvant depending on the likelihood of a recurrence and the overall survival rate, the use of genetic platforms is advised by the National Comprehensive Cancer Network (NCCN) and the European Society for Medical Oncology (ESMO) released recommendations. Relapses are less common in the central nervous system (CNS) and visceral regions, and more common at the bone level (Orrantia-Borunda et al. 2022).

Luminal Subtype B

Luminal B cancers are higher graded and have a poorer prognosis than *luminal A* tumors. They express Ki67 significantly (more than 20%) and are ER-positive, maybe PR negative. These tumors are often classified as having an intermediate to high histopathological grade. Chemotherapy and hormonal treatment together could be helpful in treating them. A worse prognosis results from the increased growth rate in comparison to *luminal A*, which is facilitated by the raised Ki67 levels. Ten to twenty percent of all tumors are classified as *luminal B* tumors, and they are characterized by enhanced expression of genes involved in cell division and growth as well as a comparatively low level of estrogen receptor expression. This grouping, which exhibits a greater sensitivity to hormone treatment, is representative of luminal tumors with the least favorable prognosis. In comparison to the *luminal A* group, this subgroup of luminal tumors has a higher reactivity to hormone treatment and chemotherapy, which indicates that they have the least favorable prognosis. *Luminal B* tumors have a higher risk of visceral recurrence in spite of frequent bone recurrence, and the survival time between identification and relapse is reduced (Orrantia-Borunda et al. 2022).

HER2 Subtype

Ten to fifteen percent of BC fall into the *HER2-positive* group, which is defined by increased human epidermal growth 2 expression and the absence of PR and ER. The development of HER2-targeted drugs has improved the prognosis, and these tumors grow faster than luminal types. The *HER2-positive* subtype is distinguished by its quick development and increased aggressiveness. Two different groupings are identified within this subtype: luminal HER2 (PR+, E+, HER2+, and Ki-67: 15–30%) and HER2-enriched (E−, HER2+, PR−, Ki-67 >30%). The prognosis for both subgroups is worse than that of luminal tumors, requiring the use of specific

Table 4 Characteristics of subtypes of BC

	TNBC	HER2	Luminal B	Luminal A	References
Frequency (%)	15	20	15	50	Barzaman et al. (2020), Poudel et al. (2019)
ER	No	Some cases	Yes	Yes	*Luminal A Breast Cancer and Molecular Assays: A Review ǀ The Oncologist ǀ Oxford Academic* (n.d.)
PR	No	Some cases	Some cases	Yes	Barzaman et al. (2020)
HER	No	Yes	No	No	*HER2-Positive Breast Cancer ǀ Drugs* (n.d.)
Ki67	High	High	Some cases	Some cases	(*Assessment of Ki67 in Breast Cancer: Updated Recommendations From the International Ki67 in Breast Cancer Working Group ǀ JNCI: Journal of the National Cancer Institute ǀ Oxford Academic* (n.d.)
Mutations	P53 and BRCA1	p53	BRCA2	No	Karami and Mehdipour (2013), *Mutant P53 in Breast Cancer: Potential as a Therapeutic Target and Biomarker ǀ Breast Cancer Research and Treatment* (n.d.)
Prognosis	Bad	Middle/bad	Middle	Good	Ahern et al. (2014)
Therapy	Chemo/experimental	Hormonal/chemo/herceptin	Hormonal/chemo	Hormonal	*Understanding Resistance to Tamoxifen in Hormone Receptor–Positive Breast Cancer ǀ Clinical Chemistry ǀ Oxford Academic* (n.d.)

medications that target the HER2/neu protein, such as surgery, customized chemotherapy, pertuzumab, and emtasin with trastuzumab (T-DM1), as well as medications that inhibit tyrosine kinase such as lapatinib and neratinib. Their response rate to chemotherapy regimens is high.

TNBC Subtype

About 20% of all instances of BC consist of the *TNBC* subtype. Most women who experience it are African American women and those under 40. Basal-like subtypes such as BL1 and BL2, mesenchymal (MES), luminal androgen receptor (LAR), claudin-low, and immunomodulatory (IM) are the other subgroups that make up the *TNBC* subtype; the first two represent fifty to seventy and twenty to thirty percent of cases, accordingly, making them the most prevalent. Furthermore, the phenotypic, pharmacological sensitivity, and clinical consequences of each of them are distinct. According to research, eighty percent of breast carcinomas (tumor inhibitor genes BRCA1 and BRCA2) exhibit aggressive behavior. The following factors affect one's chance of having *TNBC*: genetics, race, age, obesity and overweight, breastfeeding habits, and parity.

Aggression, rapid recurrence, and a higher propensity to appear in severe phases are the hallmarks of *TNBC*. It multiplies quickly, has changed DNA repair genes, and is more unstable in the genome. It is a heterogeneous, rapidly growing, poorly defined with varying prognostic subgroups, according to histology. According to immunohistochemistry (IHC), they may be further classified as basal and non-basal *TNBC*. The non-basal do not express CK5/6 cytokeratins, whereas the former are distinguished by the expression of CK5/6 cytokeratins and human epidermal growth factor receptor type 1 (EGFR1) (Orrantia-Borunda et al. 2022).

3 Molecular Biomarkers in BC

3.1 Hormone Receptor Status (ER and PR)

Certain types of BC cells require progesterone or estrogen, which the body naturally produces. Hormone receptors are unique proteins found in some cancer cells. The proliferation of cancer cells that have those receptors is accelerated by hormones that bind to them. Hormones binding to these receptors accelerate the rapid growth of cancer cells, possessing them. A pathologist examines the tumor tissue extracted during a biopsy to ascertain the hormone receptor status. Testing for the presence of hormone receptors is a basic part of treatment for all cases of BC. BC cells may express one, both, or none of these receptors. Hormone receptor–positive people have high levels of hormone receptors since they are positive for both progesterone (PR-positive) and estrogen (ER-positive) receptors. The absence of progesterone (PR-negative) and estrogen (ER-negative) receptors, which denote a shortage or absence of hormone receptors, characterizes negative tumors (*HER2-Positive Breast Cancer | Drugs* n.d., p. 2).

Physicians must determine if BC is hormone receptor positive to select the most appropriate treatment plan. If it is found that the cancer contains one or both of these hormone receptors, medical experts may employ hormone therapy medicines to reduce estrogen levels or stop it from impacting BC cells. This kind of treatment works effectively for breast tumors that are hormone receptor positive. However, it is ineffective against hormone receptor–negative tumors, which are tumors that lack both PR and ER. In all cases of invasive BC, both hormone receptors need to be inspected, either during the biopsy or after the tumor is surgically removed. At least one of these receptors is present in almost three out of every four cases of BC (*Descriptive Analysis of Estrogen Receptor (ER)-negative, Progesterone Receptor (PR)-negative, and HER2-negative Invasive Breast Cancer, the So-called Triple-negative Phenotype – Bauer – 2007 – Cancer – Wiley Online Library* 2023).

3.2 HER2 Expression Status

A protein called human epidermal growth factor receptor 2 (ERBB2 or HER2/neu), which encodes a 185 kDa glycoprotein, can be found on the exterior of certain BC cells. Among the most well-researched growth factor receptor systems in BC involves HER2. The complicated regulation of cell growth, survival, and differentiation is significantly influenced by the HER family. There are a number of ligands known to activate specific HER receptors, but none known to bind to the HER2 receptor directly. However, in response to ligand interaction, cell surface HER receptor monomers form heterodimers with different members of the HER family or homodimers with other receptors. Indeed, as compared to other homodimer/heterodimer combinations, heterodimers including HER2 seem to exhibit a particularly high signaling strength, which could account for HER2's notably important role in the oncogenic phenotype. Studies conducted on animals and in vitro have shown that oncogenic transformation, cancer, and metastasis are significantly influenced by HER2 gene amplification and protein overexpression. HER2 amplification of the gene can occur in a reasonably high percentage of several cancer forms, for example, 20–30% of breast and ovarian carcinomas. Moreover, anti-HER2 monoclonal antibodies (MAbs) block the proliferation of tumors and human breast cancer cell lines that overexpress the HER2 receptor, providing a new direction for targeted cancer treatment (Dowsett et al. 2000). Triple-positive BC is defined as BC that is positive for PR, HER2, and ER. On the other hand, BC categorized as triple-positive exhibit positivity for PR, HER2, and ER. Apart from providing information regarding BC, HER2 status may be considered a potentially measurable biomarker that provides information about an individual's overall health. HER2-positive BC make up from 10% to 20% of newly diagnosed cases (*HER2-Positive Breast Cancer | Drugs*, n.d., p. 2). HER2 status has often been assessed retroactively utilizing freshly frozen/paraffin-embedded tissue and IHC techniques that identify overexpression of the HER2 protein. To use HER2 status as a prognostic or predictive indicator in addition to direct therapy options, it is becoming increasingly crucial to determine it prospectively. To ascertain HER2 status, a plethora of methods have been employed and are presently under development; however, certain methods are more appropriate for scientific investigations than for routine patient screening (Dowsett et al. 2000). The two basic tests for HER2 status are (1) IHC, which counts the number of HER2 protein receptors in tumor cells and (2) fluorescence in situ hybridization (FISH), which counts the number of HER2 genes in cancer cells (*HER2-Positive Breast Cancer | Drugs*, n.d., p. 2). In situ hybridization (ISH) is a direct test for HER2 status; however, due to its higher failure rate, longer procedure time, and higher reagent cost when compared to IHC, numerous laboratories are currently using IHC as a screening method FISH is a confirmation procedure for HER2 IHC equivocal cases. Furthermore, HER2 protein overexpression by IHC and gene amplification by FISH have been found to be highly concordant (Couturier et al. 2000).

3.3 PIK3CA and Other Actionable Mutations

The initiation and progression of neoplastic growth are associated with the disruption of the mechanistic target of rapamycin (mTOR)/protein kinase B (AKT)/ phosphoinositide 3-kinase (PI3K) pathway. The mTOR-AKT-PI3K axis stands out as one of the frequently dysregulated pathways in BC (Naito and Urasaki 2018).

3.3.1 PIK3CA-Activating Mutations: Tier Ia

The PI3K pathway is among the extensively investigated signaling pathways, demonstrating relevance to oncogenic characteristics across various malignancies. Notably, the development of BC tumors is believed to depend on the proper functioning of the PI3K pathway. The statement arises from noting that a significant number of cases of this condition show no less than one molecular mechanism that could enhance the pathway. These mechanisms of PI3K enhancement encompass mutations within the PI3K gene, specifically in the form of PIK3CA gene mutations (Mukohara 2015).

Around forty percent of hormone receptor–positive BC cases have been recognized with mutations that activate in the PIK3CA gene, signifying a confirmed biomarker and a target that is firmly established in metastatic BC. The pivotal involvement of the PIK3CA gene, in conjunction with its downstream signaling pathways primarily implicating AKT1 and mTOR, has a major part in the development of BC tumors and may potentially contribute to the elicitation of resistance to endocrine therapies (Araki and Miyoshi 2018). Somatic mutations affecting the coding of p110α in the PIK3CA gene were initially documented in a number of solid tumors in the year 2004 (Samuels et al. 2004). The preponderance of somatic mutations in PIK3CA concentrates within two distinct "hot spots": H1047L/ H1047R in exon 20 and E545K/E542K in exon 9. Each type of these mutations has been demonstrated to exhibit gain-of-function characteristics and transformative potential. Mutations in exon 9 are situated within the p110α helical domain, wherein they have been suggested to confer the ability for p110α to evade inhibitory effects exerted by p85 through the Src-homology 2 (SH2) domain (Isakoff et al. 2005). On the contrary, mutations occurring in exon 20 are situated in close proximity to the kinase domain's activation loop. Nevertheless, it is yet unknown how precisely these mutations facilitate persistent PI3K signaling (Huang et al. 2007; Miled et al. 2007). Empirical validation has demonstrated the responsiveness of these mutations to alpelisib, a class I p110α-isoform-targeting selective inhibitor (Crimini et al. 2021).

3.3.2 PTEN-Inactivating Mutations/Loss: Tier IIa

A tumor suppressor gene located at the 10q23 region is identified as PTEN (phosphatase and tensin homolog deleted on chromosome ten), often referred to as TEP1 (tensin-like phosphatase-1)/MMAC1 (mutated in multiple advanced cancers-1) (Risinger et al. 1997). PTEN emerges among the most prevalently mutated tumor suppressor genes, exhibiting an incidence rate comparable to that of p53 in human cancer. The occurrence of mutations in both alleles of the PTEN gene is discerned throughout cancer progression across diverse tumor categories, encompassing, but not restricted to, brain, prostate, breast, endometrial cancer, and melanoma, manifesting at an approximate frequency of 50% (Simpson and Parsons 2001). Tyrosine phosphate and serine/threonine phosphate residues on protein molecules can be dephosphorylated by phosphatases, which are protein products that display sequence similarity and double specificity and are encoded by PTEN. Notably, PTEN possesses the capability to dephosphorylate inositol phospholipids as well. The data suggests dephosphorylation inositol phospholipids is not a necessary function of PTEN in its tumor suppressor role, since PTEN mutants originating from Cowden disease patients and sporadic tumors maintain protein phosphatase activity while losing this ability. Specifically, PTEN operates in the dephosphorylation of 3,4,5-triphosphate phosphatidylinositol 3,4-bisphosphate (PIP3 and PIP2), the synthesis of which is facilitated by phosphoinositide 3-kinase (PI3K) (Panigrahi et al. 2004). Cumulative evidence suggests that they are involved in AKT activation, a proto-oncogene (known as PKB). When phosphorylated AKT, or pAKT, is activated, the threonine/serine kinase AKT simultaneously promotes the progression of the cell cycle and prevents apoptosis (Franke et al. 1997). Hence, the essential function of PTEN in vivo centers around inhibiting PI3K-dependent activation of AKT. PTEN loss is categorized within tier IIa, recognizing the caveat that existing evidence is derived from data where alterations in PIK3, PTEN, and AKT are amalgamated. Moreover, there is an absence of data pertaining to the response of AKT inhibitors specifically in the context of PTEN alteration alone, and a dearth of comparative activity information with regard to PIK3 and AKT mutations (Crimini et al. 2021).

3.3.3 AKT1-Activating Mutations: Tier IIb

The initiation of AKT activation occurs through membrane localization facilitated by the interaction between the phosphatidylinositol-3,4-bisphosphate or phosphatidylinositol-3,4,5-trisphosphate and the pleckstrin homology domain (PHD). The regulatory amino acids serine 473 (Ser473) and threonine 308 (Thr308) are then phosphorylated following this (Vivanco and Sawyers 2002). The pathogenic connection between the plasma membrane and AKT is a consistent feature linking AKT to cancer. AKT's centrality in human tumors arises predominantly from the recurrence of mutations observed in enzymes responsible for the synthesis or degradation of evanescent second messenger phospholipids. Specifically, phosphatidylinositol-3,4,5-trisphosphate or its metabolite, phosphatidylinositol-3,4-bisphosphate, functions as crucial intermediaries in this context. These phospholipids, in turn, indirectly instigate the activation of AKT through the process of membrane recruitment. It is noteworthy that BC tumors in afflicted individuals frequently manifest triggering somatic mutations in phosphatidylinositol-3-OH kinase's (PI3K) catalytic subunit, PIK3CA. PI3K is the pivotal enzyme orchestrating the synthesis of D3-phosphorylated phosphoinositides, further underscoring the intricate molecular landscape associated with AKT activation in the context of BC pathogenesis (Carpten et al. 2007). As a consequence, it appears that AKT1 assumes a predominantly facilitative function in cancer growth, functioning as a mediator in an indirect manner positioned between downstream signaling molecules and mutated upstream regulatory proteins. This suggests that AKT1 serves as an intermediary component, indirectly transducing signals from upstream genetic alterations to downstream elements in the oncogenic pathway (Cantley 2002).

Notably, AKT1 mutations, recently discerned in breast, ovarian, and colorectal carcinoma, are localized within the pleckstrin homology domain (specifically E17K), as opposed to being situated in a kinase domain (Dunlap et al. 2010). The pleckstrin-homology domain assumes a central role in facilitating the localization of AKT1 to the cell membrane and its interaction with phosphatidylinositol 3-phosphate (PIP3). The activating mutation E17K induces modifications in the lipid-binding pocket, leading to heightened membrane localization and subsequent activation of AKT1. Different isoforms of AKT are linked to diverse cellular phenotypes, with AKT1 associated with cell survival and growth in an environment-dependent manner, while AKT2 is correlated with invasiveness. There exists a concern that inhibiting AKT1 may give rise to compensatory overactivation of AKT2, potentially fostering invasion and metastasis (Troxell 2012). In a nuanced examination, AKT1 exerts influence on cell proliferation by modulating proteins involved in cell cycle such as cyclin D1, p27, and p21 while concurrently inhibiting apoptosis, notably through interactions with p53. Paradoxically, AKT1 diminishes the migratory capacity of BC cells, partly achieved by the regulation of TSC2, palladin, and epithelial-mesenchymal transition (EMT)-related proteins. Conversely, AKT2 plays a major role in promoting invasion and migration, by controlling the dynamics of F-actin, EMT-associated proteins, and β-integrins. While there is evidence linking AKT3 to a negative-status estrogen receptor (ER), there are few thorough studies examining AKT3's regulatory function in important BC characteristics. Consequently, specific mutations in AKT1 and amplification of AKT2 are observed in certain BC cases, and the various AKT isoforms are linked to overall survival and therapeutic responsiveness in a manner specific to each isoform (Hinz and Jücker 2019).

3.4 Hereditary Breast Cancer Risk and *BRCA1* and *BRCA2* Mutations

Mutations in the tumor suppressor genes BRCA1 and BRCA2 significantly increase the chance of developing some epithelial malignancies, including BC. The autosomal dominant inheritance pattern of the hereditary BC syndrome, which arises from mutations in these genes, accounts for around 50% of the cancer cases linked to inherited genetic susceptibility (Mahdavi et al. 2019). The 22 exons that make up the BRCA1 gene encode a nuclear protein that weighs 220 kDa and has 1863 amino acids. The ovarian and breast tissues are among the tissues in which the BRCA1 gene is expressed. A missense substitution, an 11-base pair deletion, a stop codon, a 1-base pair insertion, an assumed regulatory mutation, and a missense substitution were among the first mutations found in the BRCA1 gene. Subsequently, it was discovered that 38 of the 63 mutations reported in a thorough screening of the BRCA1 gene were prevalent. Eight, seven, or five observations of these specific alterations were made each. Interestingly, a shortened form of the BRCA1 protein was predicted by 86% of these frequent mutations. There are currently over 1600 identified mutations in the BRCA1 gene, most of which cause frameshifts that result in missense or non-functional proteins (*BRCA Genes: The Role in Genome Stability, Cancer Stemness and Therapy Resistance – PMC*, 2023).

The BRCA2 gene is larger than the BRCA1 gene, with an open reading frame (ORF) of 10.3 kb. It codes for a 384 kDa nuclear protein. Because the proteins that the BRCA1 and BRCA2 genes encode have comparable functions, mutations in these genes cause a unique genetic propensity to BC. Six different germline variants were found in BRCA2 that were linked to BC families in the early studies. Usually, these mutations cause the transcriptional unit 17's open reading frame to be disrupted. Currently, more than 1800 BRCA2 mutations have been identified. Among these mutations are ones that result in premature protein truncation, nonsense, and frameshift deletion (Godet and Gilkes 2017).

4 Liquid Biopsies

4.1 Liquid Biopsy and Its Clinical Utility

Liquid biopsies provide a less intrusive method of obtaining information about tumors from body fluids. Blood is the most often used medium for biopsies, however, other bodily fluids can also be used. Hematological specimens can serve as a source for the isolation of extracellular nucleic acids present in the circulatory system, including circulating tumor DNA and cell-free DNA. While circulating tumor DNA only comes from tumors, cell-free DNA is the DNA that is freely flowing through the circulation and may have originated from other origins or tumors. Moreover, the isolation of circulating tumor cells (CTCs) from blood is feasible. These cells are released from neoplastic masses into the circulatory system, usually with a brief half-life of 1–2 hours before the immune system eliminates them.

CTCs can be extracted through various techniques, including methods based on immunogenicity, positive selection, negative selection, and enrichment utilizing biophysical attributes such as size and density. However, one frequently used technique for separating CTCs is immunogenicity-based enrichment. There are several methods available for positively enriching CTCs. The AdnaTest utilizes beads coated with antibodies specific to cancer and employs real-time polymerase chain reaction to discern expression patterns in cells. Magnetic-activated cell sorting is employed for cell capture, utilizing magnetic nanoparticles conjugated with antibodies. MagSweeper utilizes immune magnetic enrichment technology, employing a robotically controlled magnetic rod and beads coated with antibodies to isolate CTCs. There exist two predominant methodologies for negative enrichment in the isolation of CTCs. The EasySep system employs a technology involving the incubation of samples with antibodies specifically targeting CD45 cells, accompanied by magnetic nanoparticles. Conversely, the quadrupole magnetic separator operates as a magnetic flow cytometry system. It is designed to identify immune cells that have been magnetically labeled. In an alternative approach, the isolation of CTCs can be achieved through the application of biophysical principles. CTCs, typically exhibiting larger size compared to surrounding normal cells, can be selectively separated using centrifugation, microfiltration, and dielectrophoresis. Subsequent identification of the isolated CTCs can be accomplished through the application of DNA, RNA, or protein analysis techniques. In the assessment of circulating tumor DNA and cell-free DNA, diverse methodologies have been employed, encompassing approaches such as whole genome bisulfite sequencing, magnetics, beads, amplification, emulsion, cancer-personalized profiling by deep sequencing, tagged-amplicon deep sequencing, and whole genome and whole exome sequencing. Digital droplet polymerase chain reaction exhibits high sensitivity in discerning genomic material at exceptionally low concentrations, spanning from 0.01% to 1.0%. This attribute renders it instrumental in the identification of uncommon mutations and the quantification of copy number variants. Conversely, BEAMing, an economically efficient technique, demonstrates heightened sensitivity in screening established mutations. This methodology utilizes a composite approach involving polymerase chain reaction (PCR) and flow cytometry to detect alterations at concentrations as minimal as 0.01%. Notably, BEAMing exhibits commendable concordance with tissue testing methodologies. Tam-Seq provides a precise and sensitive analytical approach, achieving an accuracy of approximately 97%. It can discern DNA concentrations as minimal as 2% by employing primers for the tagging and identification of genomic sequences. With the use of distinctive patient sample sequence signatures and large genomic libraries, CAPP-Seq detects changes in ctDNA/cfDNA. A thorough examination of all tumor mutations is possible using whole exome sequencing, which makes it possible to identify putative tumor suppressor and oncogene genes. Whole genome sequencing (WGS) evaluates the complete genome of the tumor, identifying known harmful changes as well as variations with unclear biological relevance. While possessing the capacity to furnish an exhaustive examination of all mutations in tumors, this approach is constrained by challenges associated with quality assurance, ethical implications, financial considerations, and temporal constraints. Moreover, the interpretation of the data may pose difficulties when performed outside specialized facilities. Since WGBS-Seq offers a single cytosine measurement with extremely high precision, it is regarded as the gold standard in DNA methylation studies (Nikanjam et al. 2022). There is a wide range of clinical uses for circulating tumor DNA, cell-free DNA, and CTCs. This section covers the following topics:

(a) **Distinguishing Distinct Response Patterns to Immune Checkpoint Blockade Through Successive Liquid Biopsy Analyses**

Because immune checkpoint inhibitors can stimulate the immune system and perhaps eliminate resistant tumors, their usage in oncology is growing. Investigations have been conducted concerning circulating tumor DNA and cell-free DNA to enhance conventional imaging studies and offer initial evaluations of immune checkpoint responses. The unique reaction patterns linked to immunotherapy may make it more difficult to assess a patient's response clinically. For instance, notwithstanding a favorable clinical response to therapeutic intervention, false progression characterizes an initial perceptible enlargement of the neoplasm on diagnostic imaging. This occurrence, wherein the tumor manifests an augmented size on imaging despite undergoing regression, may be associated with inflammation stemming from immune system activity. The term "hyper-progression" refers to a condition where immune checkpoint inhibition causes tumor development to accelerate. Finding early prognostic indicators might help patients who are experiencing hyper-progression to switch to more effective treatments in a timely manner before waiting for imaging findings. To facilitate therapeutic decision-making in tandem with imaging examinations, circulating tumor DNA or cell-free DNA can provide timely and evolving assessments of response and progression in the context of immunomodulatory checkpoint therapy (Nikanjam et al. 2022).

(b) **Assessing Possibly Actionable Molecular Abnormalities**

The recognition of alterable variations in metastatic solid tumors holds paramount significance in the formulation of a precision medicine therapeutic approach. FDA has sanctioned several companion diagnostic assays utilizing blood-derived circulating tumor DNA. Among these assays is the Therascreen PIK3CA RGQ PCR kit, designed for the identification of mutations in the PIK3CA gene associated with BC. Moreover, the Foundation One liquid biopsy and Guardant360 liquid biopsy have received approval from the FDA. These diagnostic methods employ NGS techniques to analyze circulating tumor DNA. These are associated with various targeted therapies and function as auxiliary diagnostics for comprehensive molecular profiling. Patients' therapeutic decisions may be influenced by their approval, which may also open up opportunities for them to take part in clinical studies. In order to inform treatment decision-making, liquid biopsy is a useful technique that may be used in conjunction with tissue biopsy molecular profiling (Nikanjam et al. 2022).

(c) **Anticipating Early Responses and Resistance Following Targeted Therapy and Chemotherapy Through Successive Liquid Biopsy Analyses**

Because metastatic solid tumors are diverse, multiple genomic profiles may be obtained by sampling different tumor sections and disease locations. In order to ascertain the optimal therapeutic approach for patients, it is imperative to furnish a comprehensive molecular characterization of the entire tumor. Given intra-tumor heterogeneity, a single-tumor biopsy sample may result in an underestimate of the tumor genetic landscape. For a more comprehensive view, ctDNA or CTCs can be assessed as they can be released into the bloodstream from several metastatic locations. A preceding investigation demonstrated the capability of circulating tumor DNA to facilitate dynamic monitoring of polyclonal tumor progression at multiple foci. This was achieved through the concurrent analysis of circulating tumor DNA and plasma specimens in a patient diagnosed with advanced BC having distant metastases (Murtaza et al. 2020). Present research indicates that ctDNA and CTCs may offer molecular insights into the disease overall; yet, it is unclear if many metastatic lesions spread across various organs would uniformly release ctDNA and CTCs. A safer and more affordable alternative to tissue biopsies of several tumor locations, ctDNA, and CTCs may give a more thorough picture of the whole cancer's genome landscape (Nikanjam et al. 2022).

5 Fundamentals of Next-Generation Sequencing (NGS) and Genomic Profiling

The preliminary wave of sequencing the Human Genome Project (HGP) took over a decade and primarily relied on automated sequencing techniques, which Sanger initially laid out in 1977. Although the method of sequencing was highly successful in constructing early maps of the human genome, there was an urgent need to develop more reliable sequencing technologies to produce enormous amounts of data at a fast and cheaper cost. Recognizing this gap, the National Human Genome Research Institute (NHGRI) launched a financing project in 2004 with the objective of accelerating the research and development of sequencing technologies and bringing down the cost of genome sequencing. Sequencing millions of copies of DNA in parallel at significantly lower costs than traditional methods is made feasible by new technologies that offer dramatically different approaches. While alluding to these modern high-throughput techniques, the words "massive parallel sequencing" and "next-generation sequencing" have been used simultaneously (Rizzo and Buck 2012). The concepts and advancements in nanotechnology that made it possible to sequence individual DNA molecules in parallel were responsible for the notable spike in data production. NGS is distinguished by its unique combination of high throughput and single-molecule DNA sequencing capabilities, regardless of the sequencing platform (Hu et al. 2021). The main distinction is that massively parallel sequencing techniques do not separate enzymatic nucleotide incorporation from sequence ladder separation and data gathering, in contrast to Sanger sequencing.

Conceptually speaking, only three kinds of molecules are needed for DNA replication: a template strand, free bases, and a polymerase enzyme, which joins the free bases one at a time to create a new strand complementary to the template. Using unextendable, fluorescently labeled modified bases is the crucial invention that turns DNA replication into the DNA-sequencing approach at the heart of both Sanger and NGS. Only a tiny portion of bases are altered during Sanger sequencing; however, all bases are altered by NGS. Thus, the primary task of the sequencer is to arrange molecules so that the fluorescence signal can be deciphered. NGS employs positional separation to organize the fluorescent molecules instead of adopting size separation (*Understanding the Basics of NGS: From Mechanism to Variant Calling | Current Genetic Medicine Reports* n.d.). Since NGS technology carries out the enzymology and data collecting in a coordinated and sequential manner, making it possible to concurrently generate sequence data from tens of thousands to billions of templates. Because of this increased ability to generate data, the phrase "massively parallel" now describes how DNA sequencing and its uses have changed dramatically since its inception (McCombie et al. 2019).

Genome-wide NGS, a technique known as "genomic profiling," is being utilized more frequently to identify various facets of genetic heterogeneity in a variety of human disorders, including cancer (Shyr and Liu 2013). Through the deployment of NGS, the Cancer Genome Atlas (TCGA) project was a comprehensive and integrated attempt to advance the understanding of the molecular underpinnings of numerous cancer forms. The gathered data provided insights into the tight relationships between the evolutionary development of cellular processes and tumor genetics across a variety of cancer types (Akbani et al. 2014).

This is demonstrated by the widespread use of NGS in clinical settings for cancer evaluation and future prediction, as well as a significant increase in the amount of novel medication discoveries that target certain genes or/and biological pathways and are the subject of ongoing clinical trials. Moreover, the applications of NGS to generate genetic markers for precision medicine have significantly increased (Zhang et al. 2018).

5.1 Applications of NGS in Breast Cancer Research and Clinical Practice

Genomic investigations encompassing WGS for individual patients or diverse organismal cohorts, exome sequencing, focused re-sequencing endeavors, and functional inquiries emphasizing DNA-protein interactions through methodologies such as chromatin immunoprecipitation are merely a few of the numerous uses covered by second-generation sequencing that might advance the discipline of personalized medicine. Considerable endeavors have been dedicated to enhancing our understanding of BC and the genetic modifications instigating tumorigenesis. This includes elucidating mechanisms such as cell proliferation, migration, metastasis, and angiogenesis, prompted by the notable surge in the prevalence of this malady. Biochemical organizations were able to make use of the extensive knowledge gathered from the HGP to develop commercially accessible kits that are specially designed for the use of NGS in the research of BC and cancer in general. Many cancer panels are already available on the market that are intended to identify tumor-specific genomic changes by means of targeted resequencing of cancer genomes (Meldrum et al. 2011). Next-generation sequencer manufacturers make several kits that can only be used with their own gear. Two examples are the TruSeq oncology panel from Illumina and the IonAmpliseq oncology panel from Life Technologies. However, generic kits that are compatible with a variety of instrument types are available. For instance, Multiplicom developed BRCA MASTR™ Dx especially to detect genetic alterations in BRCA1 and BRCA2, associated with hereditary susceptibility to BC (PETRIC et al. 2015).

Multigene massive sequencing with highly sensitive identification of gene variants, including complicated ones, and copy number changes was made possible by the development of NGS technology. Numerous studies illustrated the benefits of employing NGS for BRCA variant discovery over more conventional techniques in terms of cost, time, and sensitivity. Crucially, the latter is crucial when deciding on a course of treatment and taking into account the potential usage of anti-PARP medications in the treatment of BC (Zelli et al. 2020).

6 Targeted Therapies and Treatment Tailoring

6.1 Locoregional Treatment of BC

Over the last three decades, there has been a substantial change in the locoregional treatment of BC. The approach has evolved from universally advocating radical mastectomy to offering a range of choices such as whole-breast irradiation (WBI) coupled with breast-conserving therapy (BCT), immediate reconstruction following mastectomy, and mastectomy as a standalone option. The intensity of the illness and each patient's preferences are now taken into account while choosing a treatment plan (Ho and Morrow 2011).

Simultaneously, our viewpoint regarding the function of locoregional management has changed. While locoregional therapy had long been the exclusive approach to treating BC, the realization that radical mastectomy did not cure many patients with BC led to the acceptance that post-surgery treatment failure is often due to disseminated tumor cells present during the operation rather than the surgical procedure's lack of radicalism. This shift in thinking supports the belief that, while local therapy is crucial for reducing problems at the original tumor site, differences in local treatment are unlikely to significantly affect BC survival—aligning with Dr. Bernard Fisher's (Fisher 1979) "systemic disease" hypothesis. The article published by the Early Breast Trialists' Collaborative Group (EBCTCG) in 2005 revealed that significant variations of more than 10–20% in local control between treatments after 5 years correlate with statistically noteworthy changes in BC survival after 15 years. This finding has led us to reconsider and affirm the significance of sufficient locoregional therapy as a crucial factor in BC survival.

6.2 Systemic Therapy

Systemic medical therapy for BC comprises three primary methods: biological therapies, chemotherapy, and endocrine therapy. Out of these, the oldest one is endocrine therapy, which is crucial for tumor patients expressing progesterone/estrogen receptors, constituting around two thirds of BC cases. Conversely, chemotherapy, the least selective option, may soon transition from reliance on progression/relapse assessment of risk to personalized administration based on the unique gene expression patterns of individual tumors. Lastly, the therapy strategy that is progressing the fastest is biological therapies, encompassing drugs that have been approved like the anti-HER2 monoclonal antibody trastuzumab and bisphosphonates, along with an expanding range of novel drugs targeting molecular factors in the cancer cell and its surroundings, the roles of which are presently under evaluation (Piccart and Cardoso 2003).

6.2.1 Early-Stage Breast Cancer

Throughout the past 80 years, considerable improvements have been observed in the systemic therapy provided to patients diagnosed with early-stage BC (Smith et al. 2011). The objectives and frameworks of clinical investigations including (neo)-adjuvants have frequently developed during this timeframe. The motivation for the transition to the post-genomic era from the pre-genomic in clinical trials stems from a more profound comprehension of the biology of the disease and an intensified curiosity in identifying molecular indicators connected to the response to treatment (Collins et al. 2003).

The current method of designing neoadjuvant trials requires a fresh set of modifications incorporating more personalized treatment strategies, encompassing both increasing and decreasing escalations and emphasizing validated endpoints that have stronger predictive links to survival outcomes (Cortazar et al. 2014). Research into novel drugs tailored to individuals with high-risk diseases is becoming more and more popular in post-neoadjuvant settings as a result of the continuous transition toward a more personalized treatment approach (Gebhart et al. 2013).

The scientific community should have easy access to large, well-annotated databases containing candidate biomarkers associated with clinical outcomes to efficiently uncover novel biomarkers to guide customized treatment (Fisher 1979). The facilitation of early data sharing from clinical trials needs to be achieved through collaborative efforts and strengthened partnerships among academic entities and the pharmaceutical industry (Agostinetto et al. 2022).

6.2.2 Advanced Breast Cancer

Neoadjuvant chemotherapy (NAC) was predominantly suggested for locally advanced BC without distant metastasis. Presently, the objectives of NAC have undergone significant transformation based on the latest clinical trial data. The goal of NAC extends beyond merely raising the rate of breast-conserving surgery; it now encompasses PM, such as the utilization of adjuvant capecitabine or T-DM1, guided by the efficacy of NAC for early BC (Shien and Iwata 2020).

The selection of a treatment regimen seeks to achieve the highest possible anticancer impact based on cancer biology. In patients who are ER negative and/or HER2 positive, the response to NAC, particularly achieving a complete response (pCR), is a prognostic factor. Postmenopausal BC patients who are ER positive are eligible for neoadjuvant hormone therapy. Prospective trials are still being conducted to evaluate the safety of skipping surgery in patients achieving pCR in response to NAC.

6.3 Hormonal Therapy

Over the past decade, we have had the good fortune to see a fruitful exploration of novel hormone agents. Although information on emerging antiestrogens is still evolving (Howell et al. 2002), intriguing developments in endocrine therapy for BC have been spurred by evidence from third-generation aromatase inhibitors. Randomized aromatase inhibitor clinical trials conducted in postmenopausal women who did not respond to tamoxifen have included a minimum of 4249 participants.

In five of these studies, the comparator was megestrol acetate (Buzdar and Howell 2001), while aminoglutethimide was the comparator in two (Gershanovich et al. 1998). The recent aromatase inhibitors demonstrated superiority in every trial across various endpoints, including response rate (RR), time to treatment failure (TTF), overall survival (OS), time to progression (TTP), or safety. The consistently positive outcomes across trials provided level 1 evidence, supporting a definitive shift in standard clinical practice. Consequently, the suggested second-line endocrine therapy after tamoxifen is represented by third-generation aromatase inhibitors drugs (exemestane, letrozole, and anastrozole) (Hamilton and Piccart 1999).

The next rational progression involved advancing these new aromatase inhibitors to the forefront of endocrine therapy for a direct comparison with tamoxifen. These trials included 2696 women in total who had only minimal exposure to tamoxifen in an adjuvant context. The combined findings from two analogous double-blind trials indicate that anastrozole increases TTP in the patient subgroup whose hormone receptors are known to be positive (Bonneterre et al. 2000).

In a more compact double-blind phase II randomized study specifically for women diagnosed with hormone receptor-positive conditions, anastrozole-treated patients exhibited improved survival.

Letrozole, however, showed impressive advantages over tamoxifen in terms of TTF, TTP, RR, and survival (for a maximum of 2 years) in a huge, randomized, double-blind trial. As the disease progressed in this experiment, half of the patients switched to the other drug, mirroring real-world clinical practice (Mouridsen et al. 2001). An open-label, randomized phase II trial comparing exemestane to tamoxifen yielded promising results, leading to the trial's expansion into an ongoing phase III research (Paridaens et al. 2000). When considering these initial findings collectively, they present level 2 data that suggests aromatase inhibitor is preferable to tamoxifen, our former "gold standard" medication. Given the exceptional safety profile of these novel endocrine drugs, it is conceivable that routine clinical practice will change before the results of the exemestane trial are released. The letrozole/tamoxifen trial's survival curves converging after 2 years could indicate that tamoxifen performs somewhat less well when given after letrozole. Regardless, the results highlight the necessity of further clinical research aimed at developing more effective "salvage" therapies if aromatase inhibitors fail. Considering the imminent integration of aromatase inhibitors into adjuvant therapy for early-stage BC, these rescue therapies gain significance. Prospective options include emerging antiestrogens like Faslodex1, proven to be comparable to anastrozole in tamoxifen-resistant cases, and biological agents (e.g., Iressa1) capable of delaying and somewhat stopping the development of hormone tolerance (Howell et al. 2002). In addition to the positive outcomes and advancements in endocrine therapy for BC, it is essential to consider the long-term side effects and impact on the quality of life associated with hormonal agents. Although the safety profile of third-generation aromatase inhibitors—letrozole, anastrozole, and exemestane—is mentioned as exceptional, it is crucial to recognize potential issues that may affect patients undergoing these treatments. Hormonal agents, particularly aromatase inhibitors such as letrozole, exemestane, and anastrozole, present both promising outcomes and potential challenges for BC patients (Nabieva and Fasching 2021). Aromatase inhibitors use has been associated with heightened risks of bone loss and fractures, raising concerns about osteoporosis development and impacting the patients' quality of life. Moreover, musculoskeletal symptoms, including joint pain and stiffness, may affect daily activities and overall well-being. Aromatase inhibitor induces a menopausal state, introducing symptoms like hot flashes, night sweats, and mood disturbances, further influencing the patients' quality of life. Tamoxifen, another widely used hormonal agent, is linked to an increased risk of uterine cancer with long-term use, posing implications for overall health. Similar to aromatase inhibitors, tamoxifen induces menopausal symptoms, affecting patient comfort. Emerging antiestrogens like Faslodex and biological agents like Iressa bring about uncertainties due to their unknown long-term effects, leaving patients in doubt regarding potential side effects and impacts on their quality of life. The mention of patients switching between drugs, as observed in trials involving letrozole, suggests challenges in adapting to different medications that may affect the continuity of care and patient well-being. The need for "salvage" therapies in case aromatase inhibitors fail underscores the uncertainty in the efficacy of subsequent treatments and potential side effects, adding complexity and potential distress to the patient's journey. Thus, while these hormonal agents offer significant advancements, the potential side effects and uncertainties surrounding emerging therapies necessitate ongoing research for a comprehensive understanding of their long-term safety profiles and their impact on patients' quality of life (Johnston and Cheung 2018).

6.4 Chemotherapy

Over the past decade, two prominent players have dominated the forefront: docetaxel and paclitaxel. Distinguished by their unique p53-independent (Fan et al. 1998) mechanism of action, these antimicrotubule drugs have been extensively investigated in various clinical trials, both as standalone treatments and with other compounds that have displayed cytotoxicity. Randomized trials have shown that among taxanes, docetaxel, when used alone, demonstrates the most well-established efficacy for patients who have already received anthracycline treatment or who are resistant to it. Docetaxel outperformed two widely used combinations (methotrexate + 5-fluorouracil (5-FU) and mitomycin + vinblastine), it also demonstrated to be equally effective as the third combination (vinorelbine + 5-FU) (Nabholtz et al. 1999). In a similar patient population, when used in conjunction with the novel antimetabolite capecitabine, docetaxel produced better results than when it was used alone (Leonard et al. 2001).

In one trial, paclitaxel showed superiority over cyclophosphamide methotrexate fluorouracil (CMF) prednisone in those who had little to no prior experience with anthracyclines. Paclitaxel, however, performed inconsistently when compared to doxorubicin, showing equivalency in one experiment and inferiority in another. Consequently, combining taxanes with anthracyclines seemed to be the next natural step. However, just one trial has reported improvements in survival thus far, and the recent treatment plans have not demonstrated consistent superiority when compared to non-taxane regimens (Jassem et al. 2001). It would be helpful to perform a meta-analysis on these data, to determine patient subgroups that may benefit most from an anthracycline-taxane mixture, and to reveal a slight but real survival advantage that could not be seen in individual trials. Meanwhile, when it comes to treating anthracycline-resistant diseases, taxanes are now preferred above alternative treatments for women whose metastases are life-threatening. However, whether they work better when used alone or in combination is still up for debate.

The actual influence of chemotherapy on the quality of life and prognosis of patients with metastatic breast cancer (MBC) is being evaluated until now, particularly when considering second and subsequent treatment lines (Carrick et al. 2009). These treatment plans consistently exhibit lower response rates and do not show a considerable effect on median survival. Presently, no "standard regimen" for second-line chemotherapy in MBC treatment has been established (Cardoso et al. 2002). A subset of women who are considerably challenging to treat include those who have received prior treatment or who have resistance to both taxanes and anthracyclines. Considering this, capecitabine stands out due to its extensively documented phase II action, an upcoming randomized European Organization for Research and Treatment of Cancer (EORTC) trial (Leonard et al. 2001) will evaluate its efficacy against the established drug vinorelbine.

Several recently developed cytotoxic drugs are undergoing clinical trials in MBC. The promising new family of anti-tubulin drugs that do not appear in preclinical models to show cross-resistance with taxanes (Bollag et al. 1995) is the epothilones and the multitargeted antifolate Alimta (Miles et al. 2001).

There may not be a significant improvement in the treatment of advanced BC with these drugs, if taxanes become an adjuvant in the following two to four years, they might prove to be indispensable. In such a situation, there would be a pressing need for active cytotoxic drugs that do not show cross-resistance would be desperately needed in such a situation, particularly for women who are having an early relapse following adjuvant chemotherapy.

6.5 Combination Therapy

6.5.1 Combined Hormone Therapy and Chemotherapy

Combined hormone therapy and chemotherapy operate through distinct mechanisms, each with its own spectrum of toxicity. It would seem promising to combine both treatment modalities in advanced BC in an attempt to improve upon the outcomes of combination chemotherapy over the previous 10 years, but this is not advised in advanced BC due to the fact that 19 trials conducted between 1977 and 1996 failed to show a discernible increase in survival associated with this strategy (Fossati et al. 1998). Notably, none of these trials used Faslodex1 or a third-generation aromatase inhibitor and were all underpowered and small. Reexamining this idea with more potent hormonal drugs and enhanced trial designs would be intriguing.

As we become aware of the crucial contribution of "optimal adjuvant hormone therapy" to increased survival of endocrine-responsive BC, the added advantage of chemotherapy needs to be carefully reviewed. In fact, there are low-risk subgroups for which the potential extra benefit may be negligible and not worth the short- and long-term adverse effects that come with it. Two complementary clinical trials were initiated by the International Breast Cancer Study Group (IBCSG) in 2003 among premenopausal women with hormonal receptor–positive BC—the tamoxifen and exemestane trial (TEXT) and suppression of ovarian function trial (SOFT) trials. The SOFT trial assessed the efficacy of ovarian ablation in young women who continue to have menstruation after adjuvant chemotherapy treatment when administered in conjunction with tamoxifen. Analogous to the first experiment, the TEXT trial evaluated the possible benefits of administering chemotherapy in addition to ovarian tamoxifen + ablation to a subset of premenopausal women for whom endocrine therapy has been designated as the principal adjuvant systemic treatment.

These trials aimed to answer questions about the role of aromatase inhibitors for women treated with ovarian function suppression (OFS) and the role of OFS for premenopausal women treated with tamoxifen. However, the trials faced challenges due to lower-than-anticipated disease-free survival (DFS) event rates and better-than-expected patient outcomes. To provide timely answers, protocol amendments in 2011 specified analyses based on chronological time and median follow-up. A combined analysis of TEXT and SOFT became the primary analysis for the aromatase inhibitor question (exemestane + OFS vs. tamoxifen + OFS), and the OFS question (tamoxifen + OFS vs. tamoxifen alone) became the primary analysis from

SOFT. The trials required additional years of follow-up to reach the targeted DFS events (Regan et al. 2013).

6.5.2 Combined Molecular Targeted Therapy and Endocrine Therapy

Combining endocrine therapy with molecular targeting compounds for the treatment of BC, especially metastatic HR+ and HER2− advanced-stage BC, involves different combination approaches, such as:

1. **PI3K/AKT/mTOR inhibitors plus hormone receptor inhibitors:** Activation of the PI3K/AKT/mTOR signaling pathway in BC led to the exploration of three CDK4/6 inhibitors and PI3K/AKT/mTOR inhibitors along with hormone therapy (fulvestrant). The study analyzed the efficacy and safety of various combinations, with abemaciclib plus fulvestrant showing the highest probability of treatment efficacy (SUCRA value of 85.29%) (Leung et al. 2021).
2. **CDK4/6 inhibitors plus hormone receptor inhibitors:** Combining endocrine therapy (AI or fulvestrant) with CDK4/6 inhibitors (palbociclib or abemaciclib) proved beneficial in increasing endocrine sensitivity. Patients in the aromatase inhibitor and CDK4/6 inhibitor combination therapy group showed a median progression-free survival (PFS) of 28 months compared to 14.9 months for the placebo plus endocrine therapy group. Similar positive outcomes were reported in various clinical trials, although not all trials were as successful (Gao et al. 2020).
3. **mTOR inhibitor plus hormone receptor inhibitors:** The combination of the mTOR inhibitor everolimus with the aromatase inhibitor letrozole was studied in patients with HR+, advanced-stage BC who had developed resistance to endocrine therapies. The trial demonstrated a significant increase in PFS from 9.0 months with letrozole alone to 22.0 months with the combination treatment. Similar positive results were observed in postmenopausal women with endocrine-resistant BC (Royce et al. 2018).
4. **PI3K inhibitors or dual PI3K/mTOR inhibitors plus AI:** Resistance to endocrine therapy in ER-negative BC, often associated with the activation of PI3K due to the lack of PTEN protein, was addressed by combining PI3K inhibitors (pilaralisib) or dual PI3K/mTOR inhibitors (voxtalisib) with the aromatase inhibitor letrozole. Clinical trials showed effectiveness in patients with HR+ and HER2− nonsteroidal aromatase inhibitor-refractory recurrent or MBC. However, challenges remain, and not all combinations yielded promising results (Cretella et al. 2018).

6.5.3 Combined Molecular Targeted Therapy and Chemotherapy

Several clinical trials are underway or have been completed to evaluate the efficacy of combination therapies for BC. Trial NCT02456857 is currently in phase II, assessing the combination of bevacizumab (VEGF inhibitor) and everolimus (mTOR inhibitors) with doxorubicin in locally advanced TNBC patients who are insensitive to standard chemotherapy. The preliminary outcomes reveal an objective response rate of 21%. Another completed phase II trial, NCT01281696, investigated the use of bevacizumab (VEGF inhibitor) in combination with etoposide and cisplatin for BC patients with brain metastases, reporting a PFS of 9.1 months and an overall survival (OS) of 10.7 months. Additionally, the ongoing phase III trial OlympiAD (NCT02000622) is exploring olaparib (PARP inhibitor) in combination with capecitabine, eribulin, or vinorelbine for metastatic breast cancer patients with a BRCA mutation, demonstrating a PFS of 7.0 months compared to 4.2 months with chemotherapy alone. These trials provide valuable insights into the potential benefits of combining targeted therapies and chemotherapy in diverse BC populations (Wang and Minden 2022).

The advantages of combined therapies include precise treatment directed toward specific tumor characteristics, synergistic or additive effects leading to prolonged survival, and reduced side effects and drug resistance compared to chemotherapeutic drugs. The specificity of combination therapies minimizes side effects, and studies have shown a reduction in drug resistance when certain drugs are combined. However, challenges such as drug resistance, risks of drug–drug interactions, and possibility of low long-term efficiency must be acknowledged. Drug resistance can still occur in some cases, and the risk of adverse effects due to drug interactions needs consideration. Additionally, the long-term efficiency of combination therapies requires further investigation, and ongoing animal studies need to advance to clinical trials for assessing effectiveness in human patients against different BC subtypes. While molecular combination therapy shows promise in improving BC outcomes, and numerous preclinical and clinical studies have indicated that it can be more effective than single-drug therapy, optimization and comprehensive personalized treatments are essential for addressing the challenges and ensuring the best possible results in the future (Mokhtari et al. 2017).

7 Multidrug Resistance in Breast Cancer

7.1 Overview of Multidrug Resistance in Breast Cancer

Multidrug resistance (MDR) stands as the predominant cause of mortality in patients with cancer, having primary causative factors encompassing heightened xenobiotic metabolism, increased ability to repair DNA, enhanced gene amplifying mutations, and greater drug permeability. Notably, changes in medication objectives constitute a significant contributory factor to the manifestation of MDR (Musyuni et al. 2022).

Chemoresistance poses a substantial challenge in treating BC therapeutically, with a notable occurrence of initially responsive tumors relapsing and acquiring resistance to multiple anticancer agents characterized by diverse structures and mechanisms of action. The precise attributes defining resistance to chemotherapy and the potential engagement of genes linked to drug resistance, notably those involved in the transport of anticancer agents, remain elusive and require further elucidation. The intricate mechanisms underlying resistance phenomena, including genetic factors contributing to altered transport processes impacting the efficacy of anticancer drugs, have not been fully characterized and necessitate in-depth investigation for a comprehensive understanding. Enhanced comprehension of the molecular mechanisms underpinning resistance to chemotherapy is imperative to formulate efficacious therapeutic approaches aimed at surmounting multidrug resistance (Perez 2009).

Resistance to drugs can manifest through diverse mechanisms. One mechanism involves an elevation in the rate of activity of ATP-dependent outflow pumps, leading to diminished intracellular drug concentrations. Noteworthy agents linked to this form of resistance encompass vincristine, paclitaxel, vinblastine, doxorubicin, and daunorubicin (Ambudkar et al. 1999). Alternatively, resistance may stem from a reduction in cellular drug uptake, particularly in the case of water-soluble drugs that may bind to transporters responsible for nutrient transport, resulting in inadequate accumulation within the cell. This mechanism is implicated in resistance to drugs such as cisplatin, 8-azaguanine, and 5-fluorouracil (Shen et al. 2000). The stimulation of controlled detoxification systems, such as cytochrome P450 dual activity oxidases, along with an increase in DNA repair processes is another major factor that leads to drug resistance. (Lowe et al. 1993). In addition, resistance may result from altered apoptotic pathways brought on by malignant transformation, changes made to the apoptotic pathway while receiving chemotherapy, or adjustments made to cell cycle processes that trigger checkpoints and prevent the start of apoptosis (Liu et al. 2001). Additional factors contributing to drug resistance encompass limitations in drug penetration, modifications in the capacity to initiate prodrugs, and adjustments to drug targets.

A notable advancement transpired through the categorization of diseases predicated on the activity patterns of PR, ER, and HER2 (Bauer et al. 2010). Despite ongoing efforts to establish a new genome-driven classification, the observed heterogeneity resulting from the presently accepted classification may account for variations in tumor-proliferative capacity and resistance to therapy (Dawson et al. 2013).

As of now, it is clear that malignant cells that overexpress ATP-binding cassette (ABC) transporters of efflux actively expel pharmaceutical molecules, reducing their intracellular level and therapeutic effectiveness while simultaneously boosting drug exposure in cells that are healthy (Allen et al. 2000; Campling et al. 1997). A thorough examination of the literature underscores the pivotal role played by this mechanism in governing the survival and progression of cancer, exerting its influence either directly or indirectly through non-genomic mediated pathways. ABCB subfamily member 1 (ABCB1, also known as multidrug susceptibility protein 1/MDR1/permeability glycoprotein/P-glycoprotein), ABCG subfamily member 2 (ABCG2, also known as BC resistance protein/BCRP), and ABC subfamily C member 1 (ABCC1, also known as MDR-associated protein 1/MRP1) are among the frequently reported ABC transporters in the scientific literature. These pumps of efflux are believed to contribute to the development of drug resistance as they are found to be present in various human cancers, including BC (He et al. 2021). MDR signaling is most likely tightly associated with other molecular processes associated with drug resistance, including the dysregulation of survival/apoptotic pathways and the suppression of DNA repair. Thus, improvements in cancer biology, in conjunction with knowledge of the fundamental mechanisms controlling tumor growth and resistance—especially the functions of well-known ABC members like the P-glycoprotein (P-gp), multidrug resistance-associated protein (MRP), and breast carcinoma resistance protein (BCRP)—hold potential for discovering better cancer treatment alternatives (Real et al. 2002; van Jaarsveld et al. 2014).

Determining the precise involvement of ABC transporters in MDR in BC has proven challenging due to the intricacies of the underlying mechanisms. Studies examining the expression of these transporters in breast carcinoma cells and tumor specimens have frequently shown conflicting results, and differences in experimental design have made it difficult to make direct comparisons between studies. Despite indications from several clinical studies suggesting an association between heightened levels of ABC transporters in tumors and the progression of malignancy, a distinct correlation between levels of expression related to patient outcomes or tumor vulnerability to treatment. Given the significant amount of ABC carriers and the amount of overlap in their activities, understanding their interaction contributions to MDR may need a thorough evaluation of combined expression levels and functionality. All things considered, further thorough investigation is necessary to completely clarify the part ABC carriers play in cancer progression and MDR. A more thorough comprehension of this intricate and ever-changing system is necessary in order to develop treatment approaches that circumvent MDR.

Additionally, effective approaches to inhibit components of MDR are needed to augment the efficacy of commonly utilized chemotherapeutic agents (Wind and Holen 2011).

7.2 Treatment Approach for Overcoming Multidrug Resistance in Breast Cancer Based on Genomic Insights

It is becoming more likely that future treatment strategies will be impacted by the molecular makeup of each patient's tumor as well as patient features. Enhanced molecular profiling, facilitated by contemporary methodologies for extracting genomic information from breast tumors, will facilitate more precise and personalized treatment modalities (Buolamwini 1999; Perez et al. 2004). The incorporation of advanced systems, such as DNA microarrays and proteomics, coupled with the utilization of in vitro–identified methods for sensitizing cells to chemotherapy (e.g., specific taxane-based MDR reversal agents, proteasome inhibitors, and siRNA/RNAi), will facilitate the refinement of tailored therapies targeted at the specific molecular anomalies present in a given cancer (O'Driscoll and Clynes 2006).

Recent advancements in the field of Molecular Biology have ushered in a new era of anticancer medications designed to specifically attack genes or genetic abnormalities implicated in cell growth, apoptosis, and angiogenesis within cancerous cells. Prominent examples encompass trastuzumab (Herceptin), an antibody-based targeted therapy specifically designed for metastatic BC exhibiting overexpression of HER2/neu, and cetuximab (Erbitux), employed in the management of metastatic colorectal cancer characterized by overexpression of EGFR, with an ongoing phase III investigation for its potential application in BC (Esteva 2004). Molecular targeting medicines have shown to be a more effective treatment option than traditional chemotherapy thus far, particularly for patients with recurrent or advanced illness. Concurrently, various vaccines targeting the HER family of proteins have been developed and are presently under investigation as potential therapies for BC. Further extensive investigations are now imperative to determine appropriate combinations of targeted therapies, and potentially vaccines, with anti-cancer drugs tailored to specific sub-groups of patients (Ko et al. 2003). PM represents a notable advancement within the domain of oncology. The shift toward PM arises from the inadequacies of the conventional "one-size-fits-all" anticancer model, which is plagued by high toxicity and has proven ineffective in treating cancer patients. Nevertheless, it is imperative to acknowledge that PM is still in its nascent stages of development (Musyuni et al. 2022). When clinical endpoints or particular biomarkers are used as the basis, the clinical management of cancer is made simpler. Once a single genetic relationship for a medicine is known, prescribing drugs based on genotype becomes simple. Still, it is critical to recognize that there are several genotypes and medications with unknown drug interactions.

Research by (Martinelli et al. 2019) examined a number of nanotechnology-based strategies for treating cancer patients with MDR. This study clarified the use of P-gp pumps, which function by removing chemicals that are either natural or foreign from cells, including chemotherapeutic medications. IHC has emerged as a promising and efficacious technique for the identification of specific biomarkers. The successful exemplification of this approach is evident in the case of "HER2," which has been found to be a valuable biomarker in the context of BC (Planes-Laine et al. 2019). Furthermore, methods such as single-nucleotide polymorphisms (SNPs), DNA microarrays, reverse transcription-polymerase chain reaction (RT-PCR), NanoString technology, genome hybridization arrays, and DNA microarrays can be used to track variations in gene expression (Cava et al. 2015).

There has been a noticeable increase in the need for new diagnostic and treatment strategies in PM. Notably, there has been a growing emphasis on new areas of scientific research, including the investigation of fluid biopsies and patient-derived xenografts, or "PDX," because of their noninvasive nature (Zhang et al. 2017). Non-invasive techniques provide a thorough knowledge of cancer, in which cell-free DNA (cfDNA) and CTCs, or cancerous cells that originate in the vascular system and circulate in the blood, are essential elements. While the examination of CTCs presents difficulties and requires careful work, the results have been useful in defining the course of tumor growth and determining how well-suggested medications work for certain patients (Marrugo-Ramírez et al. 2018).

NGS has been discussed by Morash et al. 2018 as an effective PM technique for cancer treatment. NGS technology is gaining interest in cancer for its ability to quickly uncover both common and unusual genetic changes by enabling the accurate and speedy sequencing of several genes at the same time. Comprehending drug response and pharmacotherapeutic impact analysis are aided by this. PGx Seq was shown to be a novel tool for detecting single variants in pharmacogenetics in a study by Gulilat et al. 2019, providing usefulness in the developing field of PM.

Research conducted by Wang et al. 2017 examined many innovative approaches to combat MDR in cancer therapy. This study shed light on several inhibitor molecules (small in nature) that can overcome treatment resistance in patients suffering from cancer. With medication resistance and tumor recurrence being driven by both phenotypic and genotypic heterogeneity, targeted and combination treatments are becoming more and more important in scientific studies.

8 Translating Breast Cancer Genomics into Precision Medicine Using Artificial Intelligence

In the realm of cancer genomics, a new era is characterized by extensive datasets, which are a result of the widespread accessibility of genomic data facilitated by NGS methods and the explosive expansion of medical literature. Addressing challenges associated with the scalability and high dimensionality of datasets is on the rise through the incorporation of AI methodologies, encompassing deep learning (DL), natural language processing (NLP), and machine learning (ML). This integration seeks to convert extensive data into clinically relevant knowledge, marking a noteworthy advancement in the establishment of PM (Xu et al. 2019).

Considerable expansion in the domain of personalized medicine has been achieved through advancements in AI and ML, along with the availability of robust cloud infrastructure for large-scale storage of data and medical information integration (Syrjala 2018). Enhanced diagnostic efficacy, particularly in the identification of MBC, melanoma, and diverse ocular disorders, has been demonstrated by the application of DL algorithms to imaging in medicine, surpassing the capabilities of expert pathologists and dermatologists (De Fauw et al. 2018; Ehteshami Bejnordi et al. 2017; Haenssle et al. 2018). Furthermore, a crucial role is played by AI in fostering innovations in pharmacogenomics and liquid biopsies, reshaping practices related to monitoring and cancer screening, and improving the precision of predicting unfavorable incidents and patient results (Low et al. 2018).

Advancements in sophisticated AI methodologies and solutions within the field of cancer genomics have been catalyzed by the accessibility of genotype-phenotype data and multi-omics data, obtained from genome-wide association studies (GWAS) and insights derived from data mining (Szymczak et al. 2009; Telenti et al. 2018). The ability to provide personalized care through the PM paradigm is facilitated by these advancements. PM, an emerging paradigm for disease treatment and prevention, relies on the unique genetic constitution, environmental influences, and lifestyle decisions of individual patients (Kalinin et al. 2018). PM involves two separate methodologies, namely stratified medicine and personalized medicine. Stratified medicine entails assessing the effectiveness of a drug within a defined cohort of patients characterized by a specific molecular alteration. On the other hand, personalized medicine evaluates the efficacy of individualized treatment concepts in improving outcomes across the entire population.

The genesis of this approach is rooted in instances of treatment failures, primarily associated with tumor heterogeneity and the resurgence of gene mutations not addressed by the administered therapeutic interventions (Bettaieb et al. 2017). AI systems exhibit the capacity to discern variations in drug responses at an individual level, offering recommendations informed by discerned patterns from extensive datasets, both public and proprietary. Consequently, these AI-driven capabilities stand to push the boundaries of personalized medicine, particularly in the context of cancer genomics (Lin et al. 2018).

BC stands as a prominent cause of mortality among women globally, emphasizing the critical role of early detection in significantly enhancing the well-being of millions of individuals. Recognizing the imperative of devising a solution or framework for early detection and diagnosis, a surge in research activities has transpired, propelled by recent advancements in robust AI algorithms, specifically those rooted in DL. The confluence of potent AI algorithms, the availability of hardware capable of executing and training these intricate algorithms, and access to extensive datasets requisite for AI training have fueled this research direction (Shah et al. 2022).

For the diagnosis of early-stage BC, different imaging modalities, such as breast thermography (BT), magnetic resonance imaging (MRI), mammograms (MGs), positron emission tomography (PET), ultrasound (US), histopathology (HP), and computed tomography (CT), are utilized. In the field of breast pathology, various functions, such as the identification and categorization of cancer, histologic grading, detection of lymph node (LN) metastasis, quantification of biomarkers, and prediction of genetic abnormalities like BRCA mutations, have been addressed through the utilization of AI algorithms (Ahn et al. 2023). Significant progress in anomaly detection, especially in the context of diagnosing BC, has been achieved through the utilization of AI technology for automated analysis of medical images (Domingues et al. 2020; Kozegar et al. 2020; Moghbel and Mashohor 2013; Murtaza et al. 2020; Saha et al. 2018).

In this domain, AI algorithms are categorized into two groups based on their methodology for interpreting data and extracting information from images.

1. **Conventional ML algorithms:**
   This category includes algorithms that depend on handcrafted features. The research of AI algorithms which acquire knowledge through learning without explicit programming is encompassed by ML (Samuel 1959).
2. **DL/deep neural network (DNN) algorithms:**
   Raw data is processed and information is extracted from significant regions through the mathematical optimization of classification by these algorithms. DL, a subfield of ML, relates to AI algorithms that develop models from data at various levels of conceptualization, minimizing the need for features that are handcrafted (LeCun et al. 2015). A sequence diagram of the analysis of BC genomics with AI algorithms is depicted in Fig. 3.

Robust outcomes in BC analysis have been demonstrated by both categories of AI algorithms. However, recent frameworks predominantly utilize DL algorithms because of their superior performance. This preference is rooted in the ability of DL

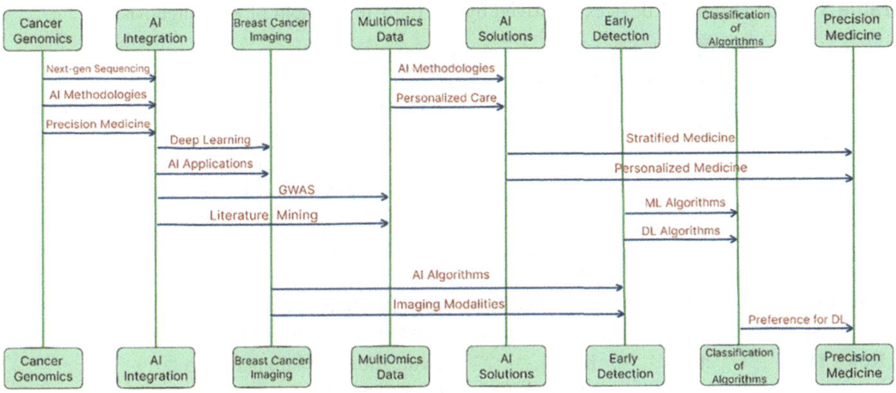

Fig. 3 Analysis of BC genomics with AI algorithms

algorithms to autonomously learn the most significant data representations without substantial human intervention, setting them apart in the domain of BC detection and analysis (Burt et al. 2018; Sharma and Mehra 2020).

The conventional "one-size-fits-all" paradigm signifies a universal approach to imaging and treatment protocols applied to patients. In contrast, PM embodies a philosophy centered on tailored strategies directed at specific cohorts of individuals, determined by their genetic, phenotypic, and environmental attributes, as well as those of their respective diseases. The amalgamation of findings from imaging, encompassing genomic, pathologic, and clinical characteristics, is progressively gaining significance in the context of BC.

The simultaneous introduction of advanced imaging modalities has augmented the diagnostic arsenal available to radiologists, providing them with enhanced tools and extensive image datasets for analysis and interpretation. The incorporation of an AI-based workflow into breast imaging facilitates the amalgamation of diverse data streams, culminating in potent multidisciplinary systems. This integration has the potential to pave the way toward personalized, patient-specific medicine, marking a transformative trajectory in the approach to BC care (Sheth and Giger 2020). Table 5 provides a comprehensive overview of AI devices that have received approval from the FDA for their application in the domain of BC diagnosis.

9 Conclusion

9.1 Achievements and Advancements in Precision Medicine for Breast Cancer

While the understanding that cancer arises from uncontrolled cell proliferation due to somatic mutations has contributed minimally to its treatment, the potential to improve patient outcomes remains unclear. Relying solely on mutations may not be

Table 5 Summary of AI devices approved by the FDA in the field of BC diagnosis

Algorithm utilizing AI	FDA approval year	Company name	Sign or signal	References
PowerLook Tomo detection	2017	iCAD, Inc.	Evaluation of malignancy in breast lesions and identification of potentially concerning breast lesions	Potnis et al. (2022)
QuantX	2017	Quantitative insights	BC diagnosis	Chen et al. (2021)
cmTriage	2019	CureMetrix	Identification of potentially concerning breast lesions	
Koios DS breast	2019	Koios medical	Evaluation of malignancy in breast lesions	
ProFound AI software V2.1	2019	iCAD	Evaluation of malignancy in breast lesions	
Genius AI detection	2020	Hologic, Inc	Evaluation of malignancy in breast lesions	
Transpara	2020	ScreenPoint medical BV	Evaluation of malignancy in breast lesions	
MammoScreen	2020	Therapixel	Evaluation of malignancy in breast lesions	
HealthMammo	2020	Zebra medical vision	Identification of potentially concerning breast lesions	
Lunit INSIGHT MMG	2021	Lunit, Inc.	Identification of potentially concerning breast lesions	Potnis et al. (2022)
Saige-Q	2021	DeepHealth, Inc.	Identification of potentially concerning breast lesions	Potnis et al. (2022)

DL deep learning, *AI* artificial intelligence, *CAD* computer-aided design, *FDA* Food and Drug Administration, *CT* computed tomography

particularly fruitful, given their indirect and multifaceted impact on the regulation of other genes. It is crucial to acknowledge cancer as a disorder of cellular regulatory mechanisms. As a result, tackling the signaling characteristic of dysregulated cancer cells is a critical aspect for achieving success in precision cancer treatment. The analysis of protein networks activated by cancer will guide the development of combination therapies, improving therapeutic outcomes. This shift from the current systemic approach to a more personalized approach is expected to enhance clinical benefits for patients. However, challenges in implementing PM often arise due to the lack of specific drugs tailored to individual cases. Consequently, oncology is at the forefront of adopting PM, utilizing the molecular profiles of individuals' genomes to optimize disease management and mitigate the risks of over- and undertreatment. This is in contrast with typical chemotherapy which is based on the systemic strategy, which reduces toxicity related to chemotherapy's nonspecific activities.

A significant challenge is encountered in the PM approach for cancer due to the diverse nature of tumors and the evolving phenotype of malignant cells. This complexity hampers accurate diagnosis and precise therapies, emphasizing the limitation of relying on a single biopsy at a specific time point. The resistance of malignant cells to medication has resulted in the use of numerous cytotoxic medicines with significant side effects. However, a combination of targeted therapies and immune system activation holds promise for more effectively eliminating malignant cells. Moreover, the advent of high-throughput techniques for cell and molecular testing at the RNA/DNA and protein levels brings optimism to the field of PM. The analysis of transcriptome profiles from malignant cells, healthy cells, and adjacent stroma cells, may potentially unveil essential hub targets implicated in the rewiring of malignant pathways. The real-time selection of drugs tailored to disease-specific protein clusters is now feasible theoretically within the framework of PM. Additionally, the continuous advancement of biosensor technology suggests that liquid biopsy could enable noninvasive cancer detection and surveillance in real time.

Despite the apparent increase in expenses, the pursuit of personalized medicine in translational medicine is deemed necessary not only from an ethical standpoint for patients but also to address the substantial costs associated with prescribing ineffective drug therapies in refractory cases. Moreover, the social costs connected to cytotoxic therapies with suboptimal clinical outcomes and debilitating side effects underscore the importance of embracing personalized medicine. The European Medicines Agency standard on the evaluation of anticancer medications already advocates for the early integration of biomarker diagnostic techniques in clinical research, solidifying the ongoing shift toward the PM strategy for cancer. International collaborative initiatives aimed at assessing the PM approach for BC treatment, exemplified by the AURORA umbrella research, which examines the impact of various drugs on diverse molecular anomalies within single or multiple tumors have already been initiated. As stratified oncology approaches the era of PM, there is a pressing requirement to incorporate extensive genomic and clinical data into actionable data to inform clinical decisions.

9.2 Future Directions and the Evolving Landscape of Breast Cancer Treatment

The resistance observed against conventional treatments serves as a notable catalyst in the exploration and advancement of novel BC drugs or the implementation of fresh strategies with existing medications. The effective utilization of combined therapies, involving diverse classes of drugs, has proven successful in establishing optimal prognoses, offering patients the chance to maximize benefits while minimizing or eliminating recurrence, resistance, and toxic effects. Additionally, investigations into cellular pathways have prompted the examination of new agents aimed at targeting or inhibiting tumorigenesis. This reasoning led to investigations into clinically significant biomarkers like CDK4/6 expression and PIK3CA mutations (Vasan et al. 2019). Recently approved by the FDA, PIK3 inhibitors represent the latest advancements in BC treatments. Ongoing research is unveiling new categories of drugs that offer enhanced benefits for BC treatment. The intricate relationship between the cell cycle and cancer biology presents numerous factors influencing the development of more effective agents and biomarkers with heightened predictive values (Le Du et al. 2013). The concept of employing genomics to drive PM for MBC is appealing (Pfefferle et al. 2016). However, there is currently a lack of evidence supporting its efficacy in improving patient outcomes. This lack of success can be ascribed to both scientific and logistical obstacles. An important hurdle in BC research is the incomplete characterization of key drivers, with the exception of ER, ERBB2, PIK3CA, and AKT1 (Xia et al. 2007). The development of comprehensive gene catalogs, along with methodologies for evaluating pathway activation, may aid in identifying such drivers in the near future. Understanding the activated protein network and the genomic landscape of tumors is crucial for guiding the development of combination medicines that maximize therapeutic effects (Stephens et al. 2012). Addressing logistical and operational challenges is equally crucial. In this context, a shift toward personalized medicine, rather than a stratified medicine approach, holds promise for improving patient management.

Acknowledgment Gratitude is extended by the authors to the Centre for Biotechnology and Bioinformatics, Dibrugarh University, Dibrugarh, Assam, for providing essential resources during the preparation of the manuscript. The authors also acknowledge the DBT e-Library Consortium (DeLCON) for granting access to some of the important journals used in writing this manuscript. Additionally, the support of this project by the Deanship of Scientific Research at Imam Mohammad Ibn Saud Islamic University (IMSIU) is appreciated by the authors.

Conflict of Interest The authors declare that they do not have any conflicts of interest to report.

References

Agostinetto, E., Gligorov, J., & Piccart, M. (2022). Systemic therapy for early-stage breast cancer: learning from the past to build the future. Nat Rev Clin Oncol, 19(12), Article 12. https://doi.org/10.1038/s41571-022-00687-1

Ahern TP, Lash TL, Damkier P, Christiansen PM, Cronin-Fenton DP (2014) Statins and breast cancer prognosis: evidence and opportunities. Lancet Oncol 15(10):e461–e468. https://doi.org/10.1016/S1470-2045(14)70119-6

Ahn JS, Shin S, Yang S-A, Park EK, Kim KH, Cho SI, Ock C-Y, Kim S (2023) Artificial intelligence in breast cancer diagnosis and personalized medicine. J Breast Cancer 26(5):405. https://doi.org/10.4048/jbc.2023.26.e45

Akbani R, Ng K-S, Werner HM, Zhang F, Ju Z, Liu W, Yang J-Y, Lu Y, Weinstein JN, Mills GB (2014) Abstract 4262: a pan-cancer proteomic analysis of The Cancer Genome Atlas (TCGA) project. Cancer Res 74(19_Supplement):4262. https://doi.org/10.1158/1538-7445.AM2014-4262

Allen JD, Brinkhuis RF, van Deemter L, Wijnholds J, Schinkel AH (2000) Extensive contribution of the multidrug transporters P-glycoprotein and Mrp1 to basal drug resistance. Cancer Res 60(20):5761–5766

Ambudkar SV, Dey S, Hrycyna CA, Ramachandra M, Pastan I, Gottesman MM (1999) Biochemical, cellular, and pharmacological aspects of the multidrug transporter. Annu Rev Pharmacol Toxicol 39(1):361–398

Araki K, Miyoshi Y (2018) Mechanism of resistance to endocrine therapy in breast cancer: the important role of PI3K/Akt/mTOR in estrogen receptor-positive, HER2-negative breast cancer. Breast Cancer 25(4):392–401. https://doi.org/10.1007/s12282-017-0812-x

Aronson SJ, Rehm HL (2015) Building the foundation for genomics in precision medicine. Nature 526(7573):336–342. https://doi.org/10.1038/nature15816

Ashley EA (2016) Towards precision medicine. Nat Rev Genet 17(9):507–522. https://doi.org/10.1038/nrg.2016.86

Assessment of Ki67 in Breast Cancer: Updated Recommendations From the International Ki67 in Breast Cancer Working Group | JNCI: Journal of the National Cancer Institute | Oxford Academic (n.d.) Retrieved November 29, 2023, from https://academic.oup.com/jnci/article/113/7/808/6053794?login=false

Barzaman K, Karami J, Zarei Z, Hosseinzadeh A, Kazemi MH, Moradi-Kalbolandi S, Safari E, Farahmand L (2020) Breast cancer: biology, biomarkers, and treatments. Int Immunopharmacol 84:106535. https://doi.org/10.1016/j.intimp.2020.106535

Bauer K, Parise C, Caggiano V (2010) Use of ER/PR/HER2 subtypes in conjunction with the 2007 St Gallen consensus statement for early breast cancer. BMC Cancer 10:1–12

Bettaieb A, Paul C, Plenchette S, Shan J, Chouchane L, Ghiringhelli F (2017) Precision medicine in breast cancer: reality or utopia? J Transl Med 15:1–13

Bollag DM, McQueney PA, Zhu J, Hensens O, Koupal L, Liesch J, Goetz M, Lazarides E, Woods CM (1995) Epothilones, a new class of microtubule-stabilizing agents with a taxol-like mechanism of action. Cancer Res 55(11):Article 11

Bonneterre, J., Thürlimann, B., Robertson, J. F., Krzakowski, M., Mauriac, L., Koralewski, P., Vergote, I., Webster, A., Steinberg, M., & von Euler, M. (2000). Anastrozole versus tamoxifen as first-line therapy for advanced breast cancer in 668 postmenopausal women: results of the Tamoxifen or Arimidex Randomized Group Efficacy and Tolerability study. J Clin Oncol, 18(22), Article 22. https://doi.org/10.1200/JCO.2000.18.22.3748

BRCA Genes: The Role in Genome Stability, Cancer Stemness and Therapy Resistance—PMC (2023, November 25). https://www.ncbi.nlm.nih.gov/pmc/articles/PMC6548160/

Buolamwini JK (1999) Novel anticancer drug discovery. Curr Opin Chem Biol 3(4):500–509

Burt JR, Torosdagli N, Khosravan N, RaviPrakash H, Mortazi A, Tissavirasingham F, Hussein S, Bagci U (2018) Deep learning beyond cats and dogs: recent advances in diagnosing breast cancer with deep neural networks. Br J Radiol 91(1089):20170545

Buzdar A, Howell A (2001) Advances in aromatase inhibition: clinical efficacy and tolerability in the treatment of breast cancer. Clin Cancer Res 7(9):Article 9

Campling BG, Young LC, Baer KA, Lam Y-M, Deeley RG, Cole SP, Gerlach JH (1997) Expression of the MRP and MDR1 multidrug resistance genes in small cell lung cancer. Clin Cancer Res 3(1):115–122

Cantley LC (2002) The phosphoinositide 3-kinase pathway. Science 296(5573):1655–1657. https://doi.org/10.1126/science.296.5573.1655

Cardoso F, Di Leo A, Lohrisch C, Bernard C, Ferreira F, Piccart MJ (2002) Second and subsequent lines of chemotherapy for metastatic breast cancer: what did we learn in the last two decades? Ann Oncol 13(2):Article 2. https://doi.org/10.1093/annonc/mdf101

Carpten JD, Faber AL, Horn C, Donoho GP, Briggs SL, Robbins CM, Hostetter G, Boguslawski S, Moses TY, Savage S, Uhlik M, Lin A, Du J, Qian Y-W, Zeckner DJ, Tucker-Kellogg G, Touchman J, Patel K, Mousses S et al (2007) A transforming mutation in the pleckstrin homology domain of AKT1 in cancer. Nature 448(7152):439–444. https://doi.org/10.1038/nature05933

Carrick S, Parker S, Thornton C, Ghersi D, Simes J, Wilcken N (2009) Single agent versus combination chemotherapy for metastatic breast cancer. Cochrane Database Syst Rev 2021(5):Article 5. https://doi.org/10.1002/14651858.CD003372.pub3

Casaubon JT, Kashyap S, Regan JP (2017) BRCA1 and BRCA2 mutations
Cava C, Bertoli G, Castiglioni I (2015) Integrating genetics and epigenetics in breast cancer: biological insights, experimental, computational methods and therapeutic potential. BMC Syst Biol 9(1):62. https://doi.org/10.1186/s12918-015-0211-x
Chaurasia M, Singh R, Sur S, Flora SJS (2023) A review of FDA approved drugs and their formulations for the treatment of breast cancer. Front Pharmacol 14:1184472
Chen Z, Lin L, Wu C, Li C, Xu R, Sun Y (2021) Artificial intelligence for assisting cancer diagnosis and treatment in the era of precision medicine. Cancer Commun 41(11):1100–1115. https://doi.org/10.1002/cac2.12215
Chiavenna SM, Jaworski JP, Vendrell A (2017) State of the art in anti-cancer mAbs. J Biomed Sci 24:1–12
Cinkaya A, Akin M, Sengul A (2016) Evaluation of treatment outcomes of triple-negative breast cancer. J Cancer Res Ther 12(1):150–154
Cobleigh MA, Vogel CL, Tripathy D, Robert NJ, Scholl S, Fehrenbacher L, . . . Slamon DJ (1999) Multinational study of the efficacy and safety of humanized anti-HER2 monoclonal antibody in women who have HER2-overexpressing metastatic breast cancer that has progressed after chemotherapy for metastatic disease. J Clin Oncol 17(9):2639–2639
Collins FS, Morgan M, Patrinos A (2003) The human genome project: lessons from large-scale biology. Science 300(5617):Article 5617. https://doi.org/10.1126/science.1084564
Cortazar P, Zhang L, Untch M, Mehta K, Costantino JP, Wolmark N, Bonnefoi H, Cameron D, Gianni L, Valagussa P, Swain SM, Prowell T, Loibl S, Wickerham DL, Bogaerts J, Baselga J, Perou C, Blumenthal G, Blohmer J et al (2014) Pathological complete response and long-term clinical benefit in breast cancer: the CTNeoBC pooled analysis. Lancet 384(9938):Article 9938. https://doi.org/10.1016/S0140-6736(13)62422-8
Couturier J, Vincent-Salomon A, Nicolas A, Beuzeboc P, Mouret E, Zafrani B, Sastre-Garau X (2000) Strong correlation between results of fluorescent in situ hybridization and immunohistochemistry for the assessment of the ERBB2 (HER-2/neu) gene status in breast carcinoma. Mod Pathol 13(11):Article 11. https://doi.org/10.1038/modpathol.3880228
Cretella D, Ravelli A, Fumarola C, La Monica S, Digiacomo G, Cavazzoni A, Alfieri R, Biondi A, Generali D, Bonelli M, Petronini PG (2018) The anti-tumor efficacy of CDK4/6 inhibition is enhanced by the combination with PI3K/AKT/mTOR inhibitors through impairment of glucose metabolism in TNBC cells. J Exp Clin Cancer Res 37(1):72. https://doi.org/10.1186/s13046-018-0741-3
Crimini E, Repetto M, Aftimos P, Botticelli A, Marchetti P, Curigliano G (2021) Precision medicine in breast cancer: from clinical trials to clinical practice. Cancer Treat Rev 98: 102223. https://doi.org/10.1016/j.ctrv.2021.102223
Dawson S-J, Rueda OM, Aparicio S, Caldas C (2013) A new genome-driven integrated classification of breast cancer and its implications. EMBO J 32(5):617–628
De Fauw J, Ledsam JR, Romera-Paredes B, Nikolov S, Tomasev N, Blackwell S, Askham H, Glorot X, O'Donoghue B, Visentin D, Van Den Driessche G, Lakshminarayanan B, Meyer C, Mackinder F, Bouton S, Ayoub K, Chopra R, King D, Karthikesalingam A et al (2018) Clinically applicable deep learning for diagnosis and referral in retinal disease. Nat Med 24(9):1342–1350. https://doi.org/10.1038/s41591-018-0107-6
Descriptive analysis of estrogen receptor (ER)-negative, progesterone receptor (PR)-negative, and HER2-negative invasive breast cancer, the so-called triple-negative phenotype—Bauer—2007—Cancer—Wiley Online Library (2023, November 25). https://acsjournals.onlinelibrary.wiley.com/doi/full/10.1002/cncr.22618
Domingues I, Pereira G, Martins P, Duarte H, Santos J, Abreu PH (2020) Using deep learning techniques in medical imaging: a systematic review of applications on CT and PET. Artif Intell Rev 53:4093–4160
Dowsett M, Cooke T, Ellis I, Gullick WJ, Gusterson B, Mallon E, Walker R (2000) Assessment of HER2 status in breast cancer: why, when and how? Eur J Cancer 36(2):170–176. https://doi.org/10.1016/S0959-8049(99)00264-6

Dunlap J, Le C, Shukla A, Patterson J, Presnell A, Heinrich MC, Corless CL, Troxell ML (2010) Phosphatidylinositol-3-kinase and AKT1 mutations occur early in breast carcinoma. Breast Cancer Res Treat 120(2):409–418. https://doi.org/10.1007/s10549-009-0406-1

Ehteshami Bejnordi B, Veta M, Johannes Van Diest P, Van Ginneken B, Karssemeijer N, Litjens G, Van Der Laak JAWM, the CAMELYON16 Consortium, Hermsen M, Manson QF, Balkenhol M, Geessink O, Stathonikos N, Van Dijk MC, Bult P, Beca F, Beck AH, Wang D, Khosla A et al (2017) Diagnostic assessment of deep learning algorithms for detection of lymph node metastases in women with breast cancer. JAMA 318(22):2199. https://doi.org/10.1001/jama.2017.14585

Esteva FJ (2004) Monoclonal antibodies, small molecules, and vaccines in the treatment of breast cancer. Oncologist 9(S3):4–9

Fan S, Cherney B, Reinhold W, Rucker K, O'Connor PM (1998) Disruption of p53 function in immortalized human cells does not affect survival or apoptosis after taxol or vincristine treatment. Clin Cancer Res 4(4):Article 4

Fisher B (1979) Breast-cancer management. N Engl J Med 301(6):Article 6. https://doi.org/10.1056/NEJM197908093010611

Fossati R, Confalonieri C, Torri V, Ghislandi E, Penna A, Pistotti V, Tinazzi A, Liberati A (1998) Cytotoxic and hormonal treatment for metastatic breast cancer: a systematic review of published randomized trials involving 31,510 women. J Clin Oncol 16(10):3439–3460. https://doi.org/10.1200/JCO.1998.16.10.3439

Franke TF, Kaplan DR, Cantley LC (1997) PI3K: downstream AKTion blocks apoptosis. Cell 88(4):435–437. https://doi.org/10.1016/S0092-8674(00)81883-8

Gao JJ, Cheng J, Bloomquist E, Sanchez J, Wedam SB, Singh H, Amiri-Kordestani L, Ibrahim A, Sridhara R, Goldberg KB, Theoret MR, Kluetz PG, Blumenthal GM, Pazdur R, Beaver JA, Prowell TM (2020) CDK4/6 inhibitor treatment for patients with hormone receptor-positive, HER2-negative, advanced or metastatic breast cancer: a US Food and Drug Administration pooled analysis. Lancet Oncol 21(2):250–260. https://doi.org/10.1016/S1470-2045(19)30804-6

Gebhart G, Gámez C, Holmes E, Robles J, Garcia C, Cortés M, De Azambuja E, Fauria K, Van Dooren V, Aktan G, Coccia-Portugal MA, Kim S-B, Vuylsteke P, Cure H, Eidtmann H, Baselga J, Piccart M, Flamen P, Di Cosimo S (2013) ^{18} F-FDG PET/CT for early prediction of response to neoadjuvant lapatinib, trastuzumab, and their combination in HER2-positive breast cancer: results from Neo-ALTTO. J Nucl Med 54(11):Article 11. https://doi.org/10.2967/jnumed.112.119271

Genetics (n.d.) Retrieved November 9, 2023, from https://www.breastcancer.org/risk/risk-factors/genetics

Genetics and risk of breast cancer (n.d.) Retrieved January 16, 2024, from https://www.bcna.org.au/resource-hub/articles/genetics-and-risk-of-breast-cancer/

Genetics as a Risk Factor for Breast Cancer (2024, January 16). https://www.breastcancer.org/risk/risk-factors/genetics

Gershanovich M, Chaudri HA, Campos D, Lurie H, Bonaventura A, Jeffrey M, Buzzi F, Bodrogi I, Ludwig H, Reichardt P, O'Higgins N, Romieu G, Friederich P, Lassus M (1998) Letrozole, a new oral aromatase inhibitor: Randomised trial comparing 2.5 mg daily, 0.5 mg daily and aminoglutethimide in postmenopausal women with advanced breast cancer. Ann Oncol 9(6): Article 6. https://doi.org/10.1023/A:1008226721932

Godet I, Gilkes DM (2017) BRCA1 and BRCA2 mutations and treatment strategies for breast cancer. Integr Cancer Sci Ther 4(1):10.15761/ICST.1000228

Gulilat M, Lamb T, Teft WA, Wang J, Dron JS, Robinson JF, Tirona RG, Hegele RA, Kim RB, Schwarz UI (2019) Targeted next generation sequencing as a tool for precision medicine. BMC Med Genet 12:1–17

Haenssle HA, Fink C, Schneiderbauer R, Toberer F, Buhl T, Blum A, Kalloo A, Hassen ABH, Thomas L, Enk A, Uhlmann L, Alt C, Arenbergerova M, Bakos R, Baltzer A, Bertlich I, Blum A, Bokor-Billmann T, Bowling J et al (2018) Man against machine: diagnostic performance of a deep learning convolutional neural network for dermoscopic melanoma recognition in comparison to 58 dermatologists. Ann Oncol 29(8):1836–1842. https://doi.org/10.1093/annonc/mdy166

Hamilton A, Piccart M (1999) The third-generation non-steroidal aromatase inhibitors: a review of their clinical benefits in the second-line hormonal treatment of advanced breast cancer. Ann Oncol 10(4):Article 4. https://doi.org/10.1023/A:1008368300827

He J, Fortunati E, Liu D-X, Li Y (2021) Pleiotropic roles of ABC transporters in breast cancer. Int J Mol Sci 22(6):3199

HER2-Positive Breast Cancer | Drugs (n.d.) Retrieved November 29, 2023, from https://link.springer.com/article/10.2165/00003495-200767090-00006

Hinz N, Jücker M (2019) Distinct functions of AKT isoforms in breast cancer: a comprehensive review. Cell Commun Signal 17:1–154. https://doi.org/10.1186/s12964-019-0450-3

Ho A, Morrow M (2011) The evolution of the locoregional therapy of breast cancer. Oncologist 16(10):Article 10. https://doi.org/10.1634/theoncologist.2011-0223

Howell A, Robertson JFR, Quaresma Albano J, Aschermannova A, Mauriac L, Kleeberg UR, Vergote I, Erikstein B, Webster A, Morris C (2002) Fulvestrant, formerly ICI 182,780, is as effective as Anastrozole in postmenopausal women with advanced breast cancer progressing after prior endocrine treatment. J Clin Oncol 20(16):Article 16. https://doi.org/10.1200/JCO.2002.10.057

Hu T, Chitnis N, Monos D, Dinh A (2021) Next-generation sequencing technologies: an overview. Hum Immunol 82(11):801–811. https://doi.org/10.1016/j.humimm.2021.02.012

Huang C-H, Mandelker D, Schmidt-Kittler O, Samuels Y, Velculescu VE, Kinzler KW, Vogelstein B, Gabelli SB, Amzel LM (2007) The structure of a human p110α/p85α complex elucidates the effects of oncogenic PI3Kα mutations. Science 318(5857):1744–1748. https://doi.org/10.1126/science.1150799

Isakoff SJ, Engelman JA, Irie HY, Luo J, Brachmann SM, Pearline RV, Cantley LC, Brugge JS (2005) Breast cancer–associated *PIK3CA* mutations are oncogenic in mammary epithelial cells. Cancer Res 65(23):10992–11000. https://doi.org/10.1158/0008-5472.CAN-05-2612

Jassem J, Pieńkowski T, Płuzańska A, Jelic S, Gorbunova V, Mrsic-Krmpotic Z, Berzins J, Nagykalnai T, Wigler N, Renard J, Munier S, Weil C (2001) Doxorubicin and paclitaxel versus fluorouracil, doxorubicin, and cyclophosphamide as first-line therapy for women with metastatic breast cancer: final results of a randomized phase III multicenter trial. J Clin Oncol 19(6): Article 6. https://doi.org/10.1200/JCO.2001.19.6.1707

Johnston SJ, Cheung K-L (2018) Endocrine therapy for breast cancer: a model of hormonal manipulation. Oncol Ther 6(2):–141, 156. https://doi.org/10.1007/s40487-018-0062-x

Kalinin AA, Higgins GA, Reamaroon N, Soroushmehr S, Allyn-Feuer A, Dinov ID, Najarian K, Athey BD (2018) Deep learning in pharmacogenomics: from gene regulation to patient stratification. Pharmacogenomics 19(7):629–650. https://doi.org/10.2217/pgs-2018-0008

Karami F, Mehdipour P (2013) A comprehensive focus on global spectrum of BRCA1 and BRCA2 mutations in breast cancer. Biomed Res Int 2013

Ko BK, Kawano K, Murray JL, Disis ML, Efferson CL, Kuerer HM, Peoples GE, Ioannides CG (2003) Clinical studies of vaccines targeting breast cancer. Clin Cancer Res 9(9):3222–3234

Kozegar E, Soryani M, Behnam H, Salamati M, Tan T (2020) Computer aided detection in automated 3-D breast ultrasound images: a survey. Artif Intell Rev 53:1919–1941

Larry Jameson J, Longo DL (2015) Precision medicine—personalized, problematic, and promising. Obstet Gynecol Surv 70(10):612–614. https://doi.org/10.1097/01.ogx.0000472121.21647.38

LeCun Y, Bengio Y, Hinton G (2015) Deep learning. Nature 521(7553):436–444

Le Du F, Ueno NT, Gonzalez-Angulo AM (2013) Breast cancer biomarkers: utility in clinical practice. Curr Breast Cancer Rep 5:284–292

Leonard R, Cervantes G, Lui W, Mauriac L, Miles D, Moiseyenko V, Chan Navarro C, Van Hazel G, Vukelja S, O'Shaughnessy J (2001) Survival update of so14999 a large phase III trial of capecitabine/doxetaxel combination therapy vs docetaxel monotherapy in patients with locally advanced (LABC) or metastatic breast cancer (MBC). Eur J Cancer 37:S151. https://doi.org/10.1016/S0959-8049(01)81043-1

Leung JH, Leung HWC, Wang S-Y, Huang S-S, Chan ALF (2021) Efficacy and safety of CDK4/6 and PI3K/AKT/mTOR inhibitors as second-line treatment in postmenopausal patients with hormone receptor-positive, HER-2-negative metastatic breast cancer: a network meta-analysis. Expert Opin Drug Saf 20(8):949–957. https://doi.org/10.1080/14740338.2021.1931116

Lin E, Kuo PH, Liu YL, Yu YWY, Yang AC, Tsai SJ (2018) A deep learning approach for predicting antidepressant response in major depression using clinical and genetic biomarkers. Front Psych 9:367995

Liu Y-Y, Han T-Y, Giuliano AE, Cabot MC (2001) Ceramide glycosylation potentiates cellular multidrug resistance. FASEB J 15(3):719–730

Low S, Zembutsu H, Nakamura Y (2018) Breast cancer: the translation of big genomic data to cancer precision medicine. Cancer Sci 109(3):497–506. https://doi.org/10.1111/cas.13463

Lowe SW, Ruley HE, Jacks T, Housman DE (1993) P53-dependent apoptosis modulates the cytotoxicity of anticancer agents. Cell 74(6):957–967

Lu RM, Hwang YC, Liu IJ, Lee CC, Tsai HZ, Li HJ, Wu HC (2020) Development of therapeutic antibodies for the treatment of diseases. J Biomed Sci 27:1–30

Luminal A Breast Cancer and Molecular Assays: A Review | The Oncologist | Oxford Academic (n.d.) Retrieved November 29, 2023, from https://academic.oup.com/oncolo/article/23/5/556/6439608?login=false

Mahdavi M, Nassiri M, Kooshyar MM, Vakili-Azghandi M, Avan A, Sandry R, Pillai S, Lam AK, Gopalan V (2019) Hereditary breast cancer; Genetic penetrance and current status with BRCA. J Cell Physiol 234(5):Article 5. https://doi.org/10.1002/jcp.27464

Marrugo-Ramírez J, Mir M, Samitier J (2018) Blood-based cancer biomarkers in liquid biopsy: a promising non-invasive alternative to tissue biopsy. Int J Mol Sci 19(10):2877. https://doi.org/10.3390/ijms19102877

Martinelli C, Pucci C, Ciofani G (2019) Nanostructured carriers as innovative tools for cancer diagnosis and therapy. APL Bioeng 3(1):011502. https://doi.org/10.1063/1.5079943

McCombie WR, McPherson JD, Mardis ER (2019) Next-generation sequencing technologies. Cold Spring Harb Perspect Med 9(11):Article 11. https://doi.org/10.1101/cshperspect.a036798

McGrath S, Ghersi D (2016) Building towards precision medicine: empowering medical professionals for the next revolution. BMC Med Genet 9(1):23. https://doi.org/10.1186/s12920-016-0183-8

Meldrum C, Doyle MA, Tothill RW (2011) Next-generation sequencing for cancer diagnostics: a practical perspective. Clin Biochem Rev 32(4):177

Miled N, Yan Y, Hon W-C, Perisic O, Zvelebil M, Inbar Y, Schneidman-Duhovny D, Wolfson HJ, Backer JM, Williams RL (2007) Mechanism of two classes of cancer mutations in the phosphoinositide 3-kinase catalytic subunit. Science 317(5835):239–242. https://doi.org/10.1126/science.1135394

Miles DW, Smith IE, Coleman RE, Calvert AH, Lind MJ (2001) A phase II study of pemetrexed disodium (LY231514) in patients with locally recurrent or metastatic breast cancer. Eur J Cancer 37(11):Article 11. https://doi.org/10.1016/S0959-8049(01)00117-4

Moghbel M, Mashohor S (2013) A review of computer assisted detection/diagnosis (CAD) in breast thermography for breast cancer detection. Artif Intell Rev 39:305–313

Mokhtari RB, Homayouni TS, Baluch N, Morgatskaya E, Kumar S, Das B, Yeger H (2017) Combination therapy in combating cancer. Oncotarget 8(23):38022–38043. https://doi.org/10.18632/oncotarget.16723

Morash M, Mitchell H, Beltran H, Elemento O, Pathak J (2018) The role of next-generation sequencing in precision medicine: a review of outcomes in oncology. J Personal Med 8(3):30. https://doi.org/10.3390/jpm8030030

Mouridsen H, Gershanovich M, Sun Y, Pérez-Carrión R, Boni C, Monnier A, Apffelstaedt J, Smith R, Sleeboom HP, Jänicke F, Pluzanska A, Dank M, Becquart D, Bapsy PP, Salminen E, Snyder R, Lassus M, Verbeek JA, Staffler B et al (2001) Superior efficacy of Letrozole versus tamoxifen as first-line therapy for postmenopausal women with advanced breast cancer: results of a phase III study of the international Letrozole breast cancer group. J Clin Oncol 19(10):Article 10. https://doi.org/10.1200/JCO.2001.19.10.2596

Mukohara T (2015) PI3K mutations in breast cancer: prognostic and therapeutic implications. Breast Cancer Targets Ther 111. https://doi.org/10.2147/BCTT.S60696

Murtaza G, Shuib L, Abdul Wahab AW, Mujtaba G, Mujtaba G, Nweke HF, Al-garadi MA, Zulfiqar F, Raza G, Azmi NA (2020) Deep learning-based breast cancer classification through medical imaging modalities: state of the art and research challenges. Artif Intell Rev 53:1655–1720

Musyuni P, Bai J, Sheikh A, Vasanthan KS, Jain GK, Abourehab MAS, Lather V, Aggarwal G, Kesharwani P, Pandita D (2022) Precision medicine: ray of hope in overcoming cancer multidrug resistance. Drug Resist Updat 65:100889. https://doi.org/10.1016/j.drup.2022.100889

Mutant p53 in breast cancer: Potential as a therapeutic target and biomarker | Breast Cancer Research and Treatment (n.d.) Retrieved November 29, 2023, from https://link.springer.com/article/10.1007/s10549-018-4753-7

Nabholtz J-M, Senn HJ, Bezwoda WR, Melnychuk D, Deschênes L, Douma J, Vandenberg TA, Rapoport B, Rosso R, Trillet-Lenoir V, Drbal J, Molino A, Nortier JWR, Richel DJ, Nagykalnai T, Siedlecki P, Wilking N, Genot JY, Hupperets PSGJ et al (1999) Prospective randomized trial of docetaxel versus mitomycin plus vinblastine in patients with metastatic breast cancer progressing despite previous anthracycline-containing chemotherapy. J Clin Oncol 17(5):Article 5. https://doi.org/10.1200/JCO.1999.17.5.1413

Nabieva N, Fasching PA (2021) Endocrine treatment for breast cancer patients revisited—history, standard of care, and possibilities of improvement. Cancers 13(22):Article 22. https://doi.org/10.3390/cancers13225643

Naito Y, Urasaki T (2018) Precision medicine in breast cancer. Chin Clin Oncol 7(3):29–29. https://doi.org/10.21037/cco.2018.06.04

Nikanjam M, Kato S, Kurzrock R (2022) Liquid biopsy: current technology and clinical applications. J Hematol Oncol 15(1):131. https://doi.org/10.1186/s13045-022-01351-y

O'Driscoll L, Clynes M (2006) Biomarkers and multiple drug resistance in breast cancer. Curr Cancer Drug Targets 6(5):365–384. https://doi.org/10.2174/156800906777723958

Orrantia-Borunda E, Anchondo-Nuñez P, Acuña-Aguilar LE, Gómez-Valles FO, Ramírez-Valdespino CA (2022) Subtypes of breast cancer. In: Mayrovitz HN (ed) Breast cancer. Exon Publications. http://www.ncbi.nlm.nih.gov/books/NBK583808/

Panigrahi A, Pinder S, Chan S, Paish E, Robertson J, Ellis I (2004) The role of PTEN and its signalling pathways, including AKT, in breast cancer; an assessment of relationships with other prognostic factors and with outcome. J Pathol 204(1):93–100. https://doi.org/10.1002/path.1611

Paridaens R, Dirix L, Beex L, Nooij M, Cufer T, Lohrisch C, Biganzoli L, Van Hoorebeeck I, Duchateau L, Lobelle J-P, Piccart M (2000) Promising results with Exemestane in the first-line treatment of metastatic breast cancer: a randomized phase II EORTC trial with a tamoxifen control. Clin Breast Cancer 1:S19–S21. https://doi.org/10.3816/CBC.2000.s.004

Perez EA (2009) Impact, mechanisms, and novel chemotherapy strategies for overcoming resistance to anthracyclines and taxanes in metastatic breast cancer. Breast Cancer Res Treat 114:195–201

Perez EA, Pusztai L, van De Vijver M (2004) Improving patient care through molecular diagnostics. Semin Oncol 31:14–20

Petric RC, Pop L-A, Jurj A, Raduly L, Dumitrascu D, Dragos N, Neagoe IB (2015) Next generation sequencing applications for breast cancer research. Clujul Med 88(3):Article 3. https://doi.org/10.15386/cjmed-486

Pfefferle AD, Agrawal YN, Koboldt DC, Kanchi KL, Herschkowitz JI, Mardis ER, … Perou CM (2016) Genomic profiling of murine mammary tumors identifies potential personalized drug targets for p53-deficient mammary cancers. Dis Model Mech 9(7):749–757

Piccart MJ, Cardoso F (2003) Progress in systemic therapy for breast cancer: an overview and perspectives. Eur J Cancer Suppl 1(2):Article 2. https://doi.org/10.1016/S1359-6349(03)00009-0

Planes-Laine G, Rochigneux P, Bertucci F, Chrétien A-S, Viens P, Sabatier R, Gonçalves A (2019) PD-1/PD-L1 targeting in breast cancer: the first clinical evidences are emerging—a literature review. Cancers 11(7):1033

Potnis KC, Ross JS, Aneja S, Gross CP, Richman IB (2022) Artificial intelligence in breast cancer screening: evaluation of FDA device regulation and future recommendations. JAMA Intern Med 182(12):1306–1312. https://doi.org/10.1001/jamainternmed.2022.4969

Poudel P, Nyamundanda G, Patil Y, Cheang MCU, Sadanandam A (2019) Heterocellular gene signatures reveal luminal-A breast cancer heterogeneity and differential therapeutic responses. Npj Breast Cancer 5(1):Article 1. https://doi.org/10.1038/s41523-019-0116-8

Real PJ, Sierra A, De Juan A, Segovia JC, Lopez-Vega JM, Fernandez-Luna JL (2002) Resistance to chemotherapy via Stat3-dependent overexpression of Bcl-2 in metastatic breast cancer cells. Oncogene 21(50):7611–7618

Regan MM, Pagani O, Fleming GF, Walley BA, Price KN, Rabaglio M, Maibach R, Ruepp B, Coates AS, Goldhirsch A, Colleoni M, Gelber RD, Francis PA (2013) Adjuvant treatment of premenopausal women with endocrine-responsive early breast cancer: design of the TEXT and SOFT trials. Breast 22(6):1094–1100. https://doi.org/10.1016/j.breast.2013.08.009

Research, C. for D. E. and (2024) Oncology (Cancer)/Hematologic Malignancies Approval Notifications. FDA. https://www.fda.gov/drugs/resources-information-approved-drugs/oncology-cancer-hematologic-malignancies-approval-notifications

Risinger JI, Hayes AK, Berchuck A, Barrett JC (1997) PTEN/MMAC1 mutations in endometrial cancers. Cancer Res 57(21):4736–4738

Rizzo JM, Buck MJ (2012) Key principles and clinical applications of "next-generation" DNA sequencing. Cancer Prev Res 5(7):Article 7. https://doi.org/10.1158/1940-6207.CAPR-11-0432

Robinson PN (2012) Deep phenotyping for precision medicine. Hum Mutat 33(5):777–780. https://doi.org/10.1002/humu.22080

Royce M, Bachelot T, Villanueva C, Özgüroğlu M, Azevedo SJ, Cruz FM, Debled M, Hegg R, Toyama T, Falkson C, Jeong J, Srimuninnimit V, Gradishar WJ, Arce C, Ridolfi A, Lin C, Cardoso F (2018) Everolimus plus endocrine therapy for postmenopausal women with Estrogen receptor–positive, human epidermal growth factor receptor 2–negative advanced breast cancer: a clinical trial. JAMA Oncol 4(7):977–984. https://doi.org/10.1001/jamaoncol.2018.0060

Sachdev JC, Sandoval AC, Jahanzeb M (2019) Update on precision medicine in breast cancer. In: Von Hoff DD, Han H (eds) Precision medicine in cancer therapy, vol 178. Springer International Publishing, pp 45–80. https://doi.org/10.1007/978-3-030-16391-4_2

Saha M, Chakraborty C, Racoceanu D (2018) Efficient deep learning model for mitosis detection using breast histopathology images. Comput Med Imaging Graph 64:29–40

Samuel AL (1959) Some studies in machine learning using the game of checkers. IBM J Res Dev 3(3):210–229

Samuels Y, Wang Z, Bardelli A, Silliman N, Ptak J, Szabo S, Yan H, Gazdar A, Powell SM, Riggins GJ, Willson JKV, Markowitz S, Kinzler KW, Vogelstein B, Velculescu VE (2004) High frequency of mutations of the *PIK3CA* gene in human cancers. Science 304(5670): 554–554. https://doi.org/10.1126/science.1096502

Shah SM, Khan RA, Arif S, Sajid U (2022) Artificial intelligence for breast cancer analysis: trends & directions. Comput Biol Med 142:105221

Sharma S, Mehra R (2020) Conventional machine learning and deep learning approach for multi-classification of breast cancer histopathology images—a comparative insight. J Digit Imaging 33:632–654

Shen D-W, Goldenberg S, Pastan IRA, Gottesman MM (2000) Decreased accumulation of [14c] carboplatin in human cisplatin-resistant cells results from reduced energy-dependent uptake. J Cell Physiol 183(1):108–116

Sheth D, Giger ML (2020) Artificial intelligence in the interpretation of breast cancer on MRI. J Magn Reson Imaging 51(5):1310–1324

Shien T, Iwata H (2020) Adjuvant and neoadjuvant therapy for breast cancer. Jpn J Clin Oncol 50(3):Article 3. https://doi.org/10.1093/jjco/hyz213

Shyr D, Liu Q (2013) Next generation sequencing in cancer research and clinical application. Biol Proced Online 15(1):4. https://doi.org/10.1186/1480-9222-15-4

Simpson L, Parsons R (2001) PTEN: life as a tumor suppressor. Exp Cell Res 264(1):29–41. https://doi.org/10.1006/excr.2000.5130

Smith BD, Jiang J, McLaughlin SS, Hurria A, Smith GL, Giordano SH, Buchholz TA (2011) Improvement in breast cancer outcomes over time: are older women missing out? J Clin Oncol 29(35):Article 35. https://doi.org/10.1200/JCO.2011.35.8408

Stephens PJ, Tarpey PS, Davies H, Van Loo P, Greenman C, Wedge DC, Nik-Zainal S, Martin S, Varela I, Bignell GR (2012) The landscape of cancer genes and mutational processes in breast cancer. Nature 486(7403):400–404

Stover DG, Wagle N (2015) Precision medicine in breast cancer: genes, genomes, and the future of Genomically driven treatments. Curr Oncol Rep 17(4):15. https://doi.org/10.1007/s11912-015-0438-0

Syrjala KL (2018) Opportunities for improving oncology care. Lancet Oncol 19(4):449. https://doi.org/10.1016/S1470-2045(18)30208-0

Szymczak S, Biernacka JM, Cordell HJ, González-Recio O, König IR, Zhang H, Sun YV (2009) Machine learning in genome-wide association studies. Genet Epidemiol 33(S1). https://doi.org/10.1002/gepi.20473

Telenti A, Lippert C, Chang P-C, DePristo M (2018) Deep learning of genomic variation and regulatory network data. Hum Mol Genet 27(Supplement_R1):R63–R71. https://doi.org/10.1093/hmg/ddy115

Thomas DC (2017) What does "precision medicine" have to say about prevention? Epidemiology 28(4):479–483

Troxell ML (2012) PIK3CA/AKT1 mutations in breast carcinoma: a comprehensive review of experimental and clinical studies. J Clin Exp Pathol 02(01). https://doi.org/10.4172/2161-0681.S1-002

Understanding Resistance to Tamoxifen in Hormone Receptor–Positive Breast Cancer | Clinical Chemistry | Oxford Academic (n.d.) Retrieved November 29, 2023, from https://academic.oup.com/clinchem/article/55/8/1453/5631800?login=false

Understanding the Basics of NGS: From Mechanism to Variant Calling | Current Genetic Medicine Reports (n.d.) Retrieved January 15, 2024, from https://link.springer.com/article/10.1007/s40142-015-0076-8

van Jaarsveld MT, Wouters MD, Boersma AW, Smid M, van IJcken WF, Mathijssen RH, Hoeijmakers JH, Martens JW, van Laere S, Wiemer EA (2014) DNA damage responsive microRNAs misexpressed in human cancer modulate therapy sensitivity. Mol Oncol 8(3):458–468

Vasan N, Razavi P, Johnson JL, Shao H, Shah H, Antoine A, . . . Baselga J (2019) Double PIK3CA mutations in cis increase oncogenicity and sensitivity to PI3Kα inhibitors. Science 366(6466):714–723

Vivanco I, Sawyers CL (2002) The phosphatidylinositol 3-kinase–AKT pathway in human cancer. Nat Rev Cancer 2(7):489–501. https://doi.org/10.1038/nrc839

Wang Y, Minden A (2022) Current molecular combination therapies used for the treatment of breast cancer. Int J Mol Sci 23(19):11046. https://doi.org/10.3390/ijms231911046

Wang J, Seebacher N, Shi H, Kan Q, Duan Z (2017) Novel strategies to prevent the development of multidrug resistance (MDR) in cancer. Oncotarget 8(48):84559–84571. https://doi.org/10.18632/oncotarget.19187

Wind NS, Holen I (2011) Multidrug resistance in breast cancer: from *in vitro* models to clinical studies. Int J Breast Cancer 2011:1–12. https://doi.org/10.4061/2011/967419

Xia W, Husain I, Liu L, Bacus S, Saini S, Spohn J, . . . Spector NL (2007) Lapatinib antitumor activity is not dependent upon phosphatase and tensin homologue deleted on chromosome 10 in ErbB2-overexpressing breast cancers. Cancer Res 67(3):1170–1175

Xu J, Yang P, Xue S, Sharma B, Sanchez-Martin M, Wang F, Beaty KA, Dehan E, Parikh B (2019) Translating cancer genomics into precision medicine with artificial intelligence: applications, challenges and future perspectives. Hum Genet 138(2):109–124. https://doi.org/10.1007/s00439-019-01970-5

Zelli, V., Compagnoni, C., Cannita, K., Capelli, R., Capalbo, C., Di Vito Nolfi, M., Alesse, E., Zazzeroni, F., & Tessitore, A. (2020). Applications of next generation sequencing to the analysis of familial breast/ovarian cancer. High-Throughput, 9(1), Article 1. https://doi.org/10.3390/ht9010001

Zhang W, Xia W, Lv Z, Ni C, Xin Y, Yang L (2017) Liquid biopsy for cancer: circulating tumor cells, circulating free DNA or exosomes? Cell Physiol Biochem 41(2):755–768. https://doi.org/10.1159/000458736

Zhang, J., Späth, S. S., Marjani, S. L., Zhang, W., & Pan, X. (2018). Characterization of cancer genomic heterogeneity by next-generation sequencing advances precision medicine in cancer treatment. Precis Clin Med, 1(1), 29–48. https://doi.org/10.1093/pcmedi/pby007

Recent Progress and Prospects of Immunotherapy in Multidrug-Resistant and Metastatic Breast Cancer Treatment

Md Abdus Subhan

Abstract

Although chemotherapy and mAbs are widely used in breast cancer treatment, these therapy approaches are allied with side effects and recurrence. US FDA has approved immunotherapy Keytruda in combination with chemotherapy for irresectable metastatic triple-negative breast cancer (TNBC) in 2020. Recent developments in different immunotherapy modalities have shown promise in breast cancer treatment to reduce mortality. Further, multifunctional nanoparticles for siRNA delivery strategies are prospective to make momentous advancement in breast cancer immunotherapy. Utilization of immunotherapies in combination approach may cause immune-related adverse effects. Further investigations to progress immunotherapies, combined chemo-immunotherapies, patient response to immunotherapies, and exclusion of immune-related side effects are essential.

Keywords

Antibody · Breast cancer · Immunotherapy · Metastasis · siRNA

1 Introduction

Brest cancer is the second most frequently diagnosed cancer, which represent 11.6% of all cancers with one of the highest rates of mortality and illness among the women all over the world. Breast cancer fast metastasized to brain, bone marrow, lung, and liver. Most common therapies applied to breast cancer treatment remain

Md. A. Subhan (✉)
Department of Chemistry, Shahjalal University of Science and Technology, Sylhet, Bangladesh
e-mail: subhan-che@sust.edu

© The Author(s), under exclusive license to Springer Nature Switzerland AG 2023
Interdisciplinary Cancer Research, https://doi.org/10.1007/16833_2023_136
Published online: 8 February 2023

chemotherapy, radiation therapy, and surgical removal of breast tumors. However, these therapy approaches are associated with severe side effects and drug resistance making breast cancer therapy difficult. Currently, immunotherapy is a challenge of breast tumor therapy. Immune circumvention is involving in all kinds of cancer and contributes to tumor development and progression. In healthy individuals, the body's immune system can identify abnormal cells and contribute to abolish them. However, cancer cells aspire to escape such damage through downregulation of the immune recognition or by attacking the T cell's function (Evans et al. 2018; Hanahan and Weinberg 2011; Kim et al. 2007; Frey 2015). Cancer immunotherapy emphasizes on regenerating the ability of human body to recognize and inhibit the cancer cells and reinstate the body's jeopardized immune system. Re-engaging this immune response can be reached by different approaches reliant on the type of cancer and target cell (Farkona et al. 2016).

Cancer immunotherapy is promising and prospective therapy modality for cancer patient including the woman with breast cancer. Cytokine, anticytokine, and antibody therapies are efficient in treatment of different forms of cancer. Cytotoxic T cells (such as, CD8+ T cell) are the key cytotoxic machineries; they are accountable for inhibiting anomalous, spoiled, or diseased cells in body (Krummel et al. 2016). The T cells are usually stimulated by antigen presenting cells (APCs). APCs, for example, dendritic cells (DCs) identify and internalize antigens and afterward present these antigens on their surface through major histocompatibility (MHC) receptors. MHC receptors expressing antigens interact with T-cell receptors (TCR) on CD8+ T cells to start a cytotoxic immune response (Krummel et al. 2016). Then the T cells inspect the body and release cytotoxic materials into cells presenting that antigen and induced cell death (Fig. 1) (Evans et al. 2018). Genetically engineered T cells are also efficient in inhibiting cancer cells (Buchbinder and Desai 2016; Frey 2015; Landskron et al. 2014; Iwai et al. 2017, Perica et al. 2015; Zhou 2014).

Cancer vaccines can induce the creation of antigen-specific T cells by delivering tumor antigens to APCs (Buonaguro et al. 2011; Dubensky and Reed 2010). Then APCs interact with CD8+ T cells in the spleen or lymph tissues, and initiate maturation, expansion, as well as migration procedures. This process usually needed a boost such as adjuvant delivery. Whereas the traditional adjuvants applied to boost B cell vaccines are not sufficient for supporting the activation of CD8+ T cells and novel adjuvants such as CpG (cytidine phosphate guanosine), TLR (toll-like receptor) agonists may be applied to provision cancer vaccine. An effective cancer vaccine can care antitumor immunity by promoting the stimulation of DCs, CD4+, and CD8+ T cells (Evans et al. 2018).

Tumor microenvironment (TME) can reduce T cell viability and function even having a strong defense of primed and functional T cells. T cells cytotoxicity can be inhibited by tumor cells through interactions with T cells via PD-1 (programmed cell death protein 1). Further, TME can destroy T-cell function by several mechanisms such as reduced pH, immune-repressive cytokines, or by biological hurdles like incomplete vasculature, dense extracellular matrix. Many therapeutic approaches have been utilized to restore T-cell inactivity in the TME, which allow existing

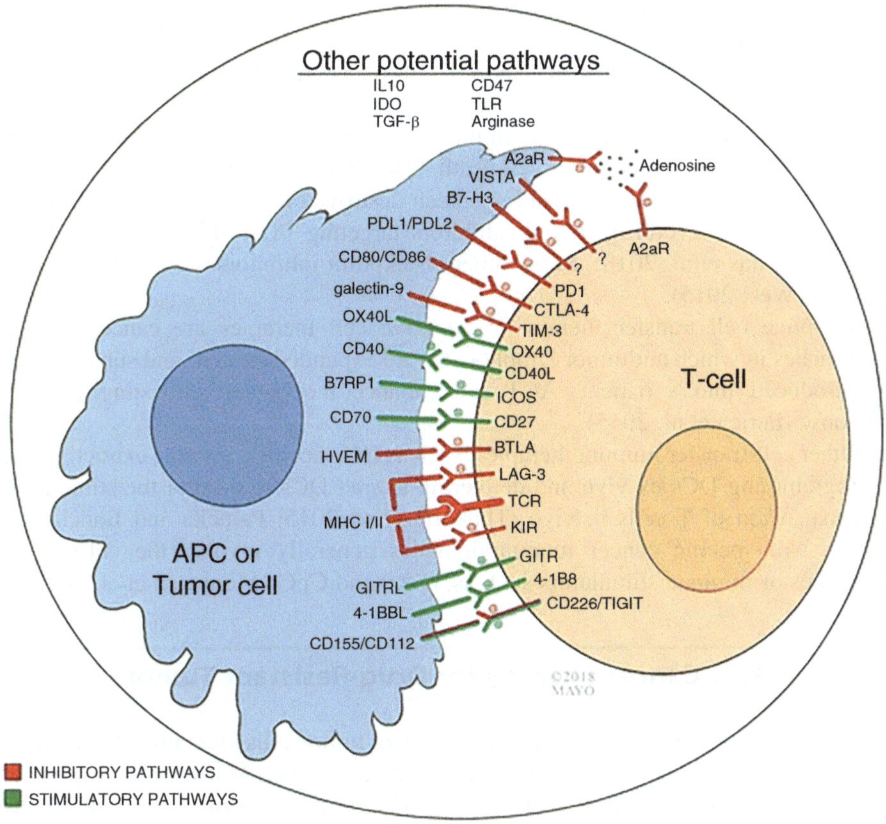

Fig. 1 Immune interactions involving antigen presenting cells or tumor cells, T cells, and tumor microenvironment. (Adapted from Marin-Acevedo et al. 2018)

activated T cells for improved performance, surveillance, and cytotoxic activity to kill cancer cells (Fig. 1) (Evans et al. 2018; Marin-Acevedo et al. 2018).

Cancer immunotherapy links body's immune system to destroy cancer cells. Many immunotherapeutic modalities have been investigated including mAbs, ICIs (immune checkpoint inhibitors), ACT (adoptive cell therapy), and non-specific cancer immunotherapy. Some immunotherapy act at the site of TME to enable immune cells to kill cancer cells. Others strengthen immunity against cancer cells enhancing the quantity of tumor-specific cytotoxic T cells at the site of tumor via cancer vaccine or adoptive cell therapy. Adjuvant-based immunotherapy usually enhances the efficiency or activation of T-cell responses by supporting pathways. Nanoparticles are suitable candidates for these immunotherapy modalities (Kim et al. 2007; Gattinoni et al. 2006; Dougan and Dougan 2017; Guo et al. 2013; Shao et al. 2015).

mAbs are engineered for targeting specific antigen. Upon binding to specific antigen, mAbs accomplish a number of crucial roles such as recruitment of immune cells, regulation of receptor or antigen functions, and local delivery of anticancer therapeutics. mAb therapy is one of the most effective cancer immunotherapies, and is often used in clinics for treating a number of cancers including breast cancer (Scott et al. 2012).

Some cancer cells overexpress immune checkpoints on the surface to disable T cells and escape immunogenic cell death. Checkpoint inhibitors empower existing anticancer immune responses that have been destroyed or neutralized by the tumor cells. There are several approved inhibitors targeting PD-1, PD-L1, and CTLA4 (Fig. 1) (Evans et al. 2018). Some other checkpoint inhibitors are in clinical trials (Fig. 1) (West 2015).

Adoptive cell transfer therapy or adoptive cell therapies are cancer therapy approaches in which antitumor lymphocytes are expanded ex vivo and subsequently re-introduced into a patient. ACT can enhance the patient's existing immune response (Perica et al. 2015).

Other cell-transfer immunotherapies, such as DC vaccines, involve extracting and reprogramming DCs ex vivo and delivering altered DCs to prompt the stimulation and expansion of T cells in vivo (Thakurta et al. 2015; Palucka and Banchereau 2012). Non-specific cancer immunotherapies generally involve the delivery of cytokines or immune stimulators such as TLR9 and CpG (Monjazeb et al. 2012).

2 Breast Cancer Therapy for Drug-Resistant Tumor

The drug resistance is the cause of insensitivity of tumor cells to treatment, which is a major factor in the unsuccessful treatment of breast cancer. Many efforts have been made to reestablish sensitivity of drugs to overwhelmed resistance of the breast cancer. The improved knowledge on resistance mechanisms and development of novel therapeutic modalities may promote clinical outcomes and circumvent the drug resistance in breast cancer (Ji et al. 2019).

In a proportion of woman with breast tumor the disease may relapse. The mechanisms in breast cancer resistance include physicochemical influences of cell membrane on drug absorption, transport, and efflux; function of membrane glycoproteins as pumps; inactivation of anticancer drugs by enzymatic action affecting metabolism; DNA repair; stemness of cancer cells; and remodeling of TME (Ji et al. 2019).

TWIST (Twist family BHLH Transcription Factor), as controller of EMT (epithelial to mesenchymal), can cause breast cancer cell susceptible to metastasis, recurrence, and resistance. TWIST can also enhance drug resistance of tumor cells. Upregulation of TWIST-1 by NF-κB enhances the chemoresistance (Lo et al. 2007; Pham et al. 2007; Vesuna et al. 2011).

IRE1 is an endoplasmic reticulum (ER) stress sensor. IRE1-RNase activity contribute to the basal production of pro-tumorigenic factors including IL-6, IL-8, CXCl-1, GM-CSF (granulocyte macrophage colony-stimulating factor), and TGFβ2 (transforming growth factor beta) in breast cancer including TNBC cells. Some chemotherapeutic anticancer drugs (e.g., paclitaxel) enhance the IRE1-RNase activity causing drug resistance and cancer relapse. Thus, inhibition of IRE1-RNase activity can increase tumor suppression and delay tumor relapse post therapy (Logue et al. 2018).

MDR1 (multidrug resistance protein 1) has crucial function in the therapy resistance of breast cancer (Gao et al. 2013). ATP-binding cassette (ABC) transporters include MDR-related protein, such as P-gp, multidrug resistance associated protein (MRP), ABCC (ATP binding cassette subfamily C member) subfamily, and breast cancer resistance protein (BCRP), such as ABCG (ATP-binding cassette subfamily G member) (Kort et al. 2015). The MDR mechanism of P-gp comprises drug efflux and enhanced DNA repair mechanism. MRP can prompt drug resistance facilitating intracellular drug elimination and amending intracellular drug distribution (Borst et al. 2000; Sodani et al. 2012). BCRP belongs to ABC transporter subfamily (Vtorushin et al. 2014; Mao and Unadkat 2014). Lung resistance protein (LRP) involves in transporting intracytoplasmic anticancer drugs to the vesicles, and then transports outside the cells and diminishes intracellular drug concentration (Ji et al. 2019).

The upstream signaling pathways such as EGFR, HER2, IGF-1R and the downstream signaling pathways including RAS/MAPK/ERK and PI3K/AKT/mTOR can cause resistance in endocrine therapy of breast tumors (Huang et al. 2011). EGFR, HER2/nue, and PI3K are allied with tamoxifen resistance, and EGFR involves resistance to gefitinib in breast cancer therapy (Huang et al. 2011).

miRNAs have an important role in amending sensitivity of breast cancer cells to anticancer drugs. miR-451 and miR-326 result in enhanced sensitivity of breast cancer cell line, MCF-7 to doxorubicin (Kovalchuk et al. 2008; Liang et al. 2010). Some miRNAs develop resistance to endocrine therapy through downregulation of the expression of Erα (He et al. 2013; Muluhngwi and Klinge 2015; Cochrane et al. 2009; Zhou et al. 2010). Further, miRNA can regulate the post translational alteration of Erα to enhance endocrine resistance (Muluhngwi and Klinge 2015). Whereas, miRNA-129-5p may reverse MDR (Luan et al. 2016), miR-218 can regulate cisplatin sensitivity in breast cancer by targeting BRCA1 (He et al. 2015). Cancer stem cells, CSCs, play crucial role in drug resistance in breast cancer (Pavlopoulou et al. 2016). The CSCs overexpress various ABC transporters including P-gp, ABCG2, ABCC1, and ABCB5. CSCs can prompt drug resistance via DNA repair and upregulation of antiapoptotic level (Zhou et al. 2016). Several signal transduction pathways including Notch play a role in drug resistance of CSCs. Over expression of LDH1 (lactate dehydrogenase 1) is also an important cause of drug resistance of CSCs (Tanei et al. 2009; Morschhauser et al., 2007; Yu et al. 2007).

Many breast cancer stem cells (BCSCs) surface markers such as, CD44, have been identified (Collina et al. 2015). The BCSCs exist in two dissimilar development statuses: mesenchymal like, EMT and epithelial-like MET (Pinto et al. 2013). In EMT, BCSCs overexpress $CD44^+$, $CD24^-$, and $EpCAM^+$ (Liu et al. 2013). However, the BCSCs with MET status express ALDH1 (aldehyde dehydrogenase 1). ALDH1 expression is an indicator of poor prognosis in breast cancer (Ginestier et al. 2007).

The BCSCs signaling pathways such as Wnt/Frizzled/β-catenin, Hippo, Notch, and Hedgehog are involved in drug resistance, which have been emphasized in many therapeutic targets (Dandawate et al. 2016). These therapy approaches can modify EMT and CSC character, which may be utilized in clinic (Kotiyal and Bhattacharya 2014).

ALDH, DNA topoisomerase, protein kinase C, dihydrofolate reductase, glutathione, and glutathione S-transferase (GST) are cell detoxifying enzymes, which are the crucial reason of MDR in cancer (Hazarika et al. 2017). GST can enhance drug resistance by inhibiting MAPK (Borrie et al. 2017). Several anticancer drugs such as Cisplatin and Adriamycin are inactivated by GST (Sharma et al. 2017). The GST can also support drug efflux activity of P-gp. GST-π may enhance metabolism of drugs, reduce effective concentration in cells, and decrease drug accumulation in target location leading to decrease in drug efficiency (Al-Harras et al. 2016). The expression of GST-π in breast cancer patient can be linked with grading and lymphatic metastasis and may be used as index for guiding therapy in clinic (Fan et al. 2014).

Topo II is the target enzyme for many drugs. These drugs can bind to Topo II and inhibit DNA damage leading to drug resistance (Ji et al. 2019). TM9SF proteins involve in adhesion, phagocytosis, and innate immunity. *TM9SF3* gene is upregulated in paclitaxel-resistant breast cancer cells (Chang et al. 2011).

Survivin protein is an inhibitor of apoptotic cancer cell death induced by anticancer drugs, which can result in MDR (Véquaud et al. 2015; Yu et al. 2015). Kinase C (PKC), DNA polymerase, and O6-alkylguanine-DNA alkyltransferase (AGT) proteins are also allied with MDR in cancer (Véquaud et al. 2015, Yu et al. 2015).

Autophagy is a type II programmed cell death, plays critical roles with autophagy-related (ATG) proteins in cancer. Tumor cells commonly use autophagy to endure various stresses, such as oncogene-induced transformation, hypoxia, endoplasmic reticulum (ER) stress, and extracellular matrix detachment. Initiation of autophagy by tumor cells also contributes to tumor dormancy, metastasis, and drug resistance to anticancer therapies. Thus, autophagy inhibitors, such as chloroquine, are promising drug candidates for breast cancer treatment to reduce drug resistance (Jain et al. 2013).

There are some intrinsic and extrinsic factors leading to cancer heterogeneity, which may cause drug resistance. Tumors consist of variety of cancer cells with diverse phenotype and genotype, which are responsible for heterogeneity among and within different cancer types. There are intra-tumoral heterogeneity as well as patient heterogeneity in cancer patients (To 2013; Gatenby et al. 2010). Besides these intrinsic factors there are some external factors that may induce drug resistance such as hypoxia, pH, paracrine signal, tumor microenvironment, etc. (Nathanson et al. 2014; Li and Dalton 2006; Ji et al. 2019). Due to this high heterogenicity of the TME and drug efflux resulting from expression of the oncogene and lack of target antigen, breast cancer treatment has become highly challenging.

Tumors are generally exposed to hypoxia. MDR1 is hypoxia-responsive gene (Comerford et al. 2002). Hypoxia-inducible factors, HIF-1 and HIF-2 can bind to the hypoxia response element (HRE) on the promoter of BCRP gene to initiate its expression, and prompt the overexpression of BCRP1 causing drug resistance

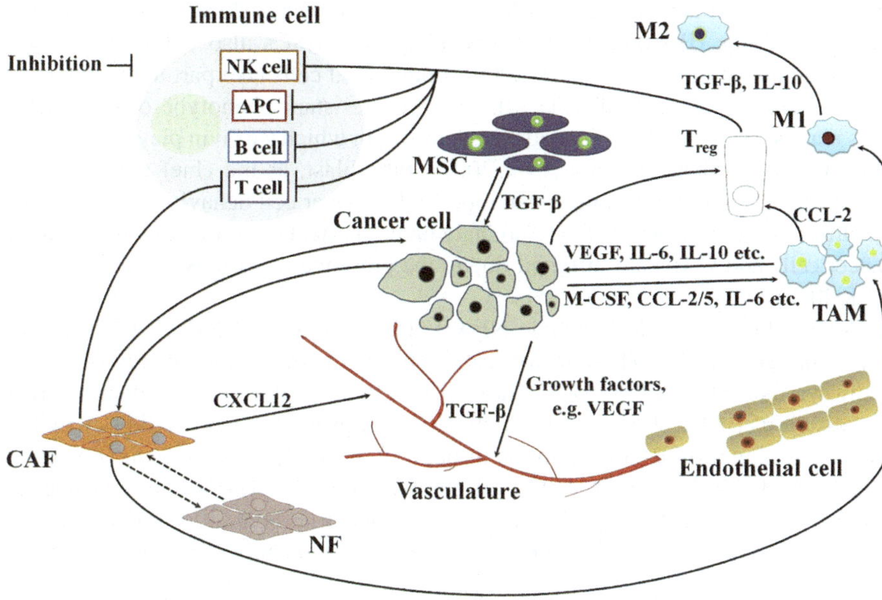

Fig. 2 Communication between cellular machineries of TME and breast cancer cells controls the response to antitumor therapy. (Adapted from Ji et al. 2019). (*CAF* cancer-associated fibroblast, *MSC* mesenchymal stem cell, *NF* normal factor, *M-CSF* macrophage colony-stimulating factor)

(Vannini et al. 2008; Martin et al. 2008). The pH adjustment bestows to the cellular mechanisms of tumors (Estrella et al. 2013). An interaction of stromal cells and breast cancer cells may result in drug resistance prompted by stromal cells (Huang et al. 2017).

The TME consists of stromal cells, extracellular matrix, and interstitial tissues (Spano and Zollo 2012). Cellular machineries of TME include TAMs (tumor-associate macrophages), mesenchymal cells, endothelial cells, and CAF (cancer-associated fibroblast) (Fig. 2) (Ansell and Vonderheide 2013). In TME macrophages defend tumor cells from cytotoxicity of anticancer drugs (Palma and Lewis 2011). TAMs and their products are involved in regulating cancer-cell behaviors in the presence of drugs (Mitchem et al. 2013). M1 TAMs are antitumor and M2 phenotypes are pro-tumor. M2-TAMs are polarized by Th2-derived cytokines such as IL4, IL10, and IL13, transforming growth factor beta (TGFβ) or prostaglandin E2 (PGE2). TAMs contribute to drug resistance and radio-protective effects. Clinical studies demonstrated that an increased number of TAMs and their M2 phenotypes are correlated with therapy failure and poor prognosis in cancer patient including breast cancer (Fig. 2) (Tang et al. 2013). TAMs are generally employed to the tumor site and polarized to M1-type at the early stages of tumor development. TAMs are converted to M2 type in the progressive stages of tumor, which exert their pro-tumor effects by blocking CD8 T cells (Fig. 2) (Tang et al. 2013). Tumor-resident mesenchymal cells are dissimilar from usual mesenchymal cells, which may reduce the drug sensitivity (Patel et al. 2010). Mesenchymal stem cells (MSCs) are resistant to anticancer drugs (Bonomi et al. 2015). However, MSCs also exhibit anticancer activity in breast tumor (Sun et al. 2009). Endothelial cells take part in angiogenesis and vasculogenesis in TME (Fig. 2). The drug-resistant phenotype of endothelial cells may be regulated by epigenetic mechanism, in which survivin plays crucial role (Virrey et al. 2008a, b; Alavi et al. 2007). Fibroblast, as the chief constituent of breast cancer microenvironment, can modify the cancer cell behavior when anticancer drugs are administered (Fig. 2) (Chaiwun et al. 2011). Interactions between the cancer cells and CAFs are different due to the presence of various phenotype (Park et al. 2015).

Soluble factors in TME include NF-κB, TGF-β, EGF and IGF II, ILGF II, etc. (Evseenko et al. 2007). These factors can influence the transcriptional regulations of pathways like PTEN/PI3K/AKT (Hartz et al. 2010). TGF-β has significant part in expression of BCRP (Ehata et al. 2010). EGF can prompt the expression of BCRP. HER2 can trigger BCRP transcription through PI3K/AKT/NF-κB signaling pathway (Zhang et al. 2011). Estradiol (E_2), in combination with TNF-α can enhance the expression of BCRP in MCF-7 cells of breast cancer (Ji et al. 2019). Further, Il-1β and TNF-α in TME may contribute to drug resistance in cancer cells (Ji et al. 2019).

3 Immunotherapy of Breast Cancer

Breast cancer is a complex disease, requiring different types of therapies. The chemotherapy, radiation therapy, and surgical resection are the most frequently used treatment options for breast cancer. However, side effects and drug resistance frequently cause therapy failure and mortality. Currently, immunotherapy is a challenge for breast cancer therapy. Different approaches of anticancer immunother-

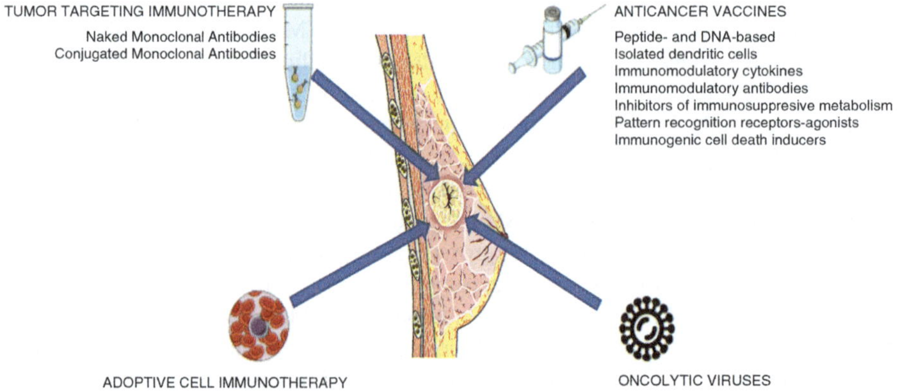

Fig. 3 Different approaches of breast cancer immunotherapy. (Adapted from Garcia-Aranda and Redondo 2019)

apy of breast tumors, including tumor-targeting immunotherapies, oncolytic viruses, anticancer vaccines, or adoptive cell immunotherapies have been developed (Fig. 3). Immunotherapies of breast cancer exhibited efficacy not only against the primary breast cancer but also effective in impeding metastasis and recurrence, complementing surgery, cytotoxic anticancer therapy and radio therapy (Garcia-Aranda and Redondo 2019).

Among the different immunotherapy approaches as monotherapy the oncolytic viruses have exhibited promising ability in targeting and killing cancer cells, activating immunotherapeutic effects in patient, and safety [Tai et al. 2019; Eissa et al. 2018; O'Bryan and Mathis 2018). This may be a promising modality to enhance efficacy in combination therapy and an excellent platform for personalized therapy of advanced and metastatic breast cancer. Many TAM-based immunotherapies for breast cancer are presently being assessed in clinical trials (Qiu et al. 2018).

Adoptive T-cell immunotherapies such as CAR T cell therapy, tumor-infiltrating lymphocyte (TIL) therapy, etc. are demonstrating promise in both pre-clinical and clinical settings. Further, nanoparticle-mediated immunotherapies are also proving valuable for breast cancer therapy for their improved anticancer immune response with minimal systemic side effects (Garcia-Aranda and Redondo 2019; Ayoub et al. 2019; Wang et al. 2014).

Several ICIs are in clinical trials, which can inhibit cancer cells (Ji et al. 2019; Sharma et al. 2017). The ICIs include nivolumab targeting PD-1, PD-L1, ipilimumab targeting CTLA-4, etc. therapies (Amaral et al. 2017). The therapeutic approach of ICIs combined with histone deacetylase inhibitor (HDACi), such as vorinostat and panobinostat, have been utilized to engulf drug resistance in clinic. CDK4/6 inhibitors are effective in exerting immunoregulatory functions in ER-positive breast cancer (Deng et al. 2018). Immunomodulator CsA and cyclosporin A can alter MDR by impeding P-gp. Immunotherapy of breast cancer is effective in combination with chemotherapy and radiation therapy and improves anticancer immune response of breast cancer (Muraro et al. 2017). However, extensive applications of immunotherapy may lead to acquired or secondary drug resistance (Ji et al. 2019). Some approved checkpoint-based immunotherapy of breast cancer is listed in Table 1 (Garcia-Aranda and Redondo 2019).

mAb-based immunotherapies of breast tumors utilize humanized mAbs (Shuptrine et al. 2012). This immunotherapy approach as monotherapy or combination therapy with other treatment modalities has contributed greatly to improve therapeutic options of breast tumors.

While mAb-based immunotherapies targeting altered proteins enhanced the result of breast cancer patients, modest response rate and development of drug resistance still persist as key obstacle for therapy achievement, and necessitate new modalities, such as antibody drug conjugate (ADC) and T-cell bispecific Abs, which are most prospective approaches for breast cancer patients. Some FDA approved mAb-based immunotherapies of breast cancer include Trastuzumab, Pertuzumab, Lapatinib, Neratinib, and Afatinib (Garcia-Aranda and Redondo 2019).

Table 1 FDA-approved immune checkpoint-based cancer immunotherapy therapies

Immune checkpoint target	Summary	Approved drugs
CTLA4 (CD152)	CTLA-4 inhibition prevents interaction with CD80/86 resulting in enhancement of T-cell function	Yervoy® (ipilimumab, Bristol Myers Squibb), FDA approved in 2011
PD-1 (CD279)	PD-1 inhibition by its ligand improves T cell-induced immune responses	Opdivo® (nivolumab, Bristol-Myers Squibb), FDA approved in 2014 Keytruda® (pembrolizumab, Merck KGaA) is a human PD-1-blocking antibody, FDA approved in 2014 Libtayo® (cemiplimab-rwlc, Sanofi S.A.), FDA approved in 2018
PD-L1 (CD274)	Immune inhibitory receptor ligands expressed by hematopoietic and non-hematopoietic cells such as T cells and B cells and different types of tumor cells	Tecentriq® (atezolizumab, Genentech Inc.), FDA approved in 2016 for use in combination with Abraxane® for the treatment of metastatic triple-negative breast cancer Bavencio® (avelumab, Merck EMD Serono), FDA approved in 2017 Imfizi® (durvalumab, AstraZeneca plc) is an anti-PD-L1 human mAb, FDA approved in 2017

MSI-H microsatellite instability high, *dMMR* mismatch repair deficient

4 Immunotherapy of Breast Cancer Utilizing Gene Delivery Approach

RNAi is fast developing as a potential therapy in the field of cancer treatment. siRNAs are used to switch off specific cancer genes that harvest damaging or irregular proteins in breast cancer. The function of siRNA can be utilized to connect cellular machinery to extinguish a corresponding sequence of mRNA that encodes a disease-causing protein. For example, *cis*-platin-induced TGIF (transforming growth factor-β (TGF-β)-induced factor) protein can be downregulated by siTGIF, which can sensitize MDA-MB-231 TNBC breast cancer cells to *cis*-platin resistance.

siRNA nanodrugs have been utilized for the downregulation of immune checkpoint proteins, transmembrane proteins, and anti-inflammatory cytokines to improve antitumor immune responses. Nanoparticle-induced targeted delivery of siRNA to CD4+ T cell lines, B lymphocytes, NK cells, and immunosuppressive cytokines have demonstrated prospects in anticancer therapy (Mei et al. 2019).

Nanomaterials have a vital part in cancer immunotherapy by assisting delivery of siRNAs (Guo et al. 2014; Horn et al. 2018). Therefore, siRNA drug-induced immunotherapy is a promising and prospective approach for breast cancer immunotherapy (Table 2) (Lin et al. 2020).

Table 2 Some selected siRNA nanodrugs for tumor immunotherapy

Cell types	Nanocarriers of siRNA	Target gene	Immunologic function
Tumor cells	PCL-PEG/PCL-PEI	PD-L1	ICI
	FA-PEI polymers	PD-L1	ICI
	Au-CGKRK nanoconjugates	DD-L1, STAT3	Antiproliferation and ICI
	Liposome-protamine-HA NPs	TGF-β	Reduce TGF-β and increase antigen-specific immune response
	ROS-sensitive NPs	TGF-β	Modify the immunosuppressive microenvironment
	HA-coated liposome	CD47	Reduce immune escape of the tumor cells
	Chitosan lactate	CD73	Reduce the immunosuppressive microenvironment of tumor
	EVs	B-catenin	Combination therapy with ICB
T cell	Lipid-coated calcium phosphate	PD-1	ICI
	PEG-PLA	CTLA-4	ICI
TAMs	Au NPs	TNF-α	Silence pro-inflammatory cytokines
	Au NPs	VEGF	Reduce the recruitment of inflammatory TAMs
	Peptide NPs	CSF-IR	Elimination of M2-like TAMs
DCs	Gold nanorods or GNRs-PEI	IDO	Enhance DCs maturation, and increase secretion of pro-inflammatory cytokines
	Cationic lipid NPs	PD-L1, PD-L2	ICI
	PEI-based NPs	PD-L1	Prompt immunosuppressive DCs to APCs
	PLGA polymeric NPs	IDO	Increase secretion of pro-inflammatory cytokines
	PLGA NPs	STAT3	Induce DCs maturation and foster antigen-cross presentation
	PLGA NPs	SOCS1	Enhance the production and release of pro-inflammatory cytokines

Multifunctional NPs for siRNA delivery offer a hybrid system, which are suitable for fighting MDR and metastatic cancer, whereas, monofunctional NPs are useful for siRNA delivery to localized cancer (Serda 2013). Multifunctional NPs for siRNA delivery approaches are likely to make noteworthy developments in cancer immunotherapy, mostly in MDR and metastatic breast cancer.

5 Conclusion and Future Perspective

Chemotherapy and mAbs are extensively utilized in the prognosis and therapy of breast cancer patient. US FDA on November 13, 2020, has approved immunotherapy Keytruda (pembrolizumab) in combination with chemotherapy for irresectable metastatic TNBC, PD-L-1-positive breast cancer. The improvement of response rate of immunotherapies of breast cancer remains a great challenge. In recent years, different immunotherapy modalities including oncolytic viruses, TAMs-targeted immunotherapies, adoptive cell immunotherapies, neoantigen-based translational immunotherapies, nanoparticle-based immunotherapies have been improved and shown promising results in breast cancer therapy. Nanoparticle-based immunotherapies are effective in reducing immune-suppressive effects of TME, drug resistance, delivering neoantigens, and adjuvants to tumor cells, and dropping the side effects of anticancer chemotherapeutics with fewer side effects. Multifunctional nanoparticles for siRNA delivery strategies are likely to produce substantial improvements in cancer immunotherapy, mainly in MDR and metastatic breast cancer treatment. Despite many successes in the development of immunotherapies of metastatic BC and TNBC cancers, there are some drawbacks that still need to be addressed for successful utilization of the immunotherapies in clinic. When immunotherapies are used in combination regimen, incidence of immune-related side effects, affecting several organs such as skin and gastrointestinal track may happen. Although, in most cases the severity of such side effects may be mild or not life threatening. Further investigations to develop new biomarker panels for patient selection, prediction of prognosis, prophecy of patient response to immunotherapies, and immune-related adverse effects are necessary for effective breast cancer therapy.

References

Alavi AS, Acevedo L, Min W, Cheresh DA (2007) Chemoresistance of endothelial cells induced by basic fibroblast growth factor depends on Raf-1-mediated inhibition of the proapoptotic kinase, ASK1. Cancer Res 67(6):2766–2772

Al-Harras MF, Houssen ME, Shaker ME, Farag K, Farouk O, Monir R, El Mahdy R, Abo-Hashem EM (2016) Polymorphisms of glutathione S-transferase π 1 and toll-like receptors 2 and 9: association with breast cancer susceptibility. Oncol Lett 11(3):2182–2188

Amaral T, Meraz-Torres F, Garbe C (2017) Immunotherapy in managing metastatic melanoma: which treatment when? Expert Opin Biol Ther 17(12):1523–1538

Ansell SM, Vonderheide RH (2013) Cellular composition of the tumor microenvironment. Am Soc Clin Oncol Educ Book 33:e91

Ayoub NM, Al-Shami KM, Yaghan RJ (2019) Immunotherapy for HER2-positive breast cancer: recent advances and combination therapeutic approaches. Breast Cancer Targets Ther 11:53–69

Bonomi A, Silini A, Vertua E et al (2015) Human amniotic mesenchymal stromal cells(hAMSCs) as potential vehicles for drug delivery in cancer therapy: an in vitro study. Stem Cell Res Ther 6(1):155

Borrie SC, Brems H, Legius E, Bagni C (2017) Cognitive dysfunctions in intellectual disabilities: the contributions of the Ras-MAPK and PI3K-AKT-mTOR pathways. Annu Rev Genomics Hum Genet 18:115–142

Borst P, Evers R, Kool M, Wijnholds J (2000) A family of drug transporters: the multi-drug resistance-associated proteins. J Natl Cancer Inst 92(16):1295–1302

Buchbinder EI, Desai A (2016) CTLA-4 and PD-1 pathways. Am J Clin Oncol 39:98–106

Buonaguro L et al (2011) Translating tumor antigens into cancer vaccine. Clin Vaccine Immunol 18:23–34

Chaiwun B, Sukhamwang N, Trakultivakorn H et al (2011) GSTPi-positive tumour microenvironment-associated fibroblasts are significantly associated with GSTPi-negative cancer cells in paired cases of primary invasive breast cancer and axillar lymph node metastases. Br J Cancer 105(8):1224–1229

Chang H, Jeung HC, Jung JJ, Kim TS, Rha SY, Chung HC (2011) Identification of genes associated with chemosensitivity to SAHA/taxane combination treatment in taxane- resistant breast cancer cell. Breast Cancer Res Treat 125(1):55–63

Cochrane DR, Spoelstra NS, Howe EN, Nordeen SK, Richer JK (2009) MicroRNA-200c mitigates invasiveness and restores sensitivity to microtubule-targeting chemotherapeutic agents. Mol Cancer Ther 8(5):1055–1066

Collina F, Bonito MD, Bergolis VL et al (2015) Prognostic value of cancer stem cells markers in triple-negative breast cancer. Biomed Res Int 2015:1–10

Comerford KM, Wallace TJ, Karhausen J, Louis NA, Montalto MC, Colgan SP (2002) Hypoxia-inducible factor-1-dependent regulation of the multidrug resistance (MDR1) gene. Cancer Res 62(12):3387–3394

Dandawate PR, Subramaniam D, Jensen RA, Anant S (2016) Targeting cancer stem cells and signaling pathways by phytochemicals: novel approach for breast cancer therapy. Semin Cancer Biol 40–44:192–208

Deng J, Wang ES, Jenkins RW et al (2018) CDK4/6 inhibition augments antitumor immunity by enhancing T-cell activation. Cancer Discov 8(2):216–233

Dougan M, Dougan SK (2017) Targeting immunotherapy to the tumor microenvironment. J Cell Biochem 118(10):3049–3054

Dubensky TW Jr, Reed SG (2010) Adjuvants for cancer vaccines. Semin Immunol 22:155–161

Ehata S, Johansson E, Katayama R et al (2010) Transforming growth factor-β decreases the cancer-initiating cell population within diffuse-type gastric carcinoma cells. Oncogene 30(14): 1693–1705

Eissa IR, Bustos-Villalobos I, Ichinose T, Matsumura S, Naoe Y, Miyajima N, Morimoto D, Mukoyama N, Zhiwen W, Tanaka M et al (2018) The current status and future prospects of oncolytic viruses in clinical trials against melanoma, glioma, pancreatic, and breast cancers. Cancers 10:356

Estrella V, Chen T, Lloyd M et al (2013) Acidity generated by the tumor microenvironment drives local invasion. Cancer Res 73(5):1524–1535

Evans ER, Bugga P, Drezek R (2018) Metallic nanoparticles for cancer immunotherapy. Mater Today (Kidlington) 21(6):673

Evseenko DA, Paxton JW, Keelan JA (2007) Independent regulation of apical and basolateral drug transporter expression and function in placental trophoblasts by cytokines, steroids, and growth factors. Drug Metab Dispos 35(4):595–601

Fan L, Strasser-Weippl K, Li J-J et al (2014) Breast cancer in China. Lancet Oncol 15(7):e279–e289

Farkona S, Diamandis EP, Blasutig IM (2016) Cancer immunotherapy: the beginning of the end of cancer? BMC Med 14:73

Frey AB (2015) Suppression of T cell responses in the tumor microenvironment. Vaccine 33: 7393–7400

Gao L, Yang Y, Song S, Hong H, Zhao X, Li D (2013) The association between geneticvariant of MDR1 gene and breast cancer risk factors in Chinese women. Int Immunopharmacol 17(1): 88–91

Garcia-Aranda M, Redondo M (2019) Immunotherapy: a challenge of breast cancer treatment. Cancers 11:1822

Gatenby RA, Gillies RJ, Brown JS (2010) Evolutionary dynamics of cancer prevention. Nat Rev Cancer 10(8):526–527

Gattinoni L et al (2006) Adoptive immunotherapy for cancer: building on success. Nat Rev Immunol 6:383–393

Ginestier C, Hur MH, Charafe-Jauffret E et al (2007) ALDH1 is a marker of normal and malignant human mammary stem cells and a predictor of poor clinical outcome. Cell Stem Cell 1(5): 555–567

Guo C et al (2013) Therapeutic cancer vaccines: past, present and future. Adv Cancer Res 119: 421–475
Guo L, Yan DD, Yang D et al (2014) Combinatorial Photothermal and Immuno cancer therapy using chitosan-coated hollow copper sulfide nanoparticles. ACS Nano 8(6):5670–5681
Hanahan D, Weinberg RA (2011) Hallmarks of cancer: the next generation. Cell 144:646–674
Hartz AMS, Madole EK, Miller DS, Bauer B (2010) Estrogen receptor signaling through phosphatase and tensin homolog/phosphoinositide 3-kinase/akt/glycogen synthase kinase 3 down-regulates blood-brain barrier breast cancer resistance protein. J Pharmacol Exp Ther 334(2): 467–476
Hazarika M, Chuk MK, Theoret MR et al (2017) U.S. FDA approval summary: nivolumab for treatment of unresectable or metastatic melanoma following progression on ipilimumab. Clin Cancer Res 23(14):3484–3488
He Y-J, Wu J-Z, Ji M-H et al (2013) miR-342 is associated with estrogen receptor-α expression and response to tamoxifen in breast cancer. Exp Ther Med 5(3):813–818
He X, Xiao X, Dong L, Wan N, Zhou Z, Deng H, Zhang X (2015) MiR-218 regulates Cisplatin chemosensitivity in breast cancer by targeting BRCA1. Tumour Biol 36(3):2065–2075
Horn L, Mansfield AS, Szczęsna A et al (2018) First-line Atezolizumab plus chemotherapy in extensive-stage small-cell lung cancer. N Engl J Med 379(23):2220–2229
Huang W-C, Chen Y-J, Li L-Y et al (2011) Nuclear translocation of epidermal growth factor receptor byakt-dependent phosphorylation enhances breast cancer-resistant protein expression in gefitinib-resistant cells. J Biol Chem 286(23):20558–20568
Huang J, Woods P, Normolle D et al (2017) Downregulation of estrogen receptor and modulation of growth of breast cancer cell lines mediated by paracrine stromal cell signals. Breast Cancer Res Treat 161(2):229–243
Iwai Y et al (2017) Cancer immunotherapies targeting the PD-1 signaling pathway. J Biomed Sci 24:26
Jain K, Paranandi KS, Sridharan S, Basu A (2013) Autophagy in breast cancer and its implications for therapy. Am J Cancer Res 3(3):251–265
Ji X, Lu Y, Tian H, Meng X, Wei M, Cho WC (2019) Chemoresistance mechanisms of breast cancer and their countermeasures. Biomed Pharmacother 114:108800
Kim R, Emi M, Tanabe K (2007) Cancer immunoediting from immune surveillance to immune escape. Immunology 121:1–14
Kort A, Durmus S, Sparidans RW, Wagenaar E, Beijnen JH, Schinkel AH (2015) Brain and testis accumulation of regorafenib is restricted by breast cancer resistance protein (BCRP/ABCG2) and P-glycoprotein (P-GP/ABCB1). Pharm Res 32(7):2205–2216
Kotiyal S, Bhattacharya S (2014) Breast cancer stem cells, EMT and therapeutic targets. Biochem Biophys Res Commun 453(1):112–116
Kovalchuk O, Filkowski J, Meservy J et al (2008) Involvement of microRNA-451 in resistance of the MCF-7 breast cancer cells to chemotherapeutic drug doxorubicin. Mol Cancer Ther 7(7): 2152–2159
Krummel MF, Bartumeus F, Gerard A (2016) T-cell migration, search strategies and mechanisms. Nat Rev Immunol 16:193–201
Landskron G et al (2014) Chronic inflammation and cytokines in the tumor microenvironment. J Immunol Res 2014:149185
Li Z-W, Dalton WS (2006) Tumor microenvironment and drug resistance in hematologic malignancies. Blood Rev 20(6):333–342
Liang Z, Wu H, Xia J et al (2010) Involvement of miR-326 in chemotherapy resistance of breast cancer through modulating expression of multidrug resistance-associated protein1. Biochem Pharmacol 79(6):817–824
Lin YX, Wang Y, Blake S, Yu M, Mei L, Wang H, Shi J (2020) RNA nanotechnology-mediated cancer immunotherapy. Theranostics 10(1):281–299
Liu S, Cong Y, Wang D et al (2013) Breast cancer stem cells transition between epithelial and mesenchymal states reflective of their normal counterparts. Stem Cell Rep 2(1):78–91

Lo H-W, Hsu S-C, Xia W, Cao X, Shih J-Y, Wei Y, Abbruzzese JL, Hortobagyi GN, Hung M-C (2007) Epidermal growth factor receptor cooperates with signal transducer and activator of transcription 3 to induce epithelial-mesench-ymal transition in cancer cells via up-regulation of TWIST gene expression. Cancer Res 67(19):9066–9076

Logue SE, McGrath EP, Cleary P et al (2018) Inhibition of IRE1 RNase activity modulates the tumor cell secretome and enhances response to chemotherapy. Nat Commun 9:3267

Luan QX, Zhang BG, Li XJ, Guo MY (2016) MiR-129-5p is downregulated in breast cancer cells partly due to promoter H3K27m3 modification and regulates epithelial-mesenchymal transition and multi-drug resistance. Eur Rev Med Pharmacol Sci 20(20):4257–4265

Mao Q, Unadkat JD (2014) Role of the breast Cancer resistance protein (BCRP/ABCG2) in drug transport—an update. AAPS J 17(1):65–82

Marin-Acevedo JA, Dholaria B, Soyano AE, Knutson KL, Chumsri S, Lou Y (2018) Next generation of immune checkpoint therapy in cancer: new developments and challenges. J Hematol Oncol 11(39):1–20

Martin CM, Ferdous A, Gallardo T et al (2008) Hypoxia-inducible Factor-2α transactivates Abcg2 and promotes cytoprotection in cardiac side population cells. Circ Res 102(9):1075–1081

Mei Y, Wang R, Jiang W et al (2019) Recent progress in nanomaterials for nucleic acid delivery in cancer immunotherapy. Biomater Sci 7(7):2640–2651

Mitchem JB, Brennan DJ, Knolhoff BL et al (2013) Targeting tumor-infiltrating macrophages decrease tumor-initiating cells, relieves immunosuppression, and improves chemotherapeutic responses. Cancer Res 73(3):1128–1141

Monjazeb AM, Hsiao H-H, Sckisel GD, Murphy WJ (2012) The role of antigen-specific and non-specific immunotherapy in the treatment of cancer. J Immunotoxicol 9:248–258

Morschhauser F, Zinzani PL, Burgess M, Sloots L, Bouafia F, Dumontet C (2007) Phase I/II trial of a P-glycoprotein inhibitor, Zosuquidar.3HCl trihydrochloride(LY335979), given orally in combination with the CHOP regimen in patients withnon-Hodgkin's lymphoma, Leuk. Lymphoma 48(4):708–715

Muluhngwi P, Klinge CM (2015) Roles for miRNAs in endocrine resistance in breast cancer. Endocr Relat Cancer 22(5):R279–R300

Muraro E, Furlan C, Avanzo M et al (2017) Local high-dose radiotherapy induces systemic immunomodulating effects of potential therapeutic relevance in oligometastatic breast cancer. Front Immunol 8(1476):1–19

Nathanson DA, Gini B, Mottahedeh J et al (2014) Targeted therapy resistance mediated by dynamic regulation of extrachromosomal mutant EGFR DNA. Science 343(6166):72–76

O'Bryan SM, Mathis JM (2018) Oncolytic virotherapy for breast cancer treatment. Curr Gene Ther 18:192–205

Palma MD, Lewis CE (2011) Cancer: macrophages limit chemotherapy. Nature 472(7343): 303–304

Palucka K, Banchereau J (2012) Cancer immunotherapy via dendritic cells. Nat Rev Cancer 12: 265–277

Park SY, Kim HM, Koo JS (2015) Differential expression of cancer-associated fibroblast-related proteins according to molecular subtype and stromal histology in breast cancer. Breast Cancer Res Treat 149(3):727–741

Patel SA, Meyer JR, Greco SJ, Corcoran KE, Bryan M, Rameshwar P (2010) Mesenchymal stem cells protect breast cancer cells through regulatory T cells: role of mesenchymal stem cell-derived TGF-β. J Immunol 184(10):5885–5894

Pavlopoulou A, Oktay Y, Vougas K, Louka M, Vorgias CE, Georgakilas AG (2016) Determinants of resistance to chemotherapy and ionizing radiation in breast cancer stem cells. Cancer Lett 380(2):485–493

Perica K et al (2015) Adoptive T cell immunotherapy for cancer. Rambam Maimonides Med J 6: e0004

Pham CG, Bubici C, Zazzeroni F, Knabb JR, Papa S, Kuntzen C, Franzoso G (2007) Upregulation of Twist-1 by NF-B blocks cytotoxicity induced by chemotherapeutic drugs. Mol Cell Biol 27(11):3920–3935

Pinto CA, Widodo E, Waltham M, Thompson EW (2013) Breast cancer stem cells and epithelial mesenchymal plasticity – implications for chemoresistance. Cancer Lett 341(1):56–62

Qiu SQ, Waaijer SJH, Zwager MC, de Vries EGE, van der Vegt B, Schroder CP (2018) Tumor-associated macrophages in breast cancer: innocent bystander or important player? Cancer Treat Rev 70:178–189

Scott AM, Wolchok JD, Old LJ (2012) Antibody therapy of cancer. Nat Rev Cancer 12:278–287

Serda RE (2013) Particle platforms for cancer immunotherapy. Int J Nanomedicine 8:1683–1696

Shao K et al (2015) Nanoparticle-based immunotherapy for cancer. ACS Nano 9:16–30

Sharma P, Hu-Lieskovan S, Wargo JA, Ribas A (2017) Primary, adaptive, and acquired resistance to cancer immunotherapy. Cell 168(4):707–723

Shuptrine C, Surana R, Weiner LM (2012) Monoclonal antibodies for the treatment of cancer. Semin Cancer Biol 22(1):3–13

Sodani K, Patel A, Kathawala RJ, Chen ZS (2012) Multidrug resistance associated proteins in multidrug resistance. Chin J Cancer 31(2):58–72

Spano D, Zollo M (2012) Tumor microenvironment: a main actor in the metastasis process. Clin Exp Metastasis 29(4):381–395

Sun B, Roh KH, Park JR, Lee SR, Park SB, Jung JW, Kang SK, Lee YS, Kang KS (2009) Therapeutic potential of mesenchymal stromal cells in a mouse breast cancer metastasis model. Cytotherapy 11(3):289–298

Tai CJ, Liu CH, Pan YC, Wong SH, Tai CJ, Richardson CD, Lin LT (2019) Chemovirotherapeutic treatment using Camptothecin enhances oncolytic measles virus-mediated killing of breast cancer cells. Sci Rep 9:6767

Tanei T, Morimoto K, Shimazu K, Kim SJ, Tanji Y, Taguchi T, Tamaki Y, Noguchi S (2009) Association of breast cancer stem cells identified by aldehyde dehydrogenase 1 expression with resistance to sequential Paclitaxel and epirubicin-based chemotherapy for breast cancers. Clin Cancer Res 15(12):4234–4241

Tang X, Mo C, Wang Y, Wei D, Xiao H (2013) Anti-tumour strategies aiming to target tumour-associated macrophages. Immunology 138(2):93–104

Thakurta DG, Sheikh NA, Fan L-Q et al (2015) Humoral immune response against nontargeted tumor antigens after treatment with Sipuleucel-T and its association with improved clinical outcome. Clin Cancer Res 21:3619–3630

To KKW (2013) MicroRNA: a prognostic biomarker and a possible druggable target for circumventing multidrug resistance in cancer chemotherapy. J Biomed Sci 20:99

Vannini I, Zoli W, Tesei A et al (2008) Role of p53 codon 72 arginine allele in cell survival in vitro and in the clinical outcome of patients with advanced breast cancer. Tumour Biol 29(3):145–151

Véquaud E, Desplanques G, Jézéquel P, Juin P, Barillé-Nion S (2015) Survivin contributes to DNA repair by homologous recombination in breast cancer cells. Breast Cancer Res Treat 155(1): 53–63

Vesuna F, Lisok A, Kimble B, Domek J, Kato Y, van der Groep P, Artemov D, Kowalski J, Carraway H, van Diest P, Raman V (2011) Twist contributes to hormoneresistance in breast cancer by downregulating estrogen receptor-α. Oncogene 31(27):3223–3234

Virrey JJ, Dong D, Stiles C et al (2008a) Stress chaperone GRP78/BiP confers Chemoresistance to tumor-associated endothelial cells. Mol Cancer Res 6(8):1268–1275

Virrey JJ, Guan S, Li W, Schönthal AH, Chen TC, Hofman FM (2008b) Increased survivin expression confers chemoresistance to tumor-associated endothelial cells. Am J Pathol 173(2):575–585

Vtorushin SV, Khristenko KY, Zavyalova MV et al (2014) The phenomenon of multi-drug resistance in the treatment of malignant tumors. Exp Oncol 36(3):144–156

Wang ZX, Cao JX, Wang M, Li D, Cui YX, Zhang XY, Liu JL, Li L (2014) Adoptive cellular immunotherapy for the treatment of patients with breast cancer: a meta-analysis. Cytotherapy 16:934–945

West HJ (2015) JAMA oncology patient page. Neoadjuvant therapy. JAMA Oncol 1:115

Yu F, Yao H, Zhu P, Zhang X, Pan Q, Gong C, Huang Y, Hu X, Su F, Lieberman J, Song E (2007) let-7 regulates self renewal and tumorigenicity of breast cancer cells. Cell 131(6):1109–1123

Yu CJ, Ou JH, Wang ML, Jialielihan N, Liu YH (2015) Elevated survivin mediated multidrug resistance and reduced apoptosis in breast cancer stem cells. J BUON 20(5):1287–1294

Zhang W, Ding W, Chen Y, Feng M, Ouyang Y, Yu Y, He Z (2011) Up-regulation of breast cancer resistance protein plays a role in HER2-mediated chemoresistance through PI3K/Akt and nuclear factor-kappa B signaling pathways in MCF7 breast cancer cells. Acta Biochim Biophys Sin Shanghai 43(8):647–653

Zhou J (2014) Advances and prospects in cancer immunotherapy. New J Sci 745808:1–13

Zhou M, Liu Z, Zhao Y et al (2010) MicroRNA-125b confers the resistance of breast cancer cells to paclitaxel through suppression of pro-apoptotic Bcl-2 antagonist killer 1 (Bak1)expression. J Biol Chem 285(28):21496–21507

Zhou ZY, Wan LL, Yang QJ, Han YL, Li D, Lu J, Guo C (2016) Nilotinib reverses ABCB1/P-glycoprotein-mediated multidrug resistance but increases cardiotoxicity of doxorubicin in a MDR xenograft model. Toxicol Lett 259:124–132

[illegible]

Targeting Myeloid Determinants of Breast Cancer

Chie Kudo-Saito and Yukinori Ozaki

Abstract

Breast cancer (BC) is the most commonly diagnosed cancer in women worldwide and the second largest cause of cancer-related deaths. Recently, numerous agents, including inhibitors targeting CDK4/6, PARP, and immune checkpoint molecules, have been clinically developed as molecular targeted therapy in the treatment of BC. However, these treatments are applicable to a limited type of BC, and are effective in only a small proportion of the treated patients, or recurrence and metastasis frequently occur in the treated patients even if receiving the treatments and effective initially. This is because tumor cells evolutionally change through the interplay with numerous components in the host, and the consequent increase of heterogeneity and complexity of both tumor cells and host environment profoundly impacts on the therapeutic efficacy. The immune system is the most influential factor in their reciprocal evolution, and the better management may be the key to successful cancer treatment. In the human immune system, myeloid cell population is the major cellular component, and its abnormality could thoroughly impact on the whole body potentially leading to treatment resistance of cancer. Thus, a deeper understanding of this large myeloid cell population may lead to successful immune control in the treatment of a broader range of types of BC, including not only TNBC but also other types. Here, we provide a recent overview of various myeloid populations in association with BC, and possible treatments that may be able to inhibit or eliminate the myeloid-derived vicious factors.

C. Kudo-Saito (✉)
Department of Immune Medicine, National Cancer Center Research Institute, Tokyo, Japan
e-mail: ckudo@ncc.go.jp

Y. Ozaki
Department of Immune Medicine, National Cancer Center Research Institute, Tokyo, Japan

Department of Breast Medical Oncology, Cancer Institute Hospital of JFCR, Tokyo, Japan

© The Author(s), under exclusive license to Springer Nature Switzerland AG 2024
Interdisciplinary Cancer Research, https://doi.org/10.1007/16833_2024_210
Published online: 28 February 2024

Keywords

Breast cancer · Dendritic cells · Eosinophils · Macrophages · Mast cells · Mesenchymal stromal cells · Myeloid cells · Neutrophils

1 Introduction

Breast cancer (BC) is the most commonly diagnosed cancer in women worldwide and the second largest cause of cancer-related deaths (Siegel et al. 2023). BC is histologically divided into invasive cancer (approximately 90%) and non-invasive cancer (approximately 10%), and the invasive cancer is further classified into four subtypes based on the molecular expression of estrogen receptor (ER) or progesterone receptor (PR), and oncogenic receptor tyrosine kinase ERBB2 (known as HER2): (1) ER/PR$^+$HER2$^-$ luminal type (60–70%), (2) ER/PR$^+$HER2$^+$ luminal HER2 type (approximately 10%), (3) ER/PR$^-$HER2$^+$ HER2-positive type (approximately 10%), and (4) ER/PR$^-$HER2$^-$ triple-negative type called "TNBC" (10–15%) (Orrantia-Borunda et al. 2022). The treatment approaches, including surgery, radiotherapy, chemotherapy, hormone therapy, molecular targeted therapy, and immunotherapy, are dependent on the phenotypes (Barzaman et al. 2020). For example, metastatic/recurrent HER2$^+$ BC is treated with chemotherapeutics and anti-HER2 mAb, including trastuzumab, pertuzumab, trastuzumab emtansine, and trastuzumab deruxtecan (Yang et al. 2023).

Recently, numerous agents have been clinically developed as molecular targeted therapy in the treatment of BC. For example, targeting the cell cycle control machinery that is necessary for tumor progression has attracted great attention as a promising strategy, and many inhibitors targeting cyclin-dependent kinases 4 and 6 (CDK4/6), including abemacicrib/LY2835219 and palbociclib/PD0332991, have been clinically approved for treating ER$^+$HER2$^-$ advanced/metastatic BC in combination with hormone/endocrine therapy (Piezzo et al. 2020). Targeting Poly (ADP-ribose) polymerase (PARP) that is an enzyme to protect cells from DNA damage for genomic stability has been also appreciated in the treatment of BC, since blocking PARP induces cell death by inducing mitotic catastrophe due to replication stress leading to loss of the DNA repair function and DNA damage accumulation (Slade 2020). Loss-of-function mutations in BRCA1 and BRCA2 genes, which encode multifunctional proteins to regulate DNA repair and chromosomal stability, are detected in approximately 5% of BC, and such cases are highly sensitive to DNA-damaging agents (Godet and Gilkes 2017). PARP inhibitors, including olaparib and talazoparib, have been clinically approved as monotherapy for HER2$^-$ BC with germline BRCA mutation (Cortesi et al. 2021). However, these treatments are applicable to a limited type of BC, and are effective in only a small proportion of the patients. In addition, even if effective initially after the treatments, recurrence and metastasis frequently occur in the patients, and such cases are usually incurable due to resistance to multimodal treatments (Loibl et al. 2021). Thus, the benefits of the treatments are actually limited to a small number of patients despite the significant contributions.

Tumor cells are known to evolutionally change through the interplay with numerous components, including stromal cells, vascular cells, and immune cells, in the host, and the consequent increase of heterogeneity and complexity of both tumor cells and host environment profoundly impacts on the therapeutic efficacy (Qin et al. 2020; Kudo-Saito et al. 2021). This suggests that targeting only the cancer side is insufficient to cure cancer, and interference of the reciprocal evolution is important to treat cancer more effectively. The immune system is the most influential factor in the underlying mechanisms, and the better management may be the key to successful cancer treatment, since only the immune system can attack and eliminate cancer cells from the host, however sometimes converts into an ally of cancer cells.

Immunotherapy has recently attracted great attention in cancer treatment due to the success of antibody drugs, which inhibit immune checkpoint (IC) signaling pathways that brake on anti-tumor immune responses (Javid et al. 2023). Clinical development of the IC inhibitory antibodies (ICIs) was slightly delayed in BC compared to other types of cancers because there are relatively fewer mutations that have been considered as an efficacy-predictive biomarker. However, ICIs targeting programmed cell death protein 1 (PD1) pembrolizumab, and the PD1 ligand (PDL1) atezolizumab have been approved in combination with chemotherapy for treating unresectable, locally advanced/metastatic PDL1$^+$ TNBC, which has higher PDL1 expression and higher tumor mutation burden (TMB) compared to those of other subtypes (Geurts and Kok 2023). This achievement seemed to change the perception of BC treatment.

However, TNBC is comprised of only 15–20% of BC, and the ICI efficacy is limited to a part of the treated patients due to innate and acquired resistance to the therapy (Zheng et al. 2023). In addition, the therapeutic efficacy for other BC subtypes except TNBC has not yet been determined despite many clinical trials for HER2$^+$ BC (Mercogliano et al. 2023). The molecular mechanisms underlying the ICI resistance have been investigated, and some factors, including abnormal MHC and IFNγ signaling mutations, have been reported (Zheng et al. 2023). However, these efforts rarely lead to practical implementation that can overcome the ICI resistance in clinical settings. Biomarkers that predict possible ICI responders have been explored, and some molecules, such as EGFR and VEGFR, have been identified in addition to PDL1 expression and TMB (El Hejjioui et al. 2023). However, these do not always correspond to clinical responses, suggesting lacking accuracy. Combination regimens with various agents, such as PARP inhibitors and angiogenesis inhibitors, have been also evaluated to improve the therapeutic efficacy in many clinical studies (Jacobs et al. 2023). However, most of the trials are still underway, and thus the efficacy remains to be determined. ICI success has been leading to development of other immunotherapies, including vaccine therapy with tumor-associated antigen peptides (Nordin et al. 2023), dendritic cells (DCs) (Qian et al. 2023), nanoparticles (Fang et al. 2022), and cell therapy with genetically engineered T cells (He et al. 2023) or NK cells (Kiaei et al. 2023) called CAR-T/ CAR-NK, for treating BC. However, most clinical trials have failed, although some of them are still underway. These suggest that immune control is very challenging in cancer therapy. A possible reason for the difficulty could be that targeting only anti-tumor effector cells is too downstream in the mechanisms that the host immunity collapses under cancer. Targeting the villains that induce and accelerate the reciprocal evolution between tumor cells and immune cells may be a favorable strategy for the eradication of refractory cancer.

In the human immune system, myeloid cell population is the major cellular component, albeit varying within tumor microenvironment. The myeloid abnormality could thoroughly impact on the whole body leading to treatment resistance of cancer. Myeloid cells produce a variety of pro-inflammatory mediators (IL1, IL6, IL8, IL18, cyclooxygenases (COXs), prostanoids, arginase, etc.), and persistent and strong stimulation with these molecules promote not only tumor progression, including genomic instability, metabolic reprograming, stromagenesis, angiogenesis, invasion, dissemination, and modification of host immunity (Singh et al. 2019), but also chronic inflammation leading to immune exhaustion and dysfunction by firmly braking the immune activation signals via inducing expression of multiple IC molecules (CTLA4, PD1, TIM3, LAG3, TIGIT, etc.) (Gao et al. 2022). These things are no exception in BC, and there are many studies showing the significant villain roles of myeloid cells in the pathogenesis of BC (Blaye et al. 2022), and the significant association between the high levels and poor prognosis in BC in clinical settings (Gonda et al. 2017). Therefore, targeting such myeloid villains has been considered as a promising strategy for improving the clinical outcomes in the treatment of BC (Fig. 1).

However, macrophages (Møs), including tumor-associated Møs (TAMs) and Mø-type myeloid-derived suppressor cells (M-MDSCs), are the focus of attention in many review papers (Tariq et al. 2017; Cha and Koo 2020), although myeloid cells are composed of heterogeneous populations and subsets, including mesenchymal stem cells (MSCs), fibroblasts, neutrophils, Møs, DCs, mast cells (MCs), eosinophils, and basophils. Despite much vigorous research on the Møs for a long time, clinical development of effective Mø-targeting therapeutics has not been rarely achieved in the practical settings. Understanding a broader range of myeloid cell populations may bring ideas for better strategy to treat BC more effectively in clinical setting. Here, we provide a recent overview of various myeloid populations in association with BC, and possible treatments that may be able to inhibit or eliminate the myeloid-derived vicious factors potentially leading to immune chaos and tumor progression.

2 MSCs

Tumor microenvironment is the closest, most influential, and essential area for cancer progression, including development, proliferation, invasion, dissemination, colonization, and survival (Anderson and Simon 2020). Although the compositions vary depending on the cancer type and individual patient, stroma is a fertile soil with numerous components, such as extracellular matrix (ECM) and multiple types of

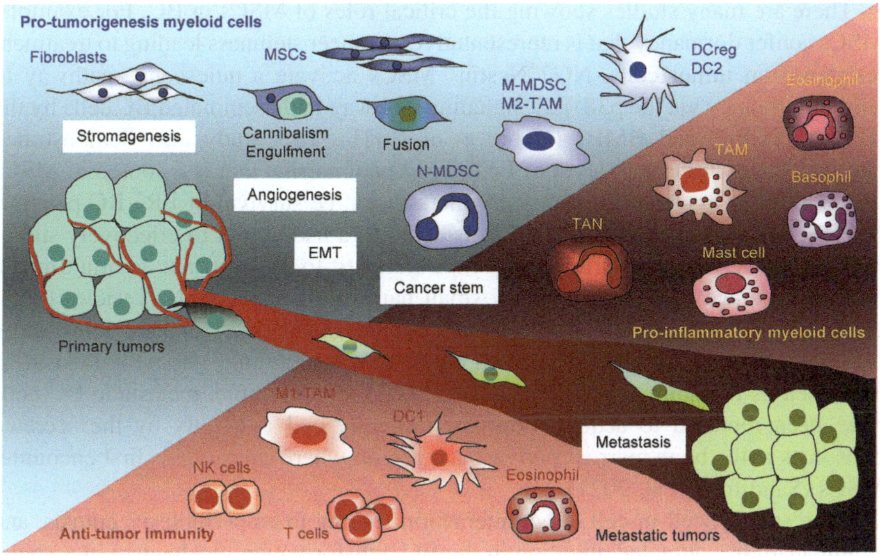

Fig. 1 Myeloid-driven mechanisms underlying breast cancer progression and metastasis. Myeloid cell populations, including mesenchymal cells, fibroblasts, neutrophils, monocytes/macrophages, dendritic cells, mast cells, eosinophils, and basophils promote tumor progression and metastasis directly and indirectly through induction of stromagenesis, angiogenesis, immune suppression, and immune dysfunction by secreting various molecules, although some subsets contribute to anti-tumor immune responses mediated by T cells and NK cells

immunomodulatory cells, and enormously supports cancer progression (Xu et al. 2022). MSCs are one of the representative stromal components. MSCs have not only stemness, including self-renewability and pluripotency to differentiate into multi-lineage cells (fibroblasts, pericytes, adipocytes, osteocytes, chondrocytes, myocytes, neurons, etc.), but also high migratory and immunosuppressive properties (Lebeau et al. 2022). MSCs highly express multiple receptors for chemokines (CCL2, CCL5, CXCL12, CXCL16, etc.) followed by expansion by multiple cytokines (TGFβ, VEGF, FGF, IGF, etc.), and acquire immunosuppressive activity upon the continuous stimulation with pro-inflammatory cytokines (IFNγ, TNF, IL1β, etc.), all of which are abundant in tumor milieu (Shi et al. 2017). Then, MSCs suppress induction of anti-tumor immunity directly by producing pro-inflammatory molecules (IL1β, IL6, IDO, PGE2, nitric oxide, etc.), and indirectly via generation and expansion of immunosuppressive cells, such as regulatory T cells (Tregs), regulatory DCs (DCregs), and myeloid-derive suppressor cells (MDSCs), in the host (Jiang and Xu 2020). Consequently, MSCs widely damage host immunity through multiple ways to extensively impair the anti-tumor functions of various immune cells (Muller et al. 2021). That is why MSCs promptly and firmly facilitate cancer progression in a direct and indirect manner, even if distributing throughout the body, not limited to the tumor microenvironment.

There are many studies showing the critical roles of MSCs in BC. For example, MSCs confer dormancy that is representative of cancer stemness leading to treatment resistance on tumor cells. $NG2^+Nestin^+$ MSCs activate a quiescence pathway to confer dormancy via the p38–p27 signaling pathway in disseminated BC cells by the secreted TGFβ2 and BMP7 (Nobre et al. 2021). This study also demonstrated clinical data that the high level of BMP7 in bone marrow that is a source of MSCs is significantly associated with longer metastasis-free survival of ER^+ BC patients. MSC-derived exosomes have been considered as a useful tool to effectively stimulate BC cells and the surrounding environments to be resistant to multimodal treatments, since exosomes are very small nano-sized extracellular vesicles (EVs) containing numerous molecules (proteins, lipids, metabolites, nucleic acids, etc.), which have a wide range of effects on development, immunity, tissue homeostasis, and diseases including cancer (Guo et al. 2022). For example, metastatic BC cells have been reported to acquire dormancy directly and indirectly by the secreted exosomes at the bone marrow perivascular region where cancer cells first encounter MSCs (Sandiford et al. 2021).

Interestingly, cell-in-cell direct interaction leading to emperipolesis, entosis, and cannibalism has attracted attention in the BC transformation mechanism in association with MSCs, although such cell eating has long attracted attention pathologically as an indicator of tumor aggressiveness (Clarke 2011). For example, BC cells after eating MSCs as cannibalism become low proliferative (dormant), but the production ability of pro-inflammatory cytokines and chemokines is remarkably augmented (Bartosh et al. 2016). Also, metastatic BC cells after eating MSCs as engulfment become more aggressive with enhanced self-renewal, migrative, invasive, and in vivo metastatic properties via increase of ZEB1 expression that is one of the regulators governing epithelial-to-mesenchymal transition (EMT) of tumor cells (Chen et al. 2019b).

We have to also pay attention to another mechanism, by which BC cells are fused with MSCs resulting in the formation of "hybrid" cells with different phenotypes and properties from the original tumor cells and MSCs. For example, human TNBC cell line HCC1806 cells spontaneously fuse with human bone marrow-derived MSCs, and become low proliferative/dormant, but highly resistant to chemotherapy via increase of CD9 expression (Ullah et al. 2019). Hybrid cells of human TNBC cell line MDA-MB-231 and MSCs isolated from metastatic tissues of BC patients also are more aggressive having higher chemoresistance and metastatic properties via increase of WNT5A expression than the original cells (Augimeri et al. 2023). Such cell fusion may increase tumor heterogeneity leading to treatment resistance.

MSC-specific markers that accurately reflect the functions have not yet been identified, and thus human MSCs are defined by several molecular expressions (CD49a, CD73, CD90, CD105, CD146, CD271, STRO1, etc.) combined with negative expression of CD11b, CD14, CD19, CD34, CD45, and CD79a, all of which are commonly expressed in other cells (Shi et al. 2017). Therefore, the only way to block the MSC effects is to target the MSC-expressing and secreting molecules. For targeting the MSC-associates, however, we need to be very careful about some points. Firstly, the sources, the pathophysiological condition, and the status (phenotypes and properties) of the MSCs used in the assays often vary between studies. Secondarily, MSCs are often cultured in vitro for a long period of time to increase the number of cells before analysis in order to ensure a sufficient number of cells for assays, since the number of MSCs that can be collected is generally very small. This means that we may not yet see the natural properties that actually exist in the body, and it may take more time to develop cancer treatments targeting MSCs.

3 Fibroblasts

MSCs differentiate into multi-lineage cells, including fibroblasts, receiving various inflammatory stimuli, such as TGFβ, PDGF, FGF, HGF, SDF, and IGF, in the tumor milieu, and such continuous stimulation converts tissue-resident and tumor-recruited fibroblasts into an inflammatory type of fibroblasts called cancer-associated fibroblasts (CAFs) in cancer (Pittenger et al. 2019). The conversion of MSCs to inflammatory CAFs has also been reported in BC research. For example, in vitro stimulation with IL1β and TNFα secreted from human BC cells induce transformation of human bone marrow-derived MSCs into CAFs accompanied by increased gene expression not only of typical CAF markers related to fibroblast proliferation and movement, but also of pro-inflammatory genes (IL6, LIF, etc.) and pro-metastatic chemokine genes (CCL2, CCL5, CXCL1, CXCL2, CXCL6, CXCL8, etc.) (Rubinstein-Achiasaf et al. 2021). Many studies have demonstrated that CAFs promote tumor development, progression, and dissemination directly and indirectly via induction of angiogenesis and immunosuppression by producing potent molecules (VEGF, FGF, PDGF, CXCL12, TGFβ, PDL1, etc.) in a variety of cancers, including BC (Salimifard et al. 2020). For example, CAFs isolated from tumor tissues of BC patients confer high proliferative and drug-resistant properties to human BC cells utilizing the secreted IL6 via the STAT3 signaling pathway (Guo et al. 2023). This study also showed clinical data that the high level of phosphorylated STAT3 expression in BC tissues is significantly associated with shorter disease-free survival of BC patients.

The intimate crosstalk between CAFs and immune cells infiltrating in the tumor tissue is essential for suppressing anti-tumor immune responses more effectively. For example, CAFs derived from tumors of BC patients recruited $CD4^+$ T cells into the tumor microenvironment utilizing the secreted CXCL12, and promoted their conversion to Tregs, which suppress proliferation of anti-tumor effector T cells, through their own expression of B7H3, CD73, and DPP4 (Costa et al. 2018). Using mouse spontaneous mammary tumor model that is genetically engineered to overexpress mammary-specific polyomavirus middle T antigen (MMTV-PyMT), it has been reported that CAFs promote tumor progression via recruiting immunosuppressive M2-type Møs by the secreted Chi3L1 (Cohen et al. 2017). Chi3L1 is a glycoprotein that regulates a broad range of biological processes, such as inflammation and tissue injury, and thus is associated with the pathogenesis of many inflammatory diseases, including liver fibrosis, asthma, arthritis, sepsis, diabetes, and cancer (Zhao et al. 2020). In cancer, Chi3L1 is known to widely support tumor progression as a multifunctional molecule that affects tumor proliferation, invasion, metastasis, angiogenesis, and immune system (Zhao et al. 2020). IL33 is another potent effector molecule that recruits and activates various inflammatory cells expressing its receptor ST2, such as neutrophils, monocytes, MCs, and eosinophils, having a major impact on cancer progression (Griesenauer and Paczesny 2017). Using the MMTV-PyMT models, CAF-secreted IL33 has been shown as a key molecule that promotes BC metastasis by orchestrating the $ST2^{+}$ cells in the metastatic lung tissues (Shani et al. 2020). This study also showed clinical data that IL33 expression is significantly higher in metastatic lung compared to other metastatic sites in BC patients. Albeit not cancer, anti-IL33 antibodies have been clinically evaluated for other inflammatory diseases, such as asthma (Wechsler et al. 2021), atopic dermatitis (Maurer et al. 2022), and COPD (Rabe et al. 2021), and it may be worthwhile to apply the antibody drugs to the treatment of BC.

Direct targeting fibroblasts/CAFs may be more effective in the treatment of BC, since fibroblasts are a major component of mammary tissues. Actually, treatment with pirfenidone, which is a drug approved for the treatment of idiopathic pulmonary fibrosis to reduce fibrosis, significantly suppressed primary tumor growth and lung metastasis, albeit in combination with doxorubicin, in mouse TNBC-like 4 T1 tumor models through reduction of cell viability and collagen production in CAFs (Takai et al. 2016). This drug may contribute to the treatment of BC. However, just like with MSC, the data often vary depending on the experimental conditions, including the sources, the pathophysiological condition, the status, and isolation methods, and the precise nature and functions of CAFs remain largely undefined. Further investigations will be necessary for the characterization of CAFs and the development of the specific drugs.

4 Neutrophils

Neutrophils are the most abundant of the immune cell components in human, and have long been considered to play only a role in innate immunity for acute host defense against microorganisms, such as bacterial and fungal pathogens (Malech et al. 2014). However, accumulating evidence suggests that neutrophils are widely involved in the pathogenesis of various diseases, including BC, and tumor-infiltrating neutrophils (called "tumor-associate neutrophils (TANs)") are known as a powerful influencer on the tumor microenvironment (Hajizadeh et al. 2021). Increase of neutrophils in the periphery and tumor tissues is frequently seen in cancer patients, and increase in the neutrophil-to-lymphocyte ratio (NLR) has been shown as a poor prognostic factor in clinical setting (Cupp et al. 2020). In BC, NLR in the periphery is particularly associated with poor survival of patients with Luminal A and $HER2^{-}$ types (Gago-Dominguez et al. 2020). In tumor tissues, CXCL1, CXCL2, CXCL3, and TGFβ produced from human TNBC cell lines (M4, MDA-MB-231, and BT549), but not ER^{+} BC cell line (BT474) recruit neutrophils (SenGupta et al. 2021). Upon receiving various stimuli (IL1β, IL6, IL8, G-CSF,

GM-CSF, CCL2, CXCL12, etc.) within the tumor milieu, the neutrophils differentiate into TANs that can produce a variety of pro-inflammatory molecules (neutrophil elastase, arginase, TNFα, IL1β, IL6, IL8, VEGF, PROK2, MMPs, TIMP1, CXCL1, CXCL2, CXCL5, etc.), all of which facilitate tumor growth and metastasis directly and indirectly via inducing immunosuppression (Carnevale et al. 2023).

It has been shown that cathepsin C (CTSC) produced from TNBC cell lines is also able to recruit neutrophils, but not Møs, via activating membrane-bound proteinase 3 that upregulates IL1β secretion into metastatic lung tissues, and the neutrophil recruitment is further facilitated by IL6 and CCL3, which are additionally secreted from the tumor cells (Xiao et al. 2021b). This study also reported that the tumor-derived CTSC induces the formation of neutrophil extracellular traps (NETs), which are extracellular fibrous scaffolds composed of its nuclear and cytoplasmic proteins, such as neutrophil elastase, myeloperoxidase, peptidylarginine deiminase 4, and MMP9 (Chen et al. 2022), for supporting the metastatic tumor growth. Interestingly, treatment with a CTSC inhibitor AZD7986 was effective in suppressing lung metastasis in mouse mammary tumor models. In this study, the degree of neutrophil infiltration and NETs in tumor tissues of BC patients was significantly associated with CTSC expression in tumor tissues and the frequency of metastasis.

NET formation is triggered by various stimuli besides CTSC within the primary and metastatic tumor milieu, and greatly contributes to cancer refractory. For example, NETs promote angiogenesis directly by the contents such as VEGF and MMP9, and indirectly via activation of endothelial cells (Chen et al. 2022). Pro-angiogenic and inflammatory molecule angiopoietins (ANGPT1 and ANGPT2) have been reported to induce NET formation (Lavoie et al. 2018). This suggests that NET formation is a key trigger to initiate and amplify angiogenesis, which provides sufficient oxygen and nutrients for tumor progression and metastasis. NETs also enhance radiation resistance in bladder cancer, albeit not BC (Shinde-Jadhav et al. 2021). In BC, close relationship between NETs and chemoresistance of metastatic tumors has been reported. For example, it has been reported that chemotherapy with cisplatin is ineffective in various mouse mammary tumor models due to the accumulation of neutrophils and NET formation in the metastatic lungs. The reasons are that the chemotherapy not only induces the production of CXCL1 and CXCL5 in tumor cells to recruit $CXCR2^{+}$ neutrophils, but also activates NLRP3 in tumor cells to induce IL1β production that promotes NET formation (Mousset et al. 2023). Sequentially the NETs confer chemoresistance on the tumor cells through the trap of latent TGFβ by the NET-derived integrin αvβ1 and sequential activation of it by the NET-derived MMP9. In this study, treatment with DNase I, NLRP3 inhibitor, or MMP9 inhibited the chemo-induced NET formation, and significantly improved the therapeutic efficacy on lung metastasis of the mouse models. They also showed clinical data that NETs were significantly increased in plasma of patients with metastatic BC who were resistant to chemotherapy, but the levels were low in chemo-responding patients. Interestingly, only a few neutrophils and no NETs were observed in the primary tumor tissues. There are several neutrophil/NET-targeted drugs having high potential for use in cancer therapy: CXCR2-specific inhibitor AZD5069 to suppress neutrophil recruitment (Ghallab et al. 2022), neutrophil elastase-specific inhibitor GW-311616 to suppress NET formation (Li et al. 2022a), and glucocorticoids including dexamethasone to reduce NET formation (Vargas et al. 2017).

Intratumoral TANs and NETs are not the only one that supports tumor progression and metastasis. Neutrophils circulating in the periphery also contact blood-circulating tumor cells (CTCs) using the expressing integrins (Mac1, ICAM1, and α3β1 integrin), and form clusters to initiate cell cycle progression and metastatic dissemination of the cancer stem-like dormant CTCs (Pereira-Veiga et al. 2022). In return, the CTCs stimulate the neutrophils to differentiate into a TAN-like tumor-supportive type that can secrete pro-inflammatory cytokines, which further enhance survival and proliferation of the CTCs. In BC, single-cell RNA sequencing analysis demonstrated that the majority of leukocytes clustering with BC cells are neutrophils in the blood of BC patients and mouse mammary tumor models, and that BC patients with the CTC-neutrophil clusters show significantly shorter progression-free survival as compared to those with unclustered CTCs (Szczerba et al. 2019). In the mechanisms, VCAM1 expressed on tumor cells is a key adhesion molecule to form clusters with neutrophils, and the CTC-attaching neutrophils release potent cytokines, such as IL1β and IL6, to confer high proliferative and metastatic properties on the partner CTCs.

CTC-neutrophil clusters in the lung vasculature were also reported using xenograft models implanted with tumor tissues of BC patients. In the mechanisms, CD11b expressed on neutrophils binds to ICAM1 expressed on tumor cells, and this binding induces release of soluble uPAR as a chemoattractant from tumor cells, followed by recruitment of neutrophils that facilitate tumor metastasis (Li et al. 2023). This study also showed clinical data that the high level of uPAR expression in tumor tissues is significantly associated with short distant metastasis-free survival and overall survival of BC. In the metastatic niches, neutrophils seem to transform into a potent nutritional supplier to the metastasized tumor cells with the help of resident mesenchymal cells. For example, $CD140a^+$ lung-resident mesenchymal cells induce increase of lipids in neutrophils by the secreted PGE2, which is a principal mediator of inflammation and a lipid mediator produced from arachidonic acid by cyclooxygenase (Cheng et al. 2021), and the lipid-laden neutrophils give its lipids to the metastasized tumor cells for facilitating its metastatic colonization in the lung (Li et al. 2020).

NETs in the periphery also contribute to BC progression and metastasis. NET formation in the peripheral circulation is induced by various molecules contained in tumor cell-derived EVs, and the NETs broadly promote the formation of cancer-associated thrombosis which is one of the serious causes of death in cancer patients, endothelial damage, organ dysfunction, and cancer metastasis (Chen et al. 2022). In the cancer metastasis mechanisms, NET-derived neutrophil elastase and MMP9 cleave and remodel laminin in the ECM, and the exposed laminin epitopes activate the integrin α3β1 and FAK/MEK/ERK signaling pathways in dormant tumor cells resulting in the cell cycle progression leading to aggressive metastatic colonization of tumor cells. NET-derived DNA also facilitates tumor motility and metastasis via binding to CCDC25 expressed on BC cells followed by activation of the ILK-β-parvin pathway (Yang et al. 2020).

Neutrophils also contribute to BC progression and metastasis indirectly by suppressing anti-tumor immune responses. For example, neutrophils suppress T-cell functions, proliferation, and IFNγ production by expressing PDL1 that is induced upon the activation by tumor-derived CCL20 (Kwantwi et al. 2021). Also, tumor-derived CCL20 expands immunosuppressive polymorphonuclear MDSCs (possibly including neutrophils), which confer cancer stemness on tumor cells by the secreted CXCL2 (Zhang et al. 2023). In this study, blocking CXCR2 with the specific antagonist SB225002 synergistically enhanced anti-tumor efficacy induced by chemotherapy with docetaxel in mouse mammary tumor models. In clinical setting, the high density of $CD66b^+$ neutrophils, which was more frequently seen in TNBC than those in the other subtypes, was significantly associated with short disease-free survival in BC patients, and Lin28B was identified as an inducer of EMT and cancer stemness in tumor cells (Lin et al. 2022). Lin28B overexpression in primary tumors promoted pre-metastatic niche formation in lung tissue via accumulation of neutrophil and the sequential conversion of the neutrophils into an immunosuppressive type with PDL2 expression (called "N2" with high expression of ARG1, CD206, Ym1, and IL6) that can suppress T-cell functions (Qi et al. 2022). In this study, co-blocking IL6 and PDL2 was significantly effective in suppressing lung metastasis in the mouse tumor models. This study also showed that the decrease of exosomal let-7s is responsible for creating the immunosuppressive metastatic niches in the lung of the tumor models, and Lin28B expression in primary tumors was significantly correlated with serum exosomal let-7s in BC patients.

Despite the accumulation of much evidence, however, neutrophil-targeted therapy remains to be clinically developed, and further clinical studies using patient samples are necessary. This may be because neutrophils are short-lived and heterogeneous leading to inaccurate analyses, and thus fundamental improvements in sampling and analysis methods may be required for further investigation.

5 Møs

Møs, which have a longer lifespan than granulocytes including neutrophils, are deeply associated with cancer refractory, and are divided into three types: classical anti-tumor types called "M1-Møs", pro-inflammatory types called tumor-associated Møs (TAMs), and immunosuppressive types called "M2-Møs" or monocytic myeloid-derived suppressor cells (M-MDSCs). Here, we summarize the role of two tumor-supportive types in BC.

5.1 Pro-inflammatory Møs

TAMs are recruited by various chemokines (CCL2, CCL3, CCL5, CXCL1, CXCL2, IL8, etc.) in the tumor microenvironment, and promote tumor development, survival, proliferation, invasion, and metastatic dissemination directly by initiating the EMT program to confer cancer stemness, and indirectly via angiogenesis and immune exhaustion using the secreted angiogenic molecules (VEGF, FGF, PDGF, MMPs, etc.) and pro-inflammatory molecules (COXs, IL1β, IL6, etc.) within the tumor milieu (Mishra et al. 2022). M-MDSCs, which are usually considered as an immunosuppressive type, have been reported to have pro-inflammatory activity in cancer. For example, MDSCs induce EMT in tumor cells by the secreted molecules (PGE2, TGFβ, EGF, and HGF), and the induced cancer stem cells further activate and expand MDSCs in the tumor milieu (Jayakumar and Bothwell 2019). MDSCs also induce angiogenesis by the secreted molecules, and induce T-cell differentiation into pro-inflammatory Th17 to further boost the MDSC-induced adverse events (Zhou et al. 2020). Thus, only slight differences in the characteristics are seen between M-MDSCs and TAMs. The M-MDSCs used in these studies may contain TAMs, or may be almost the same, although they were named by different words.

In BC as well as other types of cancers, TAMs have been reported as a major component of immune cells in tumor tissues, and many studies showed significant correlation between the high level of TAMs and poor prognosis in patients (Munir et al. 2021). It has been reported that resistance to hormone therapy of ER^+ BC is induced by crosstalk with TAM-like pro-inflammatory Møs. Møs that produce TNF and IL6 promote proliferation, invasion, and migration of ER^+ BC cells via activation of the NFκB/STAT3/ERK signaling pathways, and hyperphosphorylation of ER in tumor cells, finally leading to resistance to endocrine/hormone therapy (Castellaro et al. 2019). Recent studies have shown that TAMs modulate BC functions by releasing extracellular vesicles in addition to cytokines and chemokines. For example, TAMs enhance aerobic glycolysis and apoptosis resistance in BC cells by stabilizing HIF1α with HISLA packaged in the TAM-released EVs via blocking PHD2 and HIF1α interactions, and HISLA expression in tumor tissues is significantly correlated with tumor glycolysis and chemoresistance of BC patients (Chen et al. 2019a).

TAMs also indirectly render tumor cells refractory by cooperating with fibroblasts resident in breast tissues. For example, TAM-derived CCL18 activates $CD10^+GPR77^+$ CAF-like subset in normal breast-resident fibroblasts upon binding to a CCL18 receptor PITPNM3 that activates NFκB signaling, and stimulation with IL6 and IL8, which are released from the activated fibroblasts, confers chemoresistance on BC cells (Zeng et al. 2023). They also showed clinical data that the high level of $CCL18^+$ TAMs in tumor tissues is significantly correlated with that of $CD10^+GPR77^+$ CAFs and chemoresistance in BC patients. The same group has developed a small molecule inhibitor specific for PITPNM3, and reported the therapeutic efficacy on tumor metastasis using mouse xenograft models and organoid models (Liu et al. 2022).

TAMs negatively affect and damage the immune system, and indirectly contribute to refractoriness of BC, particularly resistance to immunotherapy (Xiao et al. 2021a). For example, the high level of $IL15R\alpha^+$ TAMs in tumor tissues is significantly associated with poor prognosis of BC patients, and the $IL15R\alpha^+$ TAMs release the IL15/IL15Rα complex that can reduce CX3CL1 expression in tumor cells followed by reduction of $CXCR3^{+}CD8^{+}$ T-cell infiltration in tumors (Zhang et al. 2022). This leads to resistance to immunotherapy with IL15 or anti-PD1 mAb, but blocking the IL15/IL15Rα complex formation with a specific peptide (called "B peptide") that they originally designed breaks the anti-PD1 resistance by increasing $CD8^{+}$ T cells in the tumors of mouse mammary tumor models. IL15 as well as IL2 has long been expected as a potent immunotherapeutics to induce and activate anti-tumor effector T cells and NK cells, while the therapeutic efficacy remains to be determined in clinical setting (Cai et al. 2023). CX3CR1 has been reported as a hallmark of anti-tumor $CD8^{+}GZMB^{+}$ T cells expanded in anti-PD1 responders after chemoimmunotherapy, albeit in melanoma (Yan et al. 2018). However, even if CTLs are recruited in the tumor microenvironment, TAMs may make them unable to attack tumors by inducing exhaustion, since it has been reported that TAMs induce immune exhaustion accompanied by PD1 upregulation through IRF8 expression (Nixon et al. 2022).

Thus, targeting TAMs may be a promising strategy to effectively treat BC. Many methodological approaches, including inhibition of TAM recruitment in tumor tissues, reduction of TAMs, and inhibition of TAM functions and its products, have been developed. In particular, there is great hope that inhibitory treatments targeting TAM-caused chronic inflammation, which facilitates cancer progression and disrupts host immunity, may be able to restrain the adverse events, resulting in the efficient elimination of cancer. For example, COXs, which produce eicosanoids including prostaglandin E2 (PGE2) and thromboxane 2 (TXA2) from arachidonic acid, are one of the most potent pro-inflammatory molecules in cancer progression mechanisms, and therapeutic efficacies induced by COX inhibitors called "non-steroidal anti-inflammatory drugs (NSAIDs)" have been shown in experimental and clinical studies on BC (Mahesh et al. 2021). A COX2-specific inhibitor celecoxib is commonly used in clinical therapy because of the fewer gastric and renal side-effects than a COX1/2 inhibitor aspirin, albeit causing cardiac side-effects upon long-term use. However, aspirin efficacy has still been expected in the treatment of cancer, including BC, and many studies have demonstrated that regular use of aspirin is associated with low risks of tumor metastasis and cancer mortality in BC patients (Chen and Holmes 2017). The reason may be that blocking both COX1 and COX2 can more broadly inhibit cancer-supportive mechanisms, including platelet aggregation, endothelial activation, tumor adhesion to endothelium, recruitment and expansion of myeloid cells, and EMT induction in tumor cells (Ma et al. 2017). Ma et al. compiled the data of aspirin effects published over a long period of time from 1989 to 2019, and concluded that regular doses of aspirin use over 3 years are associated with reduction of overall risk of BC, especially in postmenopausal women, ER^{+} tumors, and in situ BC (Ma et al. 2021). However, they suggested that further clinical studies are needed to determine the optimal doses, frequency, and duration of aspirin use for the prospective studies.

PGE2 is a COX2 product, and its receptors EP1–4 have been also pharmaceutically targeted for treating cancer, since many studies have reported aberrant expression of COX2 in 40–50% of BC, and association with tumor progression and metastasis of BC patients (Majumder et al. 2018). Among the receptors,

EP4 is most deeply associated with the COX2 activities that promote tumor invasion and migration, and enhance angiogenic and lymphangiogenic activities of TAMs via the cAMP/PKA pathway shared by EP2, and the PI3K/Akt pathway unique for EP4 (Majumder et al. 2018). After evaluating drug efficacy in mouse tumor models, EP4 antagonists, including E7046 (Hong et al. 2020), HTL0039732 (NCT05944237), and ONO-4578 (NCT03155061), and EP2/4 antagonist, including MBF-362 (NCT05940571) and TPST-1495 (NCT04344795), have been now clinically evaluated in combination with/without anti-PD1/PDL1 mAbs for solid cancer.

TAM-released MMPs are another molecule for the TAM-targeted therapy. MMPs facilitate cancer invasion and metastasis not only directly by inducing proliferation and EMT in tumor cells, but also indirectly via degradation of the ECM and generation of fibrotic stroma by activating stromal fibroblasts (Radisky et al. 2017). In BC, MMP expression is frequently upregulated in tumor tissues, and the level is associated with prognosis of patients (Kwon 2022). As promising anti-tumor efficacies induced by blocking MMPs, such as MMP9 (Juric et al. 2018) and MMP14 (Ager et al. 2015), have been shown in mouse mammary tumor models, clinical development of these inhibitors is expected in the BC therapy, albeit recent clinical trials have continued to fail probably due to a lack of specificity.

5.2 Immunosuppressive Møs

MDSCs are generated by abnormal myelopoiesis and dysdifferentiation of the progenitor cells under chronic inflammatory conditions, and acquire immunosuppressive properties upon receiving stimuli (IL4, IL6, COX2, etc.) in the host with cancer (Groth et al. 2019). MDSCs are recruited by various chemokines (CCL2, CCL5, CXCL8, CXCL12, etc.) within tumor microenvironment, and are expanded by pro-inflammatory molecules (IL1β, IL6, IL10, M-CSF, GM-CSF, COX2, PGE2, etc.) in there. In the tumor milieu under hypoxia, MDSCs are activated by HIF1α that induces ectonucleotidases, CD39 and CD73, which convert ATP to adenosine that can inhibit T-cell functions through the adenosine receptors (Jayakumar and Bothwell 2019). Using the soluble molecules (ARG1, IDO, PGE2, NO, ROS, TGFβ, IL10, etc.) and IC molecules (PDL1, VISTA, etc.), MDSCs suppress T-cell functions, and further enhance the immunosuppression by expanding immunosuppressive Tregs, regulatory B cells (Bregs), and Th17 cells. Furthermore, MDSCs induce and promote angiogenesis by secreting MMPs (MMP2, MMP13, MMP14, etc.). All these events contribute to cancer refractory. M2-Møs also suppress anti-tumor immune responses directly and indirectly using molecules similar to those of MDSCs in addition to chemokines (CCL17, CCL18, CCL22, etc.) to induce Th2 polarization (Goswami et al. 2021). As each phenotype is not clearly defined by distinct molecular expression, immunosuppressive M2-Møs may be partly contained in the MDSCs reported in the literature, and vice versa.

In BC, MDSCs/M2-Møs as well as TAMs have attracted attention as important players in shaping the refractoriness. For example, it has been reported that amplification/overexpression of KAT6A, which is a oncogenic molecule as a chromatin modifier with acetyltransferase activity, is significantly associated with tumor metastasis in TNBC, and acetylation of SMAD3 (K20 and K117) by KAT6A in tumor cells promotes BC metastasis directly by conferring cancer stemness, and indirectly by recruitment of MDSCs (Yu et al. 2021). In this study, blocking KAT6A with a specific inhibitor WM-1119 synergistically enhanced anti-PDL1 therapeutic efficacy through reduction of MDSC recruitment in the metastatic lung of mouse mammary tumor models. This study also showed clinical data that the high levels of SMAD3 acetylation and $CD11b^+$ cell accumulation in metastatic sites were significantly associated with poor prognosis of BC patients. Another group reported that MDSCs expressing complement receptors, C3aR and C5aR1, are recruited by CAFs in the metastatic niche to promote T-cell dysfunction caused by chemotherapy with doxorubicin, leading to treatment resistance (Monteran et al. 2022). In this study, combination with C3aR antagonist SB290157 or C5aR1 antagonist PMX53 was significantly effective in suppressing the chemotherapy-induced events through decrease of $PD1^+$ dysfunctional T cells in early metastatic niche, although monotherapy with the antagonists showed no effects.

MDSCs are expanded by cancer stem-like BC cells. For example, $ALDH1A1^+$ cancer stem-like BC cells induce MDSC expansion by the secreted GM-CSF followed by immunosuppression, leading to tumor progression and metastasis (Liu et al. 2021). ALDH1A1 is a cancer stem molecule that regulates the ALDH enzyme activity, and has been reported as a poor prognostic biomarker in many types of cancers, including BC (Wei et al. 2022). In the study, gemcitabine was used to eliminate MDSCs in addition to the cytotoxic activity. Combination of an ALDH1A1 inhibitor disulfiram and gemcitabine showed synergistic therapeutic efficacy through MDSC reduction and $CD8^+$ T-cell increase in tumors of mouse mammary tumor models.

Tumor suppressor gene BRCA1 is deeply associated with BC as described above, and downregulation or deficiency of BRCA1 is known to increase cancer stem-like cells (Kim et al. 2019). It has been reported that BRCA1 deficiency in mouse mammary tumor cells creates a tumor-permissive microenvironment by recruiting and expanding MDSCs locally and systemically using S100A9 and CXCL12 during tumorigenesis (Li et al. 2022b). In this study, CXCL12 knockout reduced S100A9 expression, suggesting the significant role of the S100A9-CXCL12 axis in MDSC-triggered BC initiation. Combination therapy with anti-PD1 mAb and a S100A9 inhibitor Tasquinimod or a CXCL12 inhibitor AMD3465 showed synergistic anti-tumor effects in mammary tumor models.

In clinical settings, the high frequency of MDSCs in tumor tissues and peripheral blood is significantly associated with tumor metastasis, higher stages, and poorer prognosis in BC (Bergenfelz et al. 2020). It has been reported that $HER2^-$ BC patients with the low level of M-MDSCs, rather than G-MDSCs, in peripheral blood had a clinical benefit in the treatment with pembrolizumab plus gemcitabine, albeit no clear association with other immune biomarkers (de la Cruz-Merino et al. 2022). Thus, therapeutic strategies targeting MDSCs/M2-Møs using various methodological approaches (inhibition of its recruitment, expansion, activation, etc.) are expected to be promising in the BC therapy. Conversion and/or reprogramming of MDSC/

M2-Mø could be an alternative strategy to reduce number and functionality of these cells. For example, BC cells enhance miR-182 expression in M2-like Møs by secreting TGFβ, but deficiency or deletion of the miR-182 targeting TLR4 suppresses M2 polarization and tumor development in mouse mammary tumor models (Ma et al. 2022). miR-182 is known to promote tumorigenesis and metastasis of many types of cancers, including BC (Maryam et al. 2021). Another study reported that SLIT2 treatment converts M2-Møs, which are a major component of stroma in BC tumors, into M1-Møs with high phagocytic and MMP13-producible activities for degradation of fibrosis, and also suppresses IL6 production in BC cells, resulting in suppression of fibrosis and tumor metastasis in mammary tumor models (Ahirwar et al. 2021). SLIT2 is a secreted protein that binds to a receptor Robo1, and its loss of functions by gene deletion, epigenetic inactivation, and mutations have been reported in various types of cancers, including BC (Basha et al. 2023). In the study, SLIT2 positivity was correlated with better prognosis, and low levels of $CD163^+$ M2-like Møs and fibrosis in tumor tissues of BC patients. Interestingly, a small molecule modulator (GB1275) of CD11b which is a hallmark of myeloid cells has been clinically evaluated for solid tumors. Treatment with GB1275 broadly reduces overall Møs, including TAMs, MDSCs, and M2-Møs, and the induced polarization into M1 phenotype expands potent anti-tumor effector T cells leading to tumor suppression (DeNardo et al. 2021).

MDSC/M2-Mø-secreted IDO suppresses cellular functions of anti-tumor effector T/NK cells, but induces and activates Tregs, DCregs, and MDSCs through the kynurenine pathway, suggesting a pivotal role in immune evasion of tumor cells (Girithar et al. 2023). IDO is also expressed in many other cells, such as tumor cells, vascular cells, and fibroblasts, and the high level of IDO expression in tumor tissues is associated with poor prognosis of various types of cancer, including BC, suggesting a promising target in cancer therapy. Numerous small-molecule inhibitors specific for IDO have been clinically developed, whereas most of the clinical trials were terminated, or did not proceed to the next phase, remaining only some of them, such as epacadostat (INCB024360), indoximod (NLG-8189), and INCB024360. A phase III trial with epacadostat plus pembrolizumab has ever failed due to no significant benefit in progression-free survival as compared to pembrolizumab monotherapy in unresectable or metastatic melanoma, despite the encouraging results in early phase studies (Long et al. 2019).

6 DCs

DCs are representative antigen-presenting cells, which play pivotal roles in development, maintenance, and functioning of T cells. DCs are classified into several types with different phenotypes and properties, including conventional DCs, plasmacytoid DCs (pDCs), and Langerhans cells (Balan et al. 2019). The circulating DCs express MHC class II, costimulatory molecules, and chemokine receptors (CCR7, CXCR4, etc.), and migrate into widespread tissue organs to take up pathogens via phagocytosis and endocytosis mechanisms followed by induction of immune responses within the milieu. DCs also migrate into lymphoid tissues to interact with T cells for education and activation against the specific antigens of the pathogens.

Conventional DCs are divided into two subsets, anti-tumor DC1 and immunosuppressive DC2 (Collin and Bigley 2018). DC1 expresses CD103, XCR1, and transcription factors, IRF8 and Batf3, and induce and activate potent $CD8^+$ T cells, Th1, and Th17. In contrast, DC2 expressing transcription factor IRF4 induces Th2 and Th17. DC1 may partly contain monocyte-derived DCs having inflammatory properties, and DC2 may partly contain DCregs (alternatively called tolerogenic DCs) having immunosuppressive properties. DCregs are defined by the downregulation of T-cell stimulatory molecules (MHC II, CD80, CD86, CD40, etc.), the upregulation of immunosuppressive molecules (IDO, PDL1, PDL2, ILT3, ILT4, HLA-G, IL10, IL27, TGFβ, etc.), and the tolerogenic properties, such as suppression of T-cell functions, and induction of Tregs (Ness et al. 2021). pDCs are known to play a crucial role in anti-viral and anti-tumor immune responses because of highly producing type I IFNs. However, it has been reported that pDCs become tolerogenic and dysfunctional upon receiving stimuli (TGFβ, TNFα, GM-CSF, etc.) in the host with cancer, and the high level of pDCs is associated with poor prognosis of BC cancer (Reizis 2019). Langerhans cells are mainly localized in the epidermis to promptly migrate into skin-draining lymph nodes upon receiving inflammatory stimuli, and have both immunosuppressive property to induce Tregs under the homeostatic conditions, and pro-inflammatory property to induce Th17 after barrier disruption and pathogen invasion (Doebel et al. 2017). However, DC functions in vivo remain unclear, since the number of DCs in a body is small, and DCs are often expanded by in vitro culture for the research, leading to potentially generating something different from in vivo status.

In recent research on BC, conversion of tumor-supporting DCs into anti-tumor effector types has gained attention as a therapeutic strategy to reprogram the tolerogenic environment. For example, dysfunction of pDCs is induced by lactic acid (called lactate) released from BC cells via activation of GPR81 and upregulation of MCT1 leading to tumor progression, but intratumoral injection with GPR81-mediated pathway inhibitor "galleon" and/or an MCT inhibitor "ARC" significantly suppresses tumor burden in mammary tumor models (Raychaudhuri et al. 2019). Lactate is known as a pivotal oncometabolite that drives tumor progression and metastasis, and immunosuppression in cancer, but the underlying mechanisms remain to be fully elucidated (Perez-Tomas and Perez-Guillen 2020). Another group reported that treatment with naringenin, a flavonoid widely distributed in citrus and tomato, is effective not only in suppressing autophagy and proliferation of BC cells, but also in promoting differentiation and maturation of DCs through FKBP4 downregulation, which upregulates NRF2 via nuclear translocation of NR3C1 (Xiong et al. 2021). This study also showed clinical data that high levels of FKBP4 and low levels of NRF2 correlate with poor prognosis in BC patients. It has been reported that NRF2 suppresses oxidative stress, tumor-promoting inflammation, and tumorigenesis, and regulates innate immunity and cytokine production from DCs, although there are some papers showing opposite properties to activate anabolic metabolism for cancer progression (He et al. 2020).

Vaccine therapy with tumor antigen peptides and/or DCs has been developed to potentiate DC functions showing promising results in mouse tumor models. While most clinical trials have failed, good results in a clinical trial for BC have been reported that treatment with tumor lysates, DCs, IL12/IL18-activated lymphocytes isolated from peripheral blood of BC patients showed significant reduction of the risk of relapse and metastasis through reduction of Tregs (Shevchenko et al. 2020). Nanovaccines composed of nanoparticles (liposomes, micelles, microemulsions, dendrimers, etc.) have the advantage of easy delivery of target antigens and adjuvants, and fewer toxic side effects, leading to improved antigen presentation and subsequent induction of potent anti-tumor effector T cells, and thus have attracted much attention as promising immunotherapeutics that overcomes the limitations of traditional vaccines (Fang et al. 2022). Zhang et al. developed FAP gene–engineered tumor cell–derived exosome-like nanovesicles (designated eNVs-FAP) as a nanovaccine that can promote DC maturation followed by decrease of immunosuppressive cells (M2-TAMs, MDSCs, and Tregs), but increase of effector T cells in the tumor microenvironment (Zhang et al. 2020). Subsequently, they identified miRNA-5,119 that targets mRNAs encoding multiple negative immune regulatory molecules, such as PDL1 and IDO2, in DCs by miRNA array analysis, and additionally developed DC vaccines expressing a miR-5,119 mimic that can downregulate PDL1 expression in DCs and suppress T-cell exhaustion leading to induction of potent anti-tumor immunity in mouse mammary tumor models. Another group also showed that the eNVs-FAP injection induces strong anti-tumor effects through not only promoting DC maturation, but also reduction of various immunosuppressive cells in the tumor microenvironment of various types of cancer mouse models, including BC (Hu et al. 2021).

7 MCs

MCs play multifunctional roles in regulating tissue homeostasis and immune response by secreting a variety of pro-inflammatory molecules (histamines, tryptase, chymase, heparin, lysosomal enzymes, IL1β, IL6, IL8, IL10, IL13, TNFα, PDGF, VEGF, FGF2, TGFβ, MMPs, CCL7, CCL13, etc.), and are involved in pathogenic mechanisms of many diseases, such as allergy, rheumatoid arthritis, diabetes, multiple sclerosis, and cancer (Anand et al. 2012). In cancer, blood-circulating MCs migrate into tumor tissues in response to microenvironmental chemoattractant factors (IL8, CCL2, CXCL1, CXCL10, etc.), and are subsequently activated by pro-inflammatory cytokines (SCF, IL33, PGE2, leukotriene B4, osteopontin, etc.), resulting in producing the pro-inflammatory molecules to facilitate tumor progression and metastasis directly by enhancing tumor proliferation, invasion, and metastatic dissemination, and indirectly by inducing angiogenesis and abnormal immunity, including immune tolerance and inflammation depending on the tumoral conditions (Komi and Redegeld 2020). SCF is the most potent molecule that induces tryptase and chymase upon binding to the c-kit receptor. Many clinical studies have demonstrated the significant correlations among the levels of MCs, angiogenesis and tumor progression, and prognosis of patients in various types of cancers, including BC.

In BC, MCs are also a key player in tumor progression and metastasis. For example, MCs promote tumor growth through suppression of basal gene expressions, including CK5, cMET, and EGFR, resulting in induction and activation of luminal markers, including ER and its target genes (PGR, KRT8/CK8, BCL2), in BC cells (Majorini et al. 2020). This study also showed that MC abundance is significantly seen in tumors of HER2$^+$ BC patients who relapsed after adjuvant therapy with trastuzumab, suggesting a potential biomarker to predict high relapse risk in the therapy. Interestingly, MCs in the mammary tissues appear to interact with the distant gut microbiome to promote cancer metastasis. For example, it has been reported that disruption in commensal homeostasis promotes metastatic dissemination of BC cells through promoting fibrosis, collagen deposition, and myeloid infiltration into the mammary tissue and tumor tissues (Buchta Rosean et al. 2019). The same group also reported that the commensal dysbiosis predominantly induces accumulation of CCR2$^+$ MCs in non-tumor-bearing mammary tissues, and the MCs induce fibroblast activation and collagen I deposition by secreting PDGF-B, resulting in early tumor dissemination in the mice (Feng et al. 2022). They also showed clinical data that MC abundance in mammary tissues correlates with increased collagen and metastatic recurrence in ER$^+$ BC patients.

Indeed, clinical studies have reported that the high level of MCs in tumor tissues is associated with the risk of lymph node metastasis (Floroni et al. 2022), lung metastasis (Zhang et al. 2021), low responses to treatments (Reddy et al. 2019), and short survival particularly of patients with aggressive types like TNBC (Carpenco et al. 2019). A recent study demonstrated that ANXA1, which is known as a phospholipid-binding protein having pro-tumoral properties, is a key molecule representative of aggressive phenotypes of TNBC, and its expression is associated with MC infiltration and inflammation in patients with TNBC (Okano et al. 2019). Thus, targeting MCs has been expected to contribute to the treatment of BC, and many therapeutic inhibitors targeting tyrosine kinase (gabexate, nafamostat mesylate, etc.) and various tyrosine kinases (imatinib, masitinib, sunitinib, lenvatinib, vandetanib, etc.) have been considered to reduce and inhibit MC accumulation in tumors, MC functions, and its products. However, as well as the case of other cell populations, such inhibitory treatments targeting MCs would still require caution, since there are some reports showing opposing results that increase of MCs is associated with a better prognosis in BC.

8 Eosinophils

Eosinophils play pivotal roles in maintenance of host defense and immune homeostasis, and are involved in pathogenesis of various diseases, such as allergy, asthma, parasitic and fungal infections, and autoimmune diseases, and cancer (Jackson et al. 2022). Eosinophils differentiate by receiving stimuli (IL5, IL3, GM-CSF, etc.) in the bone marrow, and migrate into damaged tissues in response to numerous chemoattractants (eotaxin1/CCL11, eotaxin2/CCL24, eotaxin3/CCL26, CCL5, CCL8, CCL7, CCL11, IL24, IL26, TSLP, etc.) secreted from various cells, including endothelial and epithelial cells, fibroblasts, adipocytes, type 2 innate lymphoid cells (ILC2s), and Th2 (Klion et al. 2020). The eosinophils are activated by various environmental stimuli, and release pathogenic molecules via distinct pathways, including classical exocytosis that releases entire granules as a fusion with plasma membrane, cytolysis that deposit intact granules directly into the tissue through a distinctive mode of cell death, and piecemeal degranulation that releases granule contents without granule through fusion with plasma membrane (Spencer et al. 2014). IL5 is a key molecule in the entire mechanisms, including eosinophil priming, activation, expansion, and survival, in paracrine and autocrine manners.

In cancer, eosinophils have been reported to exhibit two different functions, tumor-supportive and tumor-suppressive types, just like Møs and DCs (Ghaffari and Rezaei 2023). The tumor-supportive types release inflammatory molecules (IL4, IL8, IL13, CCL22, IDO, VEGF, FGF2, ROS, MMP9, etc.) to induce DNA damage and EMT in tumor cells, and matrix remodeling and angiogenesis in stroma in tumor tissues. For example, IL4 and IL13 promote M2-Mø polarization, CCL22 recruits Tregs, and VEGF induces angiogenesis in tumor tissues. In contrast, the tumor-suppressive types release various immunomodulatory molecules (granzyme A, TNFα, IL18, IFNγ, CCL5, CXCL9, CXCL10, etc.) in addition to eosinophil-specific molecules, including eosinophil-derived neurotoxin (EDN), eosinophil peroxidase (EPO), and unique acidophilic secondary granules composed of major basic protein 1 (MBP1) and a matrix composed of eosinophil cationic protein (ECP), which directly kill tumor cells. CCL5 and CXCL10 recruit activated $CD8^+$ T cells in tumor tissues.

In BC, while a relationship between the high levels of eosinophils and poor prognosis has been reported in clinical studies before, there has been recently an increase in the number of reports demonstrating its association with favorable prognosis and its important role in the anti-tumor mechanisms. For example, it has been reported that the high level of eosinophils in peripheral blood at baseline is associated with a lower risk of relapse and better prognosis in BC patients, particularly not receiving chemotherapy plus anti-HER2 therapy (Onesti et al. 2020). Decreased leukocyte counts, including eosinophils, may be the reason for less benefit in the chemo-treated patients, although another group reported that a high level of eosinophils in peripheral blood is associated with progression-free survival in TNBC, even after chemotherapy (Ghebeh et al. 2022). Grisaru-Tal et al. found accumulation of eosinophils in the metastatic lung biopsies obtained from BC patients, and showed unique anti-tumor mechanisms using mouse mammary tumor models that eosinophils are recruited to the metastatic lung tissues independently of CCR3 and eotaxins, and facilitate T-cell infiltration by being activated with TNFα and IFNγ (Grisaru-Tal et al. 2021). Another group also showed anti-metastatic roles of eosinophils using transgenic mouse models characterized by elevated IL5 and eosinophilia, or eosinophil deficiency (Cederberg et al. 2022).

There are many reports showing the anti-tumor roles of eosinophils in the therapeutic mechanisms. For example, blocking DPP4/CD26, which is highly expressed in various types of cancer, with a specific inhibitor sitagliptin significantly suppresses tumor growth, and eosinophils, but not T cells, are required for the anti-tumor efficacy (Hollande et al. 2019). The sitagliptin therapy enhanced tumoral IL33 expression followed by CCL11 induction that recruits $CCR3^+$ eosinophils, which suppress tumor growth via its degranulation, in tumors. This suggests the critical role of IL33 expressed in tumors in the eosinophil-mediated anti-tumor immunity, although the opposite tumor-supportive roles have been often reported in cancer (Griesenauer and Paczesny 2017). Interestingly, it has been demonstrated that blocking CTLA4 increases eosinophils that can induce vessel normalization in the tumor milieu of mouse mammary tumor models, contributing to the therapeutic efficacy (Zheng et al. 2020). In clinical settings, the high eosinophil levels in peripheral blood and tumor tissues are significantly associated with favorable prognosis in metastatic TNBC after nivolumab therapy (Blomberg et al. 2023). As the underlying mechanisms, eosinophils were found to be generated by IL5 secreted from $CD4^+$ T cells followed by expansion upon the anti-PD1 therapy, and sequentially accumulate in tumor tissues in response to tumor-derived IL33.

9 Basophils

Basophils circulating in peripheral blood differentiate from myeloid progenitors in response to IL3, and migrate into inflamed tissues to be activated to release a great deal of granule contents, including histamine, leukotriene C4, granzyme B, immunomodulatory cytokines (PGE2, VEGF, GM-CSF, IL4, IL6, IL13, TSLP, etc.) and chemokines (CCL3, CCL4, CCL5, IL8, etc.) via crosslinking of FceRIa and IgE-bound antigen, or upon receiving inflammatory stimuli (complement factors C5a and C3a, MBP, PAF, PAMPs, TLR ligands, extracellular nucleotides, etc.) (Chirumbolo et al. 2018). Although the similarity between basophils and mast cells has long been noticed, recent studies have demonstrated the differences between them using mouse models: For example, IL3R expression in basophils, but not mast cells, P1-RUNX and C/EBPα in basophils, and FOG1 and MITF in mast cells, although STAT5, GATA1, and GATA2 are common to both (Huang et al. 2016).

Although the roles of basophils in immune responses to bacteria and parasites, including ticks and worms, have attracted attention for many years, many studies have recently demonstrated the relationships between basophils and many other diseases, including allergy (asthma, allergic rhinitis, atopic dermatitis, nasal polyps, etc.), autoimmune diseases, fibrosis, and cancer (Miyake et al. 2022). In cancer, tumor-supportive roles have been reported more frequently, albeit still undetermined. For example, basophils recruit TAMs and Tregs, and induce Th2 and M2-Mø polarization and Treg differentiation/activation, potentially leading to tumor progression and metastasis (Chauhan et al. 2022). Angiogenic molecules (VEGF, PGE2, etc.) also facilitate the cancer refractory. In BC, however, there are very few reports on basophils, and the roles in the pathogenesis remain unclear.

10 Conclusion

The importance of immune control in treating cancer has been demonstrated through the success of ICI therapy. However, the effectiveness of immunotherapy, including ICIs, is still limited to a small number of patients, and innovative remedies are desperately needed for saving the rest of the patients. Myeloid cells are the majority of immune cells in human, and are a key upstream factor that broadly suppresses, exhausts, and impairs the anti-tumor immunity in the host with cancer, providing a great impact. Thus, targeting these myeloid cells to reduce their number and function may make it easier to achieve successful immune control, and may improve clinical outcomes for cancer treatments, including BC. In BC, ICI is currently applied to only TNBC. However, a deeper understanding of this large myeloid cell population may lead to the use of ICIs and other immunotherapeutics in a broader range of treatments of BC, including not only TNBC but also other types. However, in most clinical studies, myeloid cells, especially the short-lived and fragile granulocytes, are often not properly sampled, and the quality of such sampling has a negative impact on the quality of the output data. This means that we may not fully understand its true character. For example, in the common Ficoll method for isolating immune cells from peripheral blood, only the intermediate fraction is collected and analyzed, and the granulocytes underneath are either discarded, or most of the granulocytes in the cryopreserved specimen will die after thawing. We hope that this paper provides an opportunity to re-recognize the importance of myeloid cells in cancer progression and the need for improving sampling methods, and contributes to the development of better myeloid-targeting strategies that will lead to more effective treatment for BC.

Compliance with Ethical Standards

Disclosure of Potential Conflicts of Interest None of the authors have conflicts of interest to declare.

Research Involving Human Participants and/or Animals Not applicable.

Informed Consent Not applicable.

References

Ager EI, Kozin SV, Kirkpatrick ND, Seano G, Kodack DP, Askoxylakis V et al (2015) Blockade of MMP14 activity in murine breast carcinomas: implications for macrophages, vessels, and radiotherapy. J Natl Cancer Inst 107(4):djv017. https://doi.org/10.1093/jnci/djv017

Ahirwar DK, Charan M, Mishra S, Verma AK, Shilo K, Ramaswamy B et al (2021) Slit2 inhibits breast cancer metastasis by activating M1-like phagocytic and antifibrotic macrophages. Cancer Res 81(20):5255–5267. https://doi.org/10.1158/0008-5472.CAN-20-3909

Anand P, Singh B, Jaggi AS, Singh N (2012) Mast cells: an expanding pathophysiological role from allergy to other disorders. Naunyn Schmiedeberg's Arch Pharmacol 385(7):657–670. https://doi.org/10.1007/s00210-012-0757-8

Anderson NM, Simon MC (2020) The tumor microenvironment. Curr Biol 30(16):R921–R925. https://doi.org/10.1016/j.cub.2020.06.081

Augimeri G, Gonzalez ME, Paoli A, Eido A, Choi Y, Burman B et al (2023) A hybrid breast cancer/mesenchymal stem cell population enhances chemoresistance and metastasis. JCI Insight 8(18):e164216. https://doi.org/10.1172/jci.insight.164216

Balan S, Saxena M, Bhardwaj N (2019) Dendritic cell subsets and locations. Int Rev Cell Mol Biol 348:1–68. https://doi.org/10.1016/bs.ircmb.2019.07.004

Bartosh TJ, Ullah M, Zeitouni S, Beaver J, Prockop DJ (2016) Cancer cells enter dormancy after cannibalizing mesenchymal stem/stromal cells (MSCs). Proc Natl Acad Sci USA 113(42):E6447–E6456. https://doi.org/10.1073/pnas.1612290113

Barzaman K, Karami J, Zarei Z, Hosseinzadeh A, Kazemi MH, Moradi-Kalbolandi S et al (2020) Breast cancer: biology, biomarkers, and treatments. Int Immunopharmacol 84:106535. https://doi.org/10.1016/j.intimp.2020.106535

Basha S, Jin-Smith B, Sun C, Pi L (2023) The SLIT/ROBO pathway in liver fibrosis and cancer. Biomol Ther 13(5):785. https://doi.org/10.3390/biom13050785

Bergenfelz C, Roxa A, Mehmeti M, Leandersson K, Larsson AM (2020) Clinical relevance of systemic monocytic-MDSCs in patients with metastatic breast cancer. Cancer Immunol Immunother 69(3):435–448. https://doi.org/10.1007/s00262-019-02472-z

Blaye C, Boyer T, Peyraud F, Domblides C, Larmonier N (2022) Beyond immunosuppression: the multifaceted functions of tumor-promoting myeloid cells in breast cancers. Front Immunol 13:838040. https://doi.org/10.3389/fimmu.2022.838040

Blomberg OS, Spagnuolo L, Garner H, Voorwerk L, Isaeva OI, van Dyk E et al (2023) IL-5-producing CD4(+) T cells and eosinophils cooperate to enhance response to immune checkpoint blockade in breast cancer. Cancer Cell 41(1):106–123 e10. https://doi.org/10.1016/j.ccell.2022.11.014

Buchta Rosean C, Bostic RR, Ferey JCM, Feng TY, Azar FN, Tung KS et al (2019) Preexisting commensal Dysbiosis is a host-intrinsic regulator of tissue inflammation and tumor cell dissemination in hormone receptor-positive breast cancer. Cancer Res 79(14):3662–3675. https://doi.org/10.1158/0008-5472.CAN-18-3464

Cai M, Huang X, Huang X, Ju D, Zhu YZ, Ye L (2023) Research progress of interleukin-15 in cancer immunotherapy. Front Pharmacol 14:1184703. https://doi.org/10.3389/fphar.2023.1184703

Carnevale S, Di Ceglie I, Grieco G, Rigatelli A, Bonavita E, Jaillon S (2023) Neutrophil diversity in inflammation and cancer. Front Immunol 14:1180810. https://doi.org/10.3389/fimmu.2023.1180810

Carpenco E, Ceausu RA, Cimpean AM, Gaje PN, Saptefrati L, Fulga V et al (2019) Mast cells as an indicator and prognostic marker in molecular subtypes of breast cancer. In Vivo 33(3):743–748. https://doi.org/10.21873/invivo.11534

Castellaro AM, Rodriguez-Baili MC, Di Tada CE, Gil GA (2019) Tumor-associated macrophages induce endocrine therapy resistance in ER+ breast cancer cells. Cancers (Basel) 11(2):189. https://doi.org/10.3390/cancers11020189

Cederberg RA, Franks SE, Wadsworth BJ, So A, Decotret LR, Hall MG et al (2022) Eosinophils decrease pulmonary metastatic mammary tumor growth. Front Oncol 12:841921. https://doi.org/10.3389/fonc.2022.841921

Cha YJ, Koo JS (2020) Role of tumor-associated myeloid cells in breast cancer. Cell 9(8):1785. https://doi.org/10.3390/cells9081785

Chauhan J, Stavraka C, Grandits M, Palhares L, Josephs DH, Lacy KE et al (2022) Clinical and translational significance of basophils in patients with cancer. Cell 11(3):438. https://doi.org/10.3390/cells11030438

Chen WY, Holmes MD (2017) Role of aspirin in breast cancer survival. Curr Oncol Rep 19(7):48. https://doi.org/10.1007/s11912-017-0605-6

Chen F, Chen J, Yang L, Liu J, Zhang X, Zhang Y et al (2019a) Extracellular vesicle-packaged HIF-1alpha-stabilizing lncRNA from tumour-associated macrophages regulates aerobic glycolysis of breast cancer cells. Nat Cell Biol 21(4):498–510. https://doi.org/10.1038/s41556-019-0299-0

Chen YC, Gonzalez ME, Burman B, Zhao X, Anwar T, Tran M et al (2019b) Mesenchymal stem/stromal cell engulfment reveals metastatic advantage in breast cancer. Cell Rep 27(13):3916–3926 e5. https://doi.org/10.1016/j.celrep.2019.05.084

Chen Y, Hu H, Tan S, Dong Q, Fan X, Wang Y et al (2022) The role of neutrophil extracellular traps in cancer progression, metastasis and therapy. Exp Hematol Oncol 11(1):99. https://doi.org/10.1186/s40164-022-00345-3

Cheng H, Huang H, Guo Z, Chang Y, Li Z (2021) Role of prostaglandin E2 in tissue repair and regeneration. Theranostics 11(18):8836–8854. https://doi.org/10.7150/thno.63396

Chirumbolo S, Bjorklund G, Sboarina A, Vella A (2018) The role of basophils as innate immune regulatory cells in allergy and immunotherapy. Hum Vaccin Immunother 14(4):815–831. https://doi.org/10.1080/21645515.2017.1417711

Clarke R (2011) Cannibalism, cell survival, and endocrine resistance in breast cancer. Breast Cancer Res 13(4):311. https://doi.org/10.1186/bcr2870

Cohen N, Shani O, Raz Y, Sharon Y, Hoffman D, Abramovitz L et al (2017) Fibroblasts drive an immunosuppressive and growth-promoting microenvironment in breast cancer via secretion of Chitinase 3-like 1. Oncogene 36(31):4457–4468. https://doi.org/10.1038/onc.2017.65

Collin M, Bigley V (2018) Human dendritic cell subsets: an update. Immunology 154(1):3–20. https://doi.org/10.1111/imm.12888

Cortesi L, Rugo HS, Jackisch C (2021) An overview of PARP inhibitors for the treatment of breast cancer. Target Oncol 16(3):255–282. https://doi.org/10.1007/s11523-021-00796-4

Costa A, Kieffer Y, Scholer-Dahirel A, Pelon F, Bourachot B, Cardon M et al (2018) Fibroblast heterogeneity and immunosuppressive environment in human breast cancer. Cancer Cell 33(3):463–479 e10. https://doi.org/10.1016/j.ccell.2018.01.011

Cupp MA, Cariolou M, Tzoulaki I, Aune D, Evangelou E, Berlanga-Taylor AJ (2020) Neutrophil to lymphocyte ratio and cancer prognosis: an umbrella review of systematic reviews and meta-analyses of observational studies. BMC Med 18(1):360. https://doi.org/10.1186/s12916-020-01817-1

de la Cruz-Merino L, Gion M, Cruz J, Alonso-Romero JL, Quiroga V, Moreno F et al (2022) Pembrolizumab in combination with gemcitabine for patients with HER2-negative advanced breast cancer: GEICAM/2015-04 (PANGEA-breast) study. BMC Cancer 22(1):1258. https://doi.org/10.1186/s12885-022-10363-3

DeNardo DG, Galkin A, Dupont J, Zhou L, Bendell J (2021) GB1275, a first-in-class CD11b modulator: rationale for immunotherapeutic combinations in solid tumors. J Immunother Cancer 9(8):e003005. https://doi.org/10.1136/jitc-2021-003005

Doebel T, Voisin B, Nagao K (2017) Langerhans cells – the macrophage in dendritic cell clothing. Trends Immunol 38(11):817–828. https://doi.org/10.1016/j.it.2017.06.008

El Hejjioui B, Lamrabet S, Amrani Joutei S, Senhaji N, Bouhafa T, Malhouf MA et al (2023) New biomarkers and treatment advances in triple-negative breast cancer. Diagnostics (Basel) 13(11):1949. https://doi.org/10.3390/diagnostics13111949

Fang X, Lan H, Jin K, Gong D, Qian J (2022) Nanovaccines for cancer prevention and immunotherapy: an update review. Cancers (Basel) 14(16):3842. https://doi.org/10.3390/cancers14163842

Feng TY, Azar FN, Dreger SA, Rosean CB, McGinty MT, Putelo AM et al (2022) Reciprocal interactions between the gut microbiome and mammary tissue mast cells promote metastatic dissemination of HR+ breast tumors. Cancer Immunol Res 10(11):1309–1325. https://doi.org/10.1158/2326-6066.CIR-21-1120

Floroni E, Ceausu AR, Cosoroaba RM, Niculescu Talpos IC, Popovici RA, Gaje NP et al (2022) Mast cell density in the primary tumor predicts lymph node metastases in patients with breast cancer. Romanian J Morphol Embryol 63(1):129–135. https://doi.org/10.47162/RJME.63.1.13

Gago-Dominguez M, Matabuena M, Redondo CM, Patel SP, Carracedo A, Ponte SM et al (2020) Neutrophil to lymphocyte ratio and breast cancer risk: analysis by subtype and potential interactions. Sci Rep 10(1):13203. https://doi.org/10.1038/s41598-020-70077-z

Gao Z, Feng Y, Xu J, Liang J (2022) T-cell exhaustion in immune-mediated inflammatory diseases: new implications for immunotherapy. Front Immunol 13:977394. https://doi.org/10.3389/fimmu.2022.977394

Geurts V, Kok M (2023) Immunotherapy for metastatic triple negative breast cancer: current paradigm and future approaches. Curr Treat Options in Oncol 24(6):628–643. https://doi.org/10.1007/s11864-023-01069-0

Ghaffari S, Rezaei N (2023) Eosinophils in the tumor microenvironment: implications for cancer immunotherapy. J Transl Med 21(1):551. https://doi.org/10.1186/s12967-023-04418-7

Ghallab AM, Eissa RA, El Tayebi HM (2022) CXCR2 small-molecule antagonist combats chemoresistance and enhances immunotherapy in triple-negative breast cancer. Front Pharmacol 13:862125. https://doi.org/10.3389/fphar.2022.862125

Ghebeh H, Elshenawy MA, AlSayed AD, Al-Tweigeri T (2022) Peripheral blood eosinophil count is associated with response to chemoimmunotherapy in metastatic triple-negative breast cancer. Immunotherapy 14(4):189–199. https://doi.org/10.2217/imt-2021-0149

Girithar HN, Staats Pires A, Ahn SB, Guillemin GJ, Gluch L, Heng B (2023) Involvement of the kynurenine pathway in breast cancer: updates on clinical research and trials. Br J Cancer 129(2): 185–203. https://doi.org/10.1038/s41416-023-02245-7

Godet I, Gilkes DM (2017) BRCA1 and BRCA2 mutations and treatment strategies for breast cancer. Integr Cancer Sci Ther 4(1):10.15761. https://doi.org/10.15761/ICST.1000228

Gonda K, Shibata M, Ohtake T, Matsumoto Y, Tachibana K, Abe N et al (2017) Myeloid-derived suppressor cells are increased and correlated with type 2 immune responses, malnutrition, inflammation, and poor prognosis in patients with breast cancer. Oncol Lett 14(2):1766–1774. https://doi.org/10.3892/ol.2017.6305

Goswami KK, Bose A, Baral R (2021) Macrophages in tumor: an inflammatory perspective. Clin Immunol 232:108875. https://doi.org/10.1016/j.clim.2021.108875

Griesenauer B, Paczesny S (2017) The ST2/IL-33 axis in immune cells during inflammatory diseases. Front Immunol 8:475. https://doi.org/10.3389/fimmu.2017.00475

Grisaru-Tal S, Dulberg S, Beck L, Zhang C, Itan M, Hediyeh-Zadeh S et al (2021) Metastasis-entrained eosinophils enhance lymphocyte-mediated antitumor immunity. Cancer Res 81(21): 5555–5571. https://doi.org/10.1158/0008-5472.CAN-21-0839

Groth C, Hu X, Weber R, Fleming V, Altevogt P, Utikal J et al (2019) Immunosuppression mediated by myeloid-derived suppressor cells (MDSCs) during tumour progression. Br J Cancer 120(1):16–25. https://doi.org/10.1038/s41416-018-0333-1

Guo Y, Zhai Y, Wu L, Wang Y, Wu P, Xiong L (2022) Mesenchymal stem cell-derived extracellular vesicles: pleiotropic impacts on breast cancer occurrence, development, and therapy. Int J Mol Sci 23(6):2927. https://doi.org/10.3390/ijms23062927

Guo Z, Zhang H, Fu Y, Kuang J, Zhao B, Zhang L et al (2023) Cancer-associated fibroblasts induce growth and radioresistance of breast cancer cells through paracrine IL-6. Cell Death Discov 9(1):6. https://doi.org/10.1038/s41420-023-01306-3

Hajizadeh F, Aghebati Maleki L, Alexander M, Mikhailova MV, Masjedi A, Ahmadpour M et al (2021) Tumor-associated neutrophils as new players in immunosuppressive process of the tumor microenvironment in breast cancer. Life Sci 264:118699. https://doi.org/10.1016/j.lfs.2020.118699

He F, Antonucci L, Karin M (2020) NRF2 as a regulator of cell metabolism and inflammation in cancer. Carcinogenesis 41(4):405–416. https://doi.org/10.1093/carcin/bgaa039

He Q, Hu H, Yang F, Song D, Zhang X, Dai X (2023) Advances in chimeric antigen receptor T cells therapy in the treatment of breast cancer. Biomed Pharmacother 162:114609. https://doi.org/10.1016/j.biopha.2023.114609

Hollande C, Boussier J, Ziai J, Nozawa T, Bondet V, Phung W et al (2019) Inhibition of the dipeptidyl peptidase DPP4 (CD26) reveals IL-33-dependent eosinophil-mediated control of tumor growth. Nat Immunol 20(3):257–264. https://doi.org/10.1038/s41590-019-0321-5

Hong DS, Parikh A, Shapiro GI, Varga A, Naing A, Meric-Bernstam F et al (2020) First-in-human phase I study of immunomodulatory E7046, an antagonist of PGE(2)-receptor E-type 4 (EP4), in patients with advanced cancers. J Immunother Cancer 8(1):e000222. https://doi.org/10.1136/jitc-2019-000222

Hu S, Ma J, Su C, Chen Y, Shu Y, Qi Z et al (2021) Engineered exosome-like nanovesicles suppress tumor growth by reprogramming tumor microenvironment and promoting tumor ferroptosis. Acta Biomater 135:567–581. https://doi.org/10.1016/j.actbio.2021.09.003

Huang H, Li Y, Liu B (2016) Transcriptional regulation of mast cell and basophil lineage commitment. Semin Immunopathol 38(5):539–548. https://doi.org/10.1007/s00281-016-0562-4

Jackson DJ, Akuthota P, Roufosse F (2022) Eosinophils and eosinophilic immune dysfunction in health and disease. Eur Respir Rev 31(163):210150. https://doi.org/10.1183/16000617.0150-2021

Jacobs F, Agostinetto E, Miggiano C, De Sanctis R, Zambelli A, Santoro A (2023) Hope and hype around immunotherapy in triple-negative breast cancer. Cancers (Basel) 15(11):2933. https://doi.org/10.3390/cancers15112933

Javid H, Attarian F, Saadatmand T, Rezagholinejad N, Mehri A, Amiri H et al (2023) The therapeutic potential of immunotherapy in the treatment of breast cancer: rational strategies and recent progress. J Cell Biochem 124(4):477–494. https://doi.org/10.1002/jcb.30402

Jayakumar A, Bothwell ALM (2019) Functional diversity of myeloid-derived suppressor cells: the multitasking hydra of cancer. J Immunol 203(5):1095–1103. https://doi.org/10.4049/jimmunol.1900500

Jiang W, Xu J (2020) Immune modulation by mesenchymal stem cells. Cell Prolif 53(1):e12712. https://doi.org/10.1111/cpr.12712

Juric V, O'Sullivan C, Stefanutti E, Kovalenko M, Greenstein A, Barry-Hamilton V et al (2018) MMP-9 inhibition promotes anti-tumor immunity through disruption of biochemical and physical barriers to T-cell trafficking to tumors. PLoS One 13(11):e0207255. https://doi.org/10.1371/journal.pone.0207255

Kiaei SZF, Nouralishahi A, Ghasemirad M, Barkhordar M, Ghaffari S, Kheradjoo H et al (2023) Advances in natural killer cell therapies for breast cancer. Immunol Cell Biol 101(8):705–726. https://doi.org/10.1111/imcb.12658

Kim H, Lin Q, Yun Z (2019) BRCA1 regulates the cancer stem cell fate of breast cancer cells in the context of hypoxia and histone deacetylase inhibitors. Sci Rep 9(1):9702. https://doi.org/10.1038/s41598-019-46210-y

Klion AD, Ackerman SJ, Bochner BS (2020) Contributions of eosinophils to human health and disease. Annu Rev Pathol 15:179–209. https://doi.org/10.1146/annurev-pathmechdis-012419-032756

Komi DEA, Redegeld FA (2020) Role of mast cells in shaping the tumor microenvironment. Clin Rev Allergy Immunol 58(3):313–325. https://doi.org/10.1007/s12016-019-08753-w

Kudo-Saito C, Ozaki Y, Imazeki H, Hayashi H, Masuda J, Ozawa H et al (2021) Targeting oncoimmune drivers of cancer metastasis. Cancers (Basel) 13(3):554. https://doi.org/10.3390/cancers13030554

Kwantwi LB, Wang S, Zhang W, Peng W, Cai Z, Sheng Y et al (2021) Tumor-associated neutrophils activated by tumor-derived CCL20 (C-C motif chemokine ligand 20) promote T cell immunosuppression via programmed death-ligand 1 (PD-L1) in breast cancer. Bioengineered 12(1):6996–7006. https://doi.org/10.1080/21655979.2021.1977102

Kwon MJ (2022) Matrix metalloproteinases as therapeutic targets in breast cancer. Front Oncol 12:1108695. https://doi.org/10.3389/fonc.2022.1108695

Lavoie SS, Dumas E, Vulesevic B, Neagoe PE, White M, Sirois MG (2018) Synthesis of human neutrophil extracellular traps contributes to angiopoietin-mediated in vitro proinflammatory and proangiogenic activities. J Immunol 200(11):3801–3813. https://doi.org/10.4049/jimmunol.1701203

Lebeau G, Ah-Pine F, Daniel M, Bedoui Y, Vagner D, Frumence E et al (2022) Perivascular mesenchymal stem/stromal cells, an immune privileged niche for viruses? Int J Mol Sci 23(14):8038. https://doi.org/10.3390/ijms23148038

Li P, Lu M, Shi J, Gong Z, Hua L, Li Q et al (2020) Lung mesenchymal cells elicit lipid storage in neutrophils that fuel breast cancer lung metastasis. Nat Immunol 21(11):1444–1455. https://doi.org/10.1038/s41590-020-0783-5

Li H, Qiao C, Zhao L, Jing Q, Xue D, Zhang Y (2022a) Epigallocatechin-3-gallate reduces neutrophil extracellular trap formation and tissue injury in severe acute pancreatitis. J Leukoc Biol 112(6):1427–1443. https://doi.org/10.1002/JLB.3A0322-151R

Li J, Shu X, Xu J, Su SM, Chan UI, Mo L et al (2022b) S100A9-CXCL12 activation in BRCA1-mutant breast cancer promotes an immunosuppressive microenvironment associated with resistance to immunotherapy. Nat Commun 13(1):1481. https://doi.org/10.1038/s41467-022-29151-5

Li D, Hemati H, Park Y, Taftaf R, Zhang Y, Liu J et al (2023) ICAM-1-suPAR-CD11b axis is a novel therapeutic target for metastatic triple-negative breast cancer. Cancers (Basel) 15(10):2734. https://doi.org/10.3390/cancers15102734

Lin Z, Radaeva M, Cherkasov A, Dong X (2022) Lin28 regulates cancer cell stemness for tumour progression. Cancers (Basel) 14(19):4640. https://doi.org/10.3390/cancers14194640

Liu C, Qiang J, Deng Q, Xia J, Deng L, Zhou L et al (2021) ALDH1A1 activity in tumor-initiating cells remodels myeloid-derived suppressor cells to promote breast cancer progression. Cancer Res 81(23):5919–5934. https://doi.org/10.1158/0008-5472.CAN-21-1337

Liu Z, Shi Y, Lv L, Chen J, Jiang W, Li J et al (2022) Small molecular inhibitors reverse cancer metastasis by blockading oncogenic PITPNM3. Adv Sci (Weinh) 9(35):e2204649. https://doi.org/10.1002/advs.202204649

Loibl S, Poortmans P, Morrow M, Denkert C, Curigliano G (2021) Breast cancer. Lancet 397(10286):1750–1769. https://doi.org/10.1016/S0140-6736(20)32381-3

Long GV, Dummer R, Hamid O, Gajewski TF, Caglevic C, Dalle S et al (2019) Epacadostat plus pembrolizumab versus placebo plus pembrolizumab in patients with unresectable or metastatic melanoma (ECHO-301/KEYNOTE-252): a phase 3, randomised, double-blind study. Lancet Oncol 20(8):1083–1097. https://doi.org/10.1016/S1470-2045(19)30274-8

Ma J, Cai Z, Wei H, Liu X, Zhao Q, Zhang T (2017) The anti-tumor effect of aspirin: what we know and what we expect. Biomed Pharmacother 95:656–661. https://doi.org/10.1016/j.biopha.2017.08.085

Ma S, Guo C, Sun C, Han T, Zhang H, Qu G et al (2021) Aspirin use and risk of breast cancer: a meta-analysis of observational studies from 1989 to 2019. Clin Breast Cancer 21(6):552–565. https://doi.org/10.1016/j.clbc.2021.02.005

Ma C, He D, Tian P, Wang Y, He Y, Wu Q et al (2022) miR-182 targeting reprograms tumor-associated macrophages and limits breast cancer progression. Proc Natl Acad Sci USA 119(6):e2114006119. https://doi.org/10.1073/pnas.2114006119

Mahesh G, Anil Kumar K, Reddanna P (2021) Overview on the discovery and development of anti-inflammatory drugs: should the focus be on synthesis or degradation of PGE(2)? J Inflamm Res 14:253–263. https://doi.org/10.2147/JIR.S278514

Majorini MT, Cancila V, Rigoni A, Botti L, Dugo M, Triulzi T et al (2020) Infiltrating mast cell-mediated stimulation of estrogen receptor activity in breast cancer cells promotes the luminal phenotype. Cancer Res 80(11):2311–2324. https://doi.org/10.1158/0008-5472.CAN-19-3596

Majumder M, Nandi P, Omar A, Ugwuagbo KC, Lala PK (2018) EP4 as a therapeutic target for aggressive human breast cancer. Int J Mol Sci 19(4):1019. https://doi.org/10.3390/ijms19041019

Malech HL, Deleo FR, Quinn MT (2014) The role of neutrophils in the immune system: an overview. Methods Mol Biol 1124:3–10. https://doi.org/10.1007/978-1-62703-845-4_1

Maryam M, Naemi M, Hasani SS (2021) A comprehensive review on oncogenic miRNAs in breast cancer. J Genet 100:15. https://www.ncbi.nlm.nih.gov/pubmed/33764337

Maurer M, Cheung DS, Theess W, Yang X, Dolton M, Guttman A et al (2022) Phase 2 randomized clinical trial of astegolimab in patients with moderate to severe atopic dermatitis. J Allergy Clin Immunol 150(6):1517–1524. https://doi.org/10.1016/j.jaci.2022.08.015

Mercogliano MF, Bruni S, Mauro FL, Schillaci R (2023) Emerging targeted therapies for HER2-positive breast cancer. Cancers (Basel) 15(7):1987. https://doi.org/10.3390/cancers15071987

Mishra AK, Banday S, Bharadwaj R, Ali A, Rashid R, Kulshreshtha A et al (2022) Macrophages as a potential immunotherapeutic target in solid cancers. Vaccines (Basel) 11(1):55. https://doi.org/10.3390/vaccines11010055

Miyake K, Ito J, Karasuyama H (2022) Role of basophils in a broad spectrum of disorders. Front Immunol 13:902494. https://doi.org/10.3389/fimmu.2022.902494

Monteran L, Ershaid N, Doron H, Zait Y, Scharff Y, Ben-Yosef S et al (2022) Chemotherapy-induced complement signaling modulates immunosuppression and metastatic relapse in breast cancer. Nat Commun 13(1):5797. https://doi.org/10.1038/s41467-022-33598-x

Mousset A, Lecorgne E, Bourget I, Lopez P, Jenovai K, Cherfils-Vicini J et al (2023) Neutrophil extracellular traps formed during chemotherapy confer treatment resistance via TGF-beta activation. Cancer Cell 41(4):757–775 e10. https://doi.org/10.1016/j.ccell.2023.03.008

Muller L, Tunger A, Wobus M, von Bonin M, Towers R, Bornhauser M et al (2021) Immunomodulatory properties of mesenchymal stromal cells: an update. Front Cell Dev Biol 9:637725. https://doi.org/10.3389/fcell.2021.637725

Munir MT, Kay MK, Kang MH, Rahman MM, Al-Harrasi A, Choudhury M et al (2021) Tumor-associated macrophages as multifaceted regulators of breast tumor growth. Int J Mol Sci 22(12): 6526. https://doi.org/10.3390/ijms22126526

Ness S, Lin S, Gordon JR (2021) Regulatory dendritic cells, T cell tolerance, and dendritic cell therapy for immunologic disease. Front Immunol 12:633436. https://doi.org/10.3389/fimmu.2021.633436

Nixon BG, Kuo F, Ji L, Liu M, Capistrano K, Do M et al (2022) Tumor-associated macrophages expressing the transcription factor IRF8 promote T cell exhaustion in cancer. Immunity 55(11): 2044–2058 e5. https://doi.org/10.1016/j.immuni.2022.10.002

Nobre AR, Risson E, Singh DK, Di Martino JS, Cheung JF, Wang J et al (2021) Bone marrow NG2 (+)/nestin(+) mesenchymal stem cells drive DTC dormancy via TGFbeta2. Nat Cancer 2(3): 327–339. https://doi.org/10.1038/s43018-021-00179-8

Nordin ML, Azemi AK, Nordin AH, Nabgan W, Ng PY, Yusoff K et al (2023) Peptide-based vaccine against breast cancer: recent advances and prospects. Pharmaceuticals (Basel) 16(7): 923. https://doi.org/10.3390/ph16070923

Okano M, Oshi M, Butash AL, Katsuta E, Tachibana K, Saito K et al (2019) Triple-negative breast cancer with high levels of annexin A1 expression is associated with mast cell infiltration, inflammation, and angiogenesis. Int J Mol Sci 20(17):4197. https://doi.org/10.3390/ijms20174197

Onesti CE, Josse C, Boulet D, Thiry J, Beaumecker B, Bours V et al (2020) Blood eosinophilic relative count is prognostic for breast cancer and associated with the presence of tumor at diagnosis and at time of relapse. Onco Targets Ther 9(1):1761176. https://doi.org/10.1080/2162402X.2020.1761176

Orrantia-Borunda E, Anchondo-Nunez P, Acuna-Aguilar LE, Gomez-Valles FO, Ramirez-Valdespino CA (2022) Subtypes of breast cancer. In: Mayrovitz HN (ed) Breast cancer. Exon Publications, Brisbane (AU). https://doi.org/10.36255/exon-publications-breast-cancer-subtypes

Pereira-Veiga T, Schneegans S, Pantel K, Wikman H (2022) Circulating tumor cell-blood cell crosstalk: biology and clinical relevance. Cell Rep 40(9):111298. https://doi.org/10.1016/j.celrep.2022.111298

Perez-Tomas R, Perez-Guillen I (2020) Lactate in the tumor microenvironment: an essential molecule in cancer progression and treatment. Cancers (Basel) 12(11):3244. https://doi.org/10.3390/cancers12113244

Piezzo M, Cocco S, Caputo R, Cianniello D, Gioia GD, Lauro VD et al (2020) Targeting cell cycle in breast cancer: CDK4/6 inhibitors. Int J Mol Sci 21(18):6479. https://doi.org/10.3390/ijms21186479

Pittenger MF, Discher DE, Peault BM, Phinney DG, Hare JM, Caplan AI (2019) Mesenchymal stem cell perspective: cell biology to clinical progress. NPJ Regen Med 4:22. https://doi.org/10.1038/s41536-019-0083-6

Qi M, Xia Y, Wu Y, Zhang Z, Wang X, Lu L et al (2022) Lin28B-high breast cancer cells promote immune suppression in the lung pre-metastatic niche via exosomes and support cancer progression. Nat Commun 13(1):897. https://doi.org/10.1038/s41467-022-28438-x

Qian D, Li J, Huang M, Cui Q, Liu X, Sun K (2023) Dendritic cell vaccines in breast cancer: immune modulation and immunotherapy. Biomed Pharmacother 162:114685. https://doi.org/10.1016/j.biopha.2023.114685

Qin S, Jiang J, Lu Y, Nice EC, Huang C, Zhang J et al (2020) Emerging role of tumor cell plasticity in modifying therapeutic response. Signal Transduct Target Ther 5(1):228. https://doi.org/10.1038/s41392-020-00313-5

Rabe KF, Celli BR, Wechsler ME, Abdulai RM, Luo X, Boomsma MM et al (2021) Safety and efficacy of itepekimab in patients with moderate-to-severe COPD: a genetic association study and randomised, double-blind, phase 2a trial. Lancet Respir Med 9(11):1288–1298. https://doi.org/10.1016/S2213-2600(21)00167-3

Radisky ES, Raeeszadeh-Sarmazdeh M, Radisky DC (2017) Therapeutic potential of matrix metalloproteinase inhibition in breast cancer. J Cell Biochem 118(11):3531–3548. https://doi.org/10.1002/jcb.26185

Raychaudhuri D, Bhattacharya R, Sinha BP, Liu CSC, Ghosh AR, Rahaman O et al (2019) Lactate induces pro-tumor reprogramming in intratumoral plasmacytoid dendritic cells. Front Immunol 10:1878. https://doi.org/10.3389/fimmu.2019.01878

Reddy SM, Reuben A, Barua S, Jiang H, Zhang S, Wang L et al (2019) Poor response to neoadjuvant chemotherapy correlates with mast cell infiltration in inflammatory breast cancer. Cancer Immunol Res 7(6):1025–1035. https://doi.org/10.1158/2326-6066.CIR-18-0619

Reizis B (2019) Plasmacytoid dendritic cells: development, regulation, and function. Immunity 50(1):37–50. https://doi.org/10.1016/j.immuni.2018.12.027

Rubinstein-Achiasaf L, Morein D, Ben-Yaakov H, Liubomirski Y, Meshel T, Elbaz E et al (2021) Persistent inflammatory stimulation drives the conversion of MSCs to inflammatory CAFs that promote pro-metastatic characteristics in breast cancer cells. Cancers (Basel) 13(6):1472. https://doi.org/10.3390/cancers13061472

Salimifard S, Masjedi A, Hojjat-Farsangi M, Ghalamfarsa G, Irandoust M, Azizi G et al (2020) Cancer associated fibroblasts as novel promising therapeutic targets in breast cancer. Pathol Res Pract 216(5):152915. https://doi.org/10.1016/j.prp.2020.152915

Sandiford OA, Donnelly RJ, El-Far MH, Burgmeyer LM, Sinha G, Pamarthi SH et al (2021) Mesenchymal stem cell-secreted extracellular vesicles instruct stepwise dedifferentiation of breast cancer cells into dormancy at the bone marrow perivascular region. Cancer Res 81(6):1567–1582. https://doi.org/10.1158/0008-5472.CAN-20-2434

SenGupta S, Hein LE, Xu Y, Zhang J, Konwerski JR, Li Y et al (2021) Triple-negative breast cancer cells recruit neutrophils by secreting TGF-beta and CXCR2 ligands. Front Immunol 12:659996. https://doi.org/10.3389/fimmu.2021.659996

Shani O, Vorobyov T, Monteran L, Lavie D, Cohen N, Raz Y et al (2020) Fibroblast-derived IL33 facilitates breast cancer metastasis by modifying the immune microenvironment and driving type 2 immunity. Cancer Res 80(23):5317–5329. https://doi.org/10.1158/0008-5472.CAN-20-2116

Shevchenko JA, Khristin AA, Kurilin VV, Kuznetsova MS, Blinova DD, Starostina NM et al (2020) Autologous dendritic cells and activated cytotoxic T-cells as combination therapy for breast cancer. Oncol Rep 43(2):671–680. https://doi.org/10.3892/or.2019.7435

Shi Y, Du L, Lin L, Wang Y (2017) Tumour-associated mesenchymal stem/stromal cells: emerging therapeutic targets. Nat Rev Drug Discov 16(1):35–52. https://doi.org/10.1038/nrd.2016.193

Shinde-Jadhav S, Mansure JJ, Rayes RF, Marcq G, Ayoub M, Skowronski R et al (2021) Role of neutrophil extracellular traps in radiation resistance of invasive bladder cancer. Nat Commun 12(1):2776. https://doi.org/10.1038/s41467-021-23086-z

Siegel RL, Miller KD, Wagle NS, Jemal A (2023) Cancer statistics, 2023. CA Cancer J Clin 73(1): 17–48. https://doi.org/10.3322/caac.21763

Singh N, Baby D, Rajguru JP, Patil PB, Thakkannavar SS, Pujari VB (2019) Inflammation and cancer. Ann Afr Med 18(3):121–126. https://doi.org/10.4103/aam.aam_56_18

Slade D (2020) PARP and PARG inhibitors in cancer treatment. Genes Dev 34(5–6):360–394. https://doi.org/10.1101/gad.334516.119

Spencer LA, Bonjour K, Melo RC, Weller PF (2014) Eosinophil secretion of granule-derived cytokines. Front Immunol 5:496. https://doi.org/10.3389/fimmu.2014.00496

Szczerba BM, Castro-Giner F, Vetter M, Krol I, Gkountela S, Landin J et al (2019) Neutrophils escort circulating tumour cells to enable cell cycle progression. Nature 566(7745):553–557. https://doi.org/10.1038/s41586-019-0915-y

Takai K, Le A, Weaver VM, Werb Z (2016) Targeting the cancer-associated fibroblasts as a treatment in triple-negative breast cancer. Oncotarget 7(50):82889–82901. https://doi.org/10.18632/oncotarget.12658

Tariq M, Zhang J, Liang G, Ding L, He Q, Yang B (2017) Macrophage polarization: anti-cancer strategies to target tumor-associated macrophage in breast cancer. J Cell Biochem 118(9): 2484–2501. https://doi.org/10.1002/jcb.25895

Ullah M, Akbar A, Ng NN, Concepcion W, Thakor AS (2019) Mesenchymal stem cells confer chemoresistance in breast cancer via a CD9 dependent mechanism. Oncotarget 10(37): 3435–3450. https://doi.org/10.18632/oncotarget.26952

Vargas A, Boivin R, Cano P, Murcia Y, Bazin I, Lavoie JP (2017) Neutrophil extracellular traps are downregulated by glucocorticosteroids in lungs in an equine model of asthma. Respir Res 18(1): 207. https://doi.org/10.1186/s12931-017-0689-4

Wechsler ME, Ruddy MK, Pavord ID, Israel E, Rabe KF, Ford LB et al (2021) Efficacy and safety of itepekimab in patients with moderate-to-severe asthma. N Engl J Med 385(18):1656–1668. https://doi.org/10.1056/NEJMoa2024257

Wei Y, Li Y, Chen Y, Liu P, Huang S, Zhang Y et al (2022) ALDH1: a potential therapeutic target for cancer stem cells in solid tumors. Front Oncol 12:1026278. https://doi.org/10.3389/fonc.2022.1026278

Xiao M, He J, Yin L, Chen X, Zu X, Shen Y (2021a) Tumor-associated macrophages: critical players in drug resistance of breast cancer. Front Immunol 12:799428. https://doi.org/10.3389/fimmu.2021.799428

Xiao Y, Cong M, Li J, He D, Wu Q, Tian P et al (2021b) Cathepsin C promotes breast cancer lung metastasis by modulating neutrophil infiltration and neutrophil extracellular trap formation. Cancer Cell 39(3):423–437 e7. https://doi.org/10.1016/j.ccell.2020.12.012

Xiong H, Chen Z, Lin B, Xie B, Liu X, Chen C et al (2021) Naringenin regulates FKBP4/NR3C1/NRF2 Axis in autophagy and proliferation of breast cancer and differentiation and maturation of dendritic cell. Front Immunol 12:745111. https://doi.org/10.3389/fimmu.2021.745111

Xu M, Zhang T, Xia R, Wei Y, Wei X (2022) Targeting the tumor stroma for cancer therapy. Mol Cancer 21(1):208. https://doi.org/10.1186/s12943-022-01670-1

Yan Y, Cao S, Liu X, Harrington SM, Bindeman WE, Adjei AA et al (2018) CX3CR1 identifies PD-1 therapy-responsive CD8+ T cells that withstand chemotherapy during cancer chemoimmunotherapy. JCI Insight 3(8):e97828. https://doi.org/10.1172/jci.insight.97828

Yang L, Liu Q, Zhang X, Liu X, Zhou B, Chen J et al (2020) DNA of neutrophil extracellular traps promotes cancer metastasis via CCDC25. Nature 583(7814):133–138. https://doi.org/10.1038/s41586-020-2394-6

Yang T, Kang L, Li D, Song Y (2023) Immunotherapy for HER-2 positive breast cancer. Front Oncol 13:1097983. https://doi.org/10.3389/fonc.2023.1097983

Yu B, Luo F, Sun B, Liu W, Shi Q, Cheng SY et al (2021) KAT6A acetylation of SMAD3 regulates myeloid-derived suppressor cell recruitment, metastasis, and immunotherapy in triple-negative breast cancer. Adv Sci (Weinh) 8(20):e2100014. https://doi.org/10.1002/advs.202100014

Zeng W, Xiong L, Wu W, Li S, Liu J, Yang L et al (2023) CCL18 signaling from tumor-associated macrophages activates fibroblasts to adopt a chemoresistance-inducing phenotype. Oncogene 42(3):224–237. https://doi.org/10.1038/s41388-022-02540-2

Zhang M, Shi Y, Zhang Y, Wang Y, Alotaibi F, Qiu L et al (2020) miRNA-5119 regulates immune checkpoints in dendritic cells to enhance breast cancer immunotherapy. Cancer Immunol Immunother 69(6):951–967. https://doi.org/10.1007/s00262-020-02507-w

Zhang L, Pan J, Wang Z, Yang C, Chen W, Jiang J et al (2021) Multi-omics profiling suggesting intratumoral mast cells as predictive index of breast cancer lung metastasis. Front Oncol 11:788778. https://doi.org/10.3389/fonc.2021.788778

Zhang W, Zhang Q, Yang N, Shi Q, Su H, Lin T et al (2022) Crosstalk between IL-15Ralpha(+) tumor-associated macrophages and breast cancer cells reduces CD8(+) T cell recruitment. Cancer Commun (Lond) 42(6):536–557. https://doi.org/10.1002/cac2.12311

Zhang R, Dong M, Tu J, Li F, Deng Q, Xu J et al (2023) PMN-MDSCs modulated by CCL20 from cancer cells promoted breast cancer cell stemness through CXCL2-CXCR2 pathway. Signal Transduct Target Ther 8(1):97. https://doi.org/10.1038/s41392-023-01337-3

Zhao T, Su Z, Li Y, Zhang X, You Q (2020) Chitinase-3 like-protein-1 function and its role in diseases. Signal Transduct Target Ther 5(1):201. https://doi.org/10.1038/s41392-020-00303-7

Zheng X, Zhang N, Qian L, Wang X, Fan P, Kuai J et al (2020) CTLA4 blockade promotes vessel normalization in breast tumors via the accumulation of eosinophils. Int J Cancer 146(6):1730–1740. https://doi.org/10.1002/ijc.32829

Zheng Y, Li S, Tang H, Meng X, Zheng Q (2023) Molecular mechanisms of immunotherapy resistance in triple-negative breast cancer. Front Immunol 14:1153990. https://doi.org/10.3389/fimmu.2023.1153990

Zhou Z, Yan F, Liu O (2020) Interleukin (IL)-33: an orchestrator of immunity from host defence to tissue homeostasis. Clin Transl Immunol 9(6):e1146. https://doi.org/10.1002/cti2.1146

siRNA Loaded in Drug Delivery Nanosystems as a Strategy for Breast Cancer Treatment

Kammila Martins Nicolau Costa, Luana Beatriz Camêlo de Sousa, Demis Ferreira de Melo, Diego Paulo da Silva Lima, Bolivar Ponciano Goulart de Lima Damasceno, and João Augusto Oshiro-Júnior

Abstract

Advances in genetic sequencing techniques have enabled the knowledge of promising molecules, including siRNA (short interfering RNA), a mechanism of gene silencing directed against the mRNA strand that prevents protein synthesis and enables the formulation of drugs with greater effectiveness and directionality to the therapeutic target and less side effect. However, these molecules have low stability and need the aid of delivery systems, so they have been widely studied in conjunction with nanoparticles, structures that are up to 100 nm in size and, depending on their classification, influence the biocompatibility, biodistribution, and pharmacokinetics of the molecules. Breast cancer is one of the diseases that has been the target of these studies, due to the properties of these molecules. Thus, this chapter will bring together studies that address the siRNA carried in the main types of drug delivery nanosystems for the treatment of breast cancer.

Keywords

Chemotherapy · Drug delivery systems · Gene therapy · Oncology · RNA

K. M. N. Costa · D. F. de Melo · D. P. da Silva Lima · B. P. G. de Lima Damasceno
State University of Paraíba – UEPB, Campina Grande, Paraíba, Brazil

L. B. C. de Sousa
University Center UNIFACISA, Campina Grande, Paraíba, Brazil

J. A. Oshiro-Júnior (✉)
State University of Paraíba – UEPB, Campina Grande, Paraíba, Brazil

University Center UNIFACISA, Campina Grande, Paraíba, Brazil
e-mail: joaooshiro@yahoo.com.br

© The Author(s), under exclusive license to Springer Nature Switzerland AG 2022

Interdisciplinary Cancer Research, https://doi.org/10.1007/16833_2022_88
Published online: 30 December 2022

1 Introduction

Throughout history, one of the great enigmas has always been to understand heredity. Nowadays, we can affirm that the studies started around 1860 by Gregor Mendel, using crossbreeding and logical mathematical reasoning analysis with pea species revolutionized the understanding of the subject and founded the basic principles of genetics, the law of segregation, and the law of independent segregation (Gayon 2016; Lewis and Simpson 2021).

Moving forward in time, few significant advances were made until the 1950s, when scientist Rosalind Franklin dedicated herself to the study of the DNA molecule using her research with x-ray crystallography. Through this technique, Franklin got the famous DNA photo 51 that was used by Watson and Crick for the formulation of the DNA double-strand model, which is recognized until today (Cramer 2020). From then on, the new science of Genetics began to be founded and there were several advances. In 1970, enzymes were discovered to be able to separate genes in desired locations and insert them back into the cell in a reproducible way. In 1980, studies on gene therapy began to be done (Gonçalves and Paiva 2017).

In 1990, the Human Genome Project (HGP) was started with the goal of sequencing all nucleotide bases of genes. The project was completed in 2003 and generated several contributions to the area of genetics and molecular biology, among them, major advances in sequencing technologies, which allowed greater knowledge about the molecules of ribonucleic acids (RNAs) (Goes and Oliveira 2014). RNAs are nucleotide polymers formed by a pentose of the ribose type, a phosphate, and a nitrogenous base, which can be puerric: adenine and guanine, or primine: cytosine and uracil. RNA has a primordial function in gene expression, acting as an intermediary between the information encoded by DNA and the proteins, in addition to promoting mechanisms for their creation (Vieira 2016).

RNA may be coding or non-coding (ncRNA), only about 2% of RNAs are coding, among them, the main is the messenger RNA (mRNA), which has the function of carrying information of DNA from the nucleus to the cytoplasm of the cell for protein synthesis. The most well-known ncRNAs are ribosomal RNA (rRNA), which has the function of forming ribosomes, and transfer RNA (tRNA), responsible for transporting to the ribosomes the amino acids that will be used in the formation of proteins (Vieira 2016; Chan and Tay 2018). The ncRNAs have greater quantitative amplitude and can be classified as long (lncRNA), with more than 200 nucleotides and little-known functions, or small (sRNA), with 20–200 nucleotides and the ability to perform several functions, such as regulating gene expression, as the interference RNAs (iRNA) (Araújo 2017).

The main RNAs are miRNAs (microRNAs), endogenous RNAs, which have 21–25 nucleotides, shRNAs (short hairpin RNAs), synthetic RNAs with a structure similar to miRNAs, having an internal hybridization region similar to a hairpin, and siRNAs (short interfering RNAs), double-stranded RNA molecules, which have between 20 and 30 nucleotides, this has an advantage over the others in terms of effectiveness and delivery target specificity (Chan and Tay 2018; Hu et al. 2020). The siRNA was observed for the first time in 1990 in transgenic petunias, in which the pigment production was suppressed by these molecules; however, they were only better described in 1998 by Fire and Mello. Through studies with the nematode *Caenorhabditis elegans*, it was possible to demonstrate the interference of siRNA in gene silencing and how the process occurs. This discovery earned them the Nobel Prize in Physiology or Medicine in 2006 (Nápoles et al. 1990; Fire et al. 1998; Dong et al. 2019).

The iRNA mechanisms occur when a double-stranded RNA (dsRNA) is recognized by the dicer enzyme and cleaved in sequences of 20 to 30 nucleotides, forming, together with arginine proteins, the RISC complex (RNA-induced silencing complex), which has endonuclease activity directed against the mRNA strand, degrading it or inhibiting its translation, preventing protein synthesis (Fire et al. 1998; Li et al. 2021).

In this context, iRNAs have brought the possibility to synthesize more specific drugs with fewer side effects, due to their efficient mechanism of action to silence protein-coding genes. They can be applied to several diseases with abnormal gene expression, such as dominant genetic diseases, neurodegenerative syndromes, autoimmune diseases, viral infections, and several types of cancer. However, despite the high specificity and great therapeutic potential, the siRNA has some limitations as to delivery, has low stability, is easily degraded in tissues, short half-life, and has a negative charge, which prevents siRNA from crossing the cell membrane and reaching the therapeutic target (Singh et al. 2018; Li et al. 2021).

Thus, the incorporation of siRNA in nanostructured delivery systems is an alternative to circumvent its limitations, due to its ability to break cellular barriers and allow controlled release of the active substance, increasing the potential efficacy, accuracy, and safety of the drug and reducing possible side effects, besides increasing stability and prolonging the circulation time of the drug in the body (Mitchell et al. 2021; Verma et al. 2021).

More broadly, nanoparticles are defined as molecules with a size of up to 100 nm (nm); however, when it comes to pharmaceutical formulations, the size range between 5 and 1000 nm is considered (Lopes and Torres 2019). These particles have a wide variety and can be classified as liposomal, lipidic, polymeric, core-shell, micellar, dendrimer, nanocrystal drug, and conjugate (Mukalel et al. 2019).

The nanomaterial to be used will depend on the objective of formulating the drug, each one has different physicochemical characteristics, which interferes with the biocompatibility, biodistribution, and pharmacokinetics of the molecule. In addition to the material used, the size, morphology, concentration, surface chemistry, and release and loading capabilities of the system must be evaluated (Mukalel et al. 2019; Sun et al. 2019; Chowdhury et al. 2021). In this sense, this chapter aims to investigate the use of siRNA nanoparticles loaded into drug delivery nanosystems as a strategy for the treatment of breast cancer. Thus, we used articles published in renowned journals in the period that addressed the theme.

2 Nanoparticles Used Clinically for the Treatment of Breast Cancer

In oncology, nanoparticles have been targeted due to their ability to accumulate in tumor tissues and maintain drug concentration at the site during treatment, enabling the reduction of chemotherapeutic toxicity and potentiating and directing the therapeutic effect, avoiding the possibility of adjacent cells or tissues being affected during treatment (Zhang et al. 2019; Vallabani et al. 2019; Sun et al. 2019).

The first nanosystem drug approved by the US Food and Drug Administration (FDA) was Doxil®, a doxorubicin-based liposomal formulation, in 1995 (Barenholz 2012). Doxorubicin is an antineoplastic drug from the anthracycline class, discovered in the 1960s, used for the treatment of several types of cancer, including breast cancer, however, it has limited therapeutic activity and high gastrointestinal and cardiac toxicity. In order to end these limitations, liposomal formulations based on doxorubicin have been developed, as is the case of Doxil® (Franco et al. 2018).

Besides Doxil®, there are currently other FDA-approved nanoparticles used for the adjuvant treatment of breast cancer, such as Abraxane®, approved in 2005, composed of paclitaxel, an antimicrotubule agent, associated with albumin nanoparticles, and Genexol-PM, and several others are being studied in clinical trials (Ganesan et al. 2021).

2.1 Polymeric Nanosystems

Polymeric nanoparticles is the term used to describe particles with average size between 1 and 1000 nm that can be designed to carry biologically active compounds trapped inside a polymeric core or adsorbed on its surface (Mengarda et al. 2022). This term includes nanocapsules and nanospheres, which can be distinguished according to their morphology (Zielinska et al. 2020). The use of polymeric nanoparticles in the development of new therapeutic alternatives, such as the use of siRNA, for breast cancer relies on the ability to modify polymers to obtain distinct physicochemical and therapeutic characteristics, such as targeting tumor tissues (Sartaj et al. 2021). Furthermore, this nanosystem allows the combination of other antitumor therapeutic strategies with siRNA inhibition of gene expression, such as chemo and immunotherapy (Bahreyni et al. 2020).

In this context, monomethoxypolyethylene glycol-co-glycolic acid-poly-L-lysine (mPEG-PLGA-PLL) nanoparticles coated with monoclonal antibodies blocking programmed cell death receptor ligand 1 (PD-L1), a checkpoint inhibitor, containing siRNA targeting CD155, a glycoprotein overexpressed in breast cancer cells related to metastasis, immune response modulation and poor prognosis (Triki et al. 2019), were developed by Chen et al. (2021) as a therapeutic alternative to inhibit the progression and metastasis of triple-negative breast cancer. The nanoparticles developed had an average size of approximately 172 nm and zeta potential of about +0.38 mV, with in vitro release of 38.34% of siRNA in the first 24 h and 85.2% in 120 h. Moreover, this nanosystem was able to prevent progression and metastasis against triple-negative breast carcinoma cell line 4T1 in an in vivo model. The mechanism of this activity is based, initially, on the binding of PD-L1-blocking monoclonal antibodies to their antigens on the surfaces of 4T1 cells, which plays a role in directing the nanoparticles and increasing the immune activity against tumor cells, since PD-L1 prevents CD8+ T lymphocytes in advanced stages from exerting their cytotoxic activity. Moreover, the blocking of CD155 gene expression by siRNA prevents 4T1 cells from expressing this glycoprotein, which leads to increased activity of early-stage CD8+ T lymphocytes against tumor cells.

PD-L1 was also targeted by Gong et al. (2021) of polylactic-co-glycolic acid (PLGA) nanoparticles coated with macrophages biomimetic membrane and breast cancer cell line 4T1 cells containing CO_2-conjugated metformin, an antidiabetic drug that promotes PD-L1 degradation via activation of adenosine monophosphate-activated protein kinase, and siRNA targeting fibrinogen type I-like protein (FGL1), a protein related to suppression of T-lymphocyte activity in tumor cells (Wang et al. 2019; Qian et al. 2021; Marcucci et al. 2020). The developed nanoparticles showed an average size of 142 nm, zeta potential of approximately −30.5 mV with selective release of metformin at lysosomal pH (86% in 24 h at pH 5.0). Furthermore, these nanoparticles were able to enhance the immune response against 4T1 cells in an in vivo model. This activity is primarily based on the biomimetic membrane that covers the nanoparticles, which promotes easy internalization of the nanosystem by 4T1 cells. Once inside the cells at lysosomal pH ranges, CO_2 conjugated with metformin is released and causes rupture of the nanoparticle, which entails rapid release of metformin and siRNA, the former with the action of degrading PD-L1 and the latter with the inhibition of FGL1 gene expression, resulting in improved immunosuppression in the tumor microenvironment.

The coating of polymeric nanoparticles with tumor cell membranes was also performed by Chen et al. (2019) in the development of PLGA nanoparticles coated with cell membranes from cervical adenocarcinoma (Hela) and breast cancer (MDA-MB-231) cell lines containing doxorubicin (DOX) and siRNA targeting PD-L1. These nanoparticles showed an average size of about 110 nm with in vitro antitumor activity against MDA-MB-231 cells. The mechanism responsible for this activity is related to the presence of the tumor cell membrane coating the PLGA nanoparticle, since it facilitates its recognition and incorporation into the tumor cells, mediated by adhesion proteins such as cadherins and catenins, which are overexpressed in breast cancer cell lines such as MDA-MB-231 (Wang et al. 2019). Inside the cells, the nanoparticles promote the release of DOX and siRNA, which act together to generate apoptosis of tumor cells through cytotoxic mechanisms and inhibition of tumor immunosuppression.

Nanoparticles can also be modified by functionalization with polyethylene glycol (PEG), which has the main purpose of making the nanoparticles stealthy against the host immune system. This process prolongs the circulation time of the nanoparticles by reducing the clearance, and increases their stability in colloidal suspension mediated by the steric repulsion of the PEG chains (Ahmed and Badr-Eldin 2020; Han et al. 2015). To this purpose, Jin et al. (2021) developed polyethyleneimine-poly lactic acid (PEI-PLA) nanoparticles coated with polyethylene glycol-poly

L-aspartic acid (PEG-PAsp) with the co-release of paclitaxel (PTX) and siRNA against vascular endothelial growth factor (VEGF), which acts in several solid tumors in the process of angiogenesis (Modi and Kulkarni 2019), for the treatment of breast cancer. The developed nanosystem showed an average size of 85 nm, zeta potential of approximately +5 mV with in vivo antitumor activity against 4T1 cells. The mechanism behind this activity lies in the ability of the PEG-PAsp-coated nanoparticles to remain stable in the circulation of the animal model and to be internalized by the tumor cells. Once inside, PTX and siRNA-VEGF are released, which inhibit proliferation and angiogenesis, respectively, resulting in tumor cell apoptosis.

Molecules isolated from plant species can also be used in association with siRNA to develop polymeric nanosystems with antitumor activity. In this light, Ding et al. (2018) developed tumor peptide-coated hyaluronic acid (HA) nanoparticles for cell penetration containing galato-3-O-(-)-epigallocatechin (EGCG), a major component found in green tea that has properties to form complexes with DNA and RNA via non-covalent interactions, protamine, a polypeptide with the ability to condense genes for transfection purposes into different cell types, and siRNA specific for connective tissue growth factor (CTGF), overexpressed in multidrug-resistant (e.g., triple-negative breast cancer) (Wang et al. 2020; Kuzuhara et al. 2006). The nanoparticles developed were arranged in nanoaggregates of approximately 80 nm, showed zeta potential of about −20 mV, release of 90% of EGCG after 12 h in lysosomal pH and with antitumor activity in vitro and in vivo against tumor cells of the MDA-MB-231 and MCF-7 lineages. This activity resides, initially, in the capacity of the cell-penetrating tumor peptide to bind to CD44 receptors (adhesion receptor) present on the tumor cell membrane and promote nanoparticles internalization. Inside the cells, this nanosystem promotes EGCG and siRNA release into the lysosomal environment mediated by HA degradation via hyaluronidases, which induce apoptosis by synergism of activities and potentiation of the cytotoxic activity of EGCG (5 times higher compared to EGCG alone).

Other approaches can be adopted to enhance the antitumor activity of siRNA conveyed in polymeric nanoparticles against breast cancer cells, such as photodynamic therapy (PDT). To this purpose, Meng et al. (2018) developed 1,2-distearoyl-sn-glyceroyl-3-phosphoethanolamine-N-[maleimide-(polyethylene glycol 5000) (DSPE-PEG) nanoparticles containing docetaxel (DTX), chlorin e6 (Ce6), a second-generation photosensitizer (Feng et al. 2019), linked to poly [(1,4-butanediol)-diacrylate-β-oligoethylenimine 600] (Ce6-PDOEI) and anti-Twist siRNA, a transcription factor for induction of epithelial-mesenchymal transition (EMT) that is related to tumor migration, invasion, and metastasis (Cao et al. 2018). The developed nanosystem showed an average size of about 78 nm, zeta potential of approximately +3 mV with release of 25% of Ce6 and around 80% of DTX and siRNA under cytosol conditions (pH 7.4 and 300 mg mL^{-1} of bovine serum albumin). Moreover, the developed nanoparticles showed spatiotemporal antitumor activity in vivo against 4T1 cells, since laser irradiation ($\lambda = 671$ nm) can be performed only at the tumor growth site. This activity is related to the capacity of PDT to generate reactive oxygen species (e.g., singlet (1O2)), which can exert cytotoxic activity directly (more superficial tumor cells with laser power of $\geq$0.45 W cm^{-2}) or in synergism with DTX and anti-Twist siRNA (deeper tumor cells with laser power of $\leq$0.22 W cm^{-2}), which prevents tumor proliferation and metastasis. Moreover, this synergism may be potentiated by accelerating the endolysosomal escape of DTX and siRNA caused by PDT inside tumor cells.

2.2 Liposomes

Liposomes are spherical nanovesicles, consisting of a single or multiple concentric lipid bilayer which encapsulates an aqueous compartment. Their size varies from a few nanometers to micrometers, but in the medical field, their size is between 50 and 450 nm (Bozzuto and Molinari 2015). Furthermore, due to their structural and compositional variety, these vesicles have a range of applications, being used in the cosmetic, food, and pharmaceutical industries (Laouini et al. 2012). Its use in the pharmaceutical area has gained prominence, because its aqueous phase can contain hydrophilic drugs as well as its phospholipid bilayer can carry lipophilic drugs. Moreover, in order to make the liposome more suitable for the human body, changes are made in its structure in the head, tail, and binding of lipid molecules, forming functionalized liposomes, such as long-circulating PEGylates, active targeting of liposomes, environmentally sensitive liposomes, and, finally, multifunctional liposomes (Li et al. 2019). Among the recent uses of liposomes is their use in alternative therapies, such as siRNA for breast cancer. Its use aims to increase the delivery of these molecules that have low stability in physiological fluids and limited permeability to membranes (Bochicchio et al. 2022).

In this context, the chemotherapeutics paclitaxel (PTX), crizotinib (CRI), and Bcl-xL siRNA were co-loaded by cationic liposomes for the treatment of breast cancer in order to improve targeting efficiency and reduce side effects. The cationic liposome particles bound to crizotinib and paclitaxel (CRI-PTX-CTL) had size 138.63 $\pm$ 1.53 nm with zeta potential around 50.90 $\pm$ 0.30 mV, with in vitro release of CRI and PTX of 64.37% and 54.71%, respectively, in 24 h. Furthermore, the cell uptake experiments demonstrated that CRI, PTX, and siRNA included in CTL are effectively taken up by MCF-7 cells and that CRI and PTX have synergistic effect. The studies also demonstrated that CTL-siRNA can inhibit Bcl-xL expression in cells and that CRI, PTX, and Bcl-xl siRNA delivered by CTL exhibit increased cytotoxicity in in vitro studies. The mechanism consists in the use of siRNA to form an RNA-induced silencing complex (RISC) leading to degradation of the target mRNA and, consequently, blocking the expression of Bcl-xl, an antiapoptotic protein considered one of the main regulators of the mitochondrial pathway, and its overexpression is associated with multiple tumor processes and resistance to radiotherapy. As a result of the inhibition, a greater number of apoptosis occurred in MCF-7 cells, 76.7% of those exposed to CRI-PTX-CTL, and 67.1% of CRI-PTX (Li et al. 2022).

Survivin was also the target of studies by Chen et al., which developed a nanosystem by performing co-delivery of PXT and anti-survivin siRNA via redox-sensitive oligopeptide liposomes (⟦LHSSG2C⟧_14) to overcome PTX resistance and improve its antitumor efficiency in the synergistic treatment of breast cancer and metastasis in 4T1 cells. The redox-sensitive liposome particles loaded with paclitaxel (PTX-SS-L) and SS-L loaded with PTX and siRNA (PTX/siRNA/SS-L) were 110.5 ± 1.5 nm and 156.4 ± 2.1 nm in size with +21.6 ± 2.5 mV and 1.4 mV zeta potential respectively. Regarding PTX release, in the PTX/siRNA/SS-L system, it was 50% in the first 4 h and more 80% in 36 h while PTX/siRNA//nS-L as control released only 25%. The study also demonstrated good cellular uptake, being 6.7 times higher compared to Taxol (control group). The cells treated with PTX/siRNA/SS-L exhibited a 51% negative survivin mRNA expression compared to the control group (Taxol); in addition, the PTX/siRNA/SS-L group showed a higher apoptotic rate (32.9%) compared to the siRNA/SS-L group (14.5%) (Chen et al. 2017a, b).

VEGF siRNA delivered by calcium phosphate nanoparticles encapsulated in polycation liposomes (PLCP) was developed for the inhibition of tumor angiogenesis in breast cancer in MCF-7 cells. It is known that angiogenesis participates directly in tumor development and metastasis, and it is necessary for promotion of this process to express VEGF which can be inhibited by siRNA. As a result, PLCP nanoparticles were observed to be 207 ± 17.5 nm in size. In vitro, PLCP-mediated VEGF siRNA was able to inhibit VEGF expression by 60–80% within 24 h. Furthermore, in vivo studies, the siRNA/PLCP/VEGF group had a 72% inhibition, with doxorubicin (DOX) a 84% inhibition, and the combination of VEGF siRNA and DOX a 91% inhibition. Regarding histological studies, the microvessel density (MVD) in the VEGF siRNA group was 58 microvessels/field of view, while the saline group 153 microvessels/field of view and 91 microvessels/field of view in the DOX group (Chen et al. 2017a, b).

Another study performed co-delivery of paclitaxel and siRNA targeting PLK1 using cationic liposome functionalized by aptamer AS1411 for synergistic anti-breast cancer effects in vivo and in vitro in MCF-7 cells (Bozzuto and Molinari 2015). Polo-like kinase 1 (PLK1) plays an important role in the initiation, maintenance, and completion of mitosis and its dysfunction can trigger cancerous transformation of cells, inhibition of PLK1 can lead to cancer cell death (Bozzuto and Molinari 2015). The particles of AS1411/Lipo-PTX-siRNA was 121.27 ± 2.51 nm and zeta potential of −23.37 ± 1.45 mV and regarding the drug release, in AS1411/LipoPTX-siPLK1 was slower than free PTX, 58% and 52% of PTX released from both Lipo-PTX-siPLK1 and AS1411/Lipo-PTX-siPLK1 at the end of 96 h of the study period. Regarding the in vivo studies, mice treated with AS1411/LipoPTX-siPLK1 had a 62.32% tumor reduction in 21 days (Yu et al. 2019).

2.3 Lipidic Nanoparticles

Nanotechnology has been advancing exponentially over the years, with treatments for various diseases, however, presenting a major focus in the area of cancer, among them, breast cancer. Among the nanosystems that are part of the FDA-accepted list, lipidic nanoparticles (LNP) are part of more than 10 nanomedicines, these being solid lipid nanoparticles and nanostructured lipid carriers. Disadvantages linked to low stability, solubility, and high cost of production of liposomes led researchers to focus their studies on the other two forms of lipidic nanoparticles. siRNA-based drugs present limitations in the delivery of the compound, since they are unstable in biological fluids, suffer decomposition, and have difficulty in penetrating the target cell, besides having a low rate of accumulation in the target tissue. Their incorporation in nanostructured materials should circumvent these disadvantages, generating an intracellular delivery in target cells (Li et al. 2018; Akinc et al. 2019; Salvi and Pawar 2019).

In this scenario, targeting a co-delivery of siRNA and lycopene to perform suppression of insulin-like growth factor-1 receptor (IGF-1R) and its ligand, since its activation is related to cancer cell proliferation and resistance to cytotoxic therapies, (Mennati et al. 2022) the incorporation of this complex into LPN was performed, which led to several modifications in the final result of the process. The encapsulation efficiency of siRNA was 99.6%. The results suggested that the use of the complex increased the induction of the apoptosis process in MCF-7 breast cancer cells, where LNP-siRNA presented 89% and the control presented 14.29% after a 36-h period, besides interrupting part of the cell cycle, when compared to the siRNA that was not incorporated, which did not present any alteration in the cycle.

This time, seeking delivery of a PEGuylated LNP containing incorporated siRNA, a Fab' antibody that acts against the heparin-binding epidermal-like growth factor (HB-EGF) was added to the system aiming for a targeted delivery to the target tissue. In addition to showing a high binding affinity for the target receptor, the delivery into the cells of MDA-MB-231 triple-negative breast cancer cell line was confirmed by microscopic imaging, as this cell line shows a high expression of HB-EGF receptors. The LNPs realized a delivery of approximately 20%, compared to approximately 3% for siRNA alone and approximately 8% for LNP alone. The average size exhibited by the particles was satisfactory being 134 ± 21.5 nm for the LNPs that were not functionalized with the antibody and 161 ± 24.4 nm for those functionalized, which did not show significant growth in the portion (Okamoto et al. 2014).

LNPs coated with human serum albumin (HSA) containing incorporated siRNA were developed by Piao et al. (2013), seeking to observe the down-regulation effect of the gene in breast cancer cells in different lineages, such as MCF-7, MDA-MB-231, and SK-BR-3. The generated particles showed a satisfactory size of 79.5 ± 5.5 nm. After 3 intravenous applications of the treatment, the tumor fluorescence was significantly decreased for MCF-7 lineage cells compared to siRNA alone which showed a gradual increase in fluorescence, however, it was observed that the presence or absence of albumin did not alter the result, leading to the belief that this effect was provided only by the LNP containing the siRNA. The total photon flux per second also showed a decrease after 7 days of treatment with the LNPs in in vivo tests performed by the group.

Okamoto et al. (2018) developed a system containing siRNA incorporated into LNPs coated with Fab' antibody for application in the treatment of triple-negative breast cancer of the MDA-MB-231 cell line. The system showed an average diameter of 167 ± 56 nm. The biodistribution results of the fluorescence-labeled particles showed that the LNPs were indeed able to deliver the siRNA at the site of action, an effect that was tested in mice with carcinoma induced with the mentioned cell line, showing a 45% decrease in cell growth at the end of the seventh day. In addition, there was a decrease in the expression of more than 80% of the PLK1 protein that is linked to cell viability through cell cycle control. As observed, the incorporation of siRNA in nanomaterials of lipidic origin provides a promising advantage when compared with control groups of the studies performed. Thus, it can be stated that the system composed of siRNA and LNP comes to be a promising alternative in the treatment of breast cancer.

Breast amyloidosis may present as primary or systemic and is characterized by the ability to mimic a malignant lesion and may even be associated with a carcinoma (Toohey et al. 2007). ONPATTRO (patisiran) was the first approved drug that has the siRNA mechanism, which came about only in the year 2018. The drug is composed of lipid nanoparticles (LNPs) and is used for the treatment of adult patients with hereditary transthyretin-mediated amyloidosis (hATTR amyloidosis) with polyneuropathy, stage 1 or 2. The approval of patisirana is of great significance in advancing the use of siRNA-based drugs. In ONPATTRO, siRNA acts by silencing the TTR gene, preventing disease progression, while nanoparticles act to ensure efficient drug delivery (Akinc et al. 2019).

2.4 Solid Lipid Nanoparticles

Solid lipid nanoparticles (SLN) are considered as the first-generation solid lipid nanoparticle carrier systems, presenting a size range of 50–1000 nm, and having a solid matrix to control the release of the encapsulated compounds. These characteristics confer stability and improve the solubility of the compounds when compared to liposomes. However, as disadvantages, SLNs have a low encapsulation capacity due to the expulsion of the compounds from their matrix, since the matrix is composed of perfect crystals, leaving a reduced space between their grooves for the deposition of these compounds, thus facilitating their elimination from the matrix (Ganesan and Narayanasamy 2017; Gordillo-Galeano and Mora-Huertas 2018).

However, even with the presence of these disadvantages, several researchers direct their studies to the application of SLN in breast cancer and present promising results. In this context, Sumera et al. (2017) mention that siRNAs are applied in cancer treatment aiming to improve the chemosensitivity of these cells, besides being characterized by presenting possible better results in cancer treatment when compared to the combination of chemotherapy and radiotherapy; however, its low intracellular uptake influences negatively on the therapeutic result of its application. For this reason, their incorporation into nanostructured systems such as SLN can increase their absorption effectively.

Therefore, di- and triblock copolymers containing siRNA and PEG were incorporated into SLN, aiming to observe the efficacy of the gene silencing strategy presented by the system, using the MDA-MB-435 breast cancer cell line. An increase in the serum stability of the system was observed compared to siRNA alone and the average particle diameter showed a size of 112.9 ± 7.1 nm. Compared to the PBS application control, SLN and siRNA alone inhibited the expression of GFP (green fluorescent protein, over-expressed in the cancer cell line) by up to 34% compared to a 73% inhibition of the system as a whole (Lee et al. 2011).

Already seeking to improve the conditions of delivery to the target tissue of the siRNA, Lobovkina et al. (2011) developed SLN by means of the nanoprecipitation technique. Presenting an application through the intradermal route in mouse paws, the group presented results of prolonged siRNA release for a period corresponding to 10–13 days, with particles with an average diameter size of approximately 150 nm. In this profile, there was an initial peak siRNA release burst of 17–22% and a portion remained unchanged at the end of the experiment, i.e., it remained trapped in the SLN. However, it was observed that while a percentage remained unchanged in vitro, the in vivo tests showed degradation of SLN and release of siRNA facilitated by lipases present in the physiological environment.

2.5 Nanostructured Lipidic Carriers

On the other hand, a second generation of lipid nanoparticle carrier systems was generated, aiming to circumvent the limitations presented by SLN, being the nanostructured lipid carriers (NLC). Considered an improved form of SLN, the NLC presents an imperfect matrix, where the grooves accommodate more efficiently the encapsulated compounds, avoiding their expulsion, conferring greater stability, and the possibility of incorporating a higher concentration of the compound. The NLC can be classified according to their structure as imperfect, amorphous, and multiple; however, regardless of their structure, they have the advantage of promoting the effect of permeability and retention (EPR), this being a strategy to vector chemotherapy leading to possible reduction of adverse effects caused during treatment (Khosa et al. 2018; Haider et al. 2020).

Thus, seeking to circumvent multidrug resistance and reduce the limitations presented in the treatment of breast cancer, the production of an NLC containing siRNA and doxorubicin was proposed. The results revealed a satisfactory protection content for siRNA with a particle size diameter of 100.6 nm. MCF-7 cells tested in the anticancer profiling assay showed resistance to doxorubicin, as expected, exhibiting a percentage of cell viability of 71.6%; however, when present, NLC increased anticancer efficacy in the system with decreased cell viability, and when present siRNA showed even better anticancer effects at most concentrations tested, achieving a cell inhibition of 64.7%. Leading to the belief that the presence of NLC in the system helps in the anticancer potential and the action of siRNA potentiates the therapeutic action of the system (Zhang and Xu 2020).

2.6 Cyclodextrin-Based Nanosystems

Cyclodextrins (CDs) are cyclic oligosaccharides composed of D-glycopyranose monomers linked by α-1,4 bonds. These cyclic structures have a toroid-shaped rearrangement, which forms a lipophilic cavity and a hydrophilic exterior (Jansook et al. 2018; Zhao et al. 2014). The presence of hydroxyls on the external face of CDs allows modifications in their structure with the addition of ligands such as polymers, which can lead to the formation of anionic, PEGlated, targeted, and anionic CDs. The latter have great importance as a delivery platform for therapeutic oligonucleotides (e.g., siRNA), since cationic CDs are able to form stable complexes with these oligonucleotides from electrostatic interactions, since they have a partial negative charge (Singh et al. 2019).

In this light, Mousazadeh et al. (2022) developed stimuli-responsive nanoparticles of polyethyleneimine-β-CD attached with folate to co-label DOX and human telomerase reverse transcriptase (hTERT)-siRNA as a potential treatment for breast cancer. The developed nanoparticles, with an average size of 100 nm and zeta potential between +20 and + 30 mV, showed selective release of DOX (>85% in 4 h) into an acidic medium characteristic of intracellular lysosomal environment (pH 5) and in vitro antitumor activity against breast cancer cells of the MDA-MB-231 and MCF-7 strains. The mechanism of action of the nanoparticles is initially through folate-mediated endocytosis, since both cell lines used have these receptors overexpressed. Once inside the tumor cells, the acidic environment allows the sustained co-release of hTERT-siRNA and DOX, which enables the effective transfection and gene suppression of h-TERT and of protein expression levels by the siRNA. Furthermore, this process induces tumor cell apoptosis and increases the antitumor activity of DOX by about two-fold and nine-fold for MDA-MB-231 and MCF-7, respectively, compared to DOX alone.

With the same purpose, Jin et al. (2017) developed polyethyleneimine-γ-CD nanoparticles with the chemical element neodymium (Nd) bound on their exterior for co-release of DOX and siRNA targeting the multidrug-resistant gene (MDR-1). This gene is related to coding for the synthesis of P-glycoprotein (P-gp), which in turn is associated with resistance to chemotherapeutic drugs (e.g., DOX, epirubicin, etoposide, paclitaxel, docetaxel) in breast cancer treatments via the efflux pump mechanism (Amiri-Kordestani et al. 2012). The nanoparticles showed an average size of 300 nm, zeta potential between +16 and + 18 mV, and selective release of DOX in lysosomal acid medium (pH 5.0), with 15.5% release in 8 h compared to 7.9% in the same period at pH 7.4. Furthermore, these nanoparticles showed antitumor activity against MCF-7/ADR breast cancer cells, a multidrug-resistant strain with high levels of P-gp expression, in in vitro and in vivo models. The activity is mediated largely by the presence of Nd, which promotes endocytosis of the nanoparticles through clathrin and caveolin which are involved in cell signaling processes. Inside the cells, Nd is responsible for reducing the level of calmodulin, a calcium-binding protein involved in vital processes in multidrug-resistant tumor cells, and of metallothionein, a protein involved in mechanisms to reduce toxicity by exogenous agents. These two mechanisms, combined with the gene suppressive action of siRNA against the MDR-1 gene, contribute to the cytotoxic action of DOX (with a decrease in IC50 from 30 μg mL^{-1} of DOX alone to 0.4 μg mL^{-1}) and, consequently, cause apoptosis.

Wu et al. (2017) also used DOX to develop polyethyleneimine-2-hydroxyprophl-γ-CD nanoparticles containing siRNA targeting the oncogene homologous 1 to erythroblastosis virus E26 (ETS1), which is overexpressed in breast cancer cells resistant to antitumor drugs and participates in some resistance mechanisms (e.g., overexpression of MDR-1) (Kars et al. 2010). Nanoparticles of the average size of about 250 nm and with zeta potential between +40 and + 50 mV were developed, with the majority release of DOX in the medium at lysosomal pH, in which it was released between 13% and 25% within 84 h. These nanoparticles showed antitumor activity against the MCF-7/ADR strain in in vitro and in vivo models. The action against the tumor cells occurs when the nanoparticle is internalized and the siRNA decreases the expression of the ETS1 and MDR-1 genes, which leads to a reduction in resistance mechanisms and favors the activity of DOX (decrease in the IC50 of isolated DOX from 42.5 μg mL^{-1} to 5.2 μg mL^{-1} when compared to nanoparticles).

Besides DOX, other antitumor drugs can be associated with siRNA and conveyed in cyclodextrin-based nanosystems for the treatment of breast cancer, such as paclitaxel (PTX). With this perspective, Chen et al. (2020) developed polyethyleneimine-β-CD nanoparticles coated with chondroitin sulfate containing siRNA targeting CD146, a surface protein overexpressed in some breast cancer subtypes (e.g., HER2 and triple negative) related to progression and poor prognosis (de Kruijff et al. 2018), and PTX. The nanoparticles developed had an average size of 200 nm and zeta potential of +35 mV, with a release of about 73% of PTX and 68% of siRNA over 72 h at pH 7.4. Furthermore, the nanoparticles showed cytotoxic and anti-metastasis activity against MBA-MD-231 breast cancer cells. This activity is related to the mechanism of inhibition of CD146 gene expression by siRNA, which reduces the synthesis of the protein and inhibits the metastatic process, in addition to the ability of PTX to inhibit proliferation, mechanisms that culminate in cell apoptosis.

CDs can also be functionalized from modification by PEG. In this perspective, Kaushal et al. (2018) developed β-CD nanoparticles modified with PEG and peptide EPPT1, which targets binding to mucin type 1 receptors (MUC-1) and which are involved with tumor invasion and metastatic potential when overexpressed, containing siRNA targeting the suppression of RhoC protein expression (anti-RhoC siRNA), which is related to progression and metastasis in breast cancer (Xu et al. 2021). The nanoparticles showed an average size below 200 nm, zeta potential between 19.4 and 28 mV with in vitro anti-metastatic activity against breast cancer cells of the MDA-MB-231 and SUM149 strains. The mechanism of this activity lies in the ability of the EPPT1 peptide to bind to MUC-1 receptors and mediate nanoparticle internalization, in addition to the stabilization of siRNA in serum caused by PEG. Once inside the tumor cells, the anti-RhoC siRNA exerts its action on RhoC gene suppression, which allows the reduction of metastatic mechanisms of tumor cells.

2.7 Micelles

Micelles are globular structures, formed by an aggregate of amphipathic molecules, that is, composed of polar and apolar parts. Generally, its structure is globular, but it can also be ellipsoidal, cylindrical, and layered. The use of micellar solutions of low molecular weight surfactants has been used for hydrophobic drug solubilization. However, these surfactants have instability and critical micellar concentration. As an alternative, polymeric micelles are being used, which have gained prominence in the oncological area as carriers of chemotherapeutic agents due to their selective targeting, solubilization, inhibition of P-glycoprotein, and alteration of the internalization route of the drug. The use of micelles as delivery vehicles for siRNA has been studied, as they are easy to design and modify in which their core serves as a drug loading dock (Gong et al. 2012).

In this regard, Yin et al. (2015) performed the co-delivery of hydrophobic paclitaxel and hydrophilic AURKA-specific siRNA by redox-sensitive micelles constructed from a hyaluronic acid-based amphiphilic conjugate (HA-ss-(OA-g-bPEI), HSOP) for effective breast cancer treatment against MDA-MB-231 cells. Aurora kinase A (AURKA) has been shown to be a good potential therapeutic target for triple-negative breast cancer (TNBC), which is a more aggressive and metastatic subtype of breast cancer. Consequently, PTX- and AURKA-specific siRNA (si-AURKA) was chosen for co-delivery and loaded with redox-sensitive HA-ss-(OA-g-bPEI) micelles against the MDA-MB-231 cell line, which is a TNBC-expressed CD44. The co-loading capabilities for siRNA and PTX along with the redox sensitivity of the micelles were then observed in detail. The particle size was around 135 nm and zeta potential +20 mV and the release of PTX was 57% in 6 h and 91% in 36 h. Regarding the in vivo studies, there was a tumor inhibition of 47.4%.

Studies have also been conducted to co-deliver autophagy inhibitor of protein 7 (ATG7) siRNA and docetaxel via a polypeptide micellar system based on a cytosolic 6 (CL)-conjugated lipoic acid (LA) for the treatment of breast cancer MCF-7 cells. This conjugated aqueous LA-CL can be self-assembled into cross-linked micelles (LA-CLss). The lysosomal autophagy degradation pathway plays a crucial role in protecting cancer cells from multiple endogenous and exogenous stresses, particularly during cancer pathogenesis. Consequently, agents involved in autophagy inhibition may have broad therapeutic applications. All particles had zeta potential between 10 and 30 mV and particle size around 200 nm. The studies showed that treatment with DTX -siATG7 had 2.5-fold and 1.7-fold higher cytotoxicity and apoptosis, respectively, in MCF-7 cells compared to treatment with DTX alone. Furthermore, the release of DTX in the presence of DTT was 84.5% and 59.4% in its absence within 48 h. Regarding in vivo studies, DTX in the co-delivery system demonstrated a 1.84-fold higher tumor inhibition compared to DTX-loaded micelles (Gong et al. 2017).

Another study produced low-density lipoprotein (LDL)-coupled micelles with dual reduction and pH sensitivity for intelligent co-delivery of paclitaxel and siRNA to MCF-7/Taxol cell-based breast tumor. Among the efflux proteins that cause multiple drug resistance (MDR) is ABCG2, being found mainly in MCF-7/AdrVp multidrug-resistant breast cancer cells, and its inhibition is the aim of the study conducted by Zhu et al. (2017). In addition, LDL was used as a ligand because it can be recognized by the receptor LDLr which is abundant in tumor membranes besides being a great vehicle for lipophilic drugs such as antitumor drugs. The particle size was 167.90 ± 1.46 nm and zeta potential −21.91 ± 1.25 mV and the release was 64.97% ± 3.51% in pH 5.3 PBS in 48 h. Cytotoxicity assays showed that on average 84% of cells survived in 48 h which shows good safety.

2.8 Dendrimers

Dendrimers are compact, hyper-amplified, symmetrical, globular polymeric macromolecules with three-dimensional architecture. These molecules are characterized by having a predictable and highly ordered structure and are synthesized through repetitive chemical reactions, their synthesis can be divergent, from the nucleus to the periphery, or convergent, from the periphery to the nucleus (Mittal et al. 2021; Dias et al. 2020). The dendrimeric structure can be divided into three main components: the nucleus, the dendrons, inner layers consisting of repetitive molecular units, and terminal groups on the surface (Janaszewska et al. 2019). The structural flexibility and adaptability of the physicochemical properties of dendrimers have been attractive for their use in conjunction with antineoplastic drugs (Chis et al. 2020).

As such, Abedi Gaballu and collaborators (2021) developed the G4/MTX-siRNA nanocomplex, an anti-HMGA2 (high mobility group A2 protein) siRNA molecule using fourth-generation (G4) polyamidoamine dendrimer (PAMAM) nanosystem functionalized with methotrexate (MTX) with the aim of reducing HMGA2 gene expression and inducing apoptosis-mediated cell death in human breast cancer cells (MCF-7 and MDA-MB-231). The HMGA2 gene influences cell cycle regulation and DNA damage repair, and it has been suggested by several studies that HMGA2 is able to inhibit the induction of apoptosis of tumor cells and is found overexpressed in different types of cancer. MTX has been used as a modifier of delivery systems due to its affinity for the folate receptor (FR), and has demonstrated oncological activity by inhibiting dihydrofolate reductase (DHFR) and consequently preventing DNA synthesis. The structured nanocomplex G4/MTX-siRNA had average size 160 ± 46.7 nm, PDI (polydispersity index) 0.29, and zeta potential −7.6 ± 3.6 mV, the release profile was tested in pH 7.4 and pH 5.5, in the more acidic pH 52.3% was released 48 h, in pH 7.4 the release in 48 h was 32%, this comportment is probably related to the ionic nature of MTX. G4/MTX-siRNA was shown to successfully internalize and accumulate in the cytoplasm of cells, causing apoptosis-mediated cell death due to HMGA2 silencing in cancer cells with FR overexpression. The G4/MTX-siRNA nanocomplex was more successful in cellular uptake than the G4-siRNA polyplex in cancer cells expressing FR, indicating that the G4/MTX association is an efficient nanosystem for HMGA2 siRNA delivery.

PAMAM dendrimers were also used by Finlay and co-workers (2015) in association with siRNA, the complex was formed by a modified third-generation amphiphilic PAMAMAM dendrimer (YTZ3–15) with TWIST1 siRNA, intending to reduce TWIST1-mediated expression of EMT (epithelial-mesenchymal transition) target genes and inhibit metastatic potential. The dendrimers ranged in size from 100 nm to 200 nm, siRNA levels decreased by 40% after 24 h and 90% after 7 days. Vimentin RNA was almost undetectable after 1 day and remained at <10% after 7 days. In in vivo mouse tests, YTZ3–15 was observed to concentrate in orthotopic breast cancer tumors.

Pan et al. (2019) developed a co-delivery system based on mixed micelles composed of a lipid-modified dendrimer copolymer and PEG 5 k-DOPE, used for co-delivery of siRNA and tumor drugs via enhanced permeability and retention (EPR) effect. The molecule had a homogeneous size of about 219 nm with a PdI around 0.187 and zeta potential 20.40 ± 3.39 mV. It was observed that after 48 h of treatment, there was a 40% reduction of the molecule and the compound facilitated cell interaction and aided in endosomal escape, and had the potential to reduce toxicity.

3 Conclusion

The discovery of the suppression of protein coding through the use of siRNA molecules has brought new treatment possibilities for several diseases, including breast cancer. However, despite the high therapeutic potential, these molecules are not very stable in the body and present some restrictions.

In this context, the development of delivery systems has been studied to elude the limitations of siRNA. The incorporation of siRNA in nanostructured delivery systems is one of the most promising alternatives to circumvent the limitations of these molecules. Most of the delivery systems observed in the study had satisfactory results regarding the antitumor activity and protection of siRNA and showed potential for decreasing adverse effects.

Micelles, polymeric nanoparticles, and solid lipid nanoparticles are the nanoparticles with the most studies and evidence of success in the use of encapsulated siRNA for breast cancer treatment, while the other nanoparticles, despite good results, are less studied. It was also observed that doxorubicin shows a high affinity for siRNA molecules and has been incorporated into various delivery systems, enhancing the antitumor effect.

References

Abedi Gaballu F, Cho WCS, Dehghan G, Zarebkohan A, Baradaran B et al (2021) Silencing of HMGA2 by siRNA loaded methotrexate functionalized polyamidoamine dendrimer for human breast cancer cell therapy. Genes 12(7):1102

Ahmed OAA, Badr-Eldin SM (2020) Biodegradable self-assembled nanoparticles of PEG-PLGA amphiphilic diblock copolymer as a promising stealth system for augmented vinpocetine brain delivery. Int J Pharm 588:119778

Akinc A, Maier MA, Manoharan M, Fitzgerald K, Jayaraman M et al (2019) The Onpattro story and the clinical translation of nanomedicines containing nucleic acid-based drugs. Nat Nanotechnol 14(12):1084–1087. https://doi.org/10.1038/s41565-019-0591-y

Amiri-Kordestani L, Basseville A, Kurdziel K, Fojo AT, Bates SE (2012) Targeting MDR in breast and lung cancer: discriminating its potential importance from the failure of drug resistance reversal studies. Drug Resist Updat 15(1–2):50–61

Araújo VCS (2017) Análise de RNAs longos não codificantes do genoma de Arabidopsis thaliana (L.) Heynh. 2017. 80 f. Dissertação (Mestrado em Genética e Biologia Molecular) – Universidade Federal de Goiás, Goiânia

Bahreyni A, Mohamud Y, Luo H (2020) Emerging nanomedicines for effective breast cancer immunotherapy. J Nanobiotechnol 18(1):1–14

Barenholz Y (2012) Doxil® the first FDA approved nano-drug: lessons learned. J Control Release 160(2):117–134

Bochicchio S, Dalmoro A, Angela Barba A, Grassi G, Lamberti G (2022) Liposomes as siRNA delivery vectors: Ingenta connect. Disponível em: https://www.ingentaconnect.com/content/ben/cdm/2014/00000015/00000009/art00004. Acesso em: 18 maio. 2022

Bozzuto G, Molinari A (2015) Liposomes as nanomedical devices. Int J Nanomedicine 10:975–999, 2 fev

Cao J, Wang X, Dai T, Wu Y, Zhang M et al (2018) Twist promotes tumor metastasis in basal-like breast cancer by transcriptionally upregulating ROR1. Theranostics 8(10):2739–2751

Chan JJ, Tay Y (2018) Noncoding RNA: RNA regulatory networks in cancer. Int J Mol Sci 19(5): 1310

Chen J, Sun X, Shao R, Xu Y, Gao J, Liang W (2017a) VEGF siRNA delivered by polycation liposome-encapsulated calcium phosphate nanoparticles for tumor angiogenesis inhibition in breast cancer. Int J Nanomedicine 12:6075, 21 ago

Chen X, Zhang Y, Tang C, Tian C, Sun Q et al (2017b) Co-delivery of paclitaxel and anti-survivin siRNA via redox-sensitive oligopeptide liposomes for the synergistic treatment of breast cancer and metastasis. Int J Pharm 529(1–2):102–115, 30 ago

Chen M, Chen M, He J (2019) Cancer cell membrane cloaking nanoparticles for targeted co-delivery of doxorubicin and PD-L1 siRNA. Artif Cells Nanomed Biotechnol [Internet] 47(1):1635–1641. Available from: https://doi.org/10.1080/21691401.2019.1608219

Chen Y, Li B, Chen X, Wu M, Ji Y, Tang G et al (2020) A supramolecular co-delivery strategy for combined breast cancer treatment and metastasis prevention. Chin Chem Lett [Internet] 31(5): 1153–1158. Available from: https://doi.org/10.1016/j.cclet.2019.06.022

Chen C, Guo Q, Fu H, Yu J, Wang L, Sun Y, et al (2021) Asynchronous blockade of PD-L1 and CD155 by polymeric nanoparticles inhibits triple-negative breast cancer progression and metastasis. Biomaterials [Internet] 275(2020):120988. Available from: https://doi.org/10.1016/j.biomaterials.2021.120988

Chis AA, Dobrea C, Morgovan C, Arseniu AM, Rus LL et al (2020) Applications and limitations of dendrimers in biomedicine. Molecules 25(17):3982

Chowdhury NK, Choudhury R, Sonawane GA, Mavinamar S, Lyu X, Pandey RP, Chang CM (2021) Nanoparticles as an effective drug delivery system in COVID-19. Biomed Pharmacother 143:112162

Cramer P (2020) Rosalind Franklin and the advent of molecular biology. Cell 182(4):787–789

De Kruijff IE, Timmermans AM, Den Bakker MA, Trapman-Jansen AM, Foekens R et al (2018) The prevalence of CD146 expression in breast cancer subtypes and its relation to outcome. Cancers (Basel) 10(5):134

Dias AP, da Silva SS, da Silva JV, Parise-Filho R, Ferreira EI, El Seud O, Giarolla J (2020) Dendrimers in the context of nanomedicine. Int J Pharm 573:118814

Ding J, Liang T, Min Q, Jiang L, Zhu JJ (2018) "Stealth and fully-laden" drug carriers: self-assembled Nanogels encapsulated with epigallocatechin Gallate and siRNA for drug-resistant breast cancer therapy. ACS Appl Mater Interfaces 10(12):9938–9948

Dong Y, Siegwart DJ, Anderson DG (2019) Strategies, design, and chemistry in siRNA delivery systems. Adv Drug Deliv Rev 144:133–147

Feng C, Zhu D, Chen L, Lu Y, Liu J, Kim NY et al (2019) Targeted delivery of chlorin E6 via redox sensitive diselenide-containing micelles for improved photodynamic therapy in cluster of differentiation 44-overexpressing breast cancer. Front Pharmacol 10:369

Finlay J, Roberts CM, Lowe G, Loeza J, Rossi JJ, Glackin CA (2015) RNA-based TWIST1 inhibition via dendrimer complex to reduce breast cancer cell metastasis. Biomed Res Int 2015:382745

Fire A, Xu S, Montgomery MK, Kostas SA, Driver SE, Mello CC (1998) Potent and specific genetic interference by double-stranded RNA in Caenorhabditis elegans. Nature 391:806–811

Franco YL, Vaidya TR, Ait-Oudhia S (2018) Anticancer and cardio-protective effects of liposomal doxorubicin in the treatment of breast cancer. Breast Cancer Targets Ther 10:131

Ganesan P, Narayanasamy D (2017) Lipid nanoparticles: different preparation techniques, characterization, hurdles, and strategies for the production of solid lipid nanoparticles and nanostructured lipid carriers for oral drug delivery. Sustain Chem Pharm 6(May):37–56. https://doi.org/10.1016/j.scp.2017.07.002

Ganesan K, Wang Y, Gao F, Liu Q, Zhang C et al (2021) Targeting engineered nanoparticles for breast cancer therapy. Pharmaceutics 13(11):1829

Gayon J (2016) From Mendel to epigenetics: history of genetics. C R Biol 339(7–8):225–230

Goes ACS, Oliveira BVX (2014) Projeto Genoma Humano: um retrato da construção do conhecimento científico sob a ótica da revista Ciência Hoje. Ciência Educ Bauru 20(03): 561–577, set

Gonçalves GAR, Paiva RMA (2017) Gene therapy: advances, challenges and perspectives. Einstein (São Paulo) 15(3):369–375

Gong J et al (2012) Polymeric micelles drug delivery system in oncology. J Control Release 159(3): 312–323

Gong C et al (2017) Co-delivery of autophagy inhibitor ATG7 siRNA and docetaxel for breast cancer treatment. J Control Release 266:272–286

Gong C, Yu X, Zhang W, Han L, Wang R, Wang Y et al (2021) Regulating the immunosuppressive tumor microenvironment to enhance breast cancer immunotherapy using pH-responsive hybrid membrane-coated nanoparticles. J Nanobiotechnol [Internet] 19(1):1–20. Available from: https://doi.org/10.1186/s12951-021-00805-8

Gordillo-Galeano A, Mora-Huertas CE (2018) Solid lipid nanoparticles and nanostructured lipid carriers: a review emphasizing on particle structure and drug release. Eur J Pharm Biopharm. Elsevier B.V. https://doi.org/10.1016/j.ejpb.2018.10.017

Haider M, Abdin SM, Kamal L, Orive G (2020) Nanostructured lipid carriers for delivery of chemotherapeutics: a review. Pharmaceutics 12(3). https://doi.org/10.3390/pharmaceutics12030288

Han X, Li Z, Sun J, Luo C, Li L, Liu Y et al (2015) Stealth CD44-targeted hyaluronic acid supramolecular nanoassemblies for doxorubicin delivery: probing the effect of uncovalent pegylation degree on cellular uptake and blood long circulation. J Control Release 197:29–40

Hu B et al (2020) Therapeutic siRNA: state of the art. Signal Transduct Target Ther 5(1):1–25

Janaszewska A et al (2019) Cytotoxicity of dendrimers. Biomol Ther 9(8):330

Jansook P, Ogawa N, Loftsson T (2018) Cyclodextrins: structure, physicochemical properties and pharmaceutical applications. Int J Pharm 535(1–2):272–284

Jin W, Wang Q, Wu M, Li Y, Tang G, Ping Y et al (2017) Lanthanide-integrated supramolecular polymeric nanoassembly with multiple regulation characteristics for multidrug-resistant cancer therapy. Biomaterials [Internet] 129:83–97. Available from: https://doi.org/10.1016/j.biomaterials.2017.03.020

Jin M, Hou Y, Quan X, Chen L, Gao Z, Huang W (2021) Smart polymeric nanoparticles with pH-responsive and Peg-detachable properties (II): co-delivery of paclitaxel and VEGF siRNA for synergistic breast cancer therapy in mice. Int J Nanomedicine 16(August):5479–5494

Kars MD, Işeri ÖD, Gündüz U (2010) Drug resistant breast cancer cells overexpress ETS1 gene. Biomed Pharmacother 64(7):458–462

Kaushal N, Tiruchinapally G, Durmaz YY, Bao L, Gilani R, Merajvey SD, ElSayed ME (2018) Synergistic inhibition of aggressive breast cancer cell migration and invasion by cytoplasmic delivery of anti-RhoC silencing RNA and presentation of EPPT1 peptide on "smart" particles. J Control Release 289(July):79–93

Khosa A, Reddi S, Sasha RN (2018) Nanostructured lipid carriers for site-specific drug delivery. Biomed Pharmacother 103(February):598–613. https://doi.org/10.1016/j.biopha.2018.04.055

Kuzuhara T, Sei Y, Yamaguchi K, Suganuma M, Fujiki H (2006) DNA and RNA as new binding targets of green tea catechins. J Biol Chem 281(25):17446–17456

Laouini A, Jaafar-Maalej C, Limayem-Blouza I, Sfar S, Charcosset C, Fessi H (2012) Preparation, characterization and applications of liposomes: state of the art. J Colloid Sci Biotechnol 1(2): 147–168

Lee SH, Mok H, Park TG (2011) Di- and triblock siRNA-PEG copolymers: PEG density effect of polyelectrolyte complexes on cellular uptake and gene silencing efficiency. Macromol Biosci 11(3):410–418. https://doi.org/10.1002/mabi.201000347

Lewis RG, Simpson B (2021) Genetics, autosomal dominant. 2020. In: StatPearls. StatPearls Publishing, Treasure Island (FL). PMID: 32491444

Li Y, Humphries B, Yang C, Wang Z (2018) Nanoparticle-mediated therapeutic agent delivery for treating metastatic breast cancer—challenges and opportunities. Nanomaterials 8(6). https://doi.org/10.3390/nano8060361

Li M, Du C, Guo N, Teng Y, Meng X et al (2019) Composition design and medical application of liposomes. Eur J Med Chem 164:640–653, 15 fev

Li F, Yu W, Zhang J, Dong Y, Ding X, Ruan X et al (2021) Spatiotemporally programmable cascade hybridization of hairpin DNA in polymeric nanoframework for precise siRNA delivery. Nat Commun 12:1138

Li M, Li S, Li Y, Li X, Yang G et al (2022) Cationic liposomes co-deliver chemotherapeutics and siRNA for the treatment of breast cancer. Eur J Med Chem 233:114198, 5 abr

Lobovkina T, Jacobson GB, Gonzalez-Gonzalez E, Hickerson RP, Leake D (2011) *In vivo* sustained release of siRNA from solid lipid nanoparticles. ACS Nano 5(12):9977–9983. https://doi.org/10.1021/nn203745n

Lopes JC, Torres MLP (2019) Nanoparticles in cancer treatment: general aspects, antineoplastic mechanisms of action and tumor applications. Revista Brasileira de Cancerologia 65(4):e-08592

Marcucci F, Romeo E, Caserta CA, Rumio C, Lefoulon F (2020) Context-dependent pharmacological effects of metformin on the immune system. Trends Pharmacol Sci 41(3):162–171

Meng Q, Meng J, Ran W, Wang J, Zhai Y, Zhang P et al (2018) Light-activated Core-Shell nanoparticles for spatiotemporally specific treatment of metastatic triple-negative breast cancer. ACS Nano 12(3):2789–2802

Mengarda AC, Iles B, João JP, de Moraes J (2022) Recent trends in praziquantel nanoformulations for helminthiasis treatment. Expert Opin Drug Deliv [Internet] 19(4):1–11. Available from: https://doi.org/10.1080/17425247.2022.2051477

Mennati A, Rostamizadeh K, Manjili HK, Fathi M, Danafar H (2022) Co-delivery of siRNA and lycopene encapsulated hybrid lipid nanoparticles for dual silencing of insulin-like growth factor 1 receptor in MCF-7 breast cancer cell line. Int J Biol Macromol 200(November 2021): 335–349. https://doi.org/10.1016/j.ijbiomac.2021.12.197

Mitchell MJ, Billingsley MM, Haley RM, Wechsle ME, Peppas NA, Langer R (2021) Engineering precision nanoparticles for drug delivery. Nat Rev Drug Discov 20(2):101–124

Mittal P, Saharan A, Verma R, Altalbawy F, Alfaidi MA (2021) Dendrimers: a new race of pharmaceutical nanocarriers. Biomed Res Int 2021:8844030

Modi SJ, Kulkarni VM (2019) Medicine in drug discovery Vascular Endothelial Growth Factor Receptor (VEGFR-2)/KDR inhibitors: medicinal chemistry perspective. Med Drug Discov [Internet] 2:100009. Available from: https://doi.org/10.1016/j.medidd.2019.100009

Mousazadeh H, Bonabi E, Zarghami N (2022) Stimulus-responsive drug/gene delivery system based on polyethylenimine cyclodextrin nanoparticles for potential cancer therapy, 118747. Carbohydr Polym [Internet] 276(October 2021). Available from: https://doi.org/10.1016/j.carbpol.2021.118747

Mukalel AJ, Riley RS, Zhang R, Mitchell MJ (2019) Nanoparticles for nucleic acid delivery: applications in cancer immunotherapy. Cancer Lett 458:102–112

Nápoles C, Lemieux C, Jorgensen R (1990) Introduction of a chimeric Chalcone synthase gene into petunia results in reversible co-suppression of homologous genes in trans. Plant Cell 2(4): 279–289

Okamoto A, Asai T, Kato H, Ando H, Minamino T, Mekada E, Oku N (2014) Antibody-modified lipid nanoparticles for selective delivery of siRNA to tumors expressing membrane-anchored form of HB-EGF. Biochem Biophys Res Commun 449(4):460–465

Okamoto A, Asai T, Hirai Y, Shimizu K, Koide H, Minamino T, Oku N (2018) Systemic administration of siRNA with anti-HB-EGF antibody-modified lipid nanoparticles for the treatment of triple-negative breast cancer. Mol Pharm 15(4):1495–1504. https://doi.org/10.1021/acs.molpharmaceut.7b01055

Pan J, Mendes LP, Yao M, Filipczak N, Garai S (2019) Polyamidoamine dendrimers-based nanomedicine for combination therapy with siRNA and chemotherapeutics to overcome multi-drug resistance. Eur J Pharm Biopharm 136:18–28

Piao L, Li H, Teng L, Yung BC, Sugimoto Y, Brueggemeier RW, Lee RJ (2013) Human serum albumin-coated lipid nanoparticles for delivery of siRNA to breast cancer. Nanomedicine 9(1): 122–129. https://doi.org/10.1016/j.nano.2012.03.008

Qian W, Zhao M, Wang R, Li H (2021) Fibrinogen-like protein 1 (PGL1): the next immune checkpoint target. J Hematol Oncol 14(147):1–17

Salvi VR And Pawar P. (2019) 'Nanostructured lipid carriers (NLC) system: a novel drug targeting carrier', J Drug Deliv Sci Technol, 51(990), pp. 255–267. doi: https://doi.org/10.1016/j.jddst.2019.02.017

Sartaj A, Qamar Z, Qizilbash FF, Annu MS, Alhakamy NA et al (2021) Polymeric nanoparticles: exploring the current drug development and therapeutic insight of breast cancer treatment and recommendations. Polymers (Basel) 13(24):4400

Singh A, Trivedi P, Jain NK (2018) Advances in siRNA delivery in cancer therapy. Artif Cells Nanomed Biotechnol 46(2):274–283

Singh RP, Hidalgo T, Cazade PA, Darcy R, Cronin MF, Dorin I et al (2019) Self-assembled cationic β-Cyclodextrin nanostructures for siRNA delivery. Mol Pharm 16(3):1358–1366

Sumera et al (2017) Docetaxel-loaded solid lipid nanoparticles: a novel drug delivery system. IET Nanobiotechnol 11(6):621–629

Sun B, Hagan CT IV, Caster J, Wang AZ (2019) Nanotechnology in radiation oncology. Hematol Oncol Clin North Am 33(6):1071–1093

Toohey JM, Ismail K, Lonergan D, Lewis CR (2007) Amyloidosis of the breast mimicking recurrence in a previously treated early breast cancer. Australas Radiol 51(6): 594–596. https://doi.org/10.1111/j.1440-1673.2007.01700.x

Triki H, Charfi S, Bouzidi L, Ben Kridis W, Daoud J, Chaabane K et al (2019) CD155 expression in human breast cancer: clinical significance and relevance to natural killer cell infiltration. Life Sci 231:116543

Vallabani NVS, Singh S, Karakoti AS (2019) Magnetic nanoparticles: current trends and future aspects in diagnostics and nanomedicine. Curr Drug Metab 20(6):457–472

Verma V, Ryan KM, Padrela L (2021) Production and isolation of pharmaceutical drug nanoparticles. Int J Pharm 603:120708

Vieira LM (2016) Classificação de RNAs não-codificadores longos intergênicos usando máquina de vetores de suporte: um estudo de caso para a cana-de-açúcar. xii, 74 f., il. Monografia (Bacharelado em Engenharia da Computação)—Universidade de Brasília, Brasília

Wang J, Sanmamed MF, Datar I, Su TT, Ji L, Sun J et al (2019) Fibrinogen-like protein 1 is a major immune inhibitory ligand of LAG-3. Cell [Internet] 176(1–2):334–347.e12. Available from: https://doi.org/10.1016/j.cell.2018.11.010

Wang H, Wang C, Zou Y, Hu J, Li Y, Cheng Y (2020) Natural polyphenols in drug delivery systems: current status and future challenges. Giant [Internet] 3:100022. Available from: https://doi.org/10.1016/j.giant.2020.100022

Wu M, Liu X, Jin W, Li Y, Li Y, Hu Q et al (2017) Targeting ETS1 with RNAi-based supramolecular nanoassemblies for multidrug-resistant breast cancer therapy. J Control Release [Internet] 253:110–121. Available from: https://doi.org/10.1016/j.jconrel.2017.03.011

Xu M, Huang S, Dong X, Chen Y, Li M et al (2021) A novel isoform of ATOH8 promotes the metastasis of breast cancer by regulating RhoC. J Mol Cell Biol 13(1):59–71

Yin T, Wang L, Yin L, Zhou J, Huo M (2015) Co-delivery of hydrophobic paclitaxel and hydrophilic AURKA specific siRNA by redox-sensitive micelles for effective treatment of breast cancer. Biomaterials 61:10–25. https://doi.org/10.1016/j.biomaterials.2015.05.022

Yu S, Bi X, Yang L, Wu S, Yu Y (2019) Co-delivery of paclitaxel and PLK1-targeted siRNA using aptamer-functionalized cationic liposome for synergistic anti-breast cancer effects *in vivo*. J Biomed Nanotechnol 15(6):1135–1148

Zhang L, Xu B (2020) Autophagy inhibitior autophagy-related 7 small interfering RNA and doxorubicin dual-loaded nanostructured lipid carrier to combat multidrug resistance. J Mater Res 35(22):3136–3143. https://doi.org/10.1557/jmr.2020.223

Zhang Y, Li M, Gao X, Chen Y, Liu T (2019) Nanotechnology in cancer diagnosis: progress, challenges and opportunities. J Hematol Oncol 12(1):1–13

Zhao F, Yin H, Li J (2014) Supramolecular self-assembly forming a multifunctional synergistic system for targeted co-delivery of gene and drug. Biomaterials 35(3):1050–1062

Zhu Wen-jing et al (2017) Low-density lipoprotein-coupled micelles with reduction and pH dual sensitivity for intelligent co-delivery of paclitaxel and siRNA to breast tumor. Int J Nanomedicine 12:3375–3393. https://www.ncbi.nlm.nih.gov/pmc/articles/PMC5413542/

Zielinska A, Carreiró F, Oliveira AM, Neves A, Pires B, Venkatesh DN et al (2020) Polymeric nanoparticles: production, characterization, toxicology and ecotoxicology. Molecules 25:3731

Nucleic Acid and Drug Dual Agent Nanoplatforms for Breast Cancer

Saffiya Habib and Moganavelli Singh

Abstract

Background: Breast cancer is among the most common cancers affecting women. Besides attempting to develop novel anticancer drugs, intense research is focused on improving the efficacy of existing drugs with the aid of nanoparticles. Other options are presented through gene therapy in which nucleic acids function as therapeutic agents. This too is reliant upon nanostructures for faithful introduction into tumor cells. Studies have shown that simultaneous introduction of nucleic acids and chemotherapeutic agents can produce synergistic antitumor effects. This report was aimed at reviewing nanoplatforms for the co-delivery of nucleic acids and neoplastic agents in the treatment of breast cancer.

Methods: A broad literature survey was conducted spanning 2010 to 2021 using Google Scholar. Key words and terms used in the search included *breast cancer*; *nucleic acid and drug co-delivery*; *siRNA drug co-delivery breast cancer*; *shRNA co-delivery breast cancer*; *miRNA co-delivery breast cancer*; and *DNA drug co-delivery breast cancer*; among others.

Results: In an attempt to meet the challenges of dual agent delivery which include stable incorporation of both therapeutic agents, avoiding early release and degradation, evasion of the reticuloendothelial system (RES), and avoiding an immune response and non-specific effects in non-target cells, several multifunctional nanodevices have been explored. These have been based on organic and inorganic nanomaterials which include lipids, polymers, peptides, proteins, mesoporous silica, gold, carbon, and magnetic materials. Common design motifs

S. Habib · M. Singh (✉)
Nano-Gene and Drug Delivery Group, Discipline of Biochemistry, University of KwaZulu-Natal, Durban, South Africa
e-mail: singhm1@ukzn.ac.za

© Springer Nature Switzerland AG 2022
Interdisciplinary Cancer Research, https://doi.org/10.1007/16833_2022_2
Published online: 20 August 2022

encompass cancer cell targeting ligands, poly(ethylene) glycol for stability, and pH- and redox-sensitive moieties for controlled release. Physical methods have also been investigated to promote deposition of therapeutic agents at tumor sites.

Conclusion: Overall, research into the application of dual agent nanotherapeutics for breast cancer has demonstrated encouraging results with respect to attenuating drug resistance, reducing metastases, and treating recalcitrant triple-negative breast cancer. The advent of a clinically viable breast cancer combination therapy is thus eagerly anticipated.

Keywords

Breast cancer · Drug delivery · Dual therapy · Nanoplatforms · Nucleic acids

1 Introduction

Breast cancer refers to a heterogeneous collection of malignancies affecting and originating in the same anatomical structure, i.e., breast tissue, most commonly from the inner lining of milk ducts or the lobules that supply the ducts with milk (Sharma et al. 2010; Weigelt et al. 2010). Breast tumors usually initiate from ductal hyperproliferation and then develop into benign tumors or even metastatic carcinomas after stimulation by carcinogenic factors which induce abnormal gene expression (Sun et al. 2017). A representation of the multistep development of breast cancer is given in Fig. 1.

Several types of tumors may develop within various regions of the breast. For example, noninvasive breast cancer is confined to the ducts, while in invasive breast cancer, tumor cells invade the surrounding fatty and connective tissues. Among these two broad categories, infiltrating ductal carcinoma accounts for 80% of all breast cancer cases (Sharma et al. 2010). Alarmingly, more than 1.5 million women are diagnosed with breast cancer globally each year. This makes it one of the most common cancers affecting women (Sun et al. 2017).

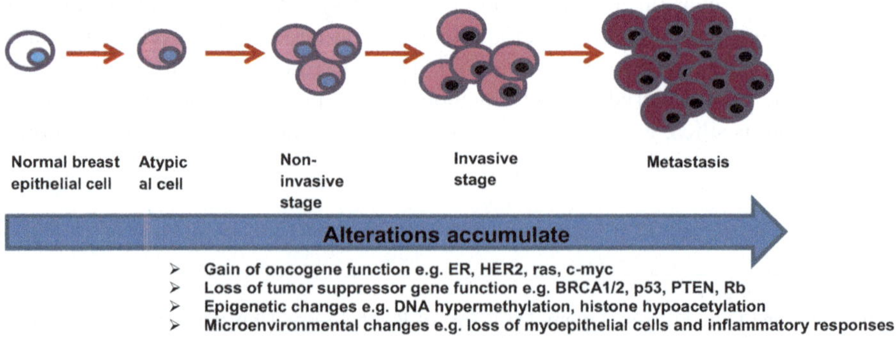

Fig. 1 Multistage process in the development and progression of breast cancer. Adapted from Rivenbark et al. (2013) and Yu and Lu (2011). Illustration created using MS Word 2010

At present two hypotheses exist to explain the origin of human breast cancer. The cancer stem cell theory is that all tumor subtypes are derived from the same progenitor cells which acquire genetic and epigenetic mutations resulting in different tumor phenotypes. The second, the stochastic theory, suggests that each tumor subtype is initiated from a single cell type (stem cell, progenitor cell, or differentiated cell) (Polyak 2007; Sgroi 2010). There are three broad subtypes, characterized based on the presence or absence of estrogen or progesterone receptors and human epidermal growth factor 2 (ERBB2): hormone receptor positive/ERBB2 negative (70% of patients), ERBB2 positive (15–20%), and triple-negative (lacking all three molecular markers; 15%) (Waks and Winer 2019).

Current treatment options are subtype and stage dependent and include a combination of surgery, cytotoxic chemotherapy, endocrine therapy, and radiation (Waks and Winer 2019) with a major emphasis on early detection for success (Sun et al. 2017). Nonetheless, it has been hypothesized that approximately 30% of newly diagnosed patients with early stage breast cancer will develop a metastasis despite receiving treatment (Morry et al. 2017; Redig and McAllister 2013).

Despite advances in understanding the pathogenesis of breast cancer, effective treatment that is free of side effects is yet to be developed. It is here that nanomedicine has the potential to make its mark. The encapsulation of conventional anticancer drugs into nanoparticles improves therapeutic efficacy by protecting drug integrity, avoiding an immune response, and prolonging retention in the body (Patra et al. 2018). The approval and widespread use of Doxil, a pegylated liposomal formulation of doxorubicin (Barenholz 2021; Barenholz 2012), is a good example, and attempts are still underway to improve nanodelivery of this drug (El Founi et al. 2021; Jafari et al. 2020; Norouzi et al. 2020). Moreover, nanodrugs may be targeted to neoplastic cells through surface modification, to reduce the therapeutic dose and prevent toxicity in healthy bystander cells (Patra et al. 2018).

Besides improving upon chemotherapy for breast cancer, nanoparticles can be used for the delivery of therapeutic nucleic acids which include small interfering RNA (siRNA), short hairpin RNA (shRNA), microRNA (miRNA), antisense oligonucleotides, and functional gene segments.

siRNA molecules are short double-stranded RNA duplexes with characteristic two nucleotide overhangs on the 3′ end. They are effector molecules of the endogenous gene silencing mechanism of RNA interference (RNAi) in which they guide the degradation of messenger RNA (mRNA) bearing a complementary sequence (Elbashir et al. 2001a, b) (Fig. 2). Cancer cells have altered gene expression signatures (Chibon 2013) to support their heightened growth requirements and to engineer the tumor microenvironment for their survival. Overexpression of genes encoding transcription factors (Xu et al. 2010), growth factor receptors (Gonzalez-Conchas et al. 2018), and angiogenesis factors (Guo et al. 2003) has been observed in breast cancer. In theory, effective silencing of any cancer-relevant gene can be achieved through successful introduction of appropriately designed siRNA.

Besides the introduction of mature siRNA molecules, siRNA can be generated in the cell through the successful delivery of expression plasmids that direct the synthesis of the target sequence in the form of short hairpin RNA (shRNA), and

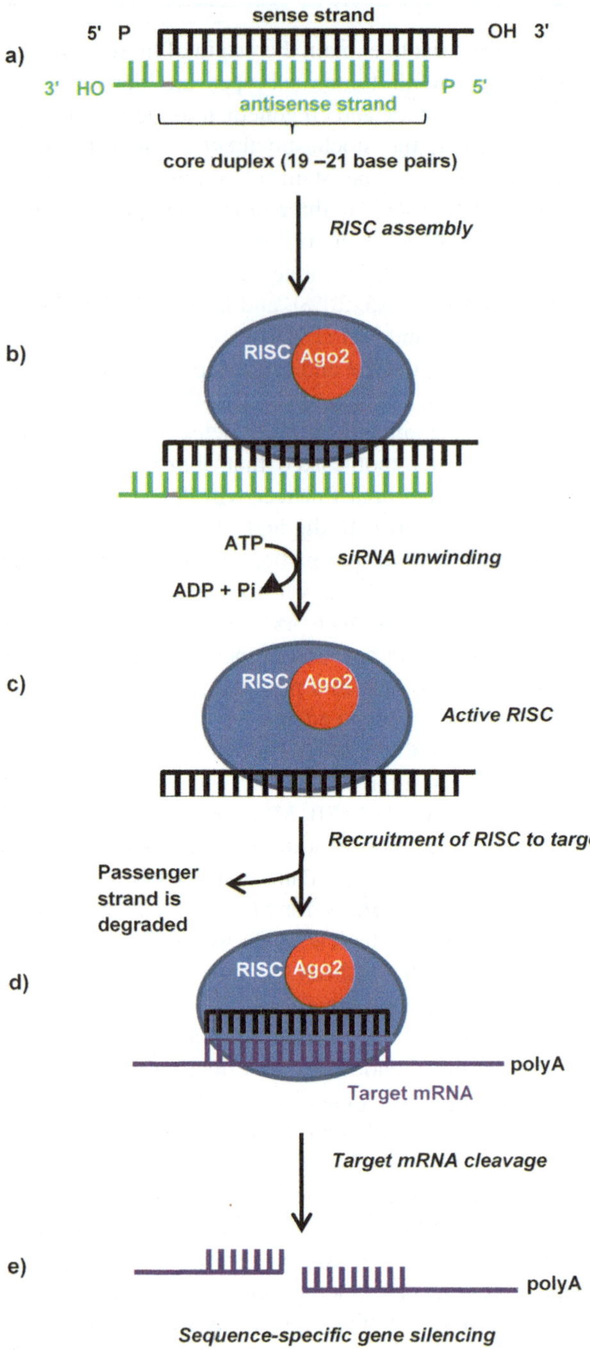

Fig. 2 siRNA-mediated gene silencing. (**a**) Schematic representation of a siRNA molecule. (**b**) The siRNA molecule associates with a network of cytoplasmic proteins containing Argonaut 2 (Ago2), and the siRNA molecule unwinds in an energy-dependent process. (**c**) Only one siRNA strand is incorporated into the active RISC, while the other is degraded. (**d**) This strand guides the

this is further processed to yield siRNA (Taxman et al. 2010). The concept of combined delivery of siRNA and drugs for cancer treatment has received attention because it achieves heightened anticancer activity as opposed to single regimen applications. Furthermore, combining siRNA/shRNA delivery with chemotherapeutics in a single delivery system is more effective than co-treatment with individual delivery systems carrying either the nucleic acid or drug alone (Yin et al. 2015). Moreover, this notion is of particular importance when facing drug resistance – siRNA molecules and shRNA expression plasmids can be designed to target genes implicated in recalcitrance, so that the chemotherapeutic agent can be effective (Dönmez and Gündüz 2011; Yang et al. 2018).

miRNA is another class of small non-coding RNA that affects endogenous gene silencing. miRNAs are single-stranded RNA molecules, averaging 22 nucleotides in length that are transcribed from miRNA genes. miRNA is recruited into RISC in which it interacts with the 3′ UTR of target mRNA leading to mRNA degradation and translational repression (Makarova et al. 2016). The deregulation of miRNA genes in breast cancer encourages proliferative signaling, resistance to apoptosis, invasion, metastasis, and angiogenesis, with different pathways in disease development being supported by the elevated expression of oncogenic miRNAs (oncomiRs) and/or reduced expression of tumor suppressor miRNAs (Loh et al. 2019). miRNA-based breast cancer therapeutics include the inhibition of oncomiRs using antisense oligonucleotides (anti-miRNAs) and the restoration of tumor suppressor miRNAs using microRNA mimics (Nagini 2017) (Fig. 3). Recently, anti-miRNA/drug and mimic/drug synergisms have been gaining attention in cancer treatment (He et al. 2020).

The introduction of functional genes such as tumor suppressors, which are either defective or downregulated in breast cancer, represents another potential treatment avenue. Furthermore, there is evidence to suggest that the restoration of such genes can attenuate resistance to drugs (Jeong et al. 2021).

Combining nucleic acids and chemotherapeutic agents in the same formulation is challenging because of their very different physicochemical properties – nucleic acids are hydrophilic, anionic, high molecular weight entities, while the most commonly used chemotherapy drugs are hydrophobic small molecules (Zhao et al. 2020). This necessitates different mechanisms of encapsulation. Nucleic acids are often electrostatically associated with cationic components of the nanoparticle, while small molecule drugs are enclosed within them by hydrophobic force, electrostatic interaction, or chemical conjugation (Dai and Tan 2015). The nanoparticle has to be engineered such that it efficiently encapsulates, protects, and delivers both types of therapeutic cargo. Kang et al. (2015) summarized the properties of the ideal co-delivery nanocarrier as one that is biocompatible, biodegradable, and stable in circulation, allows for transport of therapeutic payloads to the target sites, and

Fig. 2 (continued) recognition of mRNA bearing a complementary sequence, (**e**) which is then enzymatically cleaved to effect sequence-specific gene silencing. Illustration created using MS Word 2010

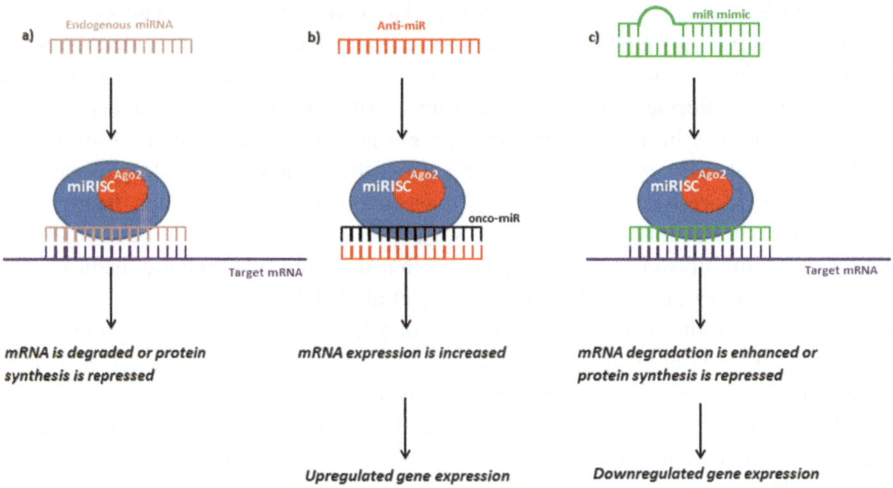

Fig. 3 miRNA-mediated gene silencing and associated therapies. (**a**) As a posttranscriptional mechanism of gene regulation in the cell, mature miRNA duplexes interact with the miRNA-induced gene silencing complex (miRISC). The antisense miRNA strand is degraded, and miRISC associated with a mature miRNA strand binds to the 3′-UTR of cognate mRNA either with perfect complementarity or with imperfect base pairing. This causes either degradation or translational repression of the targeted mRNA, which thereby inhibits synthesis of the encoded protein. (**b**) In instances of miRNA overexpression, anti-miRs also known as miRNA inhibitors may be introduced. These are small, synthetically derived RNA molecules, which have sequence complementary to mature miRNAs. In this way, they sequester target miRNAs and can suppress miRNA function. (**c**) miRNA mimics are synthetic small RNA duplexes which behave similarly to endogenous miRNAs. They can be introduced to replace the function of underexpressed miRNAs. (Adapted from Fu et al. (2019) and Paul et al. (2020). Illustration created using MS Word 2010)

possesses the ability to simultaneously transport both agents to cancer cells and release them in a controlled manner for maximum effect (Fig. 4).

For this reason nanoparticles for co-delivery are, in general, multicomponent systems based on lipids, polymers, peptides, and inorganic/organic hybrid materials. We review here nanosystems applied to the co-delivery of nucleic acids and drugs for potential treatment of breast cancer.

2 Design Features of Dual Agent Nanocarriers for Breast Cancer

An overview of some design features incorporated in dual agent nanocarriers for breast cancer treatment is given in Fig. 5.

The primary feature of any feasible dual agent nanocarrier is, as the name suggests, the capacity to serve as a depot for both nucleic acid and drug. In general, the nanoparticle is fashioned with hydrophilic cationic centers for binding to the

Fig. 4 Properties of the perfect dual agent vector. Illustration created using MS Word 2010

anionic sugar phosphate backbone of the nucleic acid as well as a core to encapsulate the drug (Dai and Tan 2015).

In addition, nanoparticle stability is an essential quality. To be considered suitable as a pharmacological agent, the nanoparticle must not aggregate upon storage and ideally retain its small size. Moreover, when introduced into the body, the nanoparticle must avoid adverse interactions with serum proteins which act as a destabilizing factor. The nanoparticle must also avoid recognition and clearance by the reticuloendothelial system. To this end stealth nanoparticles were introduced. These are designed with a hydrophilic polymer shield to sterically hinder interparticle interactions. The most common steric stabilizing agent applied to dual agent breast cancer nanodelivery is polyethylene glycol (PEG). PEG is approved for use by the Food and Drug Administration (FDA) (Alconcel et al. 2011) and has even been found to attenuate the toxicity of some nanoparticles (Ju et al. 2013). Despite it being an advantageous addition to the design of dual agent nanosystems, the "PEG dilemma" cannot be ignored. The flip side of the coin is that the PEG corona prevents the close association between the nanoparticle and cancer cell membrane that is essential for its internalization. Furthermore, PEG can hinder the association between the nanoparticle and endosomal membrane that is often necessary for endosomal escape and release of therapeutics (Zhang et al. 2015). Fortunately, this has been addressed through the design of sheddable PEG chains (Jin et al. 2021) that are detached upon exposure to stimuli such as low pH of the tumor microenvironment.

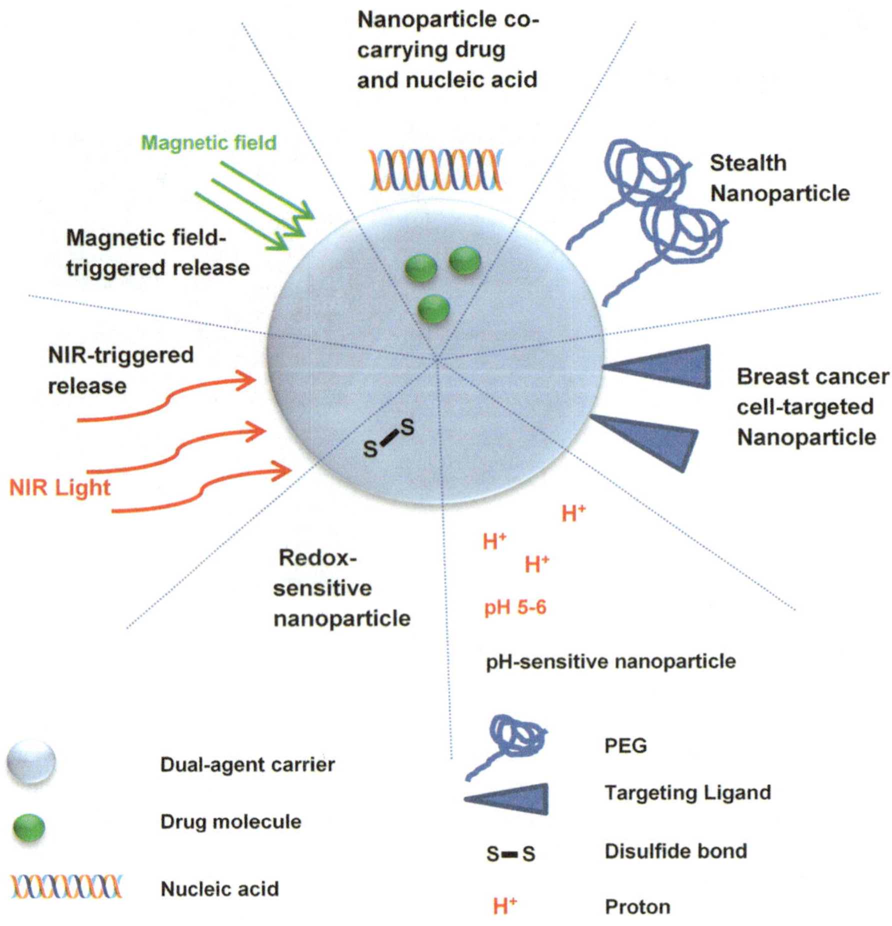

Fig. 5 An overview of some design features of nanoplatforms directed at dual agent delivery for breast cancer. Illustration created using MS Word 2010

Another important feature of the dual agent nanosystem is the ability to be taken up specifically by breast cancer cells. Not only is this important with regard to sparing healthy, normal cells but also assists in decreasing the required administered dose through increasing the effective dose of therapeutic agents to which cancer cells are exposed. In order to achieve this, the "Trojan horse" concept has come into play. This strategy exploits the natural mechanism of receptor-mediated endocytosis. Briefly a ligand to a receptor that is overexpressed on the breast cancer cell surface is displayed on the surface of the nanoparticle, and this directs its recognition and uptake by the cancer cell (Tashima 2018). Common breast cancer cell targeting ligands include folate (Jia et al. 2021; Sharma et al. 2021; Suo et al. 2017) and hyaluronate (Deng et al. 2014; Wang et al. 2016a, b; Yu et al. 2016a, b).

Other design features have been focused largely on permitting release of the nanoparticle in the cell. A common strategy is to exploit the low intracellular pH of the intracellular vesicular system. Here, the incorporation of proton sponges in nanoparticle design has gained in popularity. Briefly, a proton sponge is a molecule with buffering capacity within endosomal pH range which becomes protonated and induces an influx of chloride ions into the endosome. This causes an increase in osmotic pressure resulting in endosome rupture and release of the nanoparticle in the cell. As an example, polyethylenimine (PEI) is a common proton sponge (Zhou et al. 2016), while in lipid-based systems, the proton sponge element is built into the lipid molecule (Chen et al. 2017a, b). Besides just permitting release of the nanoparticle, the use of protonatable moieties in dual agent nanoparticle systems can be used to control release of therapeutics based on the difference between intra- and extracellular pH levels (Yu et al. 2016a, b).

The disulfide bond has also proven to be a useful element in controlling the release of therapeutics from dual agent nanocarriers (Yin et al. 2015; Zhu et al. 2017). This strategy exploits the fact that intracellular glutathione levels of tumor cells are higher than that of normal cells. Hence, within the tumor cell, the disulfide bond is reduced to release the associated therapeutic agent.

Some nanomaterials can act as NIR conversion agents for photothermal ablation of cancer cells (Li et al. 2021). This also has the added benefit of triggering release of therapeutics at the desired site by controlling the application of the external stimulus. The application of an external magnetic field can also drive release of therapeutics at the target site provided that the drug carrier possesses a strong magnetic moment. In this regard, superparamagnetic nanomaterials have been applied to dual agent delivery as well (Amani et al. 2021).

In accordance with the aforementioned design features, several classes of nanoparticles have been investigated. The advantages and disadvantages of the major nanoparticle classes that have been implicated in dual agent delivery for breast cancer are summarized in Table 1.

3 Lipid-Based Nanoparticles

Lipid-based systems have received considerable attention in nanodrug delivery due to their high temporal and thermal stability, high loading capacity, ease of preparation, low production costs, and amenability to large-scale industrial production (García-Pinel et al. 2019). Among them liposomes, bilayered phospholipid vesicles, are the most well-known and widely studied.

A novel liposome-based nanoplatform composed of a redox-sensitive cationic oligopeptide lipid with proton sponge capacity, soybean phosphatidylcholine, and cholesterol was developed for simultaneous delivery of paclitaxel and siRNA against survivin, given that survivin mediates paclitaxel resistance in breast cancer. The siRNA was electrostatically associated with the positively charged bilayer, while hydrophobic paclitaxel was contained between the bilayer. The system showed a synergistic inhibitory effect on tumor growth and pulmonary metastasis in vivo with

Table 1 Advantages and limitations of the major nanomaterials applied to breast cancer-directed dual agent delivery

Nanomaterial	Advantages	Disadvantages	References
Liposomes	Biocompatible Easy to prepare Can carry both anionic and cationic material depending on composition Tunable composition Targetable	High production cost Short shelf-life Drug leakage can occur	Yadav et al. (2017)
Polymeric nanoparticles	Biocompatible Biodegradable Long half-life High drug stability	Synthesis is complicated Purification is difficult Large-scale manufacture is problematic	Parveen and Sahoo (2008)
Micelles	Biocompatible Controlled drug release Tunable chemical and physical properties	Low drug loading capacity Dependency on critical micelle concentration	Kahraman et al. (2017)
Protein-based Nanoparticles	Biocompatible Biodegradable Good safety profile Mild conditions for preparation	High cost	Hong et al. (2020)
Gold nanoparticles	Biocompatible Small size High surface area to volume ratio Easy surface modification Green chemistry synthesis routes	Toxicity because of extended retention time in the body	Khan et al. (2014) and Sztandera et al. (2018)
Mesoporous silica nanoparticles	Biocompatible Biodegradable High surface area Large pore size Well-defined surface properties	Difficulties in preparation Scattered size distribution Formation of stable colloidal suspensions	Jafari et al. (2019b)

low toxicity (Chen et al. 2017a, b). The feasibility of using cationic liposomes for co-delivery of paclitaxel and siRNA against the polo-like kinase 1 (PLK1) gene, which is overexpressed in cancers and is associated with poor prognosis (Liu et al. 2017), was demonstrated in MCF-7 and MDA-MB-231 breast cancer cell lines (Bulbake et al. 2018). In a later study, co-delivery of the same two components via aptamer-functionalized cationic liposomes increased apoptosis and reduced angiogenesis in vitro and in vivo (Yu et al. 2019).

In keeping with the notion of cationic liposomes, 1,2-dioleoyloxy-3-trimethylammoniumpropane (DOTAP) liposomes were applied to the delivery of camptothecin and siRNA against the genes encoding chemoresistant proteins, the glucose-regulated protein 78 (GRP78) and clusterin (CLU). Gene silencing and chemosensitivity in breast cancer stem cells were more pronounced with the new system compared with the commercial lipid-based transfection agent, Lipofectamine™ 2000, and free camptothecin, respectively (Samson et al. 2019). A pH-sensitive cationic formulation containing DOTAP, dioleoylphosphatidylethanolamine, phosphatidylcholine, and cholesterol was used to encapsulate docetaxel and bind SIRT1 shRNA that targets the expression of a nuclear enzyme which deacetylates transcription factors involved in cellular regulation. The lipoplex was more effective at reducing the growth of breast tumors than the clinical counterpart, Taxotere®, and exhibited low hemo-, hepato-, and nephrotoxicity (Swami et al. 2020). Xu et al. (2019a, b) synthesized a novel peptide cationic lipid that was incorporated into a liposome for the co-delivery of resveratrol and the p53 tumor suppressor gene. The combinatory system gave stronger growth inhibition in MCF-7 cells than blank liposomes and resveratrol liposomes.

Poly-L-arginine films containing siRNA against the multiple drug resistance protein 1 (MDR1) gene were generated atop a doxorubicin-loaded hyaluronic acid-modified liposome for introduction in recalcitrant triple-negative breast cancer cells (Deng et al. 2013). The MDR1 gene encodes the multidrug resistance protein 1 or P-glycoprotein (P-gp) drug efflux transporter that acts as a protective barrier against a wide variety of substances and interferes with the bioavailability and distribution of drugs (Linardi and Natalini 2006). A single intravenous injection in animal models reduced target gene expression by almost 80% in tumors and decreased tumor volume without observable toxicity (Deng et al. 2013).

Lipid-based core-shell nanoparticles have also been used for co-delivery of RNA/drug cocktails. As an example, a protamine/chondroitin sulfate/siRNA core, with siRNA against the vascular endothelial growth factor (VEGF), was assembled. This was encapsulated in a paclitaxel-loaded cationic liposome. The surface was further modified with vapreotide, a somatostatin analogue. The nanoparticle achieved enhanced drug distribution in breast tumor tissues and tumor growth inhibition through somatostatin receptor-mediated delivery. This was accompanied by inhibition of neovascularization due to VEGF knockdown (Feng et al. 2014). In a later study, cationic albumin-coated core-shell nanoparticles were applied to the delivery of docetaxel and miRNA-34a, a potent tumor suppressor that is defective in breast cancer. The nanosystem extended the blood circulation of docetaxel, enhanced tumor accumulation of the cargo, and inhibited tumor growth and metastasis in mice bearing 4 T1 mammary tumors (Zhang et al. 2017a).

In response to the challenge to design carriers that are durable, to survive delivery, yet degradable to release their contents, Seidel et al. (2020) designed photoresponsive catanionic vesicles to co-deliver paclitaxel and siRNA against the *bcl-2* oncogene in MDA-MB-231 triple-negative breast cancer cells. The vesicles were formed spontaneously from mixing of an azobenzene-based cationic surfactant and a conventional anionic surfactant and disassociate into free surfactants with UV illumination to permit photo-triggered release of the drug and siRNA.

Lee et al. (2015) introduced a colloidal formulation of a hydrophobically associated multiple monocomplex formed from the non-ionic surfactant, pluronic F68, *bcl-2* siRNA, and benzethonium chloride. When introduced via peritumoral route in highly aggressive and resistant human breast cancer xenografts, gene silencing by siRNA and apoptosis induced by benzethonium chloride greatly suppressed tumor volume.

Solid lipid nanoparticles have also been investigated as co-carriers of genes and drugs. In a proof-of-principle study, plasmid DNA and paclitaxel were co-loaded into solid lipid nanoparticles that were modified with hyaluronic acid-acid-sensitive linker-1,2-distearoylphosphatidylethanol-amine copolymers. In vitro and in vivo introduction to breast cancer cells proved that the nanosystem could inhibit tumor growth and effect successful gene transfection (Yu et al. 2016a, b).

Recently, a calcium phosphate composite lipid nanosystem encapsulating paclitaxel and miR124 was designed to deliver its contents to triple-negative breast cancer cells via stepwise disassembly governed by the tumor microenvironment. The PEG coating was fashioned via a hydrazine bond so as to be removed in the acidic microenvironment to permit exposure of the hyaluronic acid targeting ligand. Hyaluronic acid was tethered via a disulfide bond that is reduced upon internalization to free the residue which antagonizes endogenous hyaluronic acid to reduce tumor invasiveness. Finally the paclitaxel and miR124 were sequentially released to induce significant antitumor activity in MDA-MB-231 tumor-bearing mice (Chen et al. 2021).

Lipids have also been incorporated into hybrid nanosystems for combinatorial breast cancer therapy. As an example, Sharma et al. (2021) introduced a folate-targeted lipo-polymeric nanoplex for the co-delivery of docetaxel and miR-34a. The nanoplex was taken up by MDA-MB-231 and 4 T1 breast cancer cells mainly via lipid-raft and clathrin-mediated endocytosis, improved cytotoxicity, and apoptosis, with changes in the expression of apoptotic genes compared to the free treatment group.

Besides being associated with synthetic lipid nanoplatforms, nucleic acid/drug combinations can be delivered using exosomes, which are natural vesicles and carrier agents. Cell-derived exosomes with integrin-binding potential were applied to the co-delivery of doxorubicin and cholesterol-modified tumor suppressor miRNA 159 for the treatment of triple-negative breast cancer. Dual agent delivery effectively silenced the transcription factor 7 (TCF-7) gene and showed improved anticancer activity, without adverse effects in vivo (Gong et al. 2019).

4 Polymer-Based Nanoparticles

Polymer-based nanoparticles serve as delivery agents by encapsulating the active compound within a polymeric matrix, conjugation, or adsorption to the polymer (Ansary et al. 2014; Mahapatro and Singh 2011). Polymers can be designed as linear, branched, or globular, and their size and properties can be controlled by the process used for their synthesis (Essa et al. 2020). The natural polymer, chitosan, and several synthetic polymers have been investigated as potential nucleic acid/antineoplastic co-delivery agents for breast cancer (Table 1).

4.1 Chitosan-Based Nanoparticles

Chitosan is a natural cationic polysaccharide composed of randomly distributed β-(1 → 4)-linked *N*-acetyl-D-glucosamine (GlcNAc) and D-glucosamine (GlcN). Generally derivatization is required to improve its solubility. However, its biocompatibility, biodegradability, and negligible toxicity have perpetuated interest in its biomedical application (Li et al. 2018).

Carboxymethyl dextran chitosan nanoparticles were used to encapsulate doxorubicin and electrostatically bind siRNA against IL17RB, a plasma membrane receptor that is upregulated in breast cancer. When introduced into metastatic breast cancer cells, the co-loaded nanoparticles resulted in significant silencing of *NF-kB* and *bcl-2*, both of which are targets of IL17RB/IL17B signaling, induction of apoptosis, and inhibition of migration (Alinejad et al. 2016). Trimethyl chitosan nanoparticles were used to study the impact of the co-delivery of doxorubicin and high-mobility group protein two (HMGA2) siRNA in MDA-MB-231 cells. It was found that co-delivery significantly inhibited cancer cell growth, while the siRNA knocked down the expression of HMGA2 and its downregulators, vimentin and matrix metallopeptidase-9 (MMP9), but led to overexpression of E-cadherin (Eivazy et al. 2017). Later, mucin 1-aptamer-conjugated chitosan nanoparticles were applied to the co-delivery of docetaxel and siRNA targeting c-MET (Zolbanin et al. 2018) and insulin-like growth factor receptor 1 (Jafari et al. 2019a) in mucin1-positive, SKBR3 metastatic breast cancer cells. Recently, chitosan lactate nanoparticles were functionalized with HIV-1-derived TAT peptide and hyaluronate to simultaneously deliver CD73 siRNA and doxorubicin to increase the susceptibility of cancer cells to doxorubicin-induced cell death and stimulate antitumor immune responses in 4 T1 tumor cells (Khesht et al. 2021).

In a shRNA delivery platform, a folate-modified chitosan oligosaccharide-grafted disulfide-containing PEI copolymer-based silica nanohybrid was designed to co-carry paclitaxel and P-gp shRNA to reverse paclitaxel resistance in multidrug-resistant breast cancer. The nanohybrid was able to release shRNA and the drug in a redox- and pH-responsive fashion, respectively. The nanoparticles achieved high levels of gene silencing, with a resistance reversion index of 50.59 (Jia et al. 2021).

miRNA 34a and doxorubicin were co-encapsulated in hyaluronic acid-modified chitosan nanoparticles for simultaneous delivery into breast cancer cells. It was demonstrated that co-delivery improved the antitumor activity of doxorubicin by suppressing the expression of *bcl-2*, while intracellular restoration of miR-34a inhibited cancer cell migration by targeting Notch-1 signaling (Deng et al. 2014). Also for miRNA-based therapy, anti-miR 21 and docetaxel were co-loaded into "chitosomes" formed from the self-assembly of amphiphilic chitosan derivatives for the treatment of triple-negative breast cancer (Sun et al. 2021).

4.2 Poly(lactide-co-glycolide) Nanoparticles

Synthetic polymers such as poly(lactide-co-glycolide) (PLGA) are attractive alternatives to natural polymers as these can be specifically engineered to conform to their application (Essa et al. 2020). PLGA is a copolymer synthesized by random ring-opening copolymerization of the cyclic dimers (1,4-dioxane-2,5-diones) of glycolic acid and lactic acid and has been approved by the FDA for drug delivery (Lü et al. 2009). One of the earliest examples of its application in co-drug delivery for breast cancer is that of cationic PLGA which was used to deliver doxorubicin and anti-*MDR1* siRNA to overcome multidrug resistance in MCF-7/ADR cells (Misra et al. 2014). PLGA-mediated breast cancer combinatory therapy has expanded to include a pH-responsive co-delivery platform using a hybrid biomimetic membrane-camouflaged PLGA nanoparticle for metformin and siRNA against the fibrinogen-like protein 1(FGL1) gene. Metformin, a known immuno-metabolic modulator, promotes programmed death-ligand 1 (PD-L1) degradation by activating adenosine monophosphate-activated protein kinase and blocks the inhibitory signals of PD-L1. Consequently siRNA delivery can effectively silence the FGL1 gene, promote T-cell-mediated immune responses, and enhance antitumor immunity (Gong et al. 2021).

In a breast cancer-directed, shRNA co-drug delivery platform, PLGA nanoparticles were loaded with doxorubicin and coated with alkyl-modified PEI for attachment of anti-*bcl-xl* shRNA, which targets a class of molecules that blocks cancer cell apoptosis (Ebrahimian et al. 2017). shRNA co-delivery attempts have also been coupled with diagnostics. As an example, the surface of doxorubicin-loaded PLGA nanobubbles was modified with polyethylenimine (PEI) through DCC/NHS cross-linking for condensation of shRNA bearing an anti-P-gp sequence. The nanobubbles were taken up by MCF-7/ADR cells via macropinocytosis and enhanced doxorubicin's suppressive effect on cell proliferation, possibly due to reduced efflux of the drug as a result of gene silencing. Moreover, in vitro cellular ultrasound imaging suggested that the nanobubbles could efficiently enhance ultrasound imaging of cancer cells (Yang et al. 2015).

Hyaluronic acid-coated, PEI-PLGA nanoparticles were designed for the combined delivery of doxorubicin and tumor suppressor miRNA-542-3p for the treatment of triple-negative breast cancer. The nanoparticles increased both drug uptake and cytotoxicity in MDA-MB-231 cells, which express CD44 at high levels. Furthermore, the restoration of miR-542-3p promoted cancer cell apoptosis by activating the p53 tumor suppressor and inhibiting expression of survivin (Wang et al. 2016a, b). Also for the treatment of triple-negative breast cancer, PLGA was incorporated into a novel hybrid nanovector for the delivery of anti-miR 221/222 and paclitaxel. The miR 221/222 inhibitors were co-precipitated with calcium phosphate to form cores that were coated with dioleoylphosphatidic acid. These were co-encapsulated with paclitaxel into PLGA-PEG nanoparticles to suppress miRNA 221/222 levels and significantly enhance the therapeutic efficacy of paclitaxel (Zhou et al. 2017). PLGA-PEG nanoparticles were also applied to the co-delivery of orlistat and anti-miR 21 for an enhanced apoptotic effect in triple-negative breast cancer cells (Bhargava-Shah et al. 2016). Anti-miR 21 was also co-loaded with hydroxytamoxifen using PLGA-b-PEG-COOH copolymer-based nanoparticles to enhance antiproliferative effects in estrogen-positive breast cancer cells (Devulapally et al. 2015).

4.3 Polymeric Micelles

Polymeric micelles are core/shell nanostructures formed from the self-assembly of amphiphilic copolymers. The hydrophobic polymer forms a semi-solid core which serves as a water-insoluble drug depot, while the hydrophilic polymer forms a coronal layer to delay clearance and elimination of the drug carrier (Aziz et al. 2017). Moreover, the multifunctional nature of amphiphilic copolymers lends itself to an application as demanding as dual agent delivery.

Amphiphilic hyaluronic acid-based conjugates of octandioic acid (OA) and branched PEI (bPEI) with cystamine as bioreducible linkages were assembled into polymeric micelles for the co-delivery of paclitaxel and aurora kinase A (AURKA) siRNA to triple-negative breast cancer cells. OA and bPEI formed the hydrophobic core and cationic centers for drug encapsulation and nucleic acid binding, respectively. The micelles effectively accumulated in tumors and showed better antitumor efficacy than the non-redox-sensitive co-delivery control and redox-sensitive single-drug controls (Yin et al. 2015).

Doxorubicin-loaded cores of mixed polymeric micelles were coated with chitosan, to which anti-*MDR1* siRNA was bound. The system resulted in highest doxorubicin accumulation and significantly reduced the volume of 4 T1 and 4 T1-mdr breast tumors as compared to free doxorubicin; and this improved survival times in mice (Butt et al. 2016). Micelles composed of pH-sensitive, poly(ethylene glycol)-block-poly(aminolated glycidyl methacrylate)-block-poly(2-(diisopropyl amino) ethyl methacrylate) (PEG-*b*-PAGA-*b*-PDPA) were reported. An alkylated cisplatin prodrug was encapsulated in the hydrophobic PDPA core, while p65 siRNA targeting nuclear factor kappa B (NF-kB), which stimulates cell invasion and metastasis, was loaded on the cationic PAGA interlayer. The positively charged PAGA allowed for deep penetration of tumor tissues, while protonation of the core allowed for dissociation and drug release in the early endosome (Yu et al. 2016a, b). In the same year, a comb-like amphiphilic copolymer, methoxypolyethylene glycol-*graft*-poly(l-lysine)-*block*-poly(l-phenylalanine) (mPEG-*g*-PLL-*b*-Phe), was documented for dual agent delivery. Doxorubicin was encapsulated through hydrophobic and π-π stacking interactions between drug molecules and Phe blocks, while P-gp siRNA was electrostatically bound to the PLL blocks. The dual-loaded nanomicelles demonstrated a much higher cytotoxicity in MCF-7 cells than doxorubicin-loaded nanomicelles due to their synergistic killing effect, and the blank nanomicelles showed good biocompatibility (Suo et al. 2016). Generation 4 polyamidoamine (PAMAM) was conjugated to a PEG-phospholipid copolymer which self-assembled into micelles, with anti-*MDR1* siRNA associated with PAMAM and doxorubicin loaded in the hydrophobic core, to effectively downregulate P-gp expression and reverse doxorubicin resistance in multidrug-resistant cells such as MCF-7/ADR (Pan et al. 2019). While not a micelle, methotrexate-modified PAMAM was associated with HMGA2 siRNA and induced apoptosis in MCF-7 and MDA-MB-231 cells (Abedi Gaballu et al. 2021). In another study, a pH-sensitive amphiphile, poly (β-amino ester), poly[(1,4-butanediol)-diacrylate-β-5-polyethylenimine]-block-poly[(1,4-butanediol)-diacrylate-β-5-hydroxy amylamine], was applied to the co-delivery of survivin shRNA and doxorubicin to attenuate drug resistance in MCF-7/ADR tumor-bearing mice (Tang et al. 2014).

More recently, Chen et al. (2020) demonstrated that a chondroitin sulfate-coated *β*-cyclodextrin-PEI polymer was capable of simultaneously loading paclitaxel and CD146 siRNA, which is a target of CD44 signaling in breast cancer. The nanoassembly showed strong targeting ability towards CD44-overexpressing breast cancer cells and stronger cytotoxic and anti-metastatic effects when compared with monotreatment with paclitaxel or CD146 siRNA.

A folate receptor-targeted pH-sensitive polymeric nanocomplex co-carrying VEGF siRNA and doxorubicin reduced VEGF expression by 55%, decreased tumor microvessel density, and extended lifespan of mice with breast tumor xenografts at very low doxorubicin dose (Dong et al. 2015). In keeping with the idea of receptor-mediated nanoparticle internalization via the folate receptor on breast cancer cells, Qian et al. (2015) introduced a folate-modified hydrophilic cationic star-block terpolymer, [poly(l-glutamic acid γ-hydrazide)-*b*-poly(N, N-dimethylaminopropyl methacrylamide)]3-*g*-poly(ethylene glycol), for the co-delivery of doxorubicin and anti-*bcl-2* siRNA in MCF-7 cells. PEG was incorporated via a reducible disulfide bond to permit PEG shedding for effective endosomal escape. In the same year, Xu et al. (2015) synthesized a multifunctional triblock copolymer, folate/methoxy-poly(ethylene glycol)-*block*-poly(L-glutamate-hydrazide)-*block*-poly(*N*,*N*-dimethylaminopropyl methacrylamide. Doxorubicin was covalently attached to the poly(L-glutamate-hydrazide) blocks through a pH-labile hydrazone linkage, to trigger release upon endosomal acidification; and P-gp siRNA was electrostatically associated with the cationic poly(*N*,*N*-dimethylaminopropyl methacrylamide) block. Similarly, doxorubicin and anti-*bcl-2* siRNA were loaded into micelles composed of folate-modified, pegylated poly (acrylhydrazine)-block-poly(3-dimethylaminopropyl methacrylamide)-block-poly (acrylhydrazine) to provide a synergistic antitumor effect in MCF-7 cells (Suo et al. 2017). In a proof-of-principle study, folic acid and glucose-conjugated PEG-chitosan-polycaprolactone copolymers were used to encapsulate oleic acid-coated FeCO nanoparticles, paclitaxel, and fluorescently labeled siRNA. Nanoparticles with both folic acid and glucose modifications increased drug and siRNA transfer compared to single-targeting moieties. Furthermore, the application of an external magnetic field to the breast tumor area enhanced the efficiency of paclitaxel and siRNA delivery (Amani et al. 2021).

Zhu et al. (2017) introduced a binary polymer, low-density lipoprotein-*N*-succinyl chitosan-cystamine-urocanic acid, to deliver breast cancer resistance protein siRNA and paclitaxel. Urocanic acid and cystamine were incorporated to allow for endosomal escape of polymeric micelles by the proton sponge effect and drug release in response to high intracellular glutathione concentration, respectively, while the low-density lipoprotein (LDL) permitted specific recognition by the LDL receptors that are overexpressed on tumor cells.

A recent development involved the use of pH-responsive polyethyleneimine-block-polylactic acid nanoparticles designed with a detachable PEG shield to co-deliver paclitaxel and VEGF siRNA to an orthotopic mouse model of breast cancer. Co-delivery was shown to inhibit the formation and development of new blood vessels in tumor tissues with an improved safety profile over that of single-agent nanoparticles (Jin et al. 2021).

Shen et al. (2013) simultaneously inhibited growth and pulmonary metastasis of breast cancer in vivo through co-delivery of paclitaxel and Twist shRNA using a conjugate of pluronic P85 and low-molecular-weight PEI/D-α-tocopheryl polyethylene glycol 1,000 succinate complex nanoparticles. Also for the attenuation of lung metastasis, Luo et al. (2021) recently introduced a hyaluronic acid-coated PEI-polylactic acid-lipoic acid micelle for the co-delivery of paclitaxel and siRNA against signal transducer and activator of transcription 3 (STAT3). The hyaluronic acid coat was removed by hyaluronidase in the endo/lysosome, and rapid drug release was activated in the reducing internal cellular microenvironment.

While all studies reported thus far have focused on the co-delivery of one type of siRNA and one drug, Tang et al. (2015) reported on the co-delivery of a cocktail of paclitaxel and two siRNAs, targeting Snail and Twist, which are two important transcription factors in the metastasis process of breast cancer; via pH-sensitive poly (β-amino ester) micelles. It was found that co-delivery of the siRNA duo had a more significant anti-metastasis effect than delivering a single siRNA, due to simultaneous inhibition of cancer cell motility and degradation of the extracellular matrix.

Xu et al. (2019a, b) introduced a creatine-modified amphiphilic polymer to co-carry doxorubicin and a novel bioengineered miRNA prodrug, tRNA-mir 34a, for the treatment of breast cancer metastasis. Co-delivery of doxorubicin led to more effective prodrug processing and downregulation of miRNA 34a target genes, leading to a synergistic antitumor and antimetastatic effect in vitro and in vivo. Another group co-loaded miRNA 34a and doxorubicin with a theranostic hybrid micelle composed of the diblock copolymers, polyethylenimine-polycaprolactone (PEI-PCL) and diethylenetriaminepentaacetic acid gadolinium (III) (Gd-DTPA)-conjugated polyethyleneglycol-polycaprolactone (Gd-PEG-PCL). When introduced into mice bearing MDA-MB-231 xenografts, the nanocarrier enhanced the combined efficiency of both therapeutic agents in suppressing tumor growth (Xie et al. 2019).

For DNA/drug co-delivery, micelles comprised of a novel amphiphilic bifunctional pullulan derivative containing stearic acid and low-molecular-weight (1 kDa) branched PEI were employed as carriers of p53 DNA and doxorubicin in MCF-7 cells. Co-delivery resulted in higher cytotoxicity and induced a higher apoptosis rate of tumor cells when compared with the introduction of doxorubicin and p53 as single treatments (Chen et al. 2015). Similar results were obtained with pullulan-g-desoxycholic acid-g-PEI micelles in MCF-7 cells with simultaneous delivery of the same two therapeutic agents (Chen et al. 2017a, b). In another study, amphiphilic polyethylene glycol-polyethyleneimine-poly(caprolactone)-SS-poly(caprolactone)-polyethyleneimine-polyethylene glycol (PEG-PEI-PCL-SS-PCL-PEG) was synthesized and conjugated with the T7 peptide. The polymer was assembled into micelles carrying both pORF-hTRAIL DNA and doxorubicin into breast tumor cells in vitro and in vivo. The system demonstrated redox-responsive drug release and did not induce hemolysis blood clotting and complement activation (Feng et al. 2018). More recently, Liu et al. (2021) designed a PEG-sheddable micellar system, PMet-P (cdmPEG$_{2K}$), based on polymetformin which was investigated in a doxorubicin and pIL-12 co-delivery application. When administered in an aggressive murine breast cancer (4 T1.2) model, tumor growth inhibition proved to be more effective than with monotherapy. This system also showed potential in the co-delivery of siRNA and doxorubicin.

5 Peptide and Protein-Based Nanoparticles

Protein-based nanoparticles are advantageous in drug delivery as they are stable, biodegradable, amenable to surface modification and size control, and less problematic with regard to toxicity issues. Furthermore the preparation and encapsulation methods involve mild conditions and avoid the use of toxic chemicals and organic solvents (Hong et al. 2020).

Cross-linked, reducible polypeptide micelles were applied to the co-delivery of docetaxel and autophagy-related 7 (*ATG7*) siRNA, given that autophagy plays a key role in protecting cancer cells from endogenous and exogenous stresses. The nanoassembly efficiently silenced the *ATG7* gene and suppressed docetaxel-induced autophagy for an enhanced anticancer effect (Gong et al. 2017). As a further example, protamine served as the scaffold for the assembly of epigallocatechin-3-O-gallate (EGCG) and siRNA into a nanogel that was modified with hyaluronic acid and a tumor-targeting peptide. This assembly demonstrated a 15-fold increase in cytotoxicity to drug-resistant MDA-MB-231 cells over EGCG alone, with minimal toxicity to normal tissues and organs in an animal model (Ding et al. 2018).

More recently, cationic cholesterol-peptide micelles were synthesized to encapsulate cabazitaxel and associate with siRNA against *IKBKE*, a breast cancer oncogene. Complexes were modified with hyaluronic acid to direct internalization via CD44 which is overexpressed in triple-negative breast cancer. To facilitate intracellular release of the siRNA and drug, a lysosomal cysteine protease-labile dipeptide was inserted in the cholesterol-peptide conjugate (Zhao et al. 2020). For miRNA co-drug delivery, micelles assembled from cholesterol-penetratin conjugates were prepared. When applied to the delivery of miRNA 124 and obatoclax, the nanoparticles simultaneously induced apoptosis and inhibited autophagy flux in breast cancer cells (Zhang et al. 2016).

Silk fibroin nanoparticles were loaded with doxorubicin and modified with PEI for association with survivin siRNA. The nanoparticles showed preferential accumulation at breast tumor sites, significantly silenced survivin expression, inhibited the tumor growth rate, and caused tumor cell apoptosis in a mouse model (Norouzi et al. 2021).

Lipoproteins have also featured as breast cancer-directed dual agent nanoconveyors. As an example, Wang et al. (2018a, b) used reconstituted high-density lipoprotein for the co-delivery of paclitaxel and wild-type p53. The nanostructure enhanced the cellular uptake of the drug and DNA in MCF-7 cells via scavenger receptor B-type I-mediated lysosome-independent internalization, producing high p53 expression levels, superior toxicity, and apoptosis.

6 Mesoporous Silica Nanoparticles

Mesoporous silica nanoparticles provide good opportunity to design smart drug delivery systems (Pednekar et al. 2017). They embody the porous honeycomb-like structure of silica (SiO_2) which provides high drug-loading capacity and large surface area for the attachment of functional groups (Rajani et al. 2020). Therefore, these nanoparticles are well suited for multifunctionalization (Pednekar et al. 2017) necessary for dual agent delivery.

Zhou et al. (2016) designed a multifunctional mesoporous silica nanoparticle for co-delivery of doxorubicin and siRNA against the *bcl-2* oncogene in MDA-MB-231 cells. The nanoparticle was surface decorated with a cationic, disulfide-modified, PEI-PLL copolymer. To this, a PEG-appended folic acid tumor cell-targeting moiety was attached. siRNA was electrostatically associated with the copolymer, while doxorubicin was encapsulated into the mesoporous channels. The combination of PEI and disulfide modification permitted pH- and redox-responsive drug release, respectively. Another group (Meng et al. 2013) showed that a PEI-PEG copolymer-functionalized mesoporous silica nanoparticle provided protected delivery of stably bound doxorubicin and siRNA against the P-gp drug exporter, to the tumor site in a multidrug resistant MCF-7 xenograft model. This nanoassembly attained an 8% enhanced permeability and retention effect at the tumor site. Moreover, significant P-gp silencing was achieved at heterogeneous tumor sites, and this corresponded to the regions in which doxorubicin was released intracellularly and induced apoptosis.

Wang et al. (2018a, b) presented an avenue for overcoming drug resistance in breast cancer by mesoporous silica nanoparticle-mediated regulation of Ca^{2+}. Pegylated mesoporous silica nanocapsules were loaded with doxorubicin and siRNA against the T-type Ca^{2+} channel that, when overexpressed in breast cancer, can promote intracellular proliferative pathways and provide a survival pathway after drug treatment. The synergistic anticancer effect observed in drug-resistant breast cancer cells was ascribed to the decreased T-type Ca^{2+} channel expression and Ca^{2+} concentration, G0/G1 phase cell cycle arrest, and increased intracellular drug accumulation. In the same year, an ultrathin zeolitic imidazole framework-8 film was synthesized in situ on the surface of carboxylated mesoporous silica nanoparticles for the co-delivery of anti-*bcl-2* siRNA and doxorubicin. Acidic decomposition of the film in the endo-lysosome released payloads, resulting in enhanced chemotherapeutic efficacy in multidrug-resistant cancer cells such as MCF-7/ADR (Pan et al. 2018).

Recently, redox-responsive mesoporous silica nanoparticles were functionalized with tumor-targeting DNA aptamers for the co-delivery of doxorubicin and siRNA against TIE2, a tyrosine kinase receptor for angiopoietin-1 and angiopoietin-2 that is associated with breast cancer metastasis. Controlled release of both the drug and siRNA was linked with the intracellular reducing environment, and in vivo studies demonstrated synergism in suppressing metastasis (Zhuang et al. 2021).

7 Inorganic Nanoparticles

The investigation of inorganic materials in nanomedicine stems from the search for the elusive "perfect" drug carrier. Silicon, gold, carbon-based, and magnetic nanoparticles have thus far been applied to breast cancer-directed, dual agent delivery.

Guo et al. (2019) introduced fluorescent silicon nanoparticles for imaging-guided co-delivery of doxorubicin and anti-*MDR1* siRNA. The *MDR1* gene was silenced by approximately 80%, and this enhanced the accumulation of doxorubicin molecules in drug-resistant MCF-7 cells. This, in turn, decreased the doxorubicin IC_{50} by over 35-fold.

Gold nanorods were coated with cyclodextrin-grafted PEI and then loaded with docetaxel and p65-siRNA through host-guest interaction with cyclodextrin and electrostatic binding to PEI, respectively. Laser irradiation-induced hyperthermia was used to trigger release of therapeutic components. When used to treat 4 T1 breast tumors, the combination of nanoparticles and irradiation significantly inhibited growth and lung metastasis (Wang et al. 2016a, b). The surface of mesoporous silica-coated gold nanorods was grafted with polyamidoamine dendrimers for the co-delivery of doxorubicin and anti-*bcl-2* siRNA to MCF-7 cells. The controlled release of doxorubicin was significantly enhanced under acidic conditions and the application of near-infrared laser irradiation, while the targeted siRNA was transfected into tumor cells to knock down bcl-2 protein expression (Zhang et al. 2017c).

Captopril, an anti-angiogenesis drug, was conjugated via an amide linkage to low-molecular-weight PEI. Gold nanoparticles were modified with this conjugate and complexed with anti-VEGF siRNA. The complexes suppressed angiogenesis more effectively than mono-delivery systems. Furthermore, studies on nude mice bearing MDA-MB435 tumor xenografts showed that the complexes could home-in on tumors producing strong antitumor activity (Li et al. 2015). More recently, gold nanoparticles were modified with PEI through a PEG linker to allow for assembly of anti-PLK1 siRNA on the surface, while doxorubicin was loaded via a pH-sensitive linker with a thiol group at one terminal end for controlled release. This system showed a reduced IC_{50} in 2D and 3D cultures of breast cancer cells (Shrestha et al. 2020).

Zhang et al. (2017b) designed amine dotted hollow carbon nanospheres for dual agent delivery, i.e., anti-*MDR1* siRNA and chemotherapeutics (either doxorubicin or cisplatin) to MCF-7/ADR cells. The nanospheres showed high loading capability of both siRNA and chemotherapeutics, downregulated *MDR1* expression by more than 96%, and reduced the weight of MCF-7/ADR tumors in mice by approximately 90%. Another carbon-based nanomaterial, graphene oxide, was applied to co-delivery. Polyamidoamine-functionalized graphene oxide was loaded with doxorubicin and MMP-9 shRNA for delivery into MCF-7 cells, in which the effect of co-delivery was superior to that of the single drug (Gu et al. 2017).

Yin et al. (2018) reported on a magnetic core-shell nanoparticle composed of a magnetic zinc-doped iron oxide core and a biocompatible mesoporous silica shell that was further modified with a tumor cell-targeting iRGD peptide. The nanoparticle was designed for the combined delivery of doxorubicin and Let-7a miRNA in MDA-MB-231 cells, on the basis that Let-7a sensitizes cancer cells to the action of doxorubicin by inhibiting the expression of multiple chemoresistance genes including those associated with drug efflux and DNA repair. Moreover, the magnetic nature of the nanoparticle lends itself to magnetic resonance imaging.

Nasab et al. (2021) introduced a folate and glucose dual-targeted nanohybrid fabricated from ternary magnetic copolymers based on FeCo-polyethylenimine nanoparticles and polylactic acid-polyethylene glycol. In a proof-of-principle study, the nanohybrid, carrying both siRNA and paclitaxel, induced greater anticancer effects in breast cancer cell lines than monotargeted nanohybrids.

Hydroxyapatites have also featured in breast tumor combination therapy. As an example, the design and synthesis of amine-functionalized hydroxyapatite nanoparticles to co-deliver p53 plasmid DNA and candesartan, an angiotensin II type 1 receptor blocker, was reported. Combined delivery of p53 and candesartan suppressed tumor growth in MCF-7 xenografted mice more effectively than monotherapy, and cancer cell remission was observed. These effects were associated with dual agent suppression of tumor angiogenesis (Zhao et al. 2017).

8 Other Co-delivery Systems

Li et al. (2021) recently employed molybdenum disulfide nanosheets for the co-delivery of doxorubicin and siRNA in breast cancer cell lines. Drug release was enhanced through pH sensitivity and NIR irradiation. The study demonstrated that combining chemo-, gene-, and photothermal therapy with this delivery system will achieve superior anticancer effects in drug-resistant cancer cells than doxorubicin alone.

A novel linear DNA nanostructure made up of a nucleolin-targeting AS141 aptamer, therapeutic FOXM1 aptamer, and complementary strand linker, was introduced as an epirubicin and antimiR-10b co-delivery system (Yaghoobi et al. 2021). The system showed high internalization and cytotoxicity in nucleolin-positive MCF-7 and 4 T1 breast cancer cells and inhibited tumor growth to a greater extent than monotherapy.

9 Conclusions

The design of "smart" nanodevices for nucleic acid and drug dual agent delivery holds great potential for the treatment of breast cancer. The tunable nature of nanoparticles lends itself to the incorporation of multiple design features in a single nanoplatform. The introduction of targeting ligands, such as folate and hyaluronate, for receptor-mediated cellular uptake is a common feature of breast cancer-directed dual agent nanomedicines. Other common devices include PEGylation for stability, extended circulation time, and RES avoidance, as well as pH- and redox-responsive features to permit controlled release of payloads. In addition, dual agent nanodelivery may be aided with physical methods such as near-infrared irradiation and magnetic guidance to promote deposition at breast tumor localities.

The studies covered in this report clearly demonstrate the utility of nucleic acid and drug dual agent nanotherapy in attenuating drug resistance in breast cancer. In theory, any gene implicated in drug resistance can be silenced provided that the appropriate siRNA sequence or shRNA expression plasmid is designed. This versatility provides opportunity for a "multiple hit" approach in treating breast cancer. In addition, the usefulness of dual agent nanotherapy in preventing metastases, which is one of the main causes of mortality in breast cancer patients, has also come to the fore. Encouragingly, several dual agent nanosystems have shown promise in the treatment of triple-negative breast cancer, which is highly aggressive, is prone to early recurrence, and responds poorly to conventional therapy.

However, there are considerations to take into account before a dual agent nanosystem for breast cancer becomes a clinical reality. One consideration is the fact that several studies are restricted to in vitro experiments, and the lack of supporting in vivo data represents a large gap in information. Furthermore, little is known regarding the overall production costs of such systems, as it is difficult to estimate from bench to commercial scale at this stage. Nonetheless, the overriding theme of success conveyed in this account and the promise it holds for the treatment of breast cancer cannot be ignored. With this in mind, it is likely that nucleic acid and drug dual agent onconanotherapeutics will emerge as a robust tool in the treatment of breast cancer in the future.

Acknowledgments The authors acknowledge the National Research Foundation (NRF) of South Africa: grant numbers, 138470 (S.H) and 129263 and 120455 (M.S). Please note that the opinions, findings, conclusions, or recommendations expressed in this publication are that of the authors and the NRF accepts no liability in this regard.

Compliance with Ethical Standards The authors declare that there is no conflict of interest.

References

Abedi Gaballu F, Cho WC-S, Dehghan G, Zarebkohan A, Baradaran B, Mansoori B, Abbaspour-Ravasjani S, Mohammadi A, Sheibani N, Aghanejad A (2021) Silencing of HMGA2 by siRNA loaded methotrexate functionalized Polyamidoamine dendrimer for human breast cancer cell therapy. Genes 12(7):1102. https://doi.org/10.3390/genes12071102

Alconcel SN, Baas AS, Maynard HD (2011) FDA-approved poly (ethylene glycol)–protein conjugate drug. Polym Chem 2(7):1442–1448. https://doi.org/10.1039/C1PY00034A

Alinejad V, Somi MH, Baradaran B, Akbarzadeh P, Atyabi F, Kazerooni H, Kafil HS, Maleki LA, Mansouri HS, Yousefi M (2016) Co-delivery of IL17RB siRNA and doxorubicin by chitosan-based nanoparticles for enhanced anticancer efficacy in breast cancer cells. Biomed Pharmacother 83:229–240. https://doi.org/10.1016/j.biopha.2016.06.037

Amani A, Alizadeh MR, Yaghoubi H, Nohtani M (2021) Novel multi-targeted nanoparticles for targeted co-delivery of nucleic acid and chemotherapeutic agents to breast cancer tissues. Mater Sci Eng C 118:111494. https://doi.org/10.1016/j.msec.2020.111494

Ansary RH, Awang MB, Rahman MM (2014) Biodegradable poly (D, L-lactic-co-glycolic acid)-based micro/nanoparticles for sustained release of protein drugs-A review. Trop J Pharm Res 13(7):1179–1190. https://doi.org/10.4314/tjpr.v13i7.24

Asyikin binti Abdul Aziz Z, Ahmad A, Hamidah Mohd-Setapar S, Hassan H, Lokhat D, Amjad Kamal M (2017) Recent advances in drug delivery of polymeric nano-micelles. Curr Drug Metab 18(1):16–29. https://doi.org/10.2174/1389200217666160921143616

Barenholz YC (2012) Doxil®—the first FDA-approved nano-drug: lessons learned. J Control Release 160(2):117–134. https://doi.org/10.1016/j.jconrel.2012.03.020

Barenholz YC (2021) Doxil-the first FDA-approved nano-drug: from an idea to a product. In: Peer D (ed) Handbook of harnessing biomaterials in nanomedicine, 2nd edn. Jenny Stanford Publishing, pp 335–398. https://doi.org/10.4032/9789814364270

Bhargava-Shah A, Foygel K, Devulapally R, Paulmurugan R (2016) Orlistat and antisense-miRNA-loaded PLGA-PEG nanoparticles for enhanced triple negative breast cancer therapy. Nanomedicine 11(3):235–247. https://doi.org/10.2217/nnm.15.193

Bulbake U, Kommineni N, Bryszewska M, Ionov M, Khan W (2018) Cationic liposomes for co-delivery of paclitaxel and anti-Plk1 siRNA to achieve enhanced efficacy in breast cancer. J Drug Deliv Sci Technol 48:253–265. https://doi.org/10.1016/j.jddst.2018.09.017

Butt AM, Amin MCIM, Katas H, Abdul Murad NA, Kesharwani JR (2016) Doxorubicin and siRNA codelivery via chitosan-coated pH-responsive mixed micellar polyplexes for enhanced cancer therapy in multidrug-resistant tumors. Mol Pharm 13(12):4179–4190. https://doi.org/10.1021/acs.molpharmaceut.6b00776

Chen L, Wang X, Ji F, Bao Y, Wang J, Wang X, Guo L, Li Y (2015) New bifunctional-pullulan-based micelles with good biocompatibility for efficient co-delivery of cancer-suppressing p53 gene and doxorubicin to cancer cells. RSC Adv 5(115):94719–94731. https://doi.org/10.1039/C5RA17139C

Chen L, Ji F, Bao Y, Xia J, Guo L, Wang J, Li Y (2017a) Biocompatible cationic pullulan-g-desoxycholic acid-g-PEI micelles used to co-deliver drug and gene for cancer therapy. Mater Sci Eng C 70:418–429. https://doi.org/10.1016/j.msec.2016.09.019

Chen X, Zhang Y, Tang C, Tian C, Sun Q, Su Z, Xue L, Yin Y, Ju C, Zhang C (2017b) Co-delivery of paclitaxel and anti-survivin siRNA via redox-sensitive oligopeptide liposomes for the synergistic treatment of breast cancer and metastasis. Int J Pharm 529(1–2):102–115. https://doi.org/10.1016/j.ijpharm.2017.06.071

Chen Y, Li B, Chen X, Wu M, Ji Y, Tang G, Ping Y (2020) A supramolecular co-delivery strategy for combined breast cancer treatment and metastasis prevention. Chin Chem Lett 31(5):1153–1158. https://doi.org/10.1016/j.cclet.2019.06.022

Chen C, Shen M, Liao H, Guo Q, Fu H, Yu J, Duan Y (2021) A paclitaxel and microRNA-124 coloaded stepped cleavable nanosystem against triple negative breast cancer. J Nanobiotechnol 19(1):55. https://doi.org/10.1186/s12951-021-00800-z

Chibon F (2013) Cancer gene expression signatures–the rise and fall? Eur J Cancer 49(8): 2000–2009. https://doi.org/10.1016/j.ejca.2013.02.021

Dai X, Tan C (2015) Combination of microRNA therapeutics with small-molecule anticancer drugs: mechanism of action and co-delivery nanocarriers. Adv Drug Deliv Rev 81:184–197. https://doi.org/10.1016/j.addr.2014.09.010

Deng ZJ, Morton SW, Ben-Akiva E, Dreaden EC, Shopsowitz KE, Hammond PT (2013) Layer-by-layer nanoparticles for systemic codelivery of an anticancer drug and siRNA for potential triple-negative breast cancer treatment. ACS Nano 7(11):9571–9584. https://doi.org/10.1021/nn4047925

Deng X, Cao M, Zhang J, Hu K, Yin Z, Zhou Z, Xiao X, Yang Y, Sheng W, Wu Y (2014) Hyaluronic acid-chitosan nanoparticles for co-delivery of MiR-34a and doxorubicin in therapy against triple negative breast cancer. Biomaterials 35(14):4333–4344. https://doi.org/10.1016/j.biomaterials.2014.02.006

Devulapally R, Sekar TV, Paulmurugan R (2015) Formulation of anti-miR-21 and 4-hydroxytamoxifen co-loaded biodegradable polymer nanoparticles and their antiproliferative effect on breast cancer cells. Mol Pharm 12(6):2080–2092. https://doi.org/10.1021/mp500852s

Ding J, Liang T, Min Q, Jiang L, Zhu J-J (2018) "Stealth and fully-laden" drug carriers: self-assembled nanogels encapsulated with epigallocatechin gallate and siRNA for drug-resistant breast cancer therapy. ACS Appl Mater Interfaces 10(12):9938–9948. https://doi.org/10.1021/acsami.7b19577

Dong D, Gao W, Liu Y, Qi X-R (2015) Therapeutic potential of targeted multifunctional nanocomplex co-delivery of siRNA and low-dose doxorubicin in breast cancer. Cancer Lett 359(2):178–186. https://doi.org/10.1016/j.canlet.2015.01.011

Dönmez Y, Gündüz U (2011) Reversal of multidrug resistance by small interfering RNA (siRNA) in doxorubicin-resistant MCF-7 breast cancer cells. Biomed Pharmacother 65(2):85–89. https://doi.org/10.1016/j.biopha.2010.12.007

Ebrahimian M, Taghavi S, Mokhtarzadeh A, Ramezani M, Hashemi M (2017) Co-delivery of doxorubicin encapsulated PLGA nanoparticles and Bcl-xL shRNA using alkyl-modified PEI into breast cancer cells. Appl Biochem Biotechnol 183(1):126–136. https://doi.org/10.1007/s12010-017-2434-3

Eivazy P, Atyabi F, Jadidi-Niaragh F, Aghebati Maleki L, Miahipour A, Abdolalizadeh J, Yousefi M (2017) The impact of the codelivery of drug-siRNA by trimethyl chitosan nanoparticles on the efficacy of chemotherapy for metastatic breast cancer cell line (MDA-MB-231). Artif Cells Nanomed Biotechnol 45(5):889–896. https://doi.org/10.1080/21691401.2016.1185727

El Founi M, Laroui H, Canup BS, Ametepe JS, Vanderesse R, Acherar S, Babin J, Ferji K, Chevalot I, Six J-L (2021) Doxorubicin intracellular release via external UV irradiation of dextran-g-poly (o-nitrobenzyl acrylate) photosensitive nanoparticles. ACS Appl Bio Mater. https://doi.org/10.1021/acsabm.0c01644

Elbashir SM, Harborth J, Lendeckel W, Yalcin A, Weber K, Tuschl T (2001a) Duplexes of 21-nucleotide RNAs mediate RNA interference in cultured mammalian cells. Nature 411(6836):494–498. https://doi.org/10.1038/35078107

Elbashir SM, Lendeckel W, Tuschl T (2001b) RNA interference is mediated by 21-and 22-nucleotide RNAs. Genes Dev 15(2):188–200. https://doi.org/10.1101/gad.862301

Essa D, Kondiah PP, Choonara YE, Pillay V (2020) The design of poly (lactide-co-glycolide) nanocarriers for medical applications. Front Bioeng Biotechnol 8:48. https://doi.org/10.3389/fbioe.2020.00048

Feng Q, Yu M-Z, Wang J-C, Hou W-J, Gao L-Y, Ma X-F, Pei X-W, Niu Y-J, Liu X-Y, Qiu C (2014) Synergistic inhibition of breast cancer by co-delivery of VEGF siRNA and paclitaxel via vapreotide-modified core–shell nanoparticles. Biomaterials 35(18):5028–5038. https://doi.org/10.1016/j.biomaterials.2014.03.012

Feng L, Yan S, Zhu Q, Chen J, Deng L, Zheng Y, Xue W, Guo R (2018) Targeted multifunctional redox-sensitive micelle co-delivery of DNA and doxorubicin for the treatment of breast cancer. J Mater Chem B 6(20):3372–3386. https://doi.org/10.1039/C8TB00748A

Fu Y, Chen J, Huang Z (2019) Recent progress in microRNA-based delivery systems for the treatment of human disease. ExRNA 1(1):24. https://doi.org/10.1186/s41544-019-0024-y

García-Pinel B, Porras-Alcalá C, Ortega-Rodríguez A, Sarabia F, Prados J, Melguizo C, López-Romero JM (2019) Lipid-based nanoparticles: application and recent advances in cancer treatment. Nanomaterials 9(4):638. https://doi.org/10.3390/nano9040638

Gong C, Hu C, Gu F, Xia Q, Yao C, Zhang L, Qiang L, Gao S, Gao Y (2017) Co-delivery of autophagy inhibitor ATG7 siRNA and docetaxel for breast cancer treatment. J Control Release 266:272–286. https://doi.org/10.1016/j.jconrel.2017.09.042

Gong C, Tian J, Wang Z, Gao Y, Wu X, Ding X, Qiang L, Li G, Han Z, Yuan Y (2019) Functional exosome-mediated co-delivery of doxorubicin and hydrophobically modified microRNA 159 for triple-negative breast cancer therapy. J Nanobiotechnol 17(1):1–18. https://doi.org/10.1186/s12951-019-0526-7

Gong C, Yu X, Zhang W, Han L, Wang R, Wang Y, Gao S, Yuan Y (2021) Regulating the immunosuppressive tumor microenvironment to enhance breast cancer immunotherapy using pH-responsive hybrid membrane-coated nanoparticles. J Nanobiotechnol 19(1):1–20. https://doi.org/10.1186/s12951-021-00805-8

Gonzalez-Conchas GA, Rodriguez-Romo L, Hernandez-Barajas D, Gonzalez-Guerrero JF, Rodriguez-Fernandez IA, Verdines-Perez A, Templeton AJ, Ocana A, Seruga B, Tannock IF (2018) Epidermal growth factor receptor overexpression and outcomes in early breast cancer: a systematic review and a meta-analysis. Cancer Treat Rev 62:1–8. https://doi.org/10.1016/j.ctrv.2017.10.008

Gu Y, Guo Y, Wang C, Xu J, Wu J, Kirk TB, Ma D, Xue W (2017) A polyamidoamine dendrimer functionalized graphene oxide for DOX and MMP-9 shRNA plasmid co-delivery. Mater Sci Eng C 70:572–585. https://doi.org/10.1016/j.msec.2016.09.035

Guo P, Fang Q, Tao H-Q, Schafer CA, Fenton BM, Ding I, Hu B, Cheng S-Y (2003) Overexpression of vascular endothelial growth factor by MCF-7 breast cancer cells promotes estrogen-independent tumor growth in vivo. Cancer Res 63(15):4684–4691

Guo D, Ji X, Peng F, Zhong Y, Chu B, Su Y, He Y (2019) Photostable and biocompatible fluorescent silicon nanoparticles for imaging-guided co-delivery of siRNA and doxorubicin to drug-resistant cancer cells. Nano-Micro Lett 11(1):1–13. https://doi.org/10.1007/s40820-019-0257-1

He W, Turkeshi A, Li X, Zhang H (2020) Progress in systemic co-delivery of microRNAs and chemotherapeutics for cancer treatment by using lipid-based nanoparticles. Ther Deliv 11(9): 591–603. https://doi.org/10.4155/tde-2020-0052

Hong S, Choi DW, Kim HN, Park CG, Lee W, Park HH (2020) Protein-based nanoparticles as drug delivery systems. Pharmaceutics 12(7):604. https://doi.org/10.3390/pharmaceutics12070604

Jafari R, Zolbanin NM, Majidi J, Atyabi F, Yousefi M, Jadidi-Niaragh F, Aghebati-Maleki L, Shanehbandi D, Zangbar M-SS, Rafatpanah H (2019a) Anti-mucin1 aptamer-conjugated chitosan nanoparticles for targeted co-delivery of docetaxel and IGF-1R siRNA to SKBR3 metastatic breast cancer cells. Iran Biomed J 23(1):21. https://doi.org/10.29252/.23.1.21

Jafari S, Derakhshankhah H, Alaei L, Fattahi A, Varnamkhasti BS, Saboury AA (2019b) Mesoporous silica nanoparticles for therapeutic/diagnostic applications. Biomed Pharmacother 109:1100–1111. https://doi.org/10.1016/j.biopha.2018.10.167

Jafari M, Sriram V, Xu Z, Harris GM, Lee J-Y (2020) Fucoidan-doxorubicin nanoparticles targeting P-selectin for effective breast cancer therapy. Carbohydr Polym 249:116837. https://doi.org/10.1016/j.carbpol.2020.116837

Jeong D, Ham J, Kim HW, Kim H, Ji HW, Yun SH, Park JE, Lee KS, Jo H, Han JH (2021) ELOVL2: a novel tumor suppressor attenuating tamoxifen resistance in breast cancer. Am J Cancer Res 11(6):2568–2589

Jia L, Li Z, Zheng D, Li Z, Zhao Z (2021) A targeted and redox/pH-responsive chitosan oligosaccharide derivatives based nanohybrids for overcoming multidrug resistance of breast cancer cells. Carbohydr Polym 251:117008. https://doi.org/10.1016/j.carbpol.2020.117008

Jin M, Hou Y, Quan X, Chen L, Gao Z, Huang W (2021) Smart polymeric nanoparticles with pH-responsive and PEG-detachable properties (II): co-delivery of paclitaxel and VEGF siRNA for synergistic breast cancer therapy in mice. Int J Nanomedicine 16:5479. https://doi.org/10.2147/IJN.S313339

Ju L, Zhang G, Zhang C, Sun L, Jiang Y, Yan C, Duerksen-Hughes PJ, Zhang X, Zhu X, Chen FF (2013) Quantum dot-related genotoxicity perturbation can be attenuated by PEG encapsulation. Mutat Res Genet Toxicol Environ Mutagen 753(1):54–64. https://doi.org/10.1016/j.mrgentox.2013.01.006

Kahraman E, Güngör S, Özsoy Y (2017) Potential enhancement and targeting strategies of polymeric and lipid-based nanocarriers in dermal drug delivery. Ther Deliv 8(11):967–985. https://doi.org/10.4155/tde-2017-0075

Kang L, Gao Z, Huang W, Jin M, Wang Q (2015) Nanocarrier-mediated co-delivery of chemotherapeutic drugs and gene agents for cancer treatment. Acta Pharm Sin B 5(3):169–175. https://doi.org/10.1016/j.apsb.2015.03.001

Khan A, Rashid R, Murtaza G, Zahra A (2014) Gold nanoparticles: synthesis and applications in drug delivery. Trop J Pharm Res 13:1169–1177. https://doi.org/10.4314/tjpr.v13i7.23

Khesht AMS, Karpisheh V, Gilan PS, Melnikova LA, Zekiy AO, Mohammadi M, Hojjat-Farsangi-M, Zolbanin NM, Mahmoodpoor A, Hassannia H (2021) Blockade of CD73 using siRNA loaded chitosan lactate nanoparticles functionalized with TAT-hyaluronate enhances doxorubicin mediated cytotoxicity in cancer cells both in vitro and in vivo. Int J Biol Macromol 186:849–863. https://doi.org/10.1016/j.ijbiomac.2021.07.034

Lee E, Oh C, Kim I-S, Kwon IC, Kim S (2015) Co-delivery of chemosensitizing siRNA and an anticancer agent via multiple monocomplexation-induced hydrophobic association. J Control Release 210:105–114. https://doi.org/10.1016/j.jconrel.2015.05.262

Li M, Li Y, Huang X, Lu X (2015) Captopril-polyethyleneimine conjugate modified gold nanoparticles for co-delivery of drug and gene in anti-angiogenesis breast cancer therapy. J Biomater Sci Polym Ed 26(13):813–827. https://doi.org/10.1080/09205063.2015.1057991

Li J, Cai C, Li J, Li J, Li J, Sun T, Wang L, Wu H, Yu G (2018) Chitosan-based nanomaterials for drug delivery. Molecules 23(10):2661. https://doi.org/10.3390/molecules23102661

Li G, Meng F, Lu T, Wei L, Pan X, Nong Z, Wei M, Liao C, Li X (2021) Functionalised molybdenum disulfide nanosheets for co-delivery of doxorubicin and siRNA for combined chemo/gene/photothermal therapy on multidrug-resistant cancer. J Pharm Pharmacol 73(8): 1128–1135. https://doi.org/10.1093/jpp/rgab059

Linardi RL, Natalini CC (2006) Multi-drug resistance (MDR1) gene and P-glycoprotein influence on pharmacokinetic and pharmacodynamic of therapeutic drugs. Ciência Rural 36(1):336–341. https://doi.org/10.1590/S0103-84782006000100056

Liu Z, Sun Q, Wang X (2017) PLK1, a potential target for cancer therapy. Transl Oncol 10(1): 22–32. https://doi.org/10.1016/j.tranon.2016.10.003

Liu Y, Sun J, Huang Y, Chen Y, Li J, Liang L, Xu J, Wan Z, Zhang B, Li Z (2021) Metformin-conjugated micellar system with intratumoral pH responsive de-shielding for co-delivery of doxorubicin and nucleic acid. Biochem Pharmacol 189:114453. https://doi.org/10.1016/j.bcp.2021.114453

Loh H-Y, Norman BP, Lai K-S, Rahman NMANA, Alitheen NBM, Osman MA (2019) The regulatory role of microRNAs in breast cancer. Int J Mol Sci 20(19):4940. https://doi.org/10.3390/ijms20194940

Lü J-M, Wang X, Marin-Muller C, Wang H, Lin PH, Yao Q, Chen C (2009) Current advances in research and clinical applications of PLGA-based nanotechnology. Expert Rev Mol Diagn 9(4): 325–341. https://doi.org/10.1586/erm.09.15

Luo K, Gao Y, Yin S, Yao Y, Yu H, Wang G, Li J (2021) Co-delivery of paclitaxel and STAT3 siRNA by a multifunctional nanocomplex for targeted treatment of metastatic breast cancer. Acta Biomater 134:649–663. https://doi.org/10.1016/j.actbio.2021.07.029

Mahapatro A, Singh DK (2011) Biodegradable nanoparticles are excellent vehicle for site directed in-vivo delivery of drugs and vaccines. J Nanobiotechnol 9(1):1–11. https://doi.org/10.1186/1477-3155-9-55

Makarova JA, Shkurnikov MU, Wicklein D, Lange T, Samatov TR, Turchinovich AA, Tonevitsky AG (2016) Intracellular and extracellular microRNA: an update on localization and biological role. Prog Histochem Cytochem 51(3–4):33–49. https://doi.org/10.1016/j.proghi.2016.06.001

Meng H, Mai WX, Zhang H, Xue M, Xia T, Lin S, Wang X, Zhao Y, Ji Z, Zink JI (2013) Codelivery of an optimal drug/siRNA combination using mesoporous silica nanoparticles to overcome drug resistance in breast cancer in vitro and in vivo. ACS Nano 7(2):994–1005. https://doi.org/10.1021/nn3044066

Misra R, Das M, Sahoo BS, Sahoo SK (2014) Reversal of multidrug resistance in vitro by co-delivery of MDR1 targeting siRNA and doxorubicin using a novel cationic poly (lactide-co-glycolide) nanoformulation. Int J Pharm 475(1–2):372–384. https://doi.org/10.1016/j.ijpharm.2014.08.056

Morry J, Ngamcherdtrakul W, Gu S, Reda M, Castro DJ, Sangvanich T, Gray JW, Yantasee W (2017) Targeted treatment of metastatic breast cancer by PLK1 siRNA delivered by an antioxidant nanoparticle platform. Mol Cancer Ther 16(4):763–772. https://doi.org/10.1158/1535-7163.MCT-16-0644

Nagini S (2017) Breast cancer: current molecular therapeutic targets and new players. Anti Cancer Agents Med Chem 17(2):152–163

Nasab SH, Amani A, Ebrahimi HA, Hamidi AA (2021) Design and preparation of a new multi-targeted drug delivery system using multifunctional nanoparticles for co-delivery of siRNA and paclitaxel. J Pharm Anal 11(2):163–173. https://doi.org/10.1016/j.jpha.2020.04.005

Norouzi M, Yathindranath V, Thliveris JA, Kopec BM, Siahaan TJ, Miller DW (2020) Doxorubicin-loaded iron oxide nanoparticles for glioblastoma therapy: a combinational approach for enhanced delivery of nanoparticles. Sci Rep 10:11292. https://doi.org/10.1038/s41598-020-68017-y

Norouzi P, Motasadizadeh H, Atyabi F, Dinarvand R, Gholami M, Farokhi M, Shokrgozar MA, Mottaghitalab F (2021) Combination therapy of breast cancer by Codelivery of doxorubicin and Survivin siRNA using Polyethylenimine modified silk fibroin nanoparticles. ACS Biomater Sci Eng 7(3):1074–1087. https://doi.org/10.1021/acsbiomaterials.0c01511

Pan Q-S, Chen T-T, Nie C-P, Yi J-T, Liu C, Hu Y-L, Chu X (2018) In situ synthesis of ultrathin ZIF-8 film-coated MSNs for codelivering Bcl 2 siRNA and doxorubicin to enhance chemotherapeutic efficacy in drug-resistant cancer cells. ACS Appl Mater Interfaces 10(39):33070–33077. https://doi.org/10.1021/acsami.8b13393

Pan J, Mendes LP, Yao M, Filipczak N, Garai S, Thakur GA, Sarisozen C, Torchilin VP (2019) Polyamidoamine dendrimers-based nanomedicine for combination therapy with siRNA and chemotherapeutics to overcome multidrug resistance. Eur J Pharm Biopharm 136:18–28. https://doi.org/10.1016/j.ejpb.2019.01.006

Parveen S, Sahoo SK (2008) Polymeric nanoparticles for cancer therapy. J Drug Target 16(2):108–123. https://doi.org/10.1080/10611860701794353

Patra JK, Das G, Fraceto LF, Campos EVR, del Pilar R-TM, Acosta-Torres LS, Diaz-Torres LA, Grillo R, Swamy MK, Sharma S (2018) Nano based drug delivery systems: recent developments and future prospects. J Nanobiotechnol 16(1):1–33. https://doi.org/10.1186/s12951-018-0392-8

Paul S, Bravo Vázquez LA, Pérez Uribe S, Roxana Reyes-Pérez P, Sharma A (2020) Current status of microRNA-based therapeutic approaches in neurodegenerative disorders. Cell 9(7):1698. https://doi.org/10.3390/cells9071698

Pednekar PP, Godiyal SC, Jadhav KR, Kadam VJ (2017) Mesoporous silica nanoparticles: a promising multifunctional drug delivery system. In: Grumezescu AM, Ficai A (eds) Nanostructures for cancer therapy. Elsevier, pp 593–621. https://doi.org/10.1016/B978-0-323-46144-3.00023-4

Polyak K (2007) Breast cancer: origins and evolution. J Clin Invest 117(11):3155–3163. https://doi.org/10.1172/JCI33295

Qian J, Xu M, Suo A, Xu W, Liu T, Liu X, Yao Y, Wang H (2015) Folate-decorated hydrophilic three-arm star-block terpolymer as a novel nanovehicle for targeted co-delivery of doxorubicin and Bcl-2 siRNA in breast cancer therapy. Acta Biomater 15:102–116. https://doi.org/10.1016/j.actbio.2014.12.018

Rajani C, Borisa P, Karanwad T, Borade Y, Patel V, Rajpoot K, Tekade RK (2020) Cancer-targeted chemotherapy: emerging role of the folate anchored dendrimer as drug delivery nanocarrier. In: Chauhan AS, Kulhari H (eds) Pharmaceutical applications of dendrimers. Elsevier, pp 151–198. https://doi.org/10.1016/B978-0-12-814527-2.00007-X

Redig AJ, McAllister SS (2013) Breast cancer as a systemic disease: a view of metastasis. J Intern Med 274(2):113–126. https://doi.org/10.1111/joim.12084

Rivenbark AG, O'Connor SM, Coleman WB (2013) Molecular and cellular heterogeneity in breast cancer: challenges for personalized medicine. Am J Pathol 183(4):1113–1124. https://doi.org/10.1016/j.ajpath.2013.08.002

Samson AAS, Park S, Kim S-Y, Min D-H, Jeon NL, Song JM (2019) Liposomal co-delivery-based quantitative evaluation of chemosensitivity enhancement in breast cancer stem cells by knock-down of GRP78/CLU. J Liposome Res 29(1):44–52. https://doi.org/10.1080/08982104.2017.1420081

Seidel ZP, Zhang X, MacMullan MA, Graham NA, Wang P, Lee CT Jr (2020) Photo-triggered delivery of siRNA and paclitaxel into breast cancer cells using Catanionic vesicles. ACS Appl Bio Mater 3(11):7388–7398. https://doi.org/10.1021/acsabm.0c00503

Sgroi DC (2010) Preinvasive breast cancer. Annu Rev Pathol 5:193–221. https://doi.org/10.1146/annurev.pathol.4

Sharma GN, Dave R, Sanadya J, Sharma P, Sharma K (2010) Various types and management of breast cancer: an overview. J Adv Pharm Technol Res 1(2):109

Sharma S, Pukale S, Sahel DK, Singh P, Mittal A, Chitkara D (2021) Folate targeted hybrid Lipo-polymeric Nanoplexes containing Docetaxel and miRNA-34a for breast cancer treatment. Mater Sci Eng C 128:112305. https://doi.org/10.1016/jmsec.2021.112305

Shen J, Sun H, Xu P, Yin Q, Zhang Z, Wang S, Yu H, Li Y (2013) Simultaneous inhibition of metastasis and growth of breast cancer by co-delivery of twist shRNA and paclitaxel using pluronic P85-PEI/TPGS complex nanoparticles. Biomaterials 34(5):1581–1590. https://doi.org/10.1016/j.biomaterials.2012.10.057

Shrestha B, Wang L, Zhang H, Hung CY, Tang L (2020) Gold nanoparticles mediated drug-gene combinational therapy for breast cancer treatment. Int J Nanomedicine 15:8109. https://doi.org/10.2147/IJN.S258625

Sun Y-S, Zhao Z, Yang Z-N, Xu F, Lu H-J, Zhu Z-Y, Shi W, Jiang J, Yao P-P, Zhu H-P (2017) Risk factors and preventions of breast cancer. Int J Biol Sci 13(11):1387. https://doi.org/10.7150/ijbs.21635

Sun X, Xu H, Huang T, Zhang C, Wu J, Luo S (2021) Simultaneous delivery of anti-miRNA and docetaxel with supramolecular self-assembled "chitosome" for improving chemosensitivity of triple negative breast cancer cells. Drug Deliv Transl Res 11(1):192–204. https://doi.org/10.1007/s13346-020-00779-4

Suo A, Qian J, Zhang Y, Liu R, Xu W, Wang H (2016) Comb-like amphiphilic polypeptide-based copolymer nanomicelles for co-delivery of doxorubicin and P-gp siRNA into MCF-7 cells. Mater Sci Eng C 62:564–573. https://doi.org/10.1016/j.msec.2016.02.007

Suo A, Qian J, Xu M, Xu W, Zhang Y, Yao Y (2017) Folate-decorated PEGylated triblock copolymer as a pH/reduction dual-responsive nanovehicle for targeted intracellular co-delivery of doxorubicin and Bcl-2 siRNA. Mater Sci Eng C 76:659–672. https://doi.org/10.1016/j.msec.2017.03.124

Swami R, Kumar Y, Chaudhari D, Katiyar SS, Kuche K, Katare PB, Banerjee SK, Jain S (2020) pH sensitive liposomes assisted specific and improved breast cancer therapy using co-delivery of SIRT1 shRNA and Docetaxel. Mater Sci Eng C 120:111664. https://doi.org/10.1016/j.msec.2020.111664

Sztandera K, Gorzkiewicz M, Klajnert-Maculewicz B (2018) Gold nanoparticles in cancer treatment. Mol Pharm 16(1):1–23. https://doi.org/10.1021/acs.molpharmaceut.8b00810

Tang S, Yin Q, Zhang Z, Gu W, Chen L, Yu H, Huang Y, Chen X, Xu M, Li Y (2014) Co-delivery of doxorubicin and RNA using pH-sensitive poly (β-amino ester) nanoparticles for reversal of multidrug resistance of breast cancer. Biomaterials 35(23):6047–6059. https://doi.org/10.1016/j.biomaterials.2014.04.025

Tang S, Yin Q, Su J, Sun H, Meng Q, Chen Y, Chen L, Huang Y, Gu W, Xu M (2015) Inhibition of metastasis and growth of breast cancer by pH-sensitive poly (β-amino ester) nanoparticles co-delivering two siRNA and paclitaxel. Biomaterials 48:1–15. https://doi.org/10.1016/j.biomaterials.2015.01.049

Tashima T (2018) Effective cancer therapy based on selective drug delivery into cells across their membrane using receptor-mediated endocytosis. Bioorg Med Chem Lett 28(18):3015–3024. https://doi.org/10.1016/j.bmcl.2018.07.012

Taxman DJ, Moore CB, Guthrie EH, Huang MT-H (2010) Short hairpin RNA (shRNA): design, delivery, and assessment of gene knockdown. In: Sioud M (ed) RNA therapeutics: methods in molecular biology. Humana Press, pp 139–156. https://doi.org/10.1007/978-1-60761-657-3_10

Waks AG, Winer EP (2019) Breast cancer treatment: a review. JAMA 321(3):288–300. https://doi.org/10.1001/jama.2018.19323

Wang D, Wang T, Xu Z, Yu H, Feng B, Zhang J, Guo C, Yin Q, Zhang Z, Li Y (2016a) Cooperative treatment of metastatic breast cancer using host–guest nanoplatform coloaded with docetaxel and siRNA. Small 12(4):488–498. https://doi.org/10.1002/smll.201502913

Wang S, Zhang J, Wang Y, Chen M (2016b) Hyaluronic acid-coated PEI-PLGA nanoparticles mediated co-delivery of doxorubicin and miR-542-3p for triple negative breast cancer therapy. Nanomed Nanotechnol Biol Med 12(2):411–420. https://doi.org/10.1016/j.nano.2015.09.014

Wang S, Liu X, Chen S, Liu Z, Zhang X, Liang X-J, Li L (2018a) Regulation of Ca2+ signaling for drug-resistant breast cancer therapy with mesoporous silica nanocapsule encapsulated doxorubicin/siRNA cocktail. ACS Nano 13(1):274–283. https://doi.org/10.1021/acsnano.8b05639

Wang W, Chen K, Su Y, Zhang J, Li M, Zhou J (2018b) Lysosome-independent intracellular drug/gene codelivery by lipoprotein-derived nanovector for synergistic apoptosis-inducing cancer-targeted therapy. Biomacromolecules 19(2):438–448. https://doi.org/10.1021/acs.biomac.7b01549

Weigelt B, Geyer FC, Reis-Filho JS (2010) Histological types of breast cancer: how special are they? Mol Oncol 4(3):192–208. https://doi.org/10.1016/j.molonc.2010.04.004

Xie X, Chen Y, Chen Z, Feng Y, Wang J, Li T, Li S, Qin X, Wu C, Zheng C (2019) Polymeric hybrid nanomicelles for cancer theranostics: an efficient and precise anticancer strategy for the codelivery of doxorubicin/miR-34a and magnetic resonance imaging. ACS Appl Mater Interfaces 11(47):43865–43878. https://doi.org/10.1021/acsami.9b14908

Xu J, Chen Y, Olopade OI (2010) MYC and breast cancer. Genes Cancer 1(6):629–640. https://doi.org/10.1177/1947601910378691

Xu M, Qian J, Suo A, Cui N, Yao Y, Xu W, Liu T, Wang H (2015) Co-delivery of doxorubicin and P-glycoprotein siRNA by multifunctional triblock copolymers for enhanced anticancer efficacy in breast cancer cells. J Mater Chem B 3(10):2215–2228. https://doi.org/10.1039/C5TB00031A

Xu J, Sun J, Ho PY, Luo Z, Ma W, Zhao W, Rathod SB, Fernandez CA, Venkataramanan R, Xie W (2019a) Creatine based polymer for codelivery of bioengineered MicroRNA and chemodrugs against breast cancer lung metastasis. Biomaterials 210:25–40. https://doi.org/10.1016/j.biomaterials.2019.04.025

Xu X, Liu A, Bai Y, Li Y, Zhang C, Cui S, Piao Y, Zhang S (2019b) Co-delivery of resveratrol and p53 gene via peptide cationic liposomal nanocarrier for the synergistic treatment of cervical cancer and breast cancer cells. J Drug Deliv Sci Technol 51:746–753. https://doi.org/10.1016/j.jddst.2018.05.008

Yadav D, Sandeep K, Pandey D, Dutta RK (2017) Liposomes for drug delivery. J Biotechnol Biomater 7(4):276. https://doi.org/10.4172/2155-952X.1000276

Yaghoobi E, Shojaee S, Ramezani M, Alibolandi M, Charbgoo F, Nameghi MA, Khatami F, Ashjaei MS, Abnous K, Taghdisi SM (2021) A novel targeted co-delivery system for transfer of epirubicin and antimiR-10b into cancer cells through a linear DNA nanostructure consisting of FOXM1 and AS1411 aptamers. J Drug Deliv Sci Technol 63:102521. https://doi.org/10.1016/j.jddst.2021.102521

Yang H, Deng L, Li T, Shen X, Yan J, Zuo L, Wu C, Liu Y (2015) Multifunctional PLGA nanobubbles as theranostic agents: combining doxorubicin and P-gp siRNA co-delivery into human breast cancer cells and ultrasound cellular imaging. J Biomed Nanotechnol 11(12): 2124–2136. https://doi.org/10.1166/jbn.2015.2168

Yang L, Tian Y, Leong WS, Song H, Yang W, Wang M, Wang X, Kong J, Shan B, Song Z (2018) Efficient and tumor-specific knockdown of MTDH gene attenuates paclitaxel resistance of breast cancer cells both in vivo and in vitro. Breast Cancer Res 20(1):1–13

Yin T, Wang L, Yin L, Zhou J, Huo M (2015) Co-delivery of hydrophobic paclitaxel and hydrophilic AURKA specific siRNA by redox-sensitive micelles for effective treatment of breast cancer. Biomaterials 61:10–25. https://doi.org/10.1016/j.biomaterials.2015.05.022

Yin PT, Pongkulapa T, Cho H-Y, Han J, Pasquale NJ, Rabie H, Kim J-H, Choi J-W, Lee K-B (2018) Overcoming chemoresistance in cancer via combined microRNA therapeutics with anticancer drugs using multifunctional magnetic core–shell nanoparticles. ACS Appl Mater Interfaces 10(32):26954–26963. https://doi.org/10.1021/acsami.8b09086

Yu D, Lu J (2011) Breast cancer multistep development. In: Schwab M (ed) Encyclopedia of cancer. Springer, Berlin Heidelberg, pp 522–526

Yu D, Li W, Zhang Y, Zhang B (2016a) Anti-tumor efficiency of paclitaxel and DNA when co-delivered by pH responsive ligand modified nanocarriers for breast cancer treatment. Biomed Pharmacother 83:1428–1435. https://doi.org/10.1016/j.biopha.2016.08.061

Yu H, Guo C, Feng B, Liu J, Chen X, Wang D, Teng L, Li Y, Yin Q, Zhang Z (2016b) Triple-layered pH-responsive micelleplexes loaded with siRNA and cisplatin prodrug for NF-Kappa B targeted treatment of metastatic breast cancer. Theranostics 6(1):14. https://doi.org/10.7150/thno.13515

Yu S, Bi X, Yang L, Wu S, Yu Y, Jiang B, Zhang A, Lan K, Duan S (2019) Co-delivery of paclitaxel and PLK1-targeted siRNA using aptamer-functionalized cationic liposome for synergistic anti-breast cancer effects in vivo. J Biomed Nanotechnol 15(6):1135–1148. https://doi.org/10.1166/jbn.2019.2751

Zhang D, Xu H, Hu M, Deng Y (2015) "PEG dilemma" for liposomes and its solving approaches. Acta Pharm Sin 50(3):252–260

Zhang N, Huang Y, Wu F, Zhao Y, Li X, Shen P, Yang L, Luo Y, Yang L, He G (2016) Codelivery of a miR-124 mimic and obatoclax by cholesterol-penetratin micelles simultaneously induces apoptosis and inhibits autophagic flux in breast cancer in vitro and in vivo. Mol Pharm 13(7): 2466–2483. https://doi.org/10.1021/acs.molpharmaceut.6b00211

Zhang L, Yang X, Lv Y, Xin X, Qin C, Han X, Yang L, He W, Yin L (2017a) Cytosolic co-delivery of miRNA-34a and docetaxel with core-shell nanocarriers via caveolae-mediated pathway for the treatment of metastatic breast cancer. Sci Rep 7(1):46186. https://doi.org/10.1038/srep46186

Zhang L, Yang X, Li Y, Zheng W, Jiang X (2017b) Hollow carbon nanospheres as a versatile platform for co-delivery of siRNA and chemotherapeutics. Carbon 121:79–89. https://doi.org/10.1016/j.carbon.2017.05.084

Zhang Q, Wang L, Jiang Y, Gao W, Wang Y, Yang X, Yang X, Liu Z (2017c) Gold nanorods with silica shell and PAMAM dendrimers for efficient photothermal therapy and low toxic codelivery of anticancer drug and siRNA. Adv Mater Interfaces 4(24):1701166. https://doi.org/10.1002/admi.201701166

Zhao L, Zhao W, Liu Y, Chen X, Wang Y (2017) Nano-hydroxyapatite-derived drug and gene co-delivery system for anti-angiogenesis therapy of breast cancer. Med Sci Monit 23:4723–4732. https://doi.org/10.12659/MSM.902538

Zhao Z, Li Y, Liu H, Jain A, Patel PV, Cheng K (2020) Co-delivery of IKBKE siRNA and cabazitaxel by hybrid nanocomplex inhibits invasiveness and growth of triple-negative breast cancer. Sci Adv 6(29):eabb0616. https://doi.org/10.1126/sciadv.abb0616

Zhou X, Chen L, Nie W, Wang W, Qin M, Mo X, Wang H, He C (2016) Dual-responsive mesoporous silica nanoparticles mediated codelivery of doxorubicin and Bcl-2 SiRNA for targeted treatment of breast cancer. J Phys Chem C 120(39):22375–22387. https://doi.org/10.1021/acs.jpcc.6b06759

Zhou Z, Kennell C, Lee J-Y, Leung Y-K, Tarapore P (2017) Calcium phosphate-polymer hybrid nanoparticles for enhanced triple negative breast cancer treatment via co-delivery of paclitaxel and miR-221/222 inhibitors. Nanomed Nanotechnol Biol Med 13(2):403–410. https://doi.org/10.1016/j.nano.2016.07.016

Zhu W-j, C-xQ S-d Y, Zhu Q-I, Chen W-I, Li F, Yuan Z-q, Liu Y, You B-g, Zhang X-n (2017) Low-density lipoprotein-coupled micelles with reduction and pH dual sensitivity for intelligent co-delivery of paclitaxel and siRNA to breast tumor. Int J Nanomedicine 12:3375. https://doi.org/10.2147/IJN.S126310

Zhuang J, Chen S, Hu Y, Yang F, Huo Q, Xie N (2021) Tumour-targeted and redox-responsive mesoporous silica nanoparticles for controlled release of doxorubicin and an siRNA against metastatic breast cancer. Int J Nanomedicine 16:1961. https://doi.org/10.2147/IJN.S278724

Zolbanin NM, Jafari R, Majidi J, Atyabi F, Yousefi M, Jadidi-Niaragh F, Aghebati-Maleki L, Shanehbandi D, Zangbar M-SS, Nayebi AM (2018) Targeted co-delivery of docetaxel and cMET siRNA for treatment of mucin1 overexpressing breast cancer cells. Adv Pharm Bull 8(3): 383–393. https://doi.org/10.15171/apb.2018.045

Oncolytic Virotherapy Against Breast Cancer

Mohammad Ejaz, Maria Rukan, Laiba Hassan, Hafiz Muhammad Waqas Munir, Muhammad Ali Syed, and Rani Faryal

Abstract

Breast cancer ranks as the second most frequently detected form of cancer on a global scale. A diverse array of conventional therapeutic regimens are available for the management of breast cancer including surgical intervention, radiotherapy, chemotherapy, hormone manipulation, and immunomodulation but over 30% of patients may progress to metastasis even after such treatments. Oncolytic viruses (OVs) are increasingly recognized as significant entities in the field of cancer therapeutics. Oncolytic viruses possess the appealing therapeutic amalgamation of selectively lysing tumor cells while concurrently stimulating the immune system, thus exhibiting potential as in-situ tumor vaccines. Additionally, it is possible to manipulate OVs in order to maximize their ability to selectively target tumors and stimulate the immune system. These engineered OVs can be easily integrated with other therapeutic agents for synergistic effects. The efficacy of oncolytic viruses has been substantiated through numerous preclinical investigations and more recently in human trials. Notably, the US Food and Drug Administration has granted approval for the utilization of talimogene laherparepvec, an oncolytic herpesvirus used to treat advanced melanoma. The chapter provides a comprehensive examination of oncolytic viruses as a potential treatment for breast cancer. It focuses on the mechanism of OVs in imparting

M. Ejaz
Department of Microbiology, Government Postgraduate College Mandian, Abbottabad, Pakistan

M. Rukan · L. Hassan · M. A. Syed (✉)
Department of Microbiology, The University of Haripur, Haripur, Pakistan

H. M. W. Munir
Department of Oncology, University of Turin, Turin, Italy

R. Faryal
Department of Microbiology, Quaid-i-Azam University, Islamabad, Pakistan

© The Author(s), under exclusive license to Springer Nature Switzerland AG 2023
Interdisciplinary Cancer Research, https://doi.org/10.1007/16833_2023_190
Published online: 14 November 2023

immunity against breast cancer and ongoing clinical studies that are being conducted to assess the efficacy and safety of these viruses in a clinical setting.

Keywords

Biomarkers · Breast cancer · Chemotherapy · Immune checkpoint inhibitor · Oncolytic virotherapy · Oncolytic viruses · Radiotherapy · Tumor microenvironment

1 Introduction

Breast cancer is one of the leading causes of malignancy in the female population, exerting a substantial impact on healthcare systems. Since the last three decades, there has been a noticeable upward trajectory in both the quantity of newly reported instances and the age-standardized rates for breast cancer. These trends serve as a cause for concern, suggesting a worrisome pattern. Breast cancer exhibits a high level of complexity and possesses multiple facets, thereby resulting in profound physiological, emotional, and socioeconomic consequences for affected individuals, their families, and the broader community (Higginson and Costantini 2008). Despite the remarkable progress made in the field of research, diagnosis, and treatment, breast cancer remains an ongoing and formidable adversary for both healthcare systems and individuals. Based on the most recent Global Cancer Statistics (GLOBOCAN) 2020 estimates, breast cancer in females has surpassed lung cancer as the cancer type with the highest incidence worldwide (Sung et al. 2021). Approximately 2.3 million new cases of breast cancer were diagnosed in 2020, representing 11.7% of all newly reported cancer cases (Lei et al. 2021). Breast cancer is a malignancy that exhibits metastatic behavior, frequently disseminating to remote anatomical sites including the bone, liver, lung, and brain. This characteristic metastatic propensity largely contributes to its challenging therapeutic management (Sun et al. 2017). Breast cancer arises due to the interplay of various extrinsic factors, such as tobacco usage, alcohol consumption, smoking, an unhealthy diet, an unfavorable lifestyle, and exposure to environmental elements like UV or ionizing and non-ionizing radiations. Additionally, intrinsic factors, including aging, genetic mutations, hormonal imbalances (estrogen, insulin, etc.), and a compromised immune system, contribute to the initiation or suppression of specific genes, ultimately resulting in abnormal cellular proliferation (Sun et al. 2017; Wu et al. 2018; Ejaz et al. 2020).

A multitude of conventional therapeutic regimens are available for the management of breast cancer, including surgical intervention, radiotherapy, chemotherapy, hormone manipulation, and immunomodulation. Despite the efficacy of currently accessible treatment modalities during the initial phases, approximately 30% of patients experience progression to metastatic breast cancer (Redig and McAllister 2013). The majority of patients upon initial diagnosis typically receive surgical intervention to excise the primary breast tumor; nevertheless, it is important to note that leftover cancer cells may persist following the procedure. The recurrence of tumors from these foci is likely indicative of the failure of following interventions such as radiation or targeted therapy to effectively eradicate leftover cells in the mammary or lymphatic tissue subsequently leading to the development of distant metastasis. The primary pharmacological interventions employed in the management of breast cancer, specifically those aimed at impeding hormone receptors or growth receptor kinases, which facilitate the proliferation of malignant cells, may lose their efficacy upon recurrence, thereby promoting the dissemination of cancerous cells as a consequence of the acquisition of drug resistance traits. Once patients reach the metastatic breast cancer (MBC) stage, characterized by the spread of cancer cells beyond the breast tissue, the administration of cytotoxic drugs, commonly known as chemotherapy, continues to be a viable and potentially optimal therapeutic approach. The chemotherapeutic regimens employed in the management of MBC exhibit remarkable potency and continue to demonstrate efficacy in eradicating malignant cells (Schneeweiss et al. 2015). Unfortunately, they also manifest dose-limiting toxicities due to the off-target exposure of pharmaceutical agents within the cellular structures of healthy tissues. Hence, individuals frequently fail to adhere to the recommended regimen of chemotherapeutic treatments as a result of the insufferable and profound adverse reactions. Furthermore, the administration of chemotherapy involves the utilization of multiple drugs in combination, aiming to optimize the pharmacologic effects. Certain drugs, such as the anthracycline derivatives doxorubicin and daunorubicin, demonstrate a cumulative toxicity phenomenon, which can hinder patients from successfully administering the necessary dosage to eradicate any remaining cancer cells (Yu et al. 2022).

Over the course of recent decades, oncolytic virotherapy has emerged as a novel biological approach in the battle against cancer. Extensive research is currently underway to explore its potential in treating a wide range of cancers, including breast cancer. Oncolytic viruses (OVs) are viruses that occur naturally or have been intentionally altered to specifically target and infect tumor cells while causing minimal harm to healthy cells (Hemminki et al. 2020). The infective capability of OVs toward tumor cells can be attributed to the expression of specific cytokines by malignant cells and the disruption of signal pathways that impair cellular antiviral defense mechanisms. Consequently, tumor cells become highly vulnerable to OVs. OVs were initially observed in individuals diagnosed with leukemia, wherein their clinical manifestations exhibited amelioration after influenza infection (Russell and Peng 2007). Subsequently, Italian doctors discovered that the administration of the rabies vaccine exhibited the capacity to impede the progression of cervical cancer. This observed phenomenon prompted scientists to hypothesize that specific viruses possess a function that confers resistance to tumor development (Jin et al. 2021). Presently, a limited number of oncolytic viruses (OVs) have received approval for clinical implementation. In 2005, the China Food and Drug Administration approved the utilization of Oncorine (H101), a genetically engineered oncolytic adenovirus, as a therapeutic intervention for head and neck cancer (Kelly and Russell 2007). Talimogenelaherparepvec (T-VEC), a type-1 herpes simplex virus (HSV-1) with oncolytic properties, received approval from the United States Food and Drug

Administration in 2015 for its application in treating melanoma lesions found in the skin and lymph nodes (Printz 2016).

The chapter offers a thorough analysis of oncolytic viruses in their capacity as a prospective therapeutic modality for breast cancer. The primary emphasis lies in elucidating the mechanisms by which oncolytic viruses confer immunity against breast cancer. Additionally, this discussion encompasses the exploration of combination therapy approaches and the ongoing clinical investigations aimed at evaluating the effectiveness and safety of these viral agents within a clinical context.

2 Biology of Breast Cancer

Breast cancer stands as the prevailing neoplasm in females across the globe. Breast cancer patients comprise up to 36% of the total population of individuals diagnosed with oncological conditions (Smolarz et al. 2022). The prevalence of malignant breast cancer is on the rise across all global areas, with the highest rates observed in industrialized nations. The developed countries account for over 50% of the total number of cases on a global scale (Bellanger et al. 2018).

The etiology of breast cancer is influenced by a complex interplay of genetic, environmental, nutritional, hormonal, and heritable factors, which collectively contribute to its pathogenesis. Risk factors encompass a spectrum of variables that contribute to the susceptibility of an individual to breast cancer. These factors include history of breast cancer, a positive familial lineage with a history of the disease, an elevated body mass index, an above-average height, engagement in tobacco use, consumption of alcoholic beverages, an early onset of menstruation, delayed onset of menopause, sedentary way of life, the absence of childbirth, and the utilization of hormone replacement therapy (Admoun and Mayrovitz 2022). The most widely recognized manifestations of hereditary breast cancer arise from mutations occurring in the high-penetrance genes BRCA1 and BRCA2. Besides them, TP53, CDH1, PTEN, STK11/LKB1, CHEK2, PALB2, ATM, MRE11, BRIP1, NBS1, FANCA, FANCC, FANCM, XRCC2, RAD50, RAD51, RAD51B, RAD51C, and RAD51D are among the genes that have been identified as having a significant impact on breast cancer susceptibility (Kleibl and Kristensen 2016).

The progression of breast cancer is impacted by signaling pathways such as PI3K/AKT and RAS/MEK/ERK, which typically protect normal cells against engaging in self-destructive mechanisms (Gonzalez-Angulo and Blumenschein Jr 2013, Paplomata and O'Regan 2014). Nevertheless, in the event of genetic mutations occurring in the genes accountable for these safeguarding mechanisms, the cellular entities experience a decline in their capacity to undergo programmed cell death when it is no longer required. Consequently, this leads to the initiation of the malignant condition known as cancer. Furthermore, an overabundance of leptin expression in the adipose tissue of the breast has been linked to elevated cellular proliferation and an augmented susceptibility to the development of cancer (Jardé et al. 2011). Moreover, exposure to estrogens has been linked to an elevated susceptibility to breast cancer and other forms of malignancy in humans. Estrogens undergo metabolic conversion, resulting in the formation of metabolites, specifically the catechol estrogen-3,4-quinones (CE-3,4-Q). These metabolites can interact with DNA, leading to the formation of depurinating adducts. These adducts are liberated from DNA to produce apurinic sites. The occurrence of errors in the base excision repair process for this particular damage has the potential to result in mutations that can trigger the development of breast, prostate, and various other forms of cancer (Cavalieri et al. 2006).

3 Oncolytic Virotherapy

The concept of oncolytic virotherapy is not a recent notion. It gained prominence in the early twentieth century due to sporadic findings indicating that specific viral infections were associated with the spontaneous regression of cancer in afflicted people. After these observations, a multitude of experiments were conducted. The potential utilization of naturally occurring viruses dates back to the early 1900s in a historical context. During the early and mid-1900s, it was observed that patients experienced temporary tumor remission after naturally acquired viral infections. An instance can be cited wherein a female individual, aged 42, who was diagnosed with acute leukemia, experienced a transient period of remission subsequent to an assumed influenza infection in the year 1896 (Larson et al. 2015). During the 1950s and 1960s, there was a notable advancement in the evaluation of the in vivo antitumor efficacy of oncolytic viruses in human subjects. In the year 1950, a cohort of 30 individuals diagnosed with epidermoid cervical carcinomas underwent treatment involving the administration of 10 distinct serotypes of adenovirus. In the study, it was observed that 65% of the patients had necrosis and cavity formation in the central region of the cancerous tissue (Liu et al. 2007). Consequently, the use of biotechnological advancements in genetically modified viruses has expedited the progress of virotherapy. The initial discovery of a thymidine kinase-deficient variant of herpes simplex virus-1 (dlsptk) with reduced neurovirulence and enhanced survival in glioma in mice was reported by Martuza et al. (1991). The initial approval of the oncolytic virus for cancer treatment, Rigvir, occurred in Latvia in the year 2004. The Rigvir treatment involves the utilization of an unaltered ECHO-7 virus, which has been specifically chosen for its efficacy in targeting melanoma (Alberts et al. 2018). In 2005, the oncolytic adenovirus H101, which possesses a deletion in the E1B-55KD gene and partial deletion in the E3 gene, obtained the distinction of being the first approved oncolytic virus in China for the treatment of head and neck cancer (Yu and Fang 2007). In October 2015, the US Food and Drug Administration (FDA) approved T-VEC for the treatment of melanoma. T-VEC is a therapeutic agent derived from the herpes simplex type-1 virus (HSV-1) that has been genetically modified to contain a human granulocyte-macrophage colony-stimulating factor (GM-CSF) gene (Perez et al. 2018). The approval of T-VEC has garnered significant interest in the field of oncolytic virotherapy. In 2021, a genetically modified form of the herpes simplex virus (HSV), known as Delytact, received approval in Japan for the treatment of malignant glioma (Otani et al. 2022). A diverse array of viruses are currently being utilized as oncolytic viruses including adenovirus, measles virus, Newcastle disease virus (NDV), vesicular stomatitis virus (VSV), reovirus, herpes simplex virus, and coxsackievirus (Tian et al. 2022).

3.1 Selectivity of Oncolytic Viruses in Targeting Cancer Cells

The progression of a neoplasm is frequently perceived as an intricate and dynamic phenomenon characterized by genetic and epigenetic modifications. These alterations encompass an unlimited potential for cellular proliferation, evasion of apoptosis, enhanced angiogenesis, infiltration of neighboring tissues, metastasis, and modifications to intercellular signaling pathways. Such modifications have the potential to make cancer cells susceptible to infection, as certain pathways utilized by the tumor are also essential for efficient antiviral defenses. When a virus successfully invades a tumor cell, the tumor microenvironment (TME) offers favorable conditions for the virus to undergo replication. Consequently, these alterations establish the requisite circumstances for oncolytic viruses to undergo selective replication within neoplastic cells (Hanahan and Weinberg 2000; Ilkow et al. 2014; Breitbach et al. 2016).

The precise and selective targeting of cancer cells is an essential prerequisite for the effective implementation of virotherapy. Indeed, numerous viruses that occur naturally, including parvovirus, measles virus, reovirus, and NDV, demonstrate an inherent inclination toward cancerous cells. However, it is necessary to genetically modify certain viruses, such as adenovirus, vesicular stomatitis virus, vaccinia virus, and HSV, to confer cancer-specific properties in them (Russell and Peng 2007; Hartkopf et al. 2012). There have been four distinct approaches employed in the engineering of oncolytic viruses to specifically target tumor cells (Fig. 1) (Carter et al. 2021). The first strategy involves the utilization of virus-specific, receptor-mediated cell targeting. This method focuses on targeting cell markers expressed in tumor cells, such as human epidermal growth factor receptor (HER2) and epidermal growth factor receptor (EGFR). The second strategy is grounded in the phenomenon of fast cell division observed in tumor cells. This characteristic results in heightened

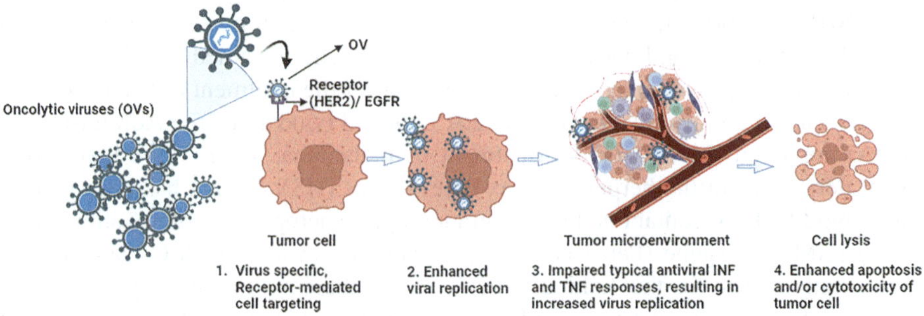

Fig. 1 Four different ways used to engineer oncolytic viruses to selectively target tumor cells

metabolic activity and replication rate, hence facilitating enhanced viral replication in comparison to normal quiescent cells (Lawler et al. 2017). Additionally, it has been observed that mutations occurring in tumor drivers or enzymes such as protein kinase R (PKR) can enhance the specificity of virus replication within tumor cells (Russell et al. 2012; Kaufman et al. 2015). Moreover, it is worth noting that a considerable number of cancer cells display impairments in their typical responses to antiviral interferon (IFN) or tumor necrosis factor (TNF), as evidenced by previous studies (Hartkopf et al. 2012; Lawler et al. 2017). These defects in the cellular response mechanisms further facilitate the reproduction of specific viruses (Russell and Peng 2007). In addition, normal cells exhibit a response to viral infection characterized by the activation of apoptosis or the inhibition of translational, transcriptional, and/or transductional targeting mechanisms. These cellular processes serve to impede the lysis of cells, potentially restricting the spread of the virus (Carter et al. 2021).

4 Mechanism of Oncolytic Virotherapy in Breast Cancer

The primary approach employed in the majority of breast cancer treatments involves the direct targeting and eradication of cancer cells, or the indirect modulation of cancer cells by disrupting the TME (Fig. 2). Oncolytic viruses have been purposefully engineered to eradicate breast cancer cells through either direct cytotoxicity or by perturbing the tumor microenvironment (Hartkopf et al. 2011; Cody and Hurst 2015; Suryawanshi et al. 2017). OVs have the potential to undergo genetic modifications, allowing for the production of diverse transgenes. These transgenes may include immune-stimulating substances and apoptosis inducers, which can augment their efficacy in eliminating cancerous cells. Furthermore, the TME can be perturbed by selectively focusing on the neoplastic mass's vascular network and mitigating the proliferative influence exerted by the TME. This is accomplished through the manipulation of OVs, wherein transgenes are introduced to selectively target specific components within the TME. Furthermore, OVs that are specifically designed for the tumor can be genetically modified to express transgenes that initiate the conversion of prodrugs, thereby confining the detrimental impacts of chemotherapy to cancerous cells and their immediate microenvironment. This approach has the potential to synergistically augment the therapeutic efficacy of conventional chemotherapy (Leveille et al. 2011; Dong et al. 2013; Suryawanshi et al. 2017). Moreover, regardless of the inherent level of oncoselectivity exhibited by a virus, it is observed that the majority of OVs possess the ability to exploit the compromised cellular pathways present in cancer cells including the antiviral protective mechanisms. Similarly, most of the cancer cells are prone to viral infections because they have abnormalities in signaling pathways that prevent them from responding normally to antiviral cytokines like IFN I/II and TNF.

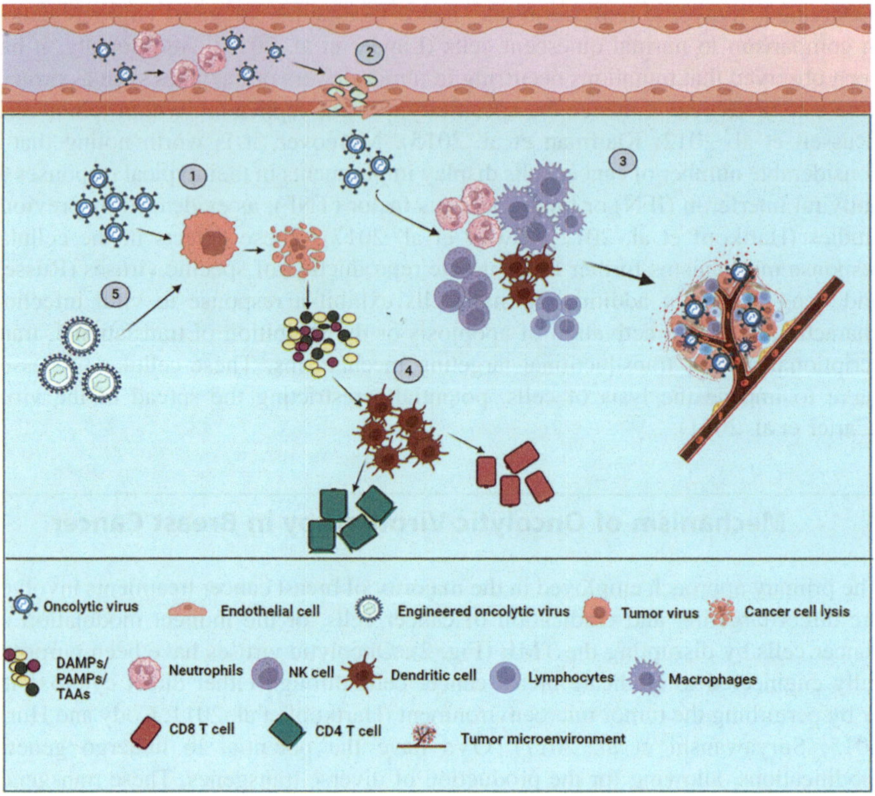

Fig. 2 Mechanisms underlying the anticancer impacts of oncolytic viruses 1: OVs can directly infiltrate and rupture cancerous cells. 2: OVs infect and rupture vascular endothelial cells, subsequently attracting neutrophils. These neutrophils induce coagulation and thrombosis, ultimately damaging blood vessels and hindering angiogenesis. 3: OVs facilitate the recruitment of dendritic cells (DCs), macrophages, neutrophils, and T cells to the tumor microenvironment, effectively shifting "cold" tumors to an immunologically active "hot" state. 4: Upon lysis, cancer cells release damage-associated molecular patterns (DAMPs), pathogen-associated molecular patterns (PAMPs), and tumor-associated antigens (TAAs), which DCs recognize, thus initiating the activation of CD4+ and CD8+ T cells to eliminate cancerous cells. 5: OVs can be equipped with transgenes targeting specific gene expressions, thereby augmenting the overall anticancer potency

4.1 Oncolytic Viruses Expressing Immune-Stimulatory Transgenes

Viruses possess the ability to elicit both innate and adaptive immune reactions in response to malignant cells, even in the absence of any genetic modifications aimed at enhancing immune stimulation. An attenuated recombinant vaccinia virus (GL-ONC1/GLV-1h68) without any immune-stimulatory transgenes can initiate a comprehensive immune response against cancer involving the production of cytokines and chemokines, ultimately leading to significant regression of breast tumor xenografts in mice (Zhang et al. 2007). The GL-ONC1/GLV-1h68 has undergone genetic modifications aimed at enhancing its specificity toward tumors by eliminating viral genes responsible for thymidine kinase (TK) and hemagglutinin. Nevertheless, there has been a notable rise in the genetic modification of OVs, intending to induce the expression of immune-stimulatory transgenes. This genetic alteration aims to enhance the immune response against cancer cells. Oncolytic viruses that carry immune-stimulatory transgenes exhibit antitumor effects by engaging various mechanisms, apart from directly lysing tumor cells. These mechanisms include (a) perturbing the tumor microenvironment that serves as a vital support system for cancer cells, (b) selectively targeting cancer cells that have not been infected within the tumor mass and at distant metastatic sites, thereby triggering an immune response against cancer, and (c) generating a sustained anticancer immune response that persists even after discontinuing virotherapy (Suryawanshi et al. 2017).

In the pursuit of exploring potential immunotherapeutic agents for different forms of cancer, researchers are currently investigating a wide range of recombinant immune-stimulatory proteins. However, the utilization of these proteins is hindered by their systemic toxicity, which restricts their administration to low doses that can be tolerated by the body. Engineered oncolytic viruses harboring genes that facilitate immune stimulation present a viable avenue for targeted delivery of these immune-stimulating proteins exclusively to malignant cells, consequently mitigating systemic toxicity. Numerous oncoviruses (OVs) have undergone genetic modifications aimed at augmenting their therapeutic potential as anticancer agents by incorporating the expression of interleukins (ILs). Some examples include an oncolytic Newcastle disease virus expressing IL-2 (Wu et al. 2016), a vesicular stomatitis virus (VSV) engineered to express IL-4 (Fernandez et al. 2002), a herpes simplex virus (HSV) engineered to express IL-12 (Liu et al. 2005), and an adenovirus engineered to express IL-24 (Sarkar et al. 2005). These different OVs share a common objective to trigger an anticancer immune response by activating and recruiting T cells, macrophages, and natural killer (NK) cells.

4.2 Oncolytic Viruses Inducing Apoptosis in Cancer Cells

An alternative strategy for augmenting the effectiveness of OVs in the treatment of breast cancer entails the induction of cellular demise through the introduction of pro-apoptotic genes into neoplastic cells. This particular approach has demonstrated a notable level of efficacy. In this instance, a genetically modified HSV with oncolytic properties was utilized to transport a pro-apoptotic gene called TNF-related apoptosis-inducing ligand (TRAIL). The results demonstrated the successful management of triple-negative breast cancer (TNBC) growth in a NU/NUnude mouse model (Zhu et al. 2013). Upon the binding of TRAIL to its specific receptor located on the surface of cancer cells, a series of events is triggered, known as the caspase cascade. This cascade is mediated by pro-caspases 8/10, which ultimately culminate in the induction of apoptosis specifically in breast cancer cells

(Gazitt 1999; Ashkenazi 2002). Moreover, the synergistic application of radiotherapy in conjunction with TRAIL therapy has exhibited enhanced regression of breast tumors. Radiotherapy enhances the susceptibility of cancer cells to undergo apoptosis when exposed to TRAIL, thereby leading to a synergistic therapeutic outcome (Chinnaiyan et al. 2000).

Similarly, it has been observed that an oncolytic adenovirus, which has been modified to carry the p53 gene, is capable of inducing cell death and restricting the growth of tumors in breast cancer and other cancer types where the p53 gene is either absent or has undergone mutations (Nielsen et al. 1997). The adenovirus, which carries the p53 gene, has demonstrated a promising safety record and holds the capacity to augment the sensitivity of malignant cells toward chemotherapy and radiotherapy. The observed phenomenon is observed to have a broad impact, even encompassing cancer types that have developed resistance. This is achieved through the restoration of the typical stress response in breast and other cancer cells that carry mutations in the p53 gene (Roth 2006). Another study was conducted by Lal and Rajala in which they engineered a recombinant measles virus that contained BNiP3, a pro-apoptotic gene derived from humans. They utilized this modified virus as an oncolytic agent and successfully showcased its capacity to trigger apoptosis in breast cancer cells when tested in a controlled laboratory setting (Lal and Rajala 2019). Furthermore, the vesicular stomatitis virus (VSV) matrix protein has recently been found to promote apoptosis via autophagy in breast cancer cells (Askari et al. 2021). Additionally, apoptin-expressing recombinant adenoviruses suppress MCF7 breast cancer cell proliferation and modulate autophagy (Chen et al. 2019).

4.3 Oncolytic Viruses Targeting Cancer Vasculature

A better comprehension of the tumor microenvironment's role in supporting tumor growth and metastasis has rendered it an appealing target for therapeutic interventions. This understanding has spurred the development of strategies aimed at various TME components, including the tumor vasculature. Disrupting tumor vasculature emerges as a potent approach to managing tumor growth and advancement. An oncolytic vaccinia virus expressing an antibody against VEGF has demonstrated efficacy in curbing tumor growth by disrupting the tumor vasculature and carrying out viral oncolysis in a mouse model simulating aggressive breast cancer types like TNBC (Gholami et al. 2014). Similarly, another engineered oncolytic VV, which expresses an antagonist of C-X-C chemokine receptor type 4 (CXCR4), can target the TME, encompassing the tumor vasculature in breast tumors. This modification has been proven to reduce tumor metastasis, leading to an improved overall survival rate (Gil et al. 2013). The CXCR4 antagonist thwarts the pro-metastatic effect mediated by the CXCR4/CXCL12 (C-X-C motif chemokine 12) axis by specifically targeting the TME (Suryawanshi et al. 2017).

4.4 Pro-drug Activation Transgene-Expressing Oncolytic Viruses

Pro-drug-activating oncolytic viruses (OVs) enable the targeted conversion of pro-drugs into their active forms exclusively within cancer cells, leading to maximal harm to cancer cells infected by the OVs. The application of this approach has been explored using oncolytic adenovirus (Dong et al. 2013), vaccinia virus (Seubert et al. 2011), and vesicular stomatitis virus (Leveille et al. 2011) in the context of breast cancer treatment. A conditionally replicating adenovirus was equipped with a suicide gene expressing the deoxyribonucleoside kinase of Drosophila melanogaster (Dm-DNK). This adenovirus carrying Dm-DNK demonstrated enhanced efficacy in treating breast cancer when used in combination with chemotherapeutic pro-drugs such as (E)-5-(2-bromovinyl)-20-deoxyuridine (Bvdu) and 20,20-difluoro-deoxycytidine (dFdC), both in laboratory settings and in living organisms (Dong et al. 2013). An oncolytic vaccinia virus was engineered to express b-galactosidase, which could activate a prodrug derived from a seco-analog of the natural antibiotic duocarmycin SA. This modified virus displayed improved therapeutic effectiveness against human breast cancer cells (Seubert et al. 2011). Another variant of the oncolytic virus, VSV-MD51, carrying the cytosine deaminase/uracil phosphoribosyl transferase (CD::UPRT) suicide gene, exhibited a synergistic effect when combined with 5-fluorocytosine (5-FC). The combination of VSV-MD51 and the 5-FC prodrug was demonstrated to affect neighboring uninfected tumor cells by releasing toxic byproducts of 5-FC, as demonstrated in laboratory experiments (Leveille et al. 2011).

5 Oncolytic Viruses

Oncolytic viruses represent a distinct category of viruses that possess the innate ability or have been modified through genetic engineering to selectively reproduce within tumor cells, thereby impeding the progression of tumors. Oncolytic viruses exhibit a remarkable ability to specifically target and infect neoplastic cells, subsequently undergoing viral replication and inducing the demise of the infected cells. This intricate phenomenon is commonly referred to as oncolysis. The subsequent dissemination of these additional infectious viruses leads to the infection of adjacent tumor cells. Simultaneously, the initiation of a tumor-specific immune response occurs, leading to additional amplification of the oncolytic effects (Carter et al. 2021). Oncolytic viruses can be categorized into two groups based on the nature of their nucleic acid composition: DNA viruses and RNA viruses.

5.1 Oncolytic DNA Viruses

5.1.1 Adenovirus

Adenoviruses exhibit a mesoscale dimension, ranging from 70 to 90 nm, and possess an icosahedral morphology. These viruses showcase a non-enveloped configuration, housing a double-stranded DNA within their structure. In humans, infections can be caused by different types of adenoviruses that exhibit immunological distinctiveness. The viruses under consideration exhibit the expression of the coxsackie-adenovirus receptor (CAR) as well as the adenoviral early genes (specifically encoded by E1A and E1B), which endows them with the capability to selectively target cells for infection. The adenovirus effectively utilizes these genes to inhibit the activity of cell cycle regulators, such as the tumor suppressor genes p53 and retinoblastoma-associated protein (pRb). These genetic elements exhibit a defensive function in typical cellular entities, yet are frequently compromised in malignant cellular entities, thereby enabling adenovirus to disrupt the regulation of cellular division. Various techniques have been utilized to manipulate adenovirus, such as the process of attenuation. The utilization of oncolytic adenovirus in clinical settings has demonstrated a limited occurrence of unfavorable occurrences. This approach introduces a unique therapeutic pathway for breast cancer owing to its precise targeting capability (Martini et al. 2020).

The investigation of ONYX-015 adenovirus, a variant lacking a specific region of E1B responsible for p53 inactivation, is currently underway in a phase I trial, in conjunction with etanercept (recombinant human TNF). Despite the inclusion of two patients diagnosed with metastatic breast cancer and several others with varying types of cancer, this trial remains in its nascent phases (Nemunaitis et al. 2007). Recent scientific investigations have brought attention to a recombinant adenovirus that exhibits a specific affinity toward E2F-1 and IL-15. E2F-1, a pivotal transcription factor governing the regulation of the cell cycle, cellular proliferation, and the development of cancer, exhibits elevated expression levels within breast cancer tissue in comparison to normal tissue. This observation implies the potential of E2F-1 as a promising therapeutic target for breast cancer treatment. Through the incorporation of the human interleukin-15 gene, which amplifies the functionality of immune cells such as macrophages and neutrophils, this adenovirus has the capacity to stimulate the generation and activation of effector T cells. Additionally, it can regulate the persistence and multiplication of memory T cells, thus leading to an immunomodulatory impact (Yan et al. 2019). Further investigations in the preclinical domain have delved into the examination of oncolytic adenoviruses, which impede the function of long non-coding RNAs (lncRNAs) or specifically target the TGF-β pathway within malignant cells (Ang et al. 2020; Li et al. 2020).

5.1.2 Herpes Simplex Virus

The herpes simplex virus, a DNA virus characterized by its double-stranded structure, harbors a genome spanning a considerable length of 152 kilobases. This organism is a member of the alpha herpes virus family and exhibits a replication process that takes place within the cellular nucleus. Significantly, HSV-1 exhibits a deficiency in its ability to induce insertional mutagenesis, making it a highly attractive candidate for potential therapeutic progress. In this particular scenario, it has been observed that human breast cancer cells exhibit a certain level of vulnerability to the herpes simplex virus type 1 (HSV-1). By making genetic alterations, it is possible to precisely adjust the virus's selectivity, thus ensuring the protection of normal cells.

Talimogene laherparepvec (T-VEC) is a highly sophisticated oncolytic virus derived from the herpes simplex virus type 1 (HSV-1), which has reached an advanced stage of commercial development in the field. T-VEC has undergone targeted genetic modifications, resulting in the elimination of two specific genes, namely ICP34.5 and ICP47. The ICP34.5 gene, which is known for its involvement in neurovirulence and its crucial role in evading the host's PKR–IFN antiviral defense mechanism, has been removed to confer cancer-specific characteristics and hinder neural infections. To augment the provocation of antitumor immune responses, the genetic material encoding the two ICP34.5 genes has been substituted with the genetic material accountable for the synthesis of granulocyte-macrophage colony-stimulating factor (GM-CSF). The absence of ICP47 leads to the display of viral antigens, which restricts infection to healthy tissues and facilitates the immune-driven eradication of cancer cells that specifically propagate the oncolytic HSV-1. The deletion of ICP47 in this context initiates the prompt activation of the US11 promoter, leading to the inhibition of PKR phosphorylation and subsequently impeding apoptosis in cancer cells (Dong et al. 2013). T-VEC has the potential to be a valuable tool when used alongside immune checkpoint inhibitors (ICIs), especially in tumors that have a low initial presence of lymphocytes and show limited response to ICIs used alone. This is particularly pertinent in instances such as breast cancer. In a phase Ib clinical trial, a cohort of 21 patients diagnosed with advanced melanoma were subjected to a therapeutic regimen involving the administration of T-VEC followed by pembrolizumab. The results of the study revealed a significant degree of efficacy exhibited by the anti-PD-1 agent, primarily attributed to its ability to modulate the tumor microenvironment (TME) (Ribas et al. 2017). The clinical trials are currently evaluating the safety and effectiveness of T-VEC, both as a monotherapy and in conjunction with paclitaxel, in a cohort of patients diagnosed with triple-negative breast cancer (TNBC) (Marra et al. 2019).

Recent observations have provided evidence that G47Δ-mIL12, a genetically modified oncolytic herpes simplex virus (oHSV), demonstrated noteworthy antitumor effects within the context of an advanced breast cancer model known as 4T1. G47Δ-mIL12, a variant that exhibits comparable genetic alterations to T-VEC, introduces ICP6 inactivation, thereby confining the replication of herpes simplex virus (HSV) exclusively to cancerous cells. This particular variant expresses murine interleukin 12 (IL-12) instead of GM-CSF. The effectiveness of this particular variant of oHSV was thoroughly examined using both in vitro and in vivo experimental approaches. In controlled laboratory conditions, it successfully infected and eliminated triple-negative breast cancer cells in both mice and humans. Upon in vivo experimentation, it was observed that the administration of G47Δ-mIL12 had a significant inhibitory effect on the proliferation of 4 T1 tumors. This inhibitory effect was found to be dependent on the presence of CD8+ T cells. Additionally, the treatment demonstrated the ability to restrict the dissemination of metastases (Ghouse et al. 2020).

The administration of HF10 in murine models exhibiting intact immune systems elicited precise immune reactions directed toward neoplastic growths, in the context of colon and breast cancer. This particular methodology additionally resulted in the emergence of resistance toward the re-implantation of malignant cells (Teshigahara et al. 2004). Similarly, HF10 is a variant of the herpes simplex virus type 1 (HSV-1) that has experienced spontaneous genetic changes and is expected to exhibit significant promise when combined with bevacizumab (Tan et al. 2015).

5.1.3 Vaccinia Virus

The vaccinia virus (VV) is a DNA virus possessing a linear, double-stranded genetic configuration, belonging to the orthopoxvirus genus of the Poxviridae family. During the typical progression of VV infection, four discrete categories of viral particles are produced. The aforementioned particles exhibit variations in their abundance, chemical makeup, spatial distribution, and roles throughout the lifecycle of the virus. These encompass (i) the virus that reaches maturity within the cell's cytoplasm (intracellular mature virus or IMV), (ii) the virus that exists outside the cell with an envelope (extracellular enveloped virus or EEV), (iii) the virus that forms an envelope within the cell (intracellular enveloped virus or IEV), and (iv) the virus that associates with the cell and possesses an envelope (cell-associated enveloped virus or CEV) (Smith et al. 2002). Cell lysis results in the release of a considerable amount of intracellular mature virus (IMV), which exhibits greater resistance in comparison to extracellular enveloped virus (EEV) and can be easily recognized by the immune system. The additional membrane of the extracellular enveloped virus (EEV), which is derived from the host cell, plays a crucial role in enhancing its ability to evade the immune system and facilitates a greater extent of spreading (Roberts and Smith 2008).

In breast cancer, it has been observed that an oncolytic vaccinia virus exhibits a notable capacity to infect tumor cells, undergo successful replication, and subsequently induce regression of the tumor. GLV-1h68, a recombinant vaccinia virus harboring three genetically engineered alterations to enable efficient visual and immunohistochemical monitoring, exhibited proficient replication and cytotoxicity against canine mammary tumor cells in both laboratory culture (in vitro) and living organism (in vivo) settings using a nude mouse model (Gentschev et al. 2009). Human breast cancer stem cells exhibiting elevated resistance to chemotherapy and irradiation, it was observed that GLV-1h68 displayed a higher degree of replication efficiency in comparison to its non-stem cell counterparts (Wang et al. 2012). When evaluated in a xenograft mouse model employing breast cancer stem cells, GLV-1 h68 demonstrated noteworthy capacity in suppressing tumor growth. This highlights the potential of this oncolytic virus as a targeted intervention against resilient cancer stem cell populations (Wang et al. 2012).

JX-929 (vvDD), alternatively referred to as the Western Reserve strain of VV, has been subject to genetic modification by selectively eliminating two distinct genes (McCart et al. 2004). Through the removal of the thymidine kinase gene, the production of viral DNA becomes dependent on the availability of thymidine triphosphate derived from vigorously proliferating cells, such as tumor cells. The ablation of the vaccinia growth factor (VGF) gene hinders the proliferation of neighboring cells (McCart et al. 2001). JX-929 utilizes a comparable recombination of the cytosine deaminase gene, enabling infected cells to undergo a conversion process where 5-fluorocytosine is transformed into 5-fluorouracil (Chalikonda et al. 2008). Furthermore, the somatostatin receptor is encompassed, enabling the observation of viral dispersion through the identification of the accumulation of radioactivity derived from 111Inpentetreotide within the infected cells (McCart et al. 2004). In another investigation, an oncolytic vaccinia virus GLV-1 h164, equipped with the capability to produce the single-chain antibody GLAF-2 targeting vascular endothelial growth factor (VEGF), was examined in the context of triple-negative breast cancer. GLV-1 h164 exhibited a notable regression in xenografts of triple-negative breast cancer tumors in comparison to the parent virus lacking GLAF-2 expression (Gholami et al. 2014).

5.2 Oncolytic RNA Viruses

5.2.1 Reovirus

The utilization of double-stranded RNA oncolytic virus, specifically reovirus, exhibits considerable potential as an innovative therapeutic approach for breast cancer. A research study has demonstrated that different cell lines of breast cancer exhibit susceptibility to infection by reovirus, whereas normal breast epithelial cells, regardless of their hormone receptor status, remain unaffected by this characteristic (Hata et al. 2008). Reoviruses demonstrate the remarkable capability to selectively target cancer stem cells (CSCs), a population of cells that typically display resistance to traditional clinical interventions. Both cancer stem cells (CSCs) and non-CSC cancer cells exhibit equivalent susceptibility to reoviruses, and the sensitivity of CSCs to reovirus therapy mirrors that of non-CSC cells (Marcato et al. 2009). Based on their observed effectiveness in eradicating cell lines of TNBC in certain studies, reoviruses may possess promising therapeutic potential against TNBC (Norman et al. 2002). Moreover, it has been observed that reoviruses exhibit the capacity to augment rates of survival and reduce instances of leptomeningeal disease in a manner that is both secure and efficacious (Yang et al. 2004). The cumulative body of evidence highlights the encouraging potential of reovirus therapy in addressing TNBC (Jin et al. 2021).

The Type 3 Dearing reovirus strain exhibits natural oncolytic properties and displays a preference for infecting tumor cells. The oncolytic potential of this particular strain has been investigated in numerous studies focusing on breast cancer cells. In a comprehensive investigation, a group of researchers conducted an experiment wherein they examined a collection of breast cancer cell lines. Intriguingly, their findings revealed that all the cell lines, irrespective of their hormone receptor status, exhibited susceptibility to reovirus infection. In stark contrast, normal breast epithelial cells did not display any susceptibility to this viral infection (Hata et al. 2008). The extensive infectivity of this phenomenon has been ascribed to the activation of the Ras pathway or the presence of mutated Ras protein within cancerous cells. In an experimental study involving mice, core biopsies extracted from a human breast cancer tumor were utilized. Notably, the administration of reovirus treatment resulted in the successful regression of the tumor (Marcato et al. 2009).

A phase II clinical trial was conducted to explore the synergistic effects of pelareorep in conjunction with paclitaxel in patients diagnosed with metastatic breast cancer. The findings suggested that the utilization of this particular combination exhibited a greater efficacy in comparison to the exclusive administration of paclitaxel (Bernstein et al. 2018). Recent preclinical investigations have revealed that the reovirus, in conjunction with anti-PD-1 therapy, elicits an immunological reaction targeting breast cancer cells (Mostafa et al. 2018).

5.2.2 Vesicular Stomatitis Virus

The vesicular stomatitis virus (VSV) is an intriguing nonpathogenic viral entity that exhibits a distinctive single-stranded negative-sense RNA configuration. The unique responsiveness of this organism to type-1 interferons (IFNs) which are known to have a beneficial effect on regular cells, combined with its lack of sensitivity toward tumor cells, enables it to efficiently proliferate within tumor cells, leading to expedited replication (Balachandran and Barber 2000). When infected with VSV, the manifestation of symptoms is observed in only a minute fraction of individuals, and the virus demonstrates a low level of pathogenicity, resulting in the absence of significant adverse outcomes. The aforementioned characteristics establish VSV as a highly favorable oncolytic virus in the field of biology. A genetically modified variant of the vesicular stomatitis virus was discovered to specifically recognize, replicate within, and eradicate breast cancer cells that exhibit ERBB2 (HER2/neu) expression (Bergman et al. 2004).

Subsequent investigations delved into the potential of VSV to mitigate the effects of TNBC. The cytotoxic impacts of recombinant vesicular stomatitis virus (VSVd51) were assessed in both murine and human TNBC cells, providing supporting evidence for its role in modulating the antitumor immune response. The administration of VSV elicited a notable therapeutic reaction by recruiting innate immune cells known as NK cells and adaptive immune cells called CD8+ T cells. The observed effect on the immune response suggests that the combination of VSV with checkpoint inhibitors may exhibit potential for the treatment of TNBC (Niavarani et al. 2020).

5.2.3 Measles Virus

Measles virus is classified as a negative-strand RNA virus within the Paramyxoviridae family, specifically belonging to the Morbillivirus genus. The measles virus has shown its potential in breast cancer treatment and is currently under investigation to effectively intervene in the progression of cancer (Engeland and Ungerechts 2021).

According to a scientific investigation, it was observed that exposure to a particular strain of the measles virus (MV) that can generate carcinoembryonic antigen (MV-CEA) resulted in the demise of MDA-MB-231 cells under controlled laboratory settings. The scientists expanded their inquiry by evaluating the efficacy of MV-CEA in combating tumors in a living organism, specifically by conducting experiments on subcutaneous MDAMB-231 xenografts. The results of their study demonstrated that MV-CEA displayed strong therapeutic efficacy against TNBC in both laboratory settings (in vitro) and living organisms (in vivo) (McDonald et al. 2006).

It has been discovered that these OVs can infiltrate cells by utilizing various receptors. These receptors include the signaling lymphocyte activation molecule (SLAM), which is found on numerous immune-associated cells, CD46 that are present on all cells with a nucleus, and the poliovirus receptor-related protein 4 (PVRL4) (Tatsuo et al. 2000; Mühlebach et al. 2011). Expanding upon this understanding, an other group of scientists conducted genetic modifications to create a modified version of the MV virus that is specifically incapable of binding to the SLAM receptor (referred to as rMV-SLAM blind). The virus exhibited a modified phenotype, demonstrating an affinity for PVRL4 as a cellular receptor while lacking the ability to utilize CD46 for infection of TNBC cells. The experimental findings revealed that the application of rMV-SLAM blind resulted in a decrease in the survival capacity of breast cancer cell lines while having no impact on the survival of SLAM-positive lymphoid cells. Additionally, the rMV-SLAM blind variant demonstrated enhanced oncolytic activity in comparison to the wildtype MV, while still ensuring safety (Sugiyama et al. 2013). In summary, these studies collectively provide compelling evidence that MV exhibits promising potential as a therapeutic candidate for addressing the challenges posed by breast cancer.

Combination therapies involving MV have garnered significant attention in the scientific community owing to the compelling evidence suggesting that such strategies have the potential to enhance the effectiveness of viral therapies. For instance, various viral strains (e.g., MV-GFP, MV-lambda, MV-s-NAP, and MV-lambda-NAP) have recently undergone experimentation involving the administration of alisertib. Alisertib, also known as MLN8237, is a compound that acts as an inhibitor of Aurora A kinase. It is noteworthy that the activity of Alisertib exhibits a synergistic effect with viral replication. The synergistic utilization of these oncolytic measles virus vectors in conjunction with alisertib exhibited a notable enhancement in the eradication of breast cancer cells when compared to treatment with the virus alone. Moreover, in certain instances, this combined approach led to the complete eradication of breast cancer cells in an in vitro setting (Iankov et al. 2015; O'Bryan and Mathis 2018).

5.2.4 Coxsackievirus

Coxsackievirus, an enterovirus possessing a positive-sense single-stranded RNA, consists of 29 distinct subtypes. Due to its limited harm to humans, the exploration of its oncolytic capabilities has been relatively limited. However, in the coming years, advancements in genetic engineering could augment the virus's selectivity for tumor cells while diminishing its pathogenicity, rendering coxsackievirus a viable oncolytic virus candidate. Specifically, Coxsackievirus B3 (CV-B3) has been genetically modified to effectively target various tumor cell types, including lung cancer and endometrial cancer (Lin et al. 2018). Meanwhile, Coxsackievirus 21 (CVA21), a naturally occurring OV without genetic alterations, has demonstrated notable oncolytic properties against multiple myeloma and melanoma (Au et al. 2007). The therapeutic potential of coxsackievirus in treating TNBC has also been verified. CVA21 has shown substantial inhibition of TNBC cells both in vitro and within mouse xenograft models (Skelding et al. 2009). Although only a limited number of studies have explored coxsackievirus's therapeutic impact on TNBC, they collectively underscore its potential capacity for treating this challenging form of breast cancer.

6 Oncolytic Virotherapy and Immune Response

The immune system's interaction with OVs elicits two distinct types of responses. One can observe a phenomenon known as restrictive response, wherein the immune system identifies OVs as pathogens, thereby triggering an antiviral reaction that hampers the viruses' capacity to replicate and spread. The adverse effects of this phenomenon can have detrimental implications for the field of oncolytic virotherapy. The former refers to a stimulatory response wherein OVs elicit the activation of immune cells. The immune cells that have been activated generate an immune response that specifically directs its efforts toward the eradication of tumors. The intricate interplay between OVs and the immune system holds profound implications for the ultimate success of therapeutic interventions (Filley and Dey 2017).

6.1 Antiviral Immune Response

One of the primary reactions elicited by the immune system in response to normal body cells involves the conventional antiviral defense mechanism targeting OVs. The findings of Shin et al. (2021) demonstrate the direct inhibitory effects of OVs on their replication and dissemination (Shin et al. 2021). The innate immune system, upon encountering oncolytic viruses, perceives them as exogenous antigens and promptly initiates the antiviral reaction. The genetic materials, namely DNA and RNA, found within organisms activate pattern recognition receptors (PRRs) that facilitate the recruitment of proteins responsible for signaling the phosphorylation of transcription factors. The initiation of an antiviral response is triggered by these factors, which stimulate the production of type I interferon (Matveeva and Chumakov 2018). Another mechanism facilitating the generation of an antiviral response entails the interaction between RNA helicases, such as RIG-I, and viral RNAs. It initiates mitochondrial signaling to elicit an antiviral response, exhibiting prion-like behavior to trigger and generate a cascade of antiviral signals (Hou et al. 2011). Specific genes associated with interferon exhibit the ability to impede viral replication, initiate the synthesis of cytokines, and recruit natural killer cells to eliminate the infected cells (Pak-Wittel et al. 2013). Natural killer cells exert a suppressive effect on viral replication and disrupt the efficacy of oncolytic viral therapy.

The innate immune system plays a crucial role in initiating the activation of the adaptive immune response through the presentation of viral antigens (Shin et al. 2021). In addition to the presence of diverse anatomical obstacles within the circulatory system, OVs face a potential hurdle in the form of antibodies specifically targeting viral entities. These antibodies may either be pre-existing as a result of prior vaccination of a patient with a closely related virus, or they may arise following the administration of antiviral antibodies aimed at neutralizing viral agents. Both of these scenarios result in a decrease in the abundance of oncolytic viruses, hinder the repeated systemic delivery of oncolytic viruses, and consequently affect the efficacy of oncolytic therapy by eliciting an immune response against the viruses (Niemann et al. 2019).

6.2 Antitumor Immune Response by Oncolytic Viruses

6.2.1 Stimulation of Immune Response Within Tumor Microenvironment (TME)

Tumors or transformed cells undergo genetic mutations, resulting in alterations of their antigens, which serve as a mechanism to evade the immune system. Therefore, they exhibit the capability to make themselves imperceptible to cells of the immune system. The phenomenon described in the literature is commonly referred to as immuno-editing (Marelli et al. 2018) The tumor microenvironment present within neoplastic growths is characterized by a low temperature, thus rendering it a frigid milieu that fosters the development of an immune-suppressed state. It hinders the immune system's ability to generate an antitumor response, thereby evading the immune system (Breitbach et al. 2016). Oncolytic viruses, in contrast, possess the capacity to elicit immune cell activation, thereby instigating the production of an antitumor response. The introduction of OVs into tumor cells facilitates the activation of the immune system, leading to the initiation of a sustained and vigorous immune response against the tumor cells (Marelli et al. 2018).

The initiation of an inflammatory response is elicited by the oncolytic viral infection of neoplastic cells. This phenomenon can be attributed to the induction of a particular form of programmed cell death, referred to as immunogenic cell death (ICD), which is triggered by viral agents. The ICD can efficiently elicit an immune system reaction against the malignant cells. It elicits chemotaxis and triggers the activation of dendritic cells, leading to the subsequent stimulation of distinct T lymphocytes (Marelli et al. 2018). The process of apoptosis in the context of immunogenic cell death is not devoid of biological activity. Instead, it triggers the activation of the endoplasmic reticulum, leading to the release of molecules known as damage-associated molecular patterns (DAMPs), which can have detrimental effects. Three distinct metabolites, namely HMGB1, calreticulin, and ATP, have been classified as DAMPs that are released as a result of immunogenic cell death (Guo et al. 2017). The cells that present antigens within the tumor microenvironment are capable of recognizing these metabolites and initiating an immune response. Moreover, the infection of OVs results in the eradication of cancer cells and the subsequent liberation of tumor-associated antigens within the TME. This mechanism facilitates the recognition of these antigens by the immune system, leading to the generation of a response and subsequent interference with the immuno-editing process.

The research conducted by Bell and his team unveils that the infection of tumors by OVs elicits the secretion of cytokines and the initiation of inflammation, thereby provoking the activation of the immune system (Bell et al. 2003). Consequently, the tumor's previous state of being "cold" can transform into a "warm" state, rendering it more prone to immune activity (Breitbach et al. 2016). Hence, the objective of oncolytic virotherapy is not solely limited to the eradication of neoplastic cells but also encompasses the activation of the immune system, which is known to be inhibited by the tumor microenvironment (Marelli et al. 2018).

6.2.2 Role of Interferon in Antitumor Response

Normal body cells initiate the immune response by producing one of the key factors such as type I interferons which is a group of proteins containing IFN-α and IFN-β (Filley and Dey 2017). These interferons within the TME activate immune cells such as NK cells, CD8+ T lymphocytes, and inflammatory cytokines which play an important role in inducing antitumor immune response (Achard et al. 2018). For instance, MV activates specific dendritic cells known as plasmacytoid DCs along with IFN-α which subsequently leads to activation of RIG-like receptors (RLRs) and Toll-like receptors (TLRs). These receptors result in the expression of tumor cell death-inducing molecules such as TNF-related apoptosis-inducing ligands (TRAIL) which are responsible for cytotoxic function (De Matos et al. 2020).

When NK cells are activated, they produce another type of protein called type II interferon (IFN-γ). This protein has an important role in inducing apoptosis, inhibiting angiogenesis, and stimulating the immune system. When stimulated, the immune system activates class II MHC molecules in dendritic cells (DCs), enhancing the phagocytic activity of macrophages and the immune responses of CD8+ T cells (Marchini et al. 2019).

The expression of class I MHC molecules in DC and the co-stimulatory molecules (CD40, CD86) can also be upregulated by the Type I IFN. These molecules maintain Th1 polarized immune response (Marchini et al. 2019). It is necessary to note that the ability of OVs to trigger the production of type I IFN at each tumor site and the effectiveness of specific OVs to activate signals for type I IFN differs greatly among various OVs. All these factors determine the capacity of OVs to trigger the acquired immune response against tumors as well as viral antigens (Achard et al. 2018).

6.2.3 Role of Macrophages in Antitumor Response

Another important factor is the population of tumor-related macrophages that determines whether the TME activates or suppresses the immune system. These macrophages can be present in two regulatory states: M1 is the immunostimulatory state in which macrophages induce inflammation and activate antitumor responses, while M2 is the immunosuppressive state in which macrophages trigger the angiogenesis and metastasis leading to suppression of antitumoral immune response (Denton et al. 2018; Marchini et al. 2019). The M1 state of macrophage is induced by lipopolysaccharides and IFN-γ, while the M2 macrophage state is triggered by cytokines IL-4. M1 macrophages promote the expression of genes encoding specific chemokines, IL-12, and TNF for the attraction and activation of dendritic cells (DCs), natural killer (NK) cells, and T cells. On the other hand, M2 macrophages generate TGF-β, IL-8, and vascular endothelial growth factor (VEGF) which promote angiogenesis (Denton et al. 2018). Tumor infection by OVs generates a higher M1 population than M2 macrophages within TME, so the chances of getting the required antitumor response by the immune system are high (Denton et al. 2016).

7 Combination Therapies in Oncolytic Virotherapy

The utilization of oncolytic viruses as a sole treatment for cancer exhibits diminished efficacy owing to a multitude of characteristics exhibited by tumors within the host's body, including the existence of preexisting immunity, the presence of inhibitory molecules, and inadequate infiltration into the TME. Hence, researchers employ the integration of oncolytic virotherapy alongside complementary treatments and antitumor agents like chemotherapy, immune checkpoint inhibitors (ICI) for immunotherapy, and radiotherapy to amplify therapeutic outcomes through the generation of synergistic effects.

7.1 Combination of Oncolytic Viruses with Chemotherapy

Several investigations have employed the amalgamation of oncolytic virotherapy alongside chemotherapy in diverse research endeavors. These studies have reported that the combination of treatments yields superior therapeutic efficacy for breast cancer compared to the use of single therapeutic regimens alone. Various chemotherapeutic agents employed in the management of breast cancer have been subjected to scrutiny in conjunction with oncolytic viruses. Notable examples encompass the alkylating cyclophosphamide drug, the cytoskeleton modifier Paclitaxel drug, the anthracycline doxorubicin drug, the autophagy inducer rapamycin drug, and additional agents as elucidated by Zou et al. (2023).

Cyclophosphamide (CPA) is an alkylating chemotherapeutic agent that is frequently employed in the medical field for the management of diverse forms of malignancies, such as breast cancer. The utilization of both CPA and oncolytic viruses in cancer therapy encompasses a dual mechanism of action. Firstly, it is observed that CPA functions as an immunosuppressant, thereby enhancing the virus's capacity to infect, replicate, and spread throughout the host organism (Ottolino-Perry et al. 2010). Furthermore, the genetic material of the virus undergoes modifications to incorporate the presence of an enzyme known as cytochrome P450.

This particular enzyme facilitates the conversion of CPA into its active form, thereby enabling the sequestration of detrimental metabolites of CPA within virus-infected cancer cells. Both of these methodologies have demonstrated efficacy in augmenting the oncolytic virus's capacity to effectively engage in the battle against cancer (Ottolino-Perry et al. 2010).

The efficacy of low-dose CPA conjugated with oncolytic viruses in the treatment of metastatic breast cancer has been observed. This particular combination of factors resulted in the reduction of a specific subset of T-cells called regulatory T-cells, which are responsible for suppressing the immune system's activity (Bramante et al. 2016). The utilization of low-dose CP in conjunction with oncolytic adenovirus has been employed in the therapeutic intervention of breast cancer, specifically targeting the TNBC to achieve synergistic outcomes (Bramante et al. 2016). The utilization of CPA has been observed in conjunction with a diverse array of oncolytic viruses, including adenovirus, reovirus, HSV, and measles virus, to induce immunosuppression and enhance the efficacy of oncolytic virotherapy (Thomas et al. 2008; Ottolino-Perry et al. 2010).

Paclitaxel, a widely utilized pharmaceutical agent, is prominently employed in the therapeutic management of breast cancer. Paclitaxel is a pharmacological agent that acts as a modifier of the cytoskeleton, specifically belonging to the taxanes class of chemotherapeutic drugs. The coadministration of this pharmaceutical compound alongside oncolytic viruses exhibited a synergistic effect on cellular toxicity when tested in a controlled laboratory environment. Furthermore, it was observed that this combined treatment led to a reduction in tumor size when evaluated in live organisms. It exerts its inhibitory effects on mitosis by impeding the proper functioning of the cytoskeleton within the cell. Apoptosis, a cellular process, can be induced by TRAIL as mediated by TRAIL receptors (Zhang et al. 2002). The co-administration of Paclitaxel, a chemotherapeutic agent, with an oncolytic virus, specifically Rhabdovirus Maraba-MG1, has been investigated to evaluate the effectiveness of this combined treatment approach in murine models of breast cancer. This combination exerted control over the growth of breast cancer by enhancing viral replication and exhibiting synergistic cytopathic effects, thereby prolonging the overall survival duration. The synergistic utilization of Paclitaxel and MG1 presents a viable therapeutic modality for individuals afflicted with refractory breast cancer (Bourgeois-Daigneault et al. 2016).

Doxorubicin, an anthracycline chemotherapeutic agent, is employed in the treatment of various malignancies, notably breast cancer. Doxorubicin exerts its deleterious effects on DNA through the inhibition of topoisomerase II activity. The coalescence of doxorubicin with oncolytic viruses exhibits a synergistic effect, resulting in an augmented level of cytotoxicity (Graat et al. 2006). In a study, the process of conjugating doxorubicin with oncolytic reovirus (reo-dox) was found to enhance the oncolytic efficacy of reovirus against metastatic breast cancer. Infection with reo-dox triggers the activation of the innate immune system, leading to the induction of DNA damage. This occurs through the activation of the double-stranded DNA break response pathway in cells infected with reo-dox. This synergistic amalgamation enhances pharmaceutical transportation while preserving viral functionality. The redox process primarily resulted in the reduction of tumor size in TNBC. Therefore, the reo-dox treatment regimen demonstrates notable efficacy as a combinatorial therapeutic approach specifically designed to target breast cancer (Berry et al. 2020).

The utilization of rapamycin, a chemotherapeutic drug known for its ability to induce autophagy, in conjunction with an adenovirus that specifically targets the overexpression of Cyclin E in human breast cancer cell lines, has demonstrated enhanced efficacy in the field of oncolytic virotherapy. The synergistic therapeutic effects observed in this combination exhibit a potent impact on breast cancer. It triggers apoptosis through the process of autophagy, facilitating the liberation of nutrients from deceased cells to support viral replication, ultimately leading to the dissemination of the virus within tumor tissues. According to Cheng et al. (2013), it has been observed that this particular phenomenon inhibits the immune response against viral infections while simultaneously augmenting the immune response against tumor development (Cheng et al. 2013).

7.2 Combination of Oncolytic Viruses with Immunotherapy Using Immune Checkpoint Inhibitors

Recently, it was found that the oncolytic viruses enhance the immune response against tumor cells within TME in breast cancer. The presence of these OVs specifically converts tumor-associated macrophages (TAMs), from the M2 state (a subtype promoting tumor growth) to the M1 state (a subtype promoting inflammation and cytotoxicity). These TAMs attract CD8+ T cells which directly target and eliminate cancer cells. In addition, the study also revealed the disadvantage of oncolytic virotherapy that the OVs also increase the level of a protein, PD-L1 within TME along with the activation of CD8+ T cells and M1 state (Kwan et al. 2021a, b). This discovery by the researchers suggests the idea that a combination of OVs with immune checkpoint inhibitors such as inhibitors targeting PD-L1 or PD-1 can eliminate this immune suppression and allow the more effective immune response to improve oncolytic virotherapy (Kwan et al. 2021a, b). Various studies have shown the synergistic therapeutic effect on breast cancer using a combination of oncolytic viruses and immune checkpoint inhibitors as given in Table 1.

7.3 Combination of Oncolytic Viruses with Radiotherapy

The combination of oncolytic virotherapy with radiotherapy is becoming more important for the treatment of breast cancer as radiotherapy is widely used for the treatment of both metastatic and curative breast cancer (Ottolino-Perry et al. 2010).

The exact way of synergistic mechanism between oncolytic virotherapy and radiotherapy is not completely identified. Many scientists suggest that it involves the combination of both therapies in which radiations cause DNA destruction and OVs inhibit DNA repairing. In addition, some researchers determined that the

Table 1 Oncolytic viruses tested against breast cancer in preclinical studies

Virus	Modification	Combination	References
Adenovirus	Fiber modified; tissue-specific promoter-driven E1A	None	Bauerschmitz et al. (2008)
Adenovirus	E1A under the control of hTERT and E1B under the control of HRE promoter; encodes IL-24	None	Zhu et al. (2012)
Adenovirus	Chimeric fiber protein; deletion in E1A gene and encodes trastuzumab	None	Liikanen et al. (2016)
Adenovirus	Modifies fiber protein; encodes TGF-β decoy protein	Anti-PD-L1, anti-CTLA-4	Xu et al. (2020)
HSV-1	Deletions in α47, γ34.5, and ICP6 genes; encodes IL-12	None	Ghouse et al. (2020)
HSV-1	Re-targeted to HER2 by trastuzumab in gD domain; deletion of γ34.5	None	Menotti et al. (2008)
Vaccinia virus	J2R gene deleted; encodes IL-15	PI3Kδ inhibitor	Ferguson et al. (2020)
Vaccinia virus	Targeted deletion of immune-modulatory genes	None	Umer et al. (2020)
Vaccinia virus	J2R replaced with an expression cassette for GM-CSF	Anti-PD-L1, anti-CTLA-4	Chaurasiya et al. (2020)
Vaccinia virus	Deleted of J2R, F14.5L, and A56R, and armed with scAb against VEGF	None	Gholami et al. (2014)
Chimeric poxvirus	Deletions in J2R and F14.5 L genes; encodes hNIS gene	Anti-PD-L1	Chaurasiya et al. (2020)
Vesicular stomatitis virus	Mutation in matrix protein	Trastuzumab emtansine	Arulanandam et al. (2020)
Maraba virus	Mutations in G and M protein	Paclitaxel	Bourgeois-Daigneault et al. (2016)
Maraba virus	Non-modified	PD-1 inhibitor and CTLA4 inhibitor	Bourgeois-Daigneault et al. (2018)
Oncolytic reovirus	Non-modified	Anti-PD-1	Mostafa et al. (2018)
Newcastle disease virus	IL-12 expression cassette inserted between M and F genes	None	Mohamed Amin et al. (2019)
Oncolytic PVSRIPO	Recombinant Polio: Rhinovirus chimera	Anti-PD1 and anti-PD-L1	Frazier et al. (2020)
Oncolytic vaccinia virus co-expressing PD-L1 inhibitor and GM-CSF	vgf gene deleted	PD-L1 inhibitor	Wang et al. (2020)

combination of radiotherapy with oncolytic virotherapy led to an increase in the replication of OVs, their reproduction, and release in TME enhancing OVs ability to destroy cancer cells (Bieler et al. 2008).

A study revealed that breast cancer was more effectively degenerated when an oncolytic virus expressing pro-apoptotic TRAIL transgene was combined with radiotherapy. Cancer cells can be sensitized to apoptosis mediated by TRAIL through radiation, leading to synergistic effects by oncolytic viruses (Chinnaiyan et al. 2000). A p53 expressing oncolytic adenovirus induced apoptosis and suppressed tumor growth in breast cancer by sensitizing cancer cells to radiotherapy. A combination of both therapies can produce a synergistic antitumor effect (Suryawanshi et al. 2017). Limited studies are available on the therapeutic effect of this combination. It is important to further investigate the synergistic effect between oncolytic virotherapy and radiotherapy.

8 Preclinical Studies of Oncolytic Viruses Against Breast Cancer

Numerous preclinical investigations conducted over the past 20 years have demonstrated encouraging antitumor properties exhibited by a diverse array of oncolytic viruses (OVs), a substantial portion of which have progressed into clinical trial stages. Talimogene Laherprepvac (T-Vec), a genetically modified herpes simplex virus that carries the gene for granulocyte-macrophage colony-stimulating factor (GM-CSF), has recently obtained approval from the FDA for its application in treating metastatic melanoma (Ejaz and Syed 2022). Breast cancer stands out as a frequently examined neoplasm in preclinical investigations, particularly in the realm of oncolytic virus exploration. In preclinical investigations, a myriad of viral species have been examined for their oncolytic potential in the context of breast cancer. Among these, adenovirus stands out as a frequently explored candidate for its oncolytic properties. Transductional targeting and transcriptional targeting are the two primary strategies employed by oncolytic adenoviruses to specifically infect and replicate within cancer cells. These strategies allow the virus to selectively target cancer cells, enhancing its efficacy as an anticancer agent (Ejaz and Syed 2022).

Bauerschmitz et al. enhanced the oncolytic efficacy of an adenovirus by specifically targeting cancer stem-like cells (CD44 + CD24) derived from pleural effusion in individuals diagnosed with breast cancer (Bauerschmitz et al. 2008). Transductional and transcriptional targeting methodologies were employed. The scientists employed a chimeric fiber protein derived from human adenovirus serotypes 3 and 5 to facilitate transductional targeting. In addition, the researchers examined multiple promoters that serve to restrict the expression of the crucial E1a gene that plays a vital role in the replication of the virus, particularly breast cancer cells. Among the various promoters that were subjected to transcriptional targeting experiments, it was observed that the promoters derived from the cyclo-oxygenase-2 and multi-drug resistance genes exhibited the highest level of activity within the BrCa CD44 + CD24 population. The adenovirus designs under the regulation of promoters effectively eliminated the targeted cells in an artificial environment and exhibited antitumor effects when tested on mice in a live setting.

In a similar vein, Liikanen and colleagues investigated an oncolytic adenovirus that employed transductional targeting through a chimeric fiber and contained a complete anti-HER2 antibody within its genetic code. The viral agent demonstrated significant efficacy in vitro against HER2-positive breast cancer cells, leading to their effective elimination. Moreover, when tested on mouse models with HER2-positive malignancies, the virus exhibited a substantial antitumor impact (Liikanen et al. 2016). In a study conducted by Xu et al., a notable investigation was carried out involving the manipulation of the adenovirus fiber protein. The objective was to introduce the Lyp-1 peptide into the protein structure. This modification resulted in a reduction in liver toxicity, while simultaneously augmenting the ability of the virus to infect breast cancer cells. It is worth noting that breast cancer cells typically exhibit a substantial presence of the Lyp-1 receptor (Xu et al. 2020). Additionally, the viral structure was engineered to incorporate a decoy molecule that mimics the transforming growth factor-beta (TGF-β) compound. In murine models of TNBC, this particular virus induced a strong immune response against the tumor and enhanced the effectiveness of anti-PD-L1 and anti-CTLA-4 treatments (Xu et al. 2020).

In addition to adenoviruses, the herpes simplex virus, and vaccinia virus are frequently investigated due to their oncolytic capabilities. In contrast to adenoviruses, herpes simplex viruses lack a distinct receptor for cellular entry, resulting in their usual focus on transcriptional regulation rather than transductional mechanisms (Akhtar and Shukla 2009). Nevertheless, there have been reports indicating the potential to redirect herpes simplex viruses from their inherent receptors to receptors that are specific to cancer cells. Menotti et al. conducted an experiment where they genetically modified a variant of the HSV-1. This engineered variant was specifically redirected to interact with the HER2 protein, thereby allowing the virus to exclusively enter cells through the HER2 receptor. Following this, the genetically altered virus exhibited oncolytic capabilities in models of ovarian and breast cancer (Menotti et al. 2008). The experimental treatment was introduced into the mice's bodies through the use of mesenchymal stromal cells, resulting in a significant reduction in the growth of the main tumor and a decrease in the spread of breast cancer to the brain (Leoni et al. 2015).

Ghouse et al. conducted an investigation involving the utilization of an oncolytic HSV in a preclinical model of TNBC. The viral genome underwent modifications to incorporate the genetic instructions for the cytokine IL-12, while simultaneously undergoing the removal of three specific genes (47, 34.5, and ICP6). These genetic alterations were implemented to enhance the virus's ability to selectively target cancer cells. The viral agent exhibited a notable decline in both the expansion of the primary tumor and the dissemination of metastatic cells within a murine model of TNBC (Ghouse et al. 2020). In addition, the IL-12 encoded by the virus played a crucial role in inhibiting the process of tumor angiogenesis (Chaurasiya and Fong 2021).

9 Clinical Trials of Oncolytic Virotherapy for Breast Cancer

In the realm of clinical trials pertaining to breast cancer, a plethora of virotherapy approaches have been scrutinized, either as independent therapeutic interventions or in conjunction with other pharmaceutical agents. The phase I trial conducted by Shahin et al. in 2012 involved the utilization of the oncolytic HSV-1 variant HF10 (Sahin et al. 2012). This investigation comprised a total of six patients exhibiting recurrent breast cancer. In the same patient, a single tumor nodule was subjected to viral injections, while another nodule underwent sham injections with a control substance. Upon administration of a maximum dose of three injections, each containing a cumulative sum of 5e05 plaque-forming units (PFU), the virus exhibited remarkable tolerance. Significantly, the process of viral replication was observed within the tumor, and in comparison to tumors that were not treated with the virus, tumors that received the virus exhibited a notable increase in the abundance of CD8+ T cells. The findings of the study unveiled that HF10 exhibited promising characteristics, thus warranting further exploration through additional clinical investigations. Nevertheless, there have been no further announcements regarding the initiation of supplementary clinical trials aimed at investigating the efficacy of breast cancer as a potential therapeutic intervention for this particular viral infection.

The findings of a phase II clinical trial involving the utilization of oncolytic reovirus, also known as pelareorep, in combination with paclitaxel among individuals diagnosed with metastatic breast cancer were documented by Bernstein et al. (2018). Intravenous administration of both the virus and paclitaxel was administered to a cohort of 74 participants as part of this Canadian trial. There was no discernible disparity in the duration of time without disease progression among patients who received solely paclitaxel and those who received the amalgamation of paclitaxel and pelareorep. However, it is noteworthy that the combined treatment notably enhanced the overall survival rate in comparison to the administration of paclitaxel alone. In an ongoing phase II trial (NCT04215146), the combination of avelumab, a monoclonal antibody that targets the PD-L1 protein, with pelareorep and paclitaxel is being investigated to further explore its potential in the context of breast cancer patients. The administration of all treatment drugs in this ongoing investigation will be conducted via the intravenous route.

Currently, a diverse array of oncolytic viruses are undergoing clinical trials to evaluate their efficacy when used in conjunction with chemotherapy and/or immunotherapy for the treatment of different manifestations of breast cancer. Nevertheless, the results of these investigations have not yet been disseminated, thereby leaving the inquiry regarding the potential enhancement of treatment outcomes for breast cancer (Chaurasiya and Fong 2021). Table 2 describes the clinical trials of ocnolytic viruses against breast cancer.

Table 2 Oncolytic viruses tested against breast cancer in clinical studies

Virus	Modification	Combination	Trial's phase	Route of administration	Country	References
Talimogene laherparepvec (HSV-1)	γ34.5-deleted encodes GM-CSF	Paclitaxel	I & II	I.T	USA	NCT02779855
Pelareorep (Reovirus)	Naturally oncolytic	Paclitaxel; Avelumab	II	I.V	USA	NCT04215146
PVSRIPO (Poliovirus)	Insertion of heterologous IRES from human rhinovirus type 2	None	I	I.T	USA	NCT03564782
Pexa-Vec (Vaccinia virus)	Deleted of J2R; encodes GM-CSF	Ipilimumab	I	I.T	France	NCT02977156
Adenovirus	Encodes HSV-TK	Valacyclovir; Pembrolizumab; radiation therapy	II	I.T	USA	NCT03004183
Pexa-Vec (Vaccinia virus)	Deleted of J2R; encodes GM-CSF	Cyclophosphamide	I & II	I.V	France	NCT02630368
TBio-6517 (Vaccinia virus)	25 kb deletion from the virus genome	Pembrolizumab	I & II	I.T	USA	NCT04301011
Talimogene laherparepvec (Herpes Simplex Virus –1)	γ34.5 deletion; encode GF-CSF	Ipilimumab; Nivolumab	I	I.T	USA	NCT04185311
Paleoreorep (Reovirus)	Naturally oncolytic	Letrozole; Atezolizumab; Trastuzumab	I	I.V	Spain	NCT04102618
CAdVEC (Adenovirus)	HER2-specific CAR-T cells	HER2-specific CAR-T cells	I	I.T	USA	NCT03740256

10 Conclusion

Oncolytic viruses, including both naturally occurring or engineered, possess the remarkable capability to undergo selective replication within tumor cells. These viruses have been extensively acknowledged for their profound capacity to induce the demise of cancerous cells. Numerous years of diligent scientific investigation have illuminated the intricate mechanism through which oncolytic viruses effectively eradicate tumors. Oncolytic viruses exhibit the ability to not only induce tumor cell death via viral replication but also possess the capacity to transform cold tumors into hot tumors by modulating the secretion of immune-related molecules. This process leads to an augmentation in the immunogenicity of tumors, as well as the facilitation of immune cell maturation, migration, and infiltration. Consequently, these mechanisms collectively contribute to the enhancement of the antitumor immune response. Furthermore, the impact of oncolytic viruses extends to the modulation of neovascularization within tumors, resulting in the inhibition of tumor growth.

OVs have emerged as a novel avenue in the field of cancer therapy, garnering significant attention from scientists worldwide. Notably, certain countries have granted clinical approval for these therapeutic agents. The realm of oncolytic virotherapy has now been validated in terms of its safety, subsequent to initial clinical setbacks, as evidenced by the endorsement of T-VEC and other substances in their final phases of trials. Despite significant advancements in mitigating the initial apprehensions, encompassing aspects such as safety and administration convenience, there remain a multitude of challenges that necessitate further attention and resolution.

The current advancement in the field of oncolytic virotherapy is boosted by a transition in emphasis toward the measure of their effectiveness. There is a growing focus on investigating targeting techniques that exhibit a higher degree of innovation. These techniques employ a combination of selective replication and targeted cancer cell destruction. Numerous valuable insights can be gleaned from diverse disciplines that employ viral vectors, such as gene therapy and viral vector vaccines, both of which are now witnessing significant advancements in clinical application. When it comes to cancer treatment, or the treatment of any specific type of cancer, a silver bullet is not likely. Therefore, it is crucial that we keep a large collection of vectors and methods to help the field overcome the many challenges of bringing these medicines into clinical practice.

References

Achard C, Surendran A, Wedge M-E, Ungerechts G, Bell J, Ilkow CS (2018) Lighting a fire in the tumor microenvironment using oncolytic immunotherapy. EBioMedicine 31:17–24

Admoun C, Mayrovitz HN (2022) The etiology of breast cancer. Exon Publications, pp 21–30

Akhtar J, Shukla D (2009) Viral entry mechanisms: cellular and viral mediators of herpes simplex virus entry. FEBS J 276:7228–7236

Alberts P, Tilgase A, Rasa A, Bandere K, Venskus D (2018) The advent of oncolytic virotherapy in oncology: the Rigvir® story. Eur J Pharmacol 837:117–126

Ang L, Guo L, Wang J, Huang J, Lou X, Zhao M (2020) Oncolytic virotherapy armed with an engineered interfering lncRNA exhibits antitumor activity by blocking the epithelial mesenchymal transition in triple-negative breast cancer. Cancer Lett 479:42–53

Arulanandam R, Taha Z, Garcia V, Selman M, Chen A, Varette O, Jirovec A, Sutherland K, Macdonald E, Tzelepis F (2020) The strategic combination of trastuzumab emtansine with oncolytic rhabdoviruses leads to therapeutic synergy. Commun Biol 3:254

Ashkenazi A (2002) Targeting death and decoy receptors of the tumour-necrosis factor superfamily. Nat Rev Cancer 2:420–430

Askari FS, Mohebbi A, Moradi A, Javid N (2021) The role of vesicular stomatitis virus matrix protein in autophagy in the breast cancer. Asian Pac J Cancer Prevent APJCP 22:249

Au GG, Lincz LF, Enno A, Shafren DR (2007) Oncolytic coxsackievirus A21 as a novel therapy for multiple myeloma. Br J Haematol 137:133–141

Balachandran S, Barber GN (2000) Vesicular stomatitis virus (VSV) therapy of tumors. IUBMB Life 50:135–138

Bauerschmitz GJ, Ranki T, Kangasniemi L, Ribacka C, Eriksson M, Porten M, Herrmann I, Ristimäki A, Virkkunen P, Tarkkanen M (2008) Tissue-specific promoters active in CD44+ CD24−/low breast cancer cells. Cancer Res 68:5533–5539

Bell JC, Lichty B, Stojdl D (2003) Getting oncolytic virus therapies off the ground. Cancer Cell 4: 7–11

Bellanger M, Zeinomar N, Tehranifar P, Terry MB (2018) Are global breast cancer incidence and mortality patterns related to country-specific economic development and prevention strategies? J Glob Oncol 4:1

Bergman I, Whitaker-Dowling P, Gao Y, Griffin JA (2004) Preferential targeting of vesicular stomatitis virus to breast cancer cells. Virology 330:24–33

Bernstein V, Ellard S, Dent S, Tu D, Mates M, Dhesy-Thind S, Panasci L, Gelmon K, Salim M, Song X (2018) A randomized phase II study of weekly paclitaxel with or without pelareorep in patients with metastatic breast cancer: final analysis of Canadian Cancer Trials Group IND. 213. Breast Cancer Res Treat 167:485–493

Berry JT, Muñoz LE, Stewart RMR, Selvaraj P, Mainou BA (2020) Doxorubicin conjugation to reovirus improves oncolytic efficacy in triple-negative breast cancer. Mol Ther -Oncolyt 18: 556–572

Bieler A, Mantwill K, Holzmüller R, Jürchott K, Kaszubiak A, Stärk S, Glockzin G, Lage H, Grosu A-L, Gansbacher B (2008) Impact of radiation therapy on the oncolytic adenovirus dl520: implications on the treatment of glioblastoma. Radiother Oncol 86:419–427

Bourgeois-Daigneault M-C, St-Germain LE, Roy DG, Pelin A, Aitken AS, Arulanandam R, Falls T, Garcia V, Diallo J-S, Bell JC (2016) Combination of paclitaxel and MG1 oncolytic virus as a successful strategy for breast cancer treatment. Breast Cancer Res 18:1–10

Bourgeois-Daigneault M-C, Roy DG, Aitken AS, El Sayes N, Martin NT, Varette O, Falls T, St-Germain LE, Pelin A, Lichty BD (2018) Neoadjuvant oncolytic virotherapy before surgery sensitizes triple-negative breast cancer to immune checkpoint therapy. Sci Transl Med 10: eaao1641

Bramante S, Koski A, Liikanen I, Vassilev L, Oksanen M, Siurala M, Heiskanen R, Hakonen T, Joensuu T, Kanerva A (2016) Oncolytic virotherapy for treatment of breast cancer, including triple-negative breast cancer. Onco Targets Ther 5:e1078057

Breitbach CJ, Lichty BD, Bell JC (2016) Oncolytic viruses: therapeutics with an identity crisis. EBioMedicine 9:31–36

Carter ME, Koch A, Lauer UM, Hartkopf AD (2021) Clinical trials of oncolytic viruses in breast cancer. Front Oncol 11:803050

Cavalieri E, Chakravarti D, Guttenplan J, Hart E, Ingle J, Jankowiak R, Muti P, Rogan E, Russo J, Santen R (2006) Catechol estrogen quinones as initiators of breast and other human cancers:

implications for biomarkers of susceptibility and cancer prevention. Biochim Biophys Acta (BBA)-Rev Cancer 1766:63–78

Chalikonda S, Kivlen MH, O'Malley ME, McCart J, Gorry M, Yin X, Brown C, Zeh H, Guo Z, Bartlett D (2008) Oncolytic virotherapy for ovarian carcinomatosis using a replication-selective vaccinia virus armed with a yeast cytosine deaminase gene. Cancer Gene Ther 15:115–125

Chaurasiya S, Fong Y (2021) Viroimmunotherapy for breast cancer: promises, problems and future directions. Cancer Gene Ther 28:757–768

Chaurasiya S, Yang A, Kang S, Lu J, Kim S-I, Park AK, Sivanandam V, Zhang Z, Woo Y, Warner SG (2020) Oncolytic poxvirus CF33-hNIS-ΔF14. 5 favorably modulates tumor immune microenvironment and works synergistically with anti-PD-L1 antibody in a triple-negative breast cancer model. Onco Targets Ther 9:1729300

Chen S, Li YQ, Yin XZ, Li SZ, Zhu YL, Fan YY, Li WJ, Cui YL, Zhao J, Li X (2019) Recombinant adenoviruses expressing apoptin suppress the growth of MCF-7 breast cancer cells and affect cell autophagy. Oncol Rep 41:2818–2832

Cheng P-H, Lian S, Zhao R, Rao X-M, McMasters KM, Zhou HS (2013) Combination of autophagy inducer rapamycin and oncolytic adenovirus improves antitumor effect in cancer cells. Virol J 10:1–13

Chinnaiyan AM, Prasad U, Shankar S, Hamstra DA, Shanaiah M, Chenevert TL, Ross BD, Rehemtulla A (2000) Combined effect of tumor necrosis factor-related apoptosis-inducing ligand and ionizing radiation in breast cancer therapy. Proc Natl Acad Sci 97:1754–1759

Cody JJ, Hurst DR (2015) Promising oncolytic agents for metastatic breast cancer treatment. Oncolyt Virother:63–73

De Matos AL, Franco LS, McFadden G (2020) Oncolytic viruses and the immune system: the dynamic duo. Mol Ther -Methods Clin Dev 17:349–358

Denton NL, Chen C-Y, Scott TR, Cripe TP (2016) Tumor-associated macrophages in oncolytic virotherapy: friend or foe? Biomedicine 4:13

Denton NL, Chen C-Y, Hutzen B, Currier MA, Scott T, Nartker B, Leddon JL, Wang P-Y, Srinivas R, Cassady KA (2018) Myelolytic treatments enhance oncolytic herpes virotherapy in models of Ewing sarcoma by modulating the immune microenvironment. Mol Ther -Oncolyt 11:62–74

Dong X, Qu W, Ma S, Zhu Z, Zheng C, He A, Karlsson A, Xu K, Zheng X (2013) Potent antitumoral effects of targeted promoter-driven oncolytic adenovirus armed with Dm-dNK for breast cancer in vitro and in vivo. Cancer Lett 328:95–103

Ejaz M, Syed MA (2022) Therapeutic cancer vaccines; past, present, and future aspects. In: Rezaei N (ed) Handbook of cancer and immunology. Springer, Cham, pp 1–21

Ejaz M, Javed S, Hamza M, Tabassum S, Abubakar M, Ullah I (2020) Fungal endophytes are effective alternatives and novel sources of anticancer drugs. Punjab Univ J Zool 35:13–24

Engeland CE, Ungerechts G (2021) Measles virus as an oncolytic immunotherapy. Cancers 13:544

Ferguson MS, Dunmall LSC, Gangeswaran R, Marelli G, Tysome JR, Burns E, Whitehead MA, Aksoy E, Alusi G, Hiley C (2020) Transient inhibition of PI3Kδ enhances the therapeutic effect of intravenous delivery of oncolytic vaccinia virus. Mol Ther 28:1263–1275

Fernandez M, Porosnicu M, Markovic D, Barber GN (2002) Genetically engineered vesicular stomatitis virus in gene therapy: application for treatment of malignant disease. J Virol 76: 895–904

Filley AC, Dey M (2017) Immune system, friend or foe of oncolytic virotherapy? Front Oncol 7: 106

Frazier VN, Holl E, Brown M, Boczkowski D, Landa K, Hwang S, Gromeier M, Nair SK (2020) Oncolytic poliovirus immunotherapy for breast cancer. J Immunol 204:249.225

Gazitt Y (1999) TRAIL is a potent inducer of apoptosis in myeloma cells derived from multiple myeloma patients and is not cytotoxic to hematopoietic stem cells. Leukemia 13:1817–1824

Gentschev I, Stritzker J, Hofmann E, Weibel S, Yu Y, Chen N, Zhang Q, Bullerdiek J, Nolte I, Szalay A (2009) Use of an oncolytic vaccinia virus for the treatment of canine breast cancer in nude mice: preclinical development of a therapeutic agent. Cancer Gene Ther 16:320–328

Gholami S, Marano A, Chen NG, Aguilar RJ, Frentzen A, Chen C-H, Lou E, Fujisawa S, Eveno C, Belin L (2014) A novel vaccinia virus with dual oncolytic and anti-angiogenic therapeutic effects against triple-negative breast cancer. Breast Cancer Res Treat 148:489–499

Ghouse SM, Nguyen H-M, Bommareddy PK, Guz-Montgomery K, Saha D (2020) Oncolytic herpes simplex virus encoding IL12 controls triple-negative breast cancer growth and metastasis. Front Oncol 10:384

Gil M, Seshadri M, Komorowski MP, Abrams SI, Kozbor D (2013) Targeting CXCL12/CXCR4 signaling with oncolytic virotherapy disrupts tumor vasculature and inhibits breast cancer metastases. Proc Natl Acad Sci 110:E1291–E1300

Gonzalez-Angulo A, Blumenschein G Jr (2013) Defining biomarkers to predict sensitivity to PI3K/Akt/mTOR pathway inhibitors in breast cancer. Cancer Treat Rev 39:313–320

Graat H, Witlox M, Schagen F, Kaspers G, Helder M, Bras J, Schaap G, Gerritsen W, Wuisman P, Van Beusechem V (2006) Different susceptibility of osteosarcoma cell lines and primary cells to treatment with oncolytic adenovirus and doxorubicin or cisplatin. Br J Cancer 94:1837–1844

Guo ZS, Liu Z, Kowalsky S, Feist M, Kalinski P, Lu B, Storkus WJ, Bartlett DL (2017) Oncolytic immunotherapy: conceptual evolution, current strategies, and future perspectives. Front Immunol 8:555

Hanahan D, Weinberg RA (2000) The hallmarks of cancer. Cell 100:57–70

Hartkopf AD, Fehm T, Wallwiener D, Lauer UM (2011) Oncolytic virotherapy of breast cancer. Gynecol Oncol 123:164–171

Hartkopf A, Fehm T, Wallwiener M, Lauer U (2012) Oncolytic viruses to treat ovarian cancer patients – a review of results from clinical trials. Tumor Diagnostik Ther 33:279–283

Hata Y, Etoh T, Inomata M, Shiraishi N, Nishizono A, Kitano S (2008) Efficacy of oncolytic reovirus against human breast cancer cells. Oncol Rep 19:1395–1398

Hemminki O, Dos Santos JM, Hemminki A (2020) Oncolytic viruses for cancer immunotherapy. J Hematol Oncol 13:1–15

Higginson IJ, Costantini M (2008) Dying with cancer, living well with advanced cancer. Eur J Cancer 44:1414–1424

Hou F, Sun L, Zheng H, Skaug B, Jiang Q-X, Chen ZJ (2011) MAVS forms functional prion-like aggregates to activate and propagate antiviral innate immune response. Cell 146:448–461

Iankov ID, Kurokawa CB, D'Assoro AB, Ingle JN, Domingo-Musibay E, Allen C, Crosby CM, Nair AA, Liu MC, Aderca I (2015) Inhibition of the Aurora A kinase augments the anti-tumor efficacy of oncolytic measles virotherapy. Cancer Gene Ther 22:438–444

Ilkow CS, Swift SL, Bell JC, Diallo J-S (2014) From scourge to cure: tumour-selective viral pathogenesis as a new strategy against cancer. PLoS Pathog 10:e1003836

Jardé T, Perrier S, Vasson M-P, Caldefie-Chézet F (2011) Molecular mechanisms of leptin and adiponectin in breast cancer. Eur J Cancer 47:33–43

Jin S, Wang Q, Wu H, Pang D, Xu S (2021) Oncolytic viruses for triple negative breast cancer and beyond. Biomarker Res 9:1–16

Kaufman HL, Kohlhapp FJ, Zloza A (2015) Oncolytic viruses: a new class of immunotherapy drugs. Nat Rev Drug Discov 14:642–662

Kelly E, Russell SJ (2007) History of oncolytic viruses: genesis to genetic engineering. Mol Ther 15:651–659

Kleibl Z, Kristensen VN (2016) Women at high risk of breast cancer: molecular characteristics, clinical presentation and management. Breast 28:136–144

Kwan A, Winder N, Atkinson E, Al-Janabi H, Allen RJ, Hughes R, Moamin M, Louie R, Evans D, Hutchinson M (2021a) Macrophages mediate the antitumor effects of the oncolytic virus HSV1716 in mammary tumors. Mol Cancer Ther 20:589–601

Kwan A, Winder N, Muthana M (2021b) Oncolytic virotherapy treatment of breast cancer: barriers and recent advances. Viruses 13:1128

Lal G, Rajala MS (2019) Combination of oncolytic measles virus armed with BNiP3, a pro-apoptotic gene and paclitaxel induces breast cancer cell death. Front Oncol 8:676

Larson C, Oronsky B, Scicinski J, Fanger GR, Stirn M, Oronsky A, Reid TR (2015) Going viral: a review of replication-selective oncolytic adenoviruses. Oncotarget 6:19976

Lawler SE, Speranza M-C, Cho C-F, Chiocca EA (2017) Oncolytic viruses in cancer treatment: a review. JAMA Oncol 3:841–849

Lei S, Zheng R, Zhang S, Wang S, Chen R, Sun K, Zeng H, Zhou J, Wei W (2021) Global patterns of breast cancer incidence and mortality: a population-based cancer registry data analysis from 2000 to 2020. Cancer Commun 41:1183–1194

Leoni V, Gatta V, Palladini A, Nicoletti G, Ranieri D, Dall'Ora M, Grosso V, Rossi M, Alviano F, Bonsi L (2015) Systemic delivery of HER2-retargeted oncolytic-HSV by mesenchymal stromal cells protects from lung and brain metastases. Oncotarget 6:34774

Leveille S, Samuel S, Goulet M, Hiscott J (2011) Enhancing VSV oncolytic activity with an improved cytosine deaminase suicide gene strategy. Cancer Gene Ther 18:435–443

Li Y, Xiao F, Zhang A, Zhang D, Nie W, Xu T, Han B, Seth P, Wang H, Yang Y (2020) Oncolytic adenovirus targeting TGF-β enhances anti-tumor responses of mesothelin-targeted chimeric antigen receptor T cell therapy against breast cancer. Cell Immunol 348:104041

Liikanen I, Tähtinen S, Guse K, Gutmann T, Savola P, Oksanen M, Kanerva A, Hemminki A (2016) Oncolytic adenovirus expressing monoclonal antibody trastuzumab for treatment of HER2-positive cancer. Mol Cancer Ther 15:2259–2269

Lin Y, Wang W, Wan J, Yang Y, Fu W, Pan D, Cai L, Cheng T, Huang X, Wang Y (2018) Oncolytic activity of a coxsackievirus B3 strain in human endometrial cancer cell lines. Virol J 15:1–12

Liu R, Varghese S, Rabkin SD (2005) Oncolytic herpes simplex virus vector therapy of breast cancer in C3 (1)/SV40 T-antigen transgenic mice. Cancer Res 65:1532–1540

Liu T-C, Galanis E, Kirn D (2007) Clinical trial results with oncolytic virotherapy: a century of promise, a decade of progress. Nat Clin Pract Oncol 4:101–117

Marcato P, Dean CA, Giacomantonio CA, Lee PW (2009) Oncolytic reovirus effectively targets breast cancer stem cells. Mol Ther 17:972–979

Marchini A, Daeffler L, Pozdeev VI, Angelova A, Rommelaere J (2019) Immune conversion of tumor microenvironment by oncolytic viruses: the protoparvovirus H-1PV case study. Front Immunol 10:1848

Marelli G, Howells A, Lemoine NR, Wang Y (2018) Oncolytic viral therapy and the immune system: a double-edged sword against cancer. Front Immunol 9:866

Marra A, Viale G, Curigliano G (2019) Recent advances in triple negative breast cancer: the immunotherapy era. BMC Med 17:1–9

Martini V, D'Avanzo F, Maggiora PM, Varughese FM, Sica A, Gennari A (2020) Oncolytic virotherapy: new weapon for breast cancer treatment. ecancermedicalscience 14:14

Martuza RL, Malick A, Markert JM, Ruffner KL, Coen DM (1991) Experimental therapy of human glioma by means of a genetically engineered virus mutant. Science 252:854–856

Matveeva OV, Chumakov PM (2018) Defects in interferon pathways as potential biomarkers of sensitivity to oncolytic viruses. Rev Med Virol 28:e2008

McCart JA, Ward JM, Lee J, Hu Y, Alexander HR, Libutti SK, Moss B, Bartlett DL (2001) Systemic cancer therapy with a tumor-selective vaccinia virus mutant lacking thymidine kinase and vaccinia growth factor genes. Cancer Res 61:8751–8757

McCart JA, Mehta N, Scollard D, Reilly RM, Carrasquillo JA, Tang N, Deng H, Miller M, Xu H, Libutti SK (2004) Oncolytic vaccinia virus expressing the human somatostatin receptor SSTR2: molecular imaging after systemic delivery using 111In-pentetreotide. Mol Ther 10:553–561

McDonald CJ, Erlichman C, Ingle JN, Rosales GA, Allen C, Greiner SM, Harvey ME, Zollman PJ, Russell SJ, Galanis E (2006) A measles virus vaccine strain derivative as a novel oncolytic agent against breast cancer. Breast Cancer Res Treat 99:177–184

Menotti L, Cerretani A, Hengel H, Campadelli-Fiume G (2008) Construction of a fully retargeted herpes simplex virus 1 recombinant capable of entering cells solely via human epidermal growth factor receptor 2. J Virol 82:10153–10161

Mohamed Amin Z, Che Ani MA, Tan SW, Yeap SK, Alitheen NB, Syed Najmuddin SUF, Kalyanasundram J, Chan SC, Veerakumarasivam A, Chia SL (2019) Evaluation of a recombinant Newcastle disease virus expressing human IL12 against human breast cancer. Sci Rep 9: 13999

Mostafa AA, Meyers DE, Thirukkumaran CM, Liu PJ, Gratton K, Spurrell J, Shi Q, Thakur S, Morris DG (2018) Oncolytic reovirus and immune checkpoint inhibition as a novel immunotherapeutic strategy for breast cancer. Cancers 10:205

Mühlebach MD, Mateo M, Sinn PL, Prüfer S, Uhlig KM, Leonard VH, Navaratnarajah CK, Frenzke M, Wong XX, Sawatsky B (2011) Adherens junction protein nectin-4 is the epithelial receptor for measles virus. Nature 480:530–533

Nemunaitis J, Senzer N, Sarmiento S, Zhang Y, Arzaga R, Sands B, Maples P, Tong A (2007) A phase I trial of intravenous infusion of ONYX-015 and enbrel in solid tumor patients. Cancer Gene Ther 14:885–893

Niavarani S-R, Lawson C, Boudaud M, Simard C, Tai L-H (2020) Oncolytic vesicular stomatitis virus – based cellular vaccine improves triple-negative breast cancer outcome by enhancing natural killer and CD8+ T-cell functionality. J Immunother Cancer 8

Nielsen LL, Dell J, Maxwell E, Armstrong L, Maneval D, Catino JJ (1997) Efficacy of p53 adenovirus-mediated gene therapy against human breast cancer xenografts. Cancer Gene Ther 4:129–138

Niemann J, Woller N, Brooks J, Fleischmann-Mundt B, Martin NT, Kloos A, Knocke S, Ernst AM, Manns MP, Kubicka S (2019) Molecular retargeting of antibodies converts immune defense against oncolytic viruses into cancer immunotherapy. Nat Commun 10:3236

Norman KL, Coffey MC, Hirasawa K, Demetrick DJ, Nishikawa SG, DiFrancesco LM, Strong JE, Lee PW (2002) Reovirus oncolysis of human breast cancer. Hum Gene Ther 13:641–652

O'Bryan SM, Mathis JM (2018) Oncolytic virotherapy for breast cancer treatment. Curr Gene Ther 18:192–205

Otani Y, Yoo JY, Shimizu T, Kurozumi K, Date I, Kaur B (2022) Implications of immune cells in oncolytic herpes simplex virotherapy for glioma. Brain Tumor Pathol 39:57–64

Ottolino-Perry K, Diallo J-S, Lichty BD, Bell JC, McCart JA (2010) Intelligent design: combination therapy with oncolytic viruses. Mol Ther 18:251–263

Pak-Wittel MA, Yang L, Sojka DK, Rivenbark JG, Yokoyama WM (2013) Interferon-γ mediates chemokine-dependent recruitment of natural killer cells during viral infection. Proc Natl Acad Sci 110:E50–E59

Paplomata E, O'Regan R (2014) The PI3K/AKT/mTOR pathway in breast cancer: targets, trials and biomarkers. Therapeutic advances in medical oncology 6:154–166

Perez MC, Miura JT, Naqvi SMH, Kim Y, Holstein A, Lee D, Sarnaik AA, Zager JS (2018) Talimogene laherparepvec (TVEC) for the treatment of advanced melanoma: a single-institution experience. Ann Surg Oncol 25:3960–3965

Printz C (2016) FDA approves new treatment of metastatic melanoma. Cancer 122:1149–1149

Redig AJ, McAllister SS (2013) Breast cancer as a systemic disease: a view of metastasis. J Intern Med 274:113–126

Ribas A, Dummer R, Puzanov I, VanderWalde A, Andtbacka RH, Michielin O, Olszanski AJ, Malvehy J, Cebon J, Fernandez E (2017) Oncolytic virotherapy promotes intratumoral T cell infiltration and improves anti-PD-1 immunotherapy. Cell 170:1109–1119.e1110

Roberts KL, Smith GL (2008) Vaccinia virus morphogenesis and dissemination. Trends Microbiol 16:472–479

Roth JA (2006) Adenovirus p53 gene therapy. Expert Opin Biol Ther 6:55–61

Russell SJ, Peng K-W (2007) Viruses as anticancer drugs. Trends Pharmacol Sci 28:326–333

Russell SJ, Peng K-W, Bell JC (2012) Oncolytic virotherapy. Nat Biotechnol 30:658–670

Sahin TT, Kasuya H, Nomura N, Shikano T, Yamamura K, Gewen T, Kanzaki A, Fujii T, Sugae T, Imai T (2012) Impact of novel oncolytic virus HF10 on cellular components of the tumor microenviroment in patients with recurrent breast cancer. Cancer Gene Ther 19:229–237

Sarkar D, Su Z-z, Vozhilla N, Park ES, Gupta P, Fisher PB (2005) Dual cancer-specific targeting strategy cures primary and distant breast carcinomas in nude mice. Proc Natl Acad Sci 102: 14034–14039

Schneeweiss A, Ruckhäberle E, Huober J (2015) Chemotherapy for metastatic breast cancer–an anachronism in the era of personalised and targeted oncological therapy? Geburtshilfe Frauenheilkd 75:574–583

Seubert C, Stritzker J, Hess M, Donat U, Sturm J, Chen N, von Hof JM, Krewer B, Tietze LF, Gentschev I (2011) Enhanced tumor therapy using vaccinia virus strain GLV-1h68 in combination with a β-galactosidase-activatable prodrug seco-analog of duocarmycin SA. Cancer Gene Ther 18:42–52

Shin DH, Nguyen T, Ozpolat B, Lang F, Alonso M, Gomez-Manzano C, Fueyo J (2021) Current strategies to circumvent the antiviral immunity to optimize cancer virotherapy. J Immunother Cancer 9:e002086

Skelding KA, Barry RD, Shafren DR (2009) Systemic targeting of metastatic human breast tumor xenografts by Coxsackievirus A21. Breast Cancer Res Treat 113:21–30

Smith GL, Vanderplasschen A, Law M (2002) The formation and function of extracellular enveloped vaccinia virus. J Gen Virol 83:2915–2931

Smolarz B, Nowak AZ, Romanowicz H (2022) Breast cancer – epidemiology, classification, pathogenesis and treatment (review of literature). Cancers 14:2569

Sugiyama T, Yoneda M, Kuraishi T, Hattori S, Inoue Y, Sato H, Kai C (2013) Measles virus selectively blind to signaling lymphocyte activation molecule as a novel oncolytic virus for breast cancer treatment. Gene Ther 20:338–347

Sun Y-S, Zhao Z, Yang Z-N, Xu F, Lu H-J, Zhu Z-Y, Shi W, Jiang J, Yao P-P, Zhu H-P (2017) Risk factors and preventions of breast cancer. Int J Biol Sci 13:1387

Sung H, Ferlay J, Siegel RL, Laversanne M, Soerjomataram I, Jemal A, Bray F (2021) Global cancer statistics 2020: GLOBOCAN estimates of incidence and mortality worldwide for 36 cancers in 185 countries. CA Cancer J Clin 71:209–249

Suryawanshi YR, Zhang T, Essani K (2017) Oncolytic viruses: emerging options for the treatment of breast cancer. Med Oncol 34:1–11

Tan G, Kasuya H, Sahin TT, Yamamura K, Wu Z, Koide Y, Hotta Y, Shikano T, Yamada S, Kanzaki A (2015) Combination therapy of oncolytic herpes simplex virus HF10 and bevacizumab against experimental model of human breast carcinoma xenograft. Int J Cancer 136:1718–1730

Tatsuo H, Ono N, Tanaka K, Yanagi Y (2000) SLAM (CDw150) is a cellular receptor for measles virus. Nature 406:893–897

Teshigahara O, Goshima F, Takao K, Kohno S, Kimata H, Nakao A, Nishiyama Y (2004) Oncolytic viral therapy for breast cancer with herpes simplex virus type 1 mutant HF 10. J Surg Oncol 85: 42–47

Thomas MA, Spencer JF, Toth K, Sagartz JE, Phillips NJ, Wold WS (2008) Immunosuppression enhances oncolytic adenovirus replication and antitumor efficacy in the Syrian hamster model. Mol Ther 16:1665–1673

Tian Y, Xie D, Yang L (2022) Engineering strategies to enhance oncolytic viruses in cancer immunotherapy. Signal Transduct Target Ther 7:117

Umer BA, Noyce RS, Franczak BC, Shenouda MM, Kelly RG, Favis NA, Desaulniers M, Baldwin TA, Hitt MM, Evans DH (2020) Deciphering the immunomodulatory capacity of oncolytic vaccinia virus to enhance the immune response to breast cancer. Cancer Immunol Res 8:618–631

Wang H, Chen NG, Minev BR, Szalay AA (2012) Oncolytic vaccinia virus GLV-1h68 strain shows enhanced replication in human breast cancer stem-like cells in comparison to breast cancer cells. J Transl Med 10:1–15

Wang G, Kang X, Chen KS, Jehng T, Jones L, Chen J, Huang XF, Chen S-Y (2020) An engineered oncolytic virus expressing PD-L1 inhibitors activates tumor neoantigen-specific T cell responses. Nat Commun 11:1395

Wu Y, He J, An Y, Wang X, Liu Y, Yan S, Ye X, Qi J, Zhu S, Yu Q (2016) Recombinant Newcastle disease virus (NDV/Anh-IL-2) expressing human IL-2 as a potential candidate for suppresses growth of hepatoma therapy. J Pharmacol Sci 132:24–30

Wu S, Zhu W, Thompson P, Hannun YA (2018) Evaluating intrinsic and non-intrinsic cancer risk factors. Nat Commun 9:3490

Xu W, Yang Y, Hu Z, Head M, Mangold KA, Sullivan M, Wang E, Saha P, Gulukota K, Helseth DL (2020) LyP-1-modified oncolytic adenoviruses targeting transforming growth factor β inhibit tumor growth and metastases and augment immune checkpoint inhibitor therapy in breast cancer mouse models. Hum Gene Ther 31:863–880

Yan Y, Xu H, Wang J, Wu X, Wen W, Liang Y, Wang L, Liu F, Du X (2019) Inhibition of breast cancer cells by targeting E2F-1 gene and expressing IL15 oncolytic adenovirus. Biosci Rep 39: BSR20190384

Yang W, Senger D, Lun X, Muzik H, Shi Z, Dyck R, Norman K, Brasher P, Rewcastle N, George D (2004) Reovirus as an experimental therapeutic for brain and leptomeningeal metastases from breast cancer. Gene Ther 11:1579–1589

Yu W, Fang H (2007) Clinical trials with oncolytic adenovirus in China. Curr Cancer Drug Targets 7:141–148

Yu J, Mu Q, Fung M, Xu X, Zhu L, Ho RJ (2022) Challenges and opportunities in metastatic breast cancer treatments: Nano-drug combinations delivered preferentially to metastatic cells may enhance therapeutic response. Pharmacol Ther 236:108108

Zhang J, Ramesh N, Chen Y, Li Y, Dilley J, Working P, Yu D-C (2002) Identification of human uroplakin II promoter and its use in the construction of CG8840, a urothelium-specific adenovirus variant that eliminates established bladder tumors in combination with docetaxel. Cancer Res 62:3743–3750

Zhang Q, Yu YA, Wang E, Chen N, Danner RL, Munson PJ, Marincola FM, Szalay AA (2007) Eradication of solid human breast tumors in nude mice with an intravenously injected light-emitting oncolytic vaccinia virus. Cancer Res 67:10038–10046

Zhu W, Wei L, Zhang H, Chen J, Qin X (2012) Oncolytic adenovirus armed with IL-24 inhibits the growth of breast cancer in vitro and in vivo. J Exp Clin Cancer Res 31:1–10

Zhu W, Zhang H, Shi Y, Song M, Zhu B, Wei L (2013) Oncolytic adenovirus encoding tumor necrosis factor-related apoptosis inducing ligand (TRAIL) inhibits the growth and metastasis of triple-negative breast cancer. Cancer Biol Ther 14:1016–1023

Zou H, Mou XZ, Zhu B (2023) Combining of oncolytic Virotherapy and other immunotherapeutic approaches in cancer: a powerful functionalization tactic. Global Chall 7:2200094

Overcoming Breast Cancer Resistance to Targeted, Chemo-, and Immuno Therapy: Mechanisms and Strategies to Overcome Resistance and Facilitate Drug Development

Adam Hermawan, Sisca Ucche, Wilfan Ibadurrahman, Kristoforus Patrana Ardi, and Chris Djerico

Abstract

Breast cancer substantially impacts female mortality, and symptoms can differ among individuals. Hence, it is crucial to tailor treatment modalities to individual patients to ensure efficient therapy. The main biomarkers for breast cancer include the estrogen receptor (ER), progesterone receptor (PR), and human epidermal growth factor receptor 2 (HER2/ERBB2), typically used to classify various types of breast cancer, such as luminal, HER2-positive (HER2+), and triple-negative breast cancer, and to determine the most suitable treatment strategies. Despite technological improvements and continuous research on drugs, including chemotherapy, targeted therapy, and immunotherapy, mortality rates for breast cancer remain markedly high owing to drug resistance. This chapter provides an in-depth examination of the fundamental mechanisms that contribute to resistance against several treatment modalities, including endocrine therapy, anti-HER2-targeted therapy, chemotherapy, and immunotherapy. Furthermore, the chapter discusses

A. Hermawan (✉)
Laboratory of Macromolecular Engineering, Department of Pharmaceutical Chemistry, Faculty of Pharmacy, Universitas Gadjah Mada Sekip Utara II, Yogyakarta, Indonesia

Cancer Chemoprevention Research Center, Faculty of Pharmacy, Universitas Gadjah Mada Sekip Utara II, Yogyakarta, Indonesia

Laboratory of Advanced Pharmaceutical Sciences. APSLC Building, Faculty of Pharmacy, Universitas Gadjah Mada Sekip Utara II, Yogyakarta, Indonesia
e-mail: adam_apt@ugm.ac.id

S. Ucche
Laboratory of Pharmacology and Toxicology, Department of Pharmacology and Clinical Pharmacy, Faculty of Pharmacy, Universitas Gadjah Mada Sekip Utara II, Yogyakarta, Indonesia

W. Ibadurrahman · K. P. Ardi · C. Djerico
Laboratory of Macromolecular Engineering, Department of Pharmaceutical Chemistry, Faculty of Pharmacy, Universitas Gadjah Mada Sekip Utara II, Yogyakarta, Indonesia

© The Author(s), under exclusive license to Springer Nature Switzerland AG 2024
Interdisciplinary Cancer Research, https://doi.org/10.1007/16833_2024_246
Published online: 9 June 2024

recent relevant advances and clinical trials to address resistance and drug development. Specifically, the application of integrated bioinformatics analysis to uncover potential anticancer therapeutics was emphasized. Accordingly, this chapter seeks to provide notable insights and suggestions to accelerate the advancement of anti-breast cancer treatments, with a specific focus on addressing the issue of treatment resistance in breast cancer.

Keywords

Breast cancer · Drug development · Drug resistance · Overcoming resistance

Abbreviations

ABC	ATP-binding cassette
ADCC	Antibody-dependent cell-mediated cytotoxicity
APC	Adenomatous polyposis coli
CDK	Cyclin-dependent kinase
CTLA4	Cytotoxic T lymphocyte–associated protein 4
CTLs	Cytotoxic T lymphocytes
DNA-PK	DNA-dependent protein kinase
DSBs	DNA double-strand breaks
EGFR	Epidermal growth factor receptor
EMT	Epithelial-to-mesenchymal transition
ER	Estrogen receptor
GSH	Glutathione
GST	Glutathione S-transferase
HER2	Human epidermal growth factor receptor 2
HLA	Human leukocyte antigen
HR	Hormone receptor
MDSCs	Myeloid-derived suppressor cells
MHC	Major histocompatibility complex
MRP2	Multidrug resistance protein 2
OS	Overall survival
pCR	Pathological complete response
PD-1/PD-L1	Programmed cell death protein 1/programmed cell death ligand 1
PFS	Progression-free survival
PR	Progesterone receptor
ROS	Reactive oxygen species
SERD	Selective estrogen receptor downregulator
SERM	Selective estrogen receptor modulator
STAT1	Signal transducer and activator of transcription 1
TAM	Tumor-associated macrophage
TMB	Tumor mutation burden
TNBC	Triple-negative breast cancer
Tregs	Regulatory T cells

1 Introduction

Breast cancer remains a notable cause of mortality among females, with several symptoms observed in affected individuals. Consequently, tailoring treatment strategies to individual patients is crucial for delivering effective therapy (Coates et al. 2015). Challenges in the diagnosis and prediction of outcomes can be attributed to the intricate nature and diverse characteristics of breast cancer (Feng et al. 2018). Notably, the diagnosis and prognosis of breast cancer heavily rely on the identification of biomarkers, which serve as indicators that indirectly point toward optimal candidates for therapeutic intervention (Nicolini et al. 2018). The main markers for breast cancer include the estrogen receptor (ER), progesterone receptor (PR), and human epidermal growth factor receptor 2 (HER2/ERBB2) (Bydoun et al. 2014). These three breast cancer markers have been extensively explored to establish the correlation between their expression and the incidence of breast cancer, as well as to identify the most suitable treatment options (Łukasiewicz et al. 2021).

Hormone receptor-positive (HR+) breast cancer accounts for approximately 70% of all cases, that is, ER positive (ER+) and/or PR positive (PR+), considered as luminal subtype, responding effectively to hormonal therapy, for instance, tamoxifen and fulvestrant (Fan et al. 2015). HER2, a transmembrane tyrosine kinase receptor from the human epidermal receptor (HER) family, substantially impacts cancer cell proliferation, differentiation, and migration and is overexpressed in approximately 25–30% of cases of breast cancer (Hsu and Hung 2016). In contrast to other breast cancer subtypes that are regulated by hormonal factors or particular proteins associated with breast tumors, triple-negative breast cancer (TBNC) is a diverse form of breast cancer characterized by the negative expression of ER, PR, and HER2, which has been associated with unfavorable clinical outcomes (Medina et al. 2020). Hence, the development of chemotherapeutic drugs capable of targeting TBNC remains an urgent priority (Hwang et al. 2019).

Despite advancements in technology and ongoing research on breast cancer, current treatment modalities, including chemotherapy, targeted therapy, and immunotherapy, have failed to substantially reduce mortality rates associated with breast cancer (McCart Reed et al. 2019), which can be attributed, at least partially, to drug resistance. Hence, ongoing research efforts aim to establish methods to identify biomarkers associated with resistance to breast cancer, as well as to create pharmaceutical interventions capable of overcoming drug resistance (Saha and Lukong 2022).

The incorporation of information technology and big data has contributed to the improvement of medication development procedures, leading to heightened efficacy and cost-effectiveness (Huang et al. 2016; Patel et al. 2020). The application of omics-based technology, accompanied by progress in data storage, data mining, and machine learning, has resulted in substantial repositories of knowledge regarding diseases, molecular pathways, and treatments (Paananen and Fortino 2020; Reel et al. 2021). Integrative bioinformatics analysis is a complex bioinformatics methodology that is predicated on the interplay between drug, gene, target, and disease networks (Poornima et al. 2016). This methodology is extensively utilized within the domain of pharmaceutical exploration and advancement (Muhammad et al. 2018).

The objective of the current chapter is to provide a comprehensive analysis of the underlying mechanisms that contribute to resistance against various treatment modalities, including endocrine therapy, anti-HER2-targeted therapy, chemotherapy, and immunotherapy. Moreover, this chapter clarifies recent advancements and clinical trials addressing resistance and drug development through the utilization of integrative bioinformatics analysis for anticancer drug discovery. Ultimately, this chapter aims to offer valuable insights and suggestions to expedite the discovery of anti-breast cancer therapeutics, with particular emphasis on the treatment resistance in breast cancer.

2 Breast Cancer Resistance Toward Endocrine, Anti-HER2, Chemotherapy, and Immunotherapy: Mechanism and Strategies to Overcome Resistance

Breast cancer therapy primarily encompasses two categories: localized therapy and systemic therapy (Moo et al. 2018). Local therapy specifically targets and eliminates cancerous cells within a confined region, such as the breast, chest wall, and lymph nodes located in the underarm region. Systemic therapeutic modalities, including hormonal therapy, chemotherapy, and targeted therapy, are increasingly prescribed and administered to individuals diagnosed with early-stage breast cancer (Teven et al. 2017). This section elucidates the underlying processes that mediate resistance to endocrine, anti-HER2, chemotherapy, and immunotherapy.

2.1 Breast Cancer Resistance Toward Endocrine Treatment

Luminal breast cancer is a prevalent subtype, accounting for approximately 30–40% of all cases of invasive breast cancer (Prat et al. 2015). The luminal subtype is distinguished by the presence of the ER+ and PR+, a lack of HER2 expression (HER2−), and a low rate of cell proliferation (Ki-67<14%) (Braun et al. 2013). Luminal breast cancer typically necessitates adjuvant therapy as a primary treatment approach, particularly hormonal therapy, rather than chemotherapy (Rossi et al. 2019). In this section, we focused on the resistance of breast cancer cells to tamoxifen and fulvestrant.

2.1.1 Tamoxifen

Tamoxifen is frequently employed as the primary hormonal treatment agent, particularly in pre- and post-menopausal females and those with ER+ status. Tamoxifen is classified as a selective estrogen receptor modulator (SERM), eliciting agonistic effects in the uterus and antagonistic effects in the breast (Y Maximov et al. 2013). In female patients with luminal breast cancer, tamoxifen was found to reduce the annual mortality rate of breast cancer by 31% (Munzone and Colleoni 2018).

However, a growing prevalence of tamoxifen resistance has been extensively documented (Clarke et al. 2015).

The Mechanism Underlying Tamoxifen Resistance in Breast Cancer

The development of resistance to tamoxifen is a substantial challenge in the clinical treatment of breast cancer (Droog et al. 2013). Tamoxifen resistance can manifest either prior to the initiation of therapy (known as de novo resistance) or after treatment commencement (referred to as acquired tamoxifen resistance) (Ali et al. 2016). Multiple molecular processes have been speculated to contribute to a role in the emergence of tamoxifen resistance, including alterations in ER signaling, stimulation of the PI3K/AKT/mTOR (PI3K) pathway, and initiation of NFκB signaling (Viedma-Rodríguez et al. 2014).

Stimulation of the PI3K/AKT/mTOR (PI3K) Pathway

Anomalies within the PI3K pathway, resulting in the emergence of breast cancer resistance to tamoxifen, have been frequently documented in breast cancer (Martini et al. 2014). According to multiple studies, tumors that display resistance to hormone therapy are characterized by increased Akt/PKB activity within the PI3K signaling pathway (Nicolini et al. 2015). Additionally, a notable intercommunication has been observed between the ER and PI3K signaling pathways (Bostner et al. 2013).

Crosstalk with the NFkB Pathway

In addition, the interplay between ER signaling and NFκB pathway was found to facilitate breast cancer cell resistance to tamoxifen. The activation of NFκB signaling markedly reduces the responsiveness of breast cancer cells resistant to tamoxifen (deGraffenried et al. 2004). The inhibition of the NFκB signaling pathway reportedly reduces tumor aggressiveness and reverses endocrine therapy resistance in patients with breast cancer (Smart et al. 2020).

Crosstalk with Epidermal Growth Factor Receptor (EGFR) Signaling

Tamoxifen resistance could be attributed to interactions between ER signaling pathways that mutually influence each other, a phenomenon commonly referred to as crosstalk with other receptors, such as the EGFR family (Arpino et al. 2008). ER signaling was found to be activated in tamoxifen-resistant breast cancer cells in an estrogen-independent manner, primarily due to increased EGFR signaling (Moerkens et al. 2014).

Recent Clinical Strategy to Overcome Tamoxifen Resistance

Various strategies have been employed to address tamoxifen resistance (Table 1).

Phosphoinositide 3-Kinase (PI3K)/Akt/Mammalian Target of Rapamycin (mTOR) Inhibitors

The PI3K/Akt/mTOR signaling pathway was targeted using a combination of inhibitors using taselisib. In patients with HR+ metastatic breast cancer, combined

Table 1 Combinatorial clinical strategies to overcome molecular mechanisms of breast cancer cell resistance to tamoxifen

No	Target	Drugs	Strategy to overcome resistance/results	References
1	PI3K/Akt/mTOR	Taselisib	Safety and affective taselisib and tamoxifen in patients with metastatic breast cancer	Baird et al. (2019)
		M2698	Tamoxifen/trastuzumab and M2698 exhibit substantial antitumor efficacy in patients with advanced breast cancer	Tsimberidou et al. (2021)
		Everolimus	Everolimus and tamoxifen in patients with metastatic breast cancer	Treilleux et al. (2015)
		TAK-228	Tamoxifen and TAK-228 in ER+ breast cancer is both safe and well tolerated	Koca et al. (2021)
2	CDK	Palbociclib	Palbociclib in combination with exemestane demonstrated promising efficacy for patients with metastatic breast cancer who had previously received tamoxifen treatment	Park et al. (2019)
		Ribociclib	Ribociclib in combination with tamoxifen has been associated with a sustained and significant improvement in overall survival (OS) when compared to tamoxifen monotherapy in pre-/perimenopausal patients	Lu et al. (2022)
3	Tyrosine kinase	Gefitinib	Gefitinib plus tamoxifen achieve a substantial improvement in progression-free survival (PFS) in patients with metastatic breast cancer	Osborne et al. (2011)
		Lapatinib	Lapatinib plus tamoxifen is safe and demonstrates clinical effectiveness in patients with metastatic breast cancer	Fumoleau et al. (2014)

CDK cyclin-dependent kinase, *ER* estrogen receptor, *PI3K/Akt/mTOR* phosphoinositide 3-kinase/protein kinase B/mammalian target of rapamycin

therapy with taselisib and tamoxifen could be administered without substantial safety concerns (Baird et al. 2019). Moreover, initial indications of therapeutic efficacy were observed in both patients with phosphatidylinositol-4,5-bisphosphate 3-kinase catalytic subunit (PIK3CA) mutants and wild-type cancer. In individuals diagnosed with advanced breast cancer exhibiting resistance to various conventional treatment modalities, a phase I study showed that M2698, an Akt inhibitor, when administered with trastuzumab or tamoxifen could elicit marked antitumor efficacy, suggesting that M2698 may be capable of overcoming treatment resistance in such patients (Tsimberidou et al. 2021). A clinical trial was conducted to evaluate the effectiveness of everolimus in conjunction with tamoxifen in patients diagnosed with metastatic breast cancer (Treilleux et al. 2015), which revealed an inverse correlation between the canonical PI3K/Akt/mTOR pathway and everolimus efficacy (Treilleux et al. 2015). A phase II clinical trial revealed that combining tamoxifen and TAK-228 as a neoadjuvant therapy for breast cancer with ER+ status was both safe and well tolerated (Koca et al. 2021).

Cyclin-Dependent Kinase (CDK) Inhibitors

The dysregulation of CDK also contributes to the breast cancer cell resistance to tamoxifen. Consequently, the development of therapeutic strategies capable of effectively counteracting tamoxifen resistance by reducing CDK activity has been explored. Palbociclib combined with exemestane has demonstrated efficacy as a viable approach to treat premenopausal individuals diagnosed with HR+, HER2− metastatic breast cancer, previously treated with tamoxifen (Park et al. 2019). In premenopausal and perimenopausal patients, a phase III randomized clinical trial revealed that the concurrent use of ribociclib and tamoxifen substantially and persistently improved overall survival (OS) when compared with tamoxifen monotherapy (Lu et al. 2022).

Tyrosine Kinase Inhibitors

Crosstalk with ErbB signaling has been identified as a contributing factor to the formation of tamoxifen resistance. Therefore, the inhibition of tyrosine kinase activity within the downstream tyrosine kinase signaling pathway could facilitate the design of effective therapeutic interventions to address tamoxifen resistance. A phase II clinical trial found that progression-free survival (PFS) was considerably increased in patients with metastatic breast cancer treated with a combination of gefitinib and tamoxifen (Osborne et al. 2011). A phase I pharmacokinetic investigation revealed the safety and clinical efficacy of concurrent tamoxifen and lapatinib administration in patients with metastatic breast cancer (Fumoleau et al. 2014).

2.1.2 Fulvestrant

Hormone therapy, including selective ER downregulators (SERD) like fulvestrant, is the treatment of choice for patients with luminal A subtype of breast cancer (Rocca et al. 2018). Upon binding to the ER, fulvestrant hinders the dimerization and nuclear translocation of the ER and further prompts the aggregation of ER in the cytoplasm, ultimately resulting in the inability of ER to respond to estradiol and the subsequent inhibition of the ER signaling pathway (Osborne and Schiff 2011). Nonetheless, the development of chemoresistance reduces the efficacy of fulvestrant therapy, resulting in tumor relapses and metastases (Andrahennadi et al. 2021).

The Mechanism Underlying Fulvestrant Resistance in Breast Cancer

The underlying mechanisms of fulvestrant resistance include various processes, such as the initiation of the PI3K/Akt signaling pathway, disturbances in cell cycle regulatory proteins, intercommunication with ErbB signaling, and induction of the NF-kB pathway.

Activation of PI3K/Akt Signaling and disturbances in cell cycle regulatory proteins

The fulvestrant resistance mechanisms in breast cancer cells include the activation of PI3K/Akt, HER2, and NFκB signaling (Huang et al. 2017). The PI3K/Akt cascade is frequently affected in breast cancer (Miller et al. 2011b). The activation of the PI3K/Akt pathway leads to upregulated ERBB2 expression, resulting in the emergence of resistance to tamoxifen and fulvestrant in breast cancer cells

(Miller et al. 2011a). Furthermore, the detection of a *PIK3CA* mutation was found as a potential contributor to the development of resistance to fulvestrant (Cristofanilli et al. 2016). Mutations in *PIK3CA*, a constituent of the PI3K complex, were found to elicit resistance to fulvestrant, as well as stimulate cellular migration and proliferation in breast cancer cells (Huang et al. 2019). Fulvestrant-resistant MCF-7 breast cancer cells exhibited overexpression of cell cycle regulatory proteins CDK6 and cyclin E2 (Giessrigl et al. 2013; Kaminska et al. 2021).

Activation of HER2 Signaling

In tumor models, the proliferation of fulvestrant-treated breast cancer in the presence of estradiol was partially facilitated by the enhanced expression of the HER2/HER3 pathway signaling (Osipo et al. 2007). In addition, fulvestrant modulates the activity of epidermal growth factor (EGF) family ligands to initiate EGFR signaling in breast cancer cells (Zhang et al. 2013). The interaction between estrogen signaling and ErbB signaling facilitates the emergence of resistance to fulvestrant in breast cancer cells via the activation of the RTKs/Akt/MTORC1 signaling (Gomes et al. 2020).

The Activation of NFκB Signaling

NFκB is a transcription factor that promotes inflammation and substantially modifies the expressions of several genes engaged in cellular proliferation, metastasis, and resistance to therapy (Rinkenbaugh and Baldwin 2016). Fulvestrant-resistant MCF-7 cells exhibited elevated levels of the p65 NFκB subunit and the upstream NFκB regulator, IkappaB kinase gamma/NFκB essential modulator (Riggins et al. 2005). In luminal breast cancer cell lines, the increased expression of receptor activator of NFkB (RANK) resulted in reduced sensitivity to fulvestrant (Gomes et al. 2020).

Recent Clinical Strategies to Overcome Fulvestrant Resistance

We next discuss recent clinical strategies to overcome fulvestrant resistance in breast cancer cells (Table 2).

PI3K/Akt/MTOR Inhibitors

The subsequent section focus on overcoming fulvestrant resistance, specifically the initiation of the PI3K/Akt/mTOR pathway. One advancement in the field of fulvestrant combination therapy involves the incorporation of PI3K inhibitors. A randomized phase II study revealed that combining fulvestrant with everolimus could improve the effectiveness of fulvestrant in postmenopausal females with ER+/PR+/HER2− metastatic breast cancer resistant to fulvestrant therapy (Kornblum et al. 2017).

Baselga et al. showed that the inclusion of buparlisib, a PI3K inhibitor, in conjunction with fulvestrant, could markedly extend PFS in the entire study population (Baselga et al. 2015). A double-blind phase III randomized clinical trial in placebo-controlled postmenopausal individuals with ER+/PR+/HER2− metastatic breast cancer demonstrated the efficacy of buparlisib combined with fulvestrant when compared with that of a placebo plus fulvestrant regimen, thereby indicating that the inhibition of PI3K, when combined with endocrine therapy, could yield

Table 2 Combinatorial clinical strategies to overcome molecular mechanisms of breast cancer cell resistance to fulvestrant

No	Target	Drugs	Strategy to overcome/results	References
1	PI3K/Akt/mTOR	Everolimus	The addition of everolimus to fulvestrant enhances the effectiveness of fulvestrant in postmenopausal women who have been diagnosed with ER+/PR+/HER2− metastatic breast cancer resistant to previous fulvestrant treatment	Kornblum et al. (2017)
		Buparlisib	Fulvestrant plus buparlisib improves PFS in postmenopausal women with advanced breast cancer that is resistant to endocrine therapy	Baselga et al. (2015) and Baselga et al. (2017)
			Fulvestrant and buparlisib afford better efficacy than fulvestrant alone in postmenopausal, ER+/PR+/HER2− advanced metastatic breast cancer	Campone et al. (2018)
		Taselisib	The combination of fulvestrant and taselisib has shown superior clinical efficacy compared to that of monotherapy in patients diagnosed with HR+ and HER2- metastatic breast cancer	Dickler et al. (2016)
		Alpelisib	Fulvestrant plus alpelisib is safe and exhibits marked therapeutic efficacy in patients with ER+ advanced breast cancer	Juric et al. (2019)
			Fulvestrant and alpelisib elicit superior outcomes to fulvestrant alone	Andrahennadi et al. (2021)
			In patients with advanced HR+ breast cancer, the combination of fulvestrant and alpelisib has resulted in improved PFS compared to that with fulvestrant monotherapy	André et al. (2021)
			Fulvestrant plus alpelisib can improve PFS and OS when compared with treatment with fulvestrant alone in patients with metastatic breast cancer (HR+/HER2−/*PIK3CA* mutations)	Leenhardt et al. (2021)
		MK-2206	MK-2206 and endocrine therapy are well tolerated in postmenopausal females with ER+ metastatic breast cancer	Ma et al. (2016)
2	CDK	Palbociclib	Fulvestrant plus palbociclib therapy can substantially improve PFS, in comparison to that with fulvestrant monotherapy, in patients with ER+ advanced breast cancer with a mutation in the *ESR1* gene	Fribbens et al. (2016)
			Fulvestrant, palbociclib, and goserelin demonstrated efficacy and favorable	Loibl et al. (2017)

(continued)

Table 2 (continued)

No	Target	Drugs	Strategy to overcome/results	References
			tolerability as a therapeutic approach for premenopausal women with HR+/HER2− advanced breast cancer that is resistant to endocrine therapy	
		Abemaciclib	Fulvestrant and abemaciclib are safe and effective for patients with HR+/HER2− advanced breast cancer	Goetz et al. (2021)
3	AurkB	Alisertib	Alisertib monotherapy is effective in patients with metastatic breast cancer that develops resistance to fulvestrant	Haddad et al. (2023)

AurkB aurora kinase B, *CDK* cyclin-dependent kinase, *ER* estrogen receptor, *HER2* human epidermal growth factor receptor 2, *HR* hormone receptor, *OS* overall survival, *PFS* progression-free survival, *PI3K/Akt/mTOR* phosphoinositide 3-kinase/protein kinase B/mammalian target of rapamycin

positive outcomes (Baselga et al. 2017). Another phase III clinical trial conducted among individuals with postmenopausal, ER+/PR+/HER2− advanced metastatic breast cancer demonstrated that OS outcomes were more favorable in the group receiving combined buparlisib and fulvestrant than those in the placebo combined with fulvestrant group (Campone et al. 2018). The co-administration of taselisib, a PI3K inhibitor, with fulvestrant in patients with HR+ and metastatic breast cancer lacking the HER2 protein afforded comparable outcomes, demonstrating clinical efficacy across the whole study population (Dickler et al. 2016).

In individuals diagnosed with ER+ advanced breast cancer, a clinical trial, phase Ib revealed that a combination of alpelisib and fulvestrant elicited a tolerable safety profile among ER+ patients. Accordingly, combining alpelisib and fulvestrant may afford more substantial therapeutic efficacy in tumors with *PIK3CA* alterations than in wild-type tumors (Juric et al. 2019). Combining fulvestrant and alpelisib in the second or subsequent line of treatment could elicit superior outcomes to the use of fulvestrant alone (Andrahennadi et al. 2021). Combined alpelisib and fulvestrant could extend PFS in patients diagnosed with advanced HR+ breast cancer (André et al. 2021). A randomized phase III trial was performed to evaluate the efficacy of combined treatment with fulvestrant and alpelisib when compared with fulvestrant monotherapy in patients with metastatic breast cancer presenting HR+/HER2− and *PIK3CA* mutations (Leenhardt et al. 2021). Combination therapy afforded a clinically meaningful PFS benefit, as well as an improvement in OS (Leenhardt et al. 2021).

The AKT protein is crucial in mediating the signaling process of the PI3K cascade (Shi et al. 2019). A phase I clinical trial investigated the efficacy of combining the AKT inhibitor MK-2206 with hormone therapy in postmenopausal females diagnosed with ER+ metastatic breast cancer (Ma et al. 2016). In a randomized phase II study, combined fulvestrant and capivasertib therapy in patients with ER+/HER2− patients with metastatic breast cancer revealed that the addition of capivasertib to fulvestrant could substantially extend the OS of participants (Howell et al. 2022).

CDK Inhibitors and Aurora Kinase B (AurkB) Inhibitors

The resistance of breast cancer cells to fulvestrant can be attributed to the disruption of cell cycle regulators. Several trials have examined the efficacy of combination therapy, including fulvestrant and CDK inhibitors. Combined administration of fulvestrant and palbociclib substantially improves PFS when compared with combined therapy with fulvestrant and placebo in ER+ advanced breast cancer patients presenting an *ESR1* gene mutation (Fribbens et al. 2016). Among premenopausal women with HR+/HER2− advanced breast cancer and previous resistance to endocrine therapy, combined administration of palbociclib, fulvestrant, and goserelin demonstrated efficacy and favorable tolerability as a therapeutic approach (Loibl et al. 2017). The effectiveness and safety of combining a CDK4/6 inhibitor, abemaciclib, with fulvestrant was examined in a population of older HR+/HER2− patients with advanced breast cancer (Goetz et al. 2021).

The roles of AurkB include the control of several biological processes, such as chromosome-microtubule interactions, cohesion, spindle stability, and cytokinesis during the cell cycle (Goldenson and Crispino 2015). The phase II randomized clinical trial (TBCRC041) reported that alisertib monotherapy affords encouraging clinical efficacy in individuals diagnosed with fulvestrant-resistant metastatic breast cancer (Haddad et al. 2023).

2.2 Resistance of Breast Cancer to HER2− Targeted Therapy

The HER2 receptor is a transmembrane tyrosine kinase and a member of the HER family, which plays a notable role in cancer cell proliferation, differentiation, and migration, with breast cancer cells exhibiting a 25–30% upregulation of this gene (Kumar et al. 2020). The extracellular domain of HER2 comprises four distinct functional domains, namely domains I, II, III, and IV (Kanthala et al. 2016), in which domain IV is located close to the transmembrane domain and has a crucial position for the stabilization of proteins associated with HER2 (Kanthala et al. 2016). According to Saxon and Lee (1999), mutations in domain IV could interfere with the phosphorylation and signaling mechanisms of heterodimers, including HER2 (Hsu and Hung 2016). HER2 expression substantially correlates with the development of breast cancer and metastatic tumors (Palmieri et al. 2007). Furthermore, breast cancer patients with HER2 overexpression exhibited a worse outcome (Aman et al. 2019), thereby suggesting that anti-HER2 therapy markedly contributes to the effective treatment of breast cancer.

The activation of HER2 induces intracellular signal transduction, which is essential for cellular growth via specific pathways, including PI3K/PKB/mTOR, Ras/Raf, and STAT (Shah and Osipo 2016). Moreover, the activation of these pathways leads to several biological activities in cancer cells, including proliferation, survival, differentiation, angiogenesis, and invasion (Gutierrez and Schiff 2011). In the following section, we focus on the resistance of HER2 + breast cancer cells toward trastuzumab and lapatinib.

2.2.1 Trastuzumab

Trastuzumab, a monoclonal antibody commonly used as a choice of initial therapy for breast cancer patients with HER2 overexpression, demonstrates a specific attraction to the extracellular domain IV of HER2 (Cho et al. 2003). Trastuzumab functions by selectively attaching to the outer part of HER2, known as the extracellular domain, leading to three distinct effects: inhibition of HER2 dimerization, stimulation of antibody-dependent cell-mediated cytotoxicity (ADCC), and induction of endocytosis-mediated HER2 degradation (Hudis 2007). The therapeutic efficacy of trastuzumab can be attributed to its ability to impede the PI3K pathway via the reduction of PTEN tyrosine phosphorylation and the augmentation of PTEN translocation in the cellular membrane, consequently resulting in AKT dephosphorylation and the subsequent inhibition of cell proliferation (Nagata et al. 2004). However, prolonged trastuzumab administration can result in the emergence of resistance, resulting in diminished efficacy (Gajria and Chandarlapaty 2011).

The Mechanism Underlying Trastuzumab Resistance in Breast Cancer

Multiple mechanisms have been identified as mediators of breast cancer cell resistance to trastuzumab; these include HER2 gene mutations impacting the binding affinity of trastuzumab, initiation of the PI3K/Akt/MTOR signaling, genetic alterations affecting cell cycle regulators, crosstalk between ER signaling and trastuzumab resistance, and evasion of ADCC (Luque-Cabal et al. 2016).

Alterations in Trastuzumab-Binding Site in the Extracellular Domain of HER2

In breast cancer cells, alterations in the binding site within the extracellular HER2 domain can result in trastuzumab resistance, and these alterations can be attributed to modifications in receptor conformation and the masking of receptors by mucin (Liu et al. 2006). In the extracellular domain of HER2, proteolytic cleavage is mediated via proteases such as ADAM10, resulting in the formation of a shortened receptor (Liu et al. 2006). This truncation prevents the binding of trastuzumab to the receptor. It is speculated that p95HER2 fragments, a specific subtype of HER2 receptors lacking the trastuzumab-binding site, are responsible for the diminished affinity of trastuzumab and subsequent resistance of breast cancer cells to this therapeutic agent (Scaltriti et al. 2007). Trastuzumab resistance in HER2+ breast cancer cells has been linked to enhanced MUC4 expression (Li et al. 2014), which, in turn, hampers the efficacy of ADCC in breast cancer cells, thereby resulting in the emergence of drug resistance (Mercogliano et al. 2017).

Activation of PI3K/Akt/MTOR Signaling

The PI3K cascade is a signal transduction system that has been implicated in the promotion of cancer growth by regulating cellular events, including cell cycle regulation, apoptosis, DNA repair, angiogenesis, and motility (Arcaro and Guerreiro

2007). Trastuzumab resistance has been associated with the dysregulation of the PI3K pathway, resulting in reduced expression of PTEN, a phosphatase enzyme, and tumor suppressor gene with a crucial function in the dephosphorylation of PIP3 to PIP2 in the PI3K/Akt cascade (Nagata et al. 2004). The development of trastuzumab resistance has been attributed to reduced PTEN levels owing to mutations within the PI3K signaling pathway (Pandolfi 2004). Additionally, excessive stimulation of the PI3K pathway resulted in the overexpression of oncogenes, including HER2 and PKB, leading to the generation of resistance to trastuzumab (Serra et al. 2011). The interaction between the ER and the HER family is recognized as a characteristic feature of breast cancer cell resistance to trastuzumab, which is mediated via the activation of the mitogen-activated protein kinase (MAPK) or the PI3K/Akt signaling (Giuliano et al. 2013).

Dysregulation of Cell Cycle Regulatory Proteins

Cell cycle and cell proliferation are known effectors of the HER2 signaling pathway (Nikolai et al. 2016). The efficacy of trastuzumab in suppressing the G1 phase of the cell cycle was demonstrated through its ability to activate p27, an inhibitor of CDK/cyclin complex (Yakes et al. 2002). Furthermore, trastuzumab-resistant breast cancer cells exhibited enhanced CDK2 activity (Scaltriti et al. 2011).

Escaping ADCC

In HER2+ breast cancer cells, prolonged trastuzumab exposure was found to elicit resistance and negatively impact ADCC (Yoshida et al. 2012). In ADCC, immune cells carrying FcγRs activity identify the Fcγ region of tumor-bound antibodies, leading to the destruction of tumor cells (Reslan et al. 2009). The therapeutic efficacy of monoclonal antibodies (mAbs) is substantially influenced by the immune response. Trastuzumab targets the HER2 and, upon binding to Fc receptors located on natural killer (NK) cells, antigen-presenting cells, or immune effector cells, it induces their activation, followed by the lysis of the antibody-coated tumor cells (Musolino et al. 2020). One putative mechanism of trastuzumab resistance is the upregulation of programmed cell death ligand 1 (PD-L1), which is facilitated via the interaction of trastuzumab with immune effector cells and subsequent stimulation of interferon (IFN)-γ production (Chaganty et al. 2018). Accordingly, anti-PD-1/-PD-L1 or anti-cytotoxic T-lymphocyte-associated protein 4 (CTLA-4) mAbs can be employed to overcome breast cancer cell resistance to trastuzumab (Ledys et al. 2021).

Recent Clinical Strategies to Overcome Trastuzumab Resistance

Recent clinical strategy to overcome trastuzumab resistance has been conducted employing antibody-drug conjugate, and combination with tyrosine kinase inhibitor, monoclonal antibody, PI3K/Akt inhibitor, PD-1/PD-L1 inhibitor, and CDK inhibitor (Table 3).

Table 3 Combinatorial clinical strategies to overcome molecular mechanisms of breast cancer cell resistance to trastuzumab

No	Target	Drugs	Strategy to overcome resistance/results	References
1	HER2	Trastuzumab emtansine and pertuzumab	The combination of trastuzumab emtansine and pertuzumab has demonstrated comparable efficacy to that of docetaxel and trastuzumab in patients diagnosed with HER2+ breast cancer	Matikas et al. (2023)
		SHR-A1201 (Trastuzumab emtansine)	SHR-A1201 demonstrates favorable tolerability and safety profiles in metastatic breast cancer characterized by HER2+	Li et al. (2023a)
		Trastuzumab-deruxtecan	Trastuzumab-deruxtecan exhibits a favorable benefit-risk profile in patients with HER2+ metastatic breast cancer refractory or resistant to trastuzumab emtansine	André et al. (2023)
			Trastuzumab-deruxtecan (DXd) or trastuzumab emtansine enhances effectiveness and tolerable adverse effects in patients with HER2+ metastatic breast cancer	Curigliano et al. (2023).
		Pertuzumab	Paclitaxel/trastuzumab/pertuzumab in patients with HER2+ breast cancer demonstrates that de-escalated neoadjuvant systemic therapy could substantially increase breast-conserving therapy rates	Weiss et al. (2023)
			Paclitaxel, trastuzumab, and pertuzumab afford higher pCR rates than endocrine therapy combined with trastuzumab and pertuzumab	Gluz et al. (2023).
			Nab-paclitaxel, trastuzumab, and pertuzumab yield a comparable pCR to that of combined trastuzumab and anthracyclines and carboplatin in patients with HER2+ primary breast cancer	Lavasani et al. (2023)
2	Tyrosine kinase	Lapatinib	Lapatinib-trastuzumab can improve pCR rates when compared with chemotherapy combined with either trastuzumab or lapatinib alone	Guarneri et al. (2021)
		Sunitinib	Sunitinib and trastuzumab elicit notable anticancer effects in individuals diagnosed with HER2+ advanced breast cancer	Bachelot et al. (2014)
		Neratinib	Neratinib plus trastuzumab exhibit favorable tolerability and promising anticancer effects among individuals with advanced HER2+ breast cancer who have previously received trastuzumab treatment	Blackwell et al. (2019)

(continued)

Table 3 (continued)

No	Target	Drugs	Strategy to overcome resistance/results	References
		Pyrotinib	Pyrotinib, trastuzumab, and chemotherapy demonstrate promising potential as a optional therapy for patients with HER2+ metastatic breast cancer	Xie et al. (2023)
3	CDK	Abemaciclib	Abemaciclib and trastuzumab, with or without fulvestrant, elicit better safety and PFS compared to those of trastuzumab plus standard-of-care chemotherapy in females with HER2+ advanced breast cancer	Tolaney et al. (2020)
		Palbociclib	Combined palbociclib and trastuzumab exhibit favorable survival outcomes in patients with HER2+ advanced breast cancer, with good tolerability	Ciruelos et al. (2020)
			In patients with advanced HER2+ breast cancer who have previously received trastuzumab and taxane therapy, palbociclib and trastuzumab emtansine reveal that palbociclib could restore sensitivity in HER2-resistant breast cancer	Haley et al. (2021)
4	PI3K/ Akt/ mTOR	Everolimus	Combined everolimus, trastuzumab, and vinorelbine can afford clinical benefit in one third of the patients with HER2+ breast cancer, with no further indications of adverse effects detected	Van Swearingen et al. (2018).
			Everolimus, letrozole, and trastuzumab exhibit a favorable safety profile and have promising anticancer efficacy in patients with HR+/HER2+ advanced breast cancers	Ballhausen et al. (2021).
		Pictilisib	Pictilisib, when combined with paclitaxel and used with or without bevacizumab, trastuzumab, or letrozole, has been observed to have a tolerable safety profile in individuals diagnosed with locally recurrent or metastatic breast cancer	Schöffski et al. (2018)
		M2698	A combination of M2698 with trastuzumab or tamoxifen exerts notable antitumor effects in patients with advanced breast cancer that has developed resistance to several conventional treatments	Tsimberidou et al. (2021).
5	PD-1/ PDL-1	Durvalumab	In patients with HER2+ metastatic breast cancer and progression despite previous anti HER2 therapies, combined durvalumab plus trastuzumab show a lack of substantial clinical activity	Chia et al. (2018)

(continued)

Table 3 (continued)

No	Target	Drugs	Strategy to overcome resistance/results	References
		Pembrolizumab	Combination with pembrolizumab shows preliminary antitumor activity in patients with previously resistance to trastuzumab	Loi et al. (2019)
			The combination of pembrolizumab with trastuzumab emtansine for the treatment of metastatic HER2+ breast cancer has shown promising efficacy	Adrienne et al. (2022)

CDK cyclin-dependent kinase, *HER2* human epidermal growth factor receptor 2, *HR* hormone receptor, *OS* overall survival, *pCR* pathological complete response, *PD-1/PDL-1* programmed cell death protein 1/programmed cell death ligand 1, *PFS* progression-free survival, *PI3K/Akt/mTOR* phosphoinositide 3-kinase/protein kinase B/mammalian target of rapamycin

Targeting HER2

Antibody-Drug Conjugate

A previous randomized phase II clinical trial, namely PREDIX HER2 trial, in HER2+ breast cancer patients demonstrated that neoadjuvant treatment with docetaxel, pertuzumab, and trastuzumab in combination with trastuzumab emtansine afforded comparable long-term effects (Matikas et al. 2023). In HER2+ metastatic breast cancer patients, a phase I clinical trial demonstrated that SHR-A1201, an antibody-drug combination, could elicit favorable tolerability and safety profiles (Li et al. 2023a). The pharmacokinetic parameters exhibited a linear pattern, while the immunogenicity outcomes aligned with the anticipated clinical expectations (Li et al. 2023a). The DESTINY-Breast02 study, a randomized, open-label, multicenter, phase III trial, reported the favorable benefit-risk profile of trastuzumab-deruxtecan in HER2+ metastatic breast cancer patients who were refractory or resistant to trastuzumab emtansine (André et al. 2023). Additionally, the DESTINY-Breast03 study, a phase III clinical trial that included patients with HER2+ metastatic breast cancer treated with either trastuzumab-deruxtecan or trastuzumab emtansine, reported enhanced effectiveness and tolerable adverse effects (Curigliano et al. 2023). These outcomes highlight the overall advantages of trastuzumab-deruxtecan therapy among patients with HER2+ metastatic breast cancer (Curigliano et al. 2023).

Development of Novel Antibody

A prospective clinical trial evaluated the effectiveness of neoadjuvant paclitaxel/trastuzumab/pertuzumab in patients with HER2+ breast cancer. De-escalated neoadjuvant systemic therapy substantially enhanced breast-conserving therapy rates (Weiss et al. 2023). The neoadjuvant WSG-TP-II randomized clinical trial investigated the effectiveness of endocrine therapy in combination with trastuzumab and pertuzumab when compared with that of de-escalated chemotherapy in patients with HR+/HER2+ early breast cancer (Gluz et al. 2023). Twelve weeks of therapy with paclitaxel, trastuzumab, and pertuzumab substantially enhanced the pathological complete response (pCR) rate, indicating superior effectiveness to endocrine therapy combined with trastuzumab and pertuzumab (Gluz et al. 2023). In addition, a recent phase II prospective open-label study revealed that the application of neoadjuvant nab-paclitaxel, trastuzumab, and pertuzumab in patients with HER2+ primary breast cancer yielded a pCR comparable with that of the combined administration of trastuzumab, anthracyclines, and carboplatin (Lavasani et al. 2023). Notably, this treatment approach correlated with a decreased incidence of treatment-related toxicities (Lavasani et al. 2023).

Tyrosine Kinase Inhibitors

According to the Cher-LOB randomized phase II study, the addition of lapatinib-trastuzumab in combination with chemotherapy elicited a higher pCR rate than the use of chemotherapy in conjunction with either trastuzumab or lapatinib alone (Guarneri et al. 2021). Accordingly, lapatinib-trastuzumab may yield more favorable outcomes than trastuzumab alone (Guarneri et al. 2021). A combination of sunitinib and trastuzumab produced notable anticancer effects in a phase II clinical trial for patients with HER2+ advanced breast cancer (Bachelot et al. 2014). A clinical phase I/II study also demonstrated the promising anticancer effects and favorable tolerability of neratinib plus trastuzumab combination therapy for individuals with advanced HER2+ breast cancer who had been previously treated with trastuzumab (Blackwell et al. 2019). Moreover, the combination of pyrotinib, trastuzumab, and chemotherapy has shown great promise as a therapeutic option for patients diagnosed with HER2+ metastatic breast cancer, based on the results of a single-arm exploratory phase II trial. Furthermore, this treatment regimen had a manageable safety profile, which is encouraging for patients and healthcare providers alike. This regimen was also found to be effective for heavily pretreated individuals with HER2+ metastatic breast cancer, particularly those without liver and/or lung metastases (Xie et al. 2023).

CDK Inhibitors

The monarcHER trial, a phase II study, employed a randomized and open-label design to investigate the efficacy of abemaciclib, a CDK inhibitor, in comparison to that of trastuzumab plus standard-of-care chemotherapy; here, the combination of trastuzumab with or without fulvestrant was associated with better results. The trial specifically targeted females with advanced breast cancer who exhibited HR+ and HER2+ characteristics (Tolaney et al. 2020). Combination therapy comprising abemaciclib and fulvestrant, in addition to trastuzumab, resulted in a significant improvement in PFS compared to that with standard chemotherapy with trastuzumab. The safety profile of this treatment approach was also found to be manageable (Tolaney et al. 2020).

The use of trastuzumab and CDK inhibitors together as a treatment method has also been studied. In the phase II clinical trial SOLTI-1303 PATRICIA, the safety and efficacy of palbociclib and trastuzumab in patients with advanced HER2+ breast cancer was evaluated. The trial revealed that palbociclib combined with trastuzumab could afford favorable survival outcomes and was well tolerated by patients

(Ciruelos et al. 2020). In a clinical trial that was conducted on patients with advanced HER2+ breast cancer, who had previously received trastuzumab and taxane therapy, the combination of PD 0332991 (palbociclib) and trastuzumab emtansine was evaluated. The results of this phase I/Ib trial showed that palbociclib could restore sensitivity in HER2-resistant breast cancer (Haley et al. 2021).

PI3K/Akt/MTOR Inhibitors

Several studies have investigated the possibility of using a combination of trastuzumab and PI3K/Akt inhibitors. In the LCCC 1025 clinical trial, the effectiveness of everolimus, trastuzumab, and vinorelbine was examined as a treatment for patients with HER2+ breast cancer with progressive brain metastases (Van Swearingen et al. 2018). Approximately one third of the participants attained a clinical benefit rate, with no additional signs of adverse effects observed (Van Swearingen et al. 2018). Combined therapy with everolimus, letrozole, and trastuzumab is a safe and effective treatment for individuals with metastatic breast cancer. Further, this therapy has shown promising anticancer efficacy in heavily treated patients with advanced HR+ and HER2+ cancers (Ballhausen et al. 2021).

A tolerable safety profile among individuals diagnosed with locally recurrent or metastatic breast cancer was demonstrated in a phase Ib clinical trial of pictilisib combined with paclitaxel, with or without bevacizumab, trastuzumab, or letrozole (Schöffski et al. 2018). Moreover, the results of a phase I clinical trial on the efficacy of M2698, a dual inhibitor that targets p70S6K and AKT, demonstrated that this drug, when administered in combination with trastuzumab or tamoxifen, can exert significant antitumor effects in patients with advanced breast cancer. This is especially significant for patients with cancer that has developed resistance to conventional treatments. Therefore, M2698 could be effective in treating patients with disease that is resistant to other forms of treatment (Tsimberidou et al. 2021).

Anti-PD-1/PDL-1 Therapy

Combination therapy with trastuzumab and immunotherapy drugs has been explored. A phase I clinical trial evaluated the effectiveness of combining durvalumab, a PD-L1 antibody, with trastuzumab for patients with HER2+ metastatic breast cancer who had progressed despite prior anti-HER2 therapies, and it showed that the combination lacked significant clinical activity (Chia et al. 2018). Moreover, a clinical phase Ib-2 trial combining trastuzumab with pembrolizumab has shown early signs of effectiveness for patients with trastuzumab-resistant disease (Loi et al. 2019). Another clinical phase Ib trial assessed the efficacy of pembrolizumab (antibody against PD-1) in conjunction with trastuzumab emtansine to treat patients with metastatic HER2+ breast cancer, revealing that this combination was well tolerated and did not pose any notable safety concerns (Adrienne et al. 2022).

2.2.2 Lapatinib

Lapatinib is a small-molecule tyrosine kinase inhibitor that has been approved to treat advanced breast cancer characterized by HER2 amplification and effectively targets both EGFR and HER2 (Tsang et al. 2011). It competitively binds to the ATP-binding site of EGFR and HER2 receptors, effectively inhibiting signal transduction mediated by both (Kim et al. 2009). The activation of HER2 signaling triggers downstream signaling pathways, including PI3K/Akt and Ras/MAPK (Tsang et al. 2011). Expression of the survivin protein, a member of the inhibitor of apoptosis family, was found to be significantly reduced upon lapatinib administration, thus promoting cell death (Xia et al. 2006). Nevertheless, a considerable percentage of patients experience disease progression owing to the occurrence of acquired resistance to lapatinib (Eustace et al. 2018).

The Mechanism Underlying Lapatinib Resistance in Breast Cancer

Breast cancer cells can become resistant to lapatinib through various mechanisms, such as mutations in the *HER2* gene within the intracellular domain, and the activation of compensatory pathways, such as PI3K/Akt/mTOR, MET, and JAK and crosstalk with ER signaling.

Mutation of HER2 Gene in the Tyrosine Kinase Intracellular Domain

Mutations in the *HER2* gene within the intracellular/tyrosine kinase domain cause lapatinib resistance. The presence of a HER2 kinase domain mutation leads to the continuous phosphorylation and activation of HER2 (Wang et al. 2006). The presence of *HER2 L755S* and *T798I* mutations was found to be correlated with elevated levels of resistance to lapatinib (Trowe et al. 2008). FOXO3A activation is responsible for the interplay between ER and HER2 signaling, resulting in an increase in ER signaling in lapatinib-treated patients with HER2+/ER+ breast cancer (Guo and Sonenshein 2004). In HER2+ breast cancers subjected to lapatinib treatment, lapatinib was found to reduce ER expression and signaling, serving as an adaptive mechanism for cell survival (Giuliano et al. 2015).

Establishment of Compensatory/Alternative Pathways

The alternative pathways are regulated by receptor tyrosine kinases (HER3, MET, and AXL) (Thery et al. 2014) and intracellular kinases, such as PI3K/Akt, mTOR, and PTK6 (Sahu and Pattanayak 2020). Augmenting the level of activated HER3 can potentially diminish the efficacy of tyrosine kinase inhibitors (Gala and Chandarlapaty 2014). Lapatinib administration could inhibit the functioning of HER2 and establish an alternative route characterized by the overexpression of PI3K/Akt and FoxO3A, inducing the overexpression of regulatory genes related to the cell cycle, apoptosis, and proliferation of malignant cells (Mukherjee et al. 2007). Various other receptor tyrosine inhibitors, including c-Met and AXL, also impact the development of resistance to lapatinib (Schoumacher and Burbridge 2017).

Recent Clinical Strategies to Overcome Lapatinib Resistance

Diverse targeted therapies, such as PI3K/Akt/mTOR inhibitors, tyrosine kinase inhibitors, endocrine therapy, and advancements in mAbs, have been explored to combat lapatinib resistance (Table 4).

Table 4 Combinatorial clinical strategies to overcome molecular mechanisms of breast cancer cell resistance to lapatinib

No	Target	Drugs	Strategy to overcome resistance/results	References
1	PI3K/Akt/mTOR	Everolimus	Patients receiving lapatinib and everolimus have exhibited good tolerance to the treatment, with a partial response (PR) observed in 17% of individuals with advanced HER2+ breast cancer	Gadgeel et al. (2013)
		MK-2206	Lapatinib and MK-2206 combined treatment tolerable in patients with advanced HER2+ breast cancer	Wisinski et al. (2016)
2	HER2	Trastuzumab	Lapatinib and trastuzumab combined treatment resulted in augmentation of the pCR rate among women with HER2+ breast cancer	Rimawi et al. (2020)
3	ER	Letrozole	The combination of letrozole and lapatinib has produced favorable responses in Asian postmenopausal women diagnosed with ER+/HER2+ breast cancer	Park et al. (2016)
		Letrozole, anastrozole, and exemestane	Combination of lapatinib and trastuzumab, together with letrozole, anastrozole, and exemestane can substantially improve PFS in postmenopausal women with HR+/HER2+ metastatic breast cancer	Johnston et al. (2021)
4	Tyrosine kinase c-MET/AXL/VEGFR	Foretinib	Combined foretinib and lapatinib afford positive outcomes in patients with HER2+ metastatic breast cancer	Chia et al. (2017)
		Afatinib	Combined afatinib as a standalone treatment or in combination with vinorelbine or paclitaxel yields objective responses in at least 18% of patients with HER2+ breast cancer that has developed resistance to lapatinib	Hickish et al. (2022)

ER estrogen receptor, *HER2* human epidermal growth factor receptor 2, *HR* hormone receptor, *pCR* pathological complete response, *PFS* progression-free survival, *PI3K/Akt/mTOR* phosphoinositide 3-kinase/protein kinase B/mammalian target of rapamycin, *VEGFR* vascular endothelial growth factor receptor

PI3K/Akt/MTOR Inhibitors

Patients with advanced HER2+ breast cancers were included in a phase I trial, called Southwest Oncology Group (SWOG) S0528, to evaluate the effects of lapatinib with the mTOR inhibitor everolimus. Here, patients responded favorably to lapatinib and everolimus; 17% experienced a partial response (Gadgeel et al. 2013). Moreover, a phase I clinical trial examined the effectiveness of a new AKT inhibitor (MK-2206) and lapatinib for the treatment of adult solid tumors. The trial additionally explored the effects of dose expansion for advanced HER2+ breast cancer and demonstrated that this combination therapy approach was tolerable and has potential clinical applications (Wisinski et al. 2016).

mABs

The utilization of trastuzumab in combination with lapatinib is employed to overcome lapatinib resistance in breast cancer. In a controlled phase II study targeting individuals diagnosed with HER2+ breast cancer, the application of a dual anti-HER2 therapy regimen, comprising lapatinib and trastuzumab, over 24 weeks, resulted in a significant improvement in the rate of pCR in women afflicted with HER2+ breast cancer, notably avoiding the use of chemotherapy (Rimawi et al. 2020).

Anti-Estrogen Therapy

To address breast cancer cell resistance, the potential of anti-estrogen therapy combined with lapatinib was explored. In a phase II clinical study involving Asian postmenopausal women diagnosed with ER+ and HER2+ breast cancer, neoadjuvant treatment combining letrozole with lapatinib was shown to have a positive impact on efficacy (Park et al. 2016). Further, the results of a phase III randomized trial indicated that the combination of lapatinib and trastuzumab, when used alongside an aromatase inhibitor (examples include letrozole, anastrozole, and exemestane), significantly enhanced PFS in postmenopausal women with HR+/HER2+ metastatic breast cancer (Johnston et al. 2021).

Tyrosine Kinase Inhibitors

The focus of lapatinib combination therapy lies in the targeting of tyrosine kinase as part of its developmental process. The effectiveness of combining lapatinib with foretinib, a compound that inhibits c-MET, AXL, and the vascular endothelial growth factor receptor (VEGFR), was evaluated in individuals with HER2+ metastatic breast cancer (Chia et al. 2017). The study reported positive outcomes, resulting in the identification of recommended phase II dosages for foretinib and lapatinib, i.e., 45 and 1000 mg taken orally once daily, respectively (Chia et al. 2017). Further findings from an open-label, multicenter, phase II study showed that the administration of afatinib, alone or in combination with vinorelbine or paclitaxel, led to objective responses in a minimum of 18% of the patients with HER2+ breast cancer who had previously shown resistance to lapatinib (Hickish et al. 2022).

2.3 Breast Cancer Resistance Toward Chemotherapy

Triple-negative breast cancer (TNBC), accounting for approximately 15–20% of breast cancer cases, is identified based on its heterogeneity and the absence of estrogen receptor (ER), progesterone receptor (PR), and HER2 expression, leading to a less favorable outcome for those diagnosed with it (Yao et al. 2017). Accordingly, the lack of distinctive molecular cancer markers poses a substantial challenge to TNBC therapy (Gluz et al. 2009). Chemotherapy has been recognized as the most effective therapeutic approach to treat TNBC (Wahba and El-Hadaad 2015).

Chemotherapy is a widely employed therapeutic approach for breast cancer, wherein cytotoxic medications are utilized to eradicate cancerous cells via either oral or systemic delivery (Fang et al. 2020). To treat breast cancer, frequently employed chemotherapeutic agents include doxorubicin, paclitaxel, cyclophosphamide, and methotrexate (Hanna and Mayden 2021). Chemotherapy is administered to diminish the tumor size prior to radiation therapy or surgery, either in a neoadjuvant capacity before surgery (Graham et al. 2015) or in an adjuvant capacity after surgery (Breugom et al. 2015). The use of neoadjuvant chemotherapy for the treatment of breast cancer has become increasingly common, especially for shrinking the primary tumor in the breast and any metastases in the axillary lymph nodes (Wang and Mao 2020). Chemotherapeutic interventions afford patients the potential to attain optimal therapeutic outcomes while concurrently mitigating or eradicating the likelihood of relapse, treatment resistance, and adverse effects. Furthermore, this approach ensures that patients are able to maintain a high quality of living (Fisusi and Akala 2019).

2.3.1 The Mechanism Underlying Doxorubicin, Paclitaxel, and Cisplatin Resistance in Breast Cancer

Doxorubicin

Doxorubicin, a type of anthracycline antibiotic, is commonly utilized in breast cancer therapy owing to its effectiveness (Kaklamani and Gradishar 2003). It is commonly administered as typical adjuvant chemotherapy in conjunction with docetaxel and cyclophosphamide (Mackey et al. 2013), and also as a neoadjuvant therapy in combination with paclitaxel (Le et al. 2023). Doxorubicin has multiple cellular-level modes of action, one of which involves the inhibition of topoisomerase II, thereby impeding the process of DNA replication (Singh et al. 2023). Doxorubicin can intercalate with DNA to hinder the process of DNA synthesis (Yang et al. 2014). Doxorubicin is known to enhance levels of reactive oxygen species (ROS), resulting in cellular damage to various components such as DNA and intracellular proteins (Meredith and Dass 2016). The development of resistance to doxorubicin in breast cancer cells is promoted by the presence of extracellular matrix proteins (Lovitt et al. 2018). A reduction in the expression of the RNA-binding protein (HuR) has been investigated as a new factor contributing to the emergence of doxorubicin resistance in breast cancer cells (Latorre et al. 2012). Further, resistance to doxorubicin in breast cancer cells can stem from the intrinsic properties of the cancer itself or from extrinsic factors that are acquired over time (Wang et al. 2019).

Drug Efflux Overexpression

Doxorubicin resistance can be caused by various processes, one of which involves upregulation of the expression of ATP-binding cassette (ABC) transporters, such as ABCB4 (Huang et al. 2018), which participate in the elimination of doxorubicin from the cytoplasm to the extracellular space (Wen et al. 2019). The emergence of doxorubicin resistance in breast cancer cells is regulated by the upregulation of the expression of genes associated with drug efflux pumps and the heightened activity of multidrug resistance protein 1 (MDR1) and breast cancer resistance protein 1 (BCRP1) pumps (Hermawan et al. 2016). Doxorubicin can effectively eradicate liver cancer stem cells by inhibiting ABCB1/MDR1 or ABCG2/BCRP (Yin et al. 2021).

Alterations in Cell Survival, Apoptosis, and Autophagy Regulatory Protein

Breast cancer resistance to doxorubicin evolves through a complicated process, including modifications to the proteins that play crucial roles in regulating cell survival, apoptosis, and autophagy. Resistin, an adipokine overexpressed in breast cancer, promotes resistance to doxorubicin-induced cell death by activating the AMPK/mTOR/ULK1 and JNK signaling pathways, leading to increased autophagy (Liu et al. 2017). Further, the MAPK/ERK pathway can act adaptively in breast cancer cells, resulting in doxorubicin resistance and supporting tumor progression to protect cancer cells (Christowitz et al. 2019).

Epithelial-to-Mesenchymal Transition (EMT)

EMT is a dynamic process during which epithelial cells undergo a transformation into mesenchymal cells (Kalluri and Neilson 2003). EMT takes place during fetal development, tissue regeneration, and wound healing. Furthermore, EMT has been linked to breast cancer progression, metastasis, and resistance to various therapeutic interventions (Smith and Bhowmick 2016). The increased resistance of breast cancer to doxorubicin is associated with EMT and involves the overexpression or activation of multiple cellular signaling pathways, including PI3K/Akt, ERK, c-Myc, NOTCH, and β-catenin, which collectively contribute to this process (Wang and Zhou 2011).

Modified of DNA Repair Mechanism

As a mechanism of action, doxorubicin induces the production of free radicals and causes oxidative damage, which plays a role in DNA damage (Cappetta et al. 2017). Loss of the adenomatous polyposis coli (*APC*) gene, a tumor suppressor, has been demonstrated to decrease the sensitivity of breast cancer cells to doxorubicin, indicating its role in chemotherapy resistance (Stefanski et al. 2019). This occurs because doxorubicin diminishes the activity of DNA damage–response proteins, including ATM, Chk1, and Chk2, thereby impeding DNA repair processes and resulting in reduced cell sensitivity to the drug (Stefanski et al. 2019). DNA-dependent protein kinase (DNA-PK) is critical for the repair of DNA double-strand breaks (DSBs), and the inhibition of DNA-PK has been shown to increase the sensitivity of breast cancer cells to doxorubicin treatment (Ciszewski et al. 2014).

Paclitaxel

Paclitaxel is frequently used as the primary therapeutic drug for breast cancer. Nonetheless, the development of resistance to this treatment presents a significant clinical challenge and contributes substantially to mortality, owing to treatment inefficacy (Abu Samaan et al. 2019). Regarding the mechanism of action, paclitaxel binds to β-tubulin, a constituent of microtubules, resulting in the formation of stable microtubules (Entwistle et al. 2008). The subsequent arrest at the mitotic checkpoint is believed to induce apoptosis, potentially via G2/M arrest, followed by the induction of apoptosis via the mitochondrial route (Dziadyk et al. 2004). In addition to its known effects on interphase microtubules, paclitaxel can induce disruption of microtubules, leading to interference with cellular development and metabolism (Murray et al. 2012). In patients with metastasis, paclitaxel frequently elicits an overall response rate ranging from 20 to 40% when utilized as second-line or salvage single-agent therapy (Perez 1998).

The development of resistance to paclitaxel involves several mechanisms, including:

The Upregulation of Transmembrane Efflux Transporters

The development of resistance to paclitaxel was found to be linked to elevated levels of efflux pump proteins, including ABC transporter (Rago et al. 2003). Silencing of *ABCB1* markedly enhanced the sensitivity of breast cancer cells to paclitaxel (Němcová-Fürstová et al. 2016). Accordingly, upregulation of *ABCB1* and *ABCG2* expression could serve as markers indicating the development of resistance to paclitaxel (Duz and Karatas 2021).

Mutations in Tubulin and Changes in the Stability of the Microtubule Network

Reportedly, cells exhibiting resistance to paclitaxel possess a mutation in the gene encoding the paclitaxel-binding area of β-tubulin, namely *Asp26Glu*, resulting in the decreased stability of their microtubule structures (Hari et al. 2006). Moreover, mutations in tubulin could modify the stability of microtubules, thereby counteracting the drug action (Huzil et al. 2007). In addition, human tumor cells can undergo spontaneous mutations in genes that encode β1-tubulin, resulting in resistance to paclitaxel; hence, individuals possessing certain polymorphisms in genes that encode β1-tubulin may necessitate larger paclitaxel doses to achieve beneficial therapeutic outcomes (Yin et al. 2010).

Decreased Activity of Important Apoptotic Proteins

In addition, paclitaxel resistance in breast cancer can be attributed to the suppression of apoptotic cell death by eRα (Sui et al. 2007). Modifications to the mitochondrial apoptotic pathway play a crucial role in the development of acquired resistance to paclitaxel in breast cancer cells (Kutuk and Letai 2008). Moreover, paclitaxel resistance in breast cancer cells is characterized by the depletion of apoptosis-promoting elements and activation of the autophagic pathway, diminishing drug efficacy (Ajabnoor et al. 2012).

Cisplatin

Cisplatin, an anticancer drug derived from platinum, is widely employed in chemotherapy regimens for the management of breast cancer (Brown et al. 2019). Cisplatin is employed to treat TNBC by inhibiting EMT (Wang et al. 2021). The formation of cisplatin-DNA adducts, which occurs through the covalent binding of cisplatin to DNA via platinum atoms and intercalation between base pairs (Fichtinger-Schepman et al. 1985), results in DNA damage and the suppression of the transcription process (Brabec 2002), in turn resulting in cell cycle arrest and apoptosis (Schärer 2005).

Whereas cisplatin initially effectively inhibits the growth of breast cancer cells, prolonged exposure to this drug leads to the development of resistance, ultimately resulting in uncontrolled tumor recurrence and proliferation (Vassilopoulos et al. 2014). Numerous molecular pathways that underlie cisplatin resistance have been comprehensively described, including diminished cisplatin accumulation within cancer cells (Schärer 2005) via various mechanisms.

Reduced Influx Transporter Expression and Enhanced Efflux Transporter Expression

Cisplatin reportedly reduces the expression levels of organic cation transporters and the copper transporter protein known as copper transport protein (Holzer et al. 2006). The downregulation of copper influx transporters (CTR1), which serve as importers of platinum-containing medications, leads to cellular resistance against cisplatin (Schoeberl et al. 2022). The protein ATP7B, classified as a P-type ATPase and an efflux transporter of copper from cells, participates in the regulation of cisplatin expulsion, thereby mitigating the cytotoxic effects associated with this drug (Safaei et al. 2008). Cisplatin resistance can also be attributed to the augmented activity of the efflux transporter, i.e., multidrug resistance protein 2 (MRP2), within the cancer cells (Kasaian et al. 2016).

Overexpression of Glutathione-S-Transferase (GST) and Glutathione (GSH)

During breast cancer therapy, cisplatin forms adducts with DNA (Basu and Krishnamurthy 2010). Cisplatin resistance is primarily linked to the intracellular upregulation of GST, essential detoxification enzymes capable of facilitating the conjugation of cisplatin with GSH (Li et al. 2017). GSH depletion (Jia et al. 2015) and inhibition of GST (Niu et al. 2022) serve as notable mechanisms for counteracting cisplatin resistance in breast cancer cells.

Impairment in DNA Repair Process

Cisplatin resistance can also be attributed to the impairment of DNA repair processes that shield cancer cells against the DNA damage induced by chemotherapeutic agents (Rocha et al. 2018). Cisplatin-resistant cells were shown to exhibit increased expression of nucleotide excision repair capacity, as demonstrated by the overexpression of DNA excision repair protein ERCC-1 (Bellmunt et al. 2007). Moreover, cisplatin-resistant cells showed increased homologous recombination capacity, as demonstrated by the overexpression of *BRCA1/BRCA2* (Imyanitov and Sokolenko 2021).

2.3.2 Recent Clinical Strategy to Overcome Doxorubicin, Paclitaxel, and Cisplatin Resistance

Recent clinical trials have examined the effectiveness of combining afatinib, lapatinib, cyclophosphamide, pembrolizumab, and atezolizumab as a strategy to overcome doxorubicin resistance in breast cancer cells (Table 5). The DAFNE (GBG-70) study examined the effect of neoadjuvant therapy with afatinib, trastuzumab, and a chemotherapy regimen (paclitaxel, epirubicin, and cyclophosphamide) in patients with locally advanced or operable breast cancer receiving taxane-anthracycline-containing chemotherapy (Hanusch et al. 2015). The applied treatment approach elicited acceptable tolerability and achieved a pCR rate comparable with that of other anti-HER2 doublets (Hanusch et al. 2015). In a phase II clinical study that focused on patients with advanced metastatic HER2+ breast cancer, all of whom were previously exposed to trastuzumab, the combined treatment approach utilizing lapatinib and pegylated liposomal doxorubicin exhibited favorable effectiveness and safety profiles. This treatment approach is particularly well suited for patients with a history of cardiac risk or brain metastases (Pircher et al. 2015).

In the TITAN study, a phase III clinical trial involving early-stage TNBC, the combination of doxorubicin/cyclophosphamide followed by anthracycline/ixabepilone treatment resulted in disease-free survival (DFS) and OS outcomes that were comparable to those with the standard doxorubicin/paclitaxel treatment for individuals with operable TNBC (Yardley et al. 2017). In a phase II clinical trial including patients with early-stage TNBC resistant to anthracycline therapy, the primary objective was successfully achieved with neoadjuvant treatment comprising atezolizumab and nab-paclitaxel (Yam et al. 2023), thereby indicating the efficacy of this intervention in this population at high-risk for disease progression (Yam et al. 2023). In patients with metastatic TNBC, a phase I clinical trial revealed that patients who had not previously received anthracycline treatment and were administered a combination therapy of pembrolizumab and doxorubicin exhibited a promising rate of response (Egelston et al. 2023). The findings from the ALICE trial, a phase IIb study conducted on patients with metastatic TNBC, indicated that the addition of atezolizumab to pegylated liposomal doxorubicin/cyclophosphamide was well tolerated, with promising signs of potential clinical effectiveness (Røssevold et al. 2022).

Clinical trials have also been undertaken to tackle the issue of paclitaxel resistance in breast cancer cells. These trials involved the use of combination therapy comprising afatinib, bevacizumab, pembrolizumab, and epirubicin. In a multicenter, open-label phase II trial, the effectiveness of afatinib, either as a standalone treatment or in combination with vinorelbine or paclitaxel, was examined in patients with HER2+ breast cancer who had previously experienced treatment failure or disease progression after receiving trastuzumab and/or lapatinib. In patients with metastatic breast cancer who had previously received HER2-targeted therapy without success, treatment with afatinib, either as a monotherapy or in combination with vinorelbine or paclitaxel, led to objective responses in at least 18% of cases (Hickish et al. 2022). In a clinical trial investigating the combination of paclitaxel and bevacizumab for the

Table 5 Combinatorial clinical strategies to overcome molecular mechanisms of breast cancer cell resistance toward doxorubicin, paclitaxel, and cisplatin

No	Drugs	Strategy to overcome resistance/results	References
1	Paclitaxel, epirubicin, and cyclophosphamide and afatinib, trastuzumab	The combination therapy demonstrates favorable tolerability and results in a pCR among patients with locally advanced or operable breast cancer undergoing chemotherapy that includes taxanes and anthracyclines	Hanusch et al. (2015)
2	Doxorubicin and trastuzumab	The use of lapatinib and pegylated liposomal doxorubicin in combination exhibits effectiveness and safety in individuals with advanced metastatic HER2+ breast cancer who had previously been treated with trastuzumab	Pircher et al. (2015)
3	Doxorubicin/ cyclophosphamide followed by ixabepilone/paclitaxel	Doxorubicin/cyclophosphamide continued with ixabepilone treatment yield DFS and OS similar to doxorubicin/ cyclophosphamide continued with paclitaxel in individuals with operable TNBC	Yardley et al. (2017)
4	Paclitaxel and atezolizumab	Combined atezolizumab and nab-paclitaxel in patients with early-stage TNBC resistant to anthracycline resulted in good efficacy	Yam et al. (2023)
5	Doxorubicin and pembrolizumab	Combined pembrolizumab and doxorubicin exhibited a promising rate of response in patients with metastatic TNBC	Egelston et al. (2023)
6	Doxorubicin/ cyclophosphamide, atezolizumab	Atezolizumab combined with pegylated liposomal doxorubicin/ cyclophosphamide was well tolerated and had signs of potential clinical efficacy	Røssevold et al. (2022)
7	Paclitaxel/vinorelbine, with or without afatinib	Afatinib as a single treatment or in combination with vinorelbine or paclitaxel results in objective responses in a minimum of 18% of the patients who have HER2+ breast cancer and have faced treatment failure with trastuzumab	Hickish et al. (2022)
8	Paclitaxel and bevacizumab	Combined paclitaxel and bevacizumab in individuals diagnosed with advanced ER+/HER2- breast cancer demonstrate favorable outcomes	Bos et al. (2023)
9	Paclitaxel, anthracycline, and pembrolizumab	Combined neoadjuvant nab-paclitaxel, anthracycline, and pembrolizumab elicits promising pCR rates in patients diagnosed with early-stage TNBC	Fasching et al. (2023)
10	Cisplatin and veliparib	Combined veliparib and cisplatin can notably enhance PFS in patients with metastatic TNBC and BRCA mutation	Rodler et al. (2023)

(continued)

Table 5 (continued)

No	Drugs	Strategy to overcome resistance/results	References
11	Cisplatin and adavosertib	Adavosertib combined with cisplatin yields an ORR threshold exceeding 30% in patients with metastatic TNBC	Keenan et al. (2021)

DFS disease-free survival, *ER* estrogen receptor, *HER2* human epidermal growth factor receptor 2, *ORR* objective response rate, *pCR* pathological complete response, *PFS* progression-free survival, *TNBC* triple-negative breast cancer

treatment of advanced ER+/HER2- breast cancer, favorable outcomes were observed, with approximately 86% of the patients achieving a 6-month PFS rate (Bos et al. 2023). A phase II study (NeoImmunoboost, AGO-B-041) of patients with early-stage TNBC demonstrated promising pCR rates following neoadjuvant chemotherapy with a combination of neoadjuvant nab-paclitaxel, anthracycline, and pembrolizumab (Fasching et al. 2023). Additionally, a phase II clinical trial (S1416) employing randomization, double-blinding, and placebo control revealed that veliparib combined with cisplatin led to a notable enhancement in PFS among individuals diagnosed with metastatic TNBC and *BRCA* mutation-associated breast cancer, specifically those with BRCA-like characteristics (Rodler et al. 2023).

Clinical trials have been conducted to overcome breast cancer resistance toward cisplatin. For example, clinical trials have evaluated the clinical efficacy and molecular response to the WEE1 inhibitor adavosertib when combined with cisplatin in patients diagnosed with metastatic TNBC (Keenan et al. 2021). In patients with metastatic TNBC who had received either no prior treatment or only one prior line of therapy, the combination of adavosertib and cisplatin did not meet the predetermined objective response rate threshold >30% (Keenan et al. 2021).

2.4 Resistance Mechanism to Immunotherapy

Immunotherapy has become more prevalent as a revolutionary approach to cancer treatment, with potential advantages compared to conventional therapies, such as chemotherapy and radiation therapy. In breast cancer, particularly in aggressive forms such as TNBC, immunotherapy has shown promising results. Immune checkpoint inhibitors represent a remarkable and revolutionary approach to breast cancer treatment, with therapies targeting pathways such as programmed cell death protein 1 (PD-1)/PD-L1 and cytotoxic T-lymphocyte-associated protein 4 (CTLA-4). These inhibitors function by restoring the ability of the immune system to recognize and attack cancer cells, representing a significant advancement in breast cancer therapy (Buchbinder and Desai 2016). However, the efficacy of these treatments varies across breast cancer subtypes, with certain subtypes exhibiting more responsiveness than others.

2.4.1 The Mechanism Underlying Immunotherapy Resistance in Breast Cancer

Resistance to immunotherapy is categorized as primary and adaptive resistance. In primary resistance, patients fail to respond to the therapy; in adaptive resistance, patients experience relapse after the initial response. Furthermore, resistance to immunotherapy in breast cancer involves a complex interplay of tumor-intrinsic and -extrinsic factors. Intrinsic factors associated with tumor immune resistance involve the loss of tumor antigen presentation, tumor mutation burden and epigenetic changes, and dysregulated signaling pathways. Meanwhile, extrinsic mechanisms of tumor immune resistance include the availability of immunosuppressive cells and altered composition of tumor microenvironment (Sharma et al. 2017).

Intrinsic Factors Related to Tumor Immune Resistance

Loss of Tumor Antigen Presentation

Mutations interfering with the proteasome, transporters involved in antigen processing, and structural alterations in the major histocompatibility complex (MHC) are capable of disrupting antigen presentation, hampering T cell activation, and ultimately allowing for tumor immune evasion. In patients with TNBC, the absence of tumor-specific MHC-I molecules has been linked to the diminished effectiveness of immunotherapy and is correlated with the presence of lymph node metastasis (Dusenbery et al. 2021; Taylor et al. 2023). Furthermore, HER2-expressing tumors can evade cytotoxic T lymphocytes (CTLs)-mediated lysis by downregulating MHC class I and antigen processing machinery (Lollini et al. 1998; Herrmann et al. 2004; Kaplan et al. 2006; Inoue et al. 2012). In patients with breast cancer, the downregulation of human leukocyte antigen (HLA)-class I expression has been associated with tumor progression and relapse, particularly in cases where estrogen, progesterone, and HER2 receptors are overexpressed (Kaneko et al. 2011).

Tumor Mutation Burden (TMB) and Epigenetic Changes

TMB refers to the quantity of somatic mutations observed per mega base of the sequenced genome. It is associated with high neoantigen load, rendering tumors more immunogenic and responsive to immune checkpoint inhibitors. However, when compared to other types of cancer, breast cancer typically exhibits a lower TMB (Zheng 2022). The APOBEC family of cytidine deaminases is notably mutated in breast cancer, which is particularly prevalent in HER2+ breast cancer cases (Roberts et al. 2013). Elevated APOBEC enrichment scores and substantial genetic heterogeneity in patients with HER2-enriched tumors could promote immune escape mechanisms and the development of a cold tumor microenvironment in breast cancer (DiMarco et al. 2022). Furthermore, elevated levels of APOBEC subunit 3B (*APOBEC3B*) mRNA expression have been associated with poorer prognosis in patients diagnosed with the luminal A subtype of breast cancer (Mao et al. 2020). In addition to *APOBEC* mutations, increased levels of *CCL18* copy number variations have been linked to a notable reduction in immune cell infiltration and a less favorable survival prognosis in patients with breast cancer (Cui et al.

2023). In addition, low *FAT3* expression was documented in the high TMB group of patients with TNBC, affording better patient survival (Gao et al. 2021).

Dysregulated Pathway

Aberrant signaling pathways are related to breast cancer resistance to immunotherapy. IFN-γ, mostly produced by NK cells and CTLs, is the predominant mediator of immune responses in the antitumor immunity responses. In response to IFN-γ binding to its receptor, downstream signaling pathways involving Janus kinases (JAK1 and JAK2) and the activation of signal transducer and activator of transcription 1 (STAT1) are initiated. This activation of IFN-γ is known to induce the expression of various inflammatory cytokines and factors, including tumor necrosis factor (TNF)-α, interleukin (IL)-1β, IL-10, and nitric oxide. Additionally, IFN-γ stimulation leads to upregulation of the expression of MHC molecules, enhances tumor immunogenicity, and augments sensitivity with respect to tumor recognition and destruction. However, prolonged IFN-γ exposure was found to persistently induce immune pressure and lead to tumor immune escape. Moreover, IFN-γ enhanced genetic instability in 4T1 mammary carcinoma cells (Takeda et al. 2017).

Tumor Extrinsic Factor Mechanism

Immunotherapy efficacy is often hampered by the existence of immunosuppressive cells within the tumor microenvironment (TME). Immunosuppressive cells, which include regulatory T cells (Tregs), myeloid-derived suppressor cells (MDSCs), and tumor-associated macrophages (TAMs) have been reported to hinder antitumor activity within the TME. High levels of CD25, the marker used to identify Tregs, were detected in aggressive primary TNBC and were related to resistance to anti-PD-1 therapy (Fattori et al. 2023). MDSCs comprise heterogenous populations of immature cells capable of suppressing the function of cytotoxic T cells (Veglia et al. 2018).

2.4.2 Recent Clinical Strategy to Overcome Immunotherapy Resistance

Research aimed at overcoming immunotherapy resistance in breast cancer is rapidly evolving, and several studies have documented the benefits of combining immunotherapy with other immunotherapy, chemotherapy, or targeted therapy (Table 6).

Pairing Immune Checkpoint Inhibitors PD-1/PD-L1 and CTLA-4

Whereas immune checkpoint inhibitors targeting PD-1/PD-L1 and CTLA-4 have achieved significant success in cancer therapy, resistance can emerge when focusing on a single checkpoint, highlighting the need for combination approaches and alternative strategies in immunotherapy. Accordingly, researchers are exploring the simultaneous blockade of multiple checkpoints. The pairing of inhibitors could afford a more robust immune response, circumventing the pathways employed by tumors to evade a single-target therapy. A combination of durvalumab and tremelimumab can elicit an enhanced immune response and activate T cells to identify cancer-specific antigens in patients with metastatic TNBC (Santa-Maria et al. 2018).

Table 6 Combinatorial clinical strategies to overcome molecular mechanisms of breast cancer cell resistance to immunotherapy

No	Target	Drugs	Mechanism/strategy to overcome resistance	References
1	CTLA-4	Tremelimumab	Durvalumab and tremelimumab increase immune response and activate CD8+ T cells, granzyme, and perforin transcription in individuals with metastatic TNBC	Santa-Maria et al. (2018)
2	Microtubule, topoisomerase DNA	Paclitaxel	Atezolizumab and nab-paclitaxel increase the PFS in patients with metastatic TNBC	Schmid et al. (2018)
		Paclitaxel, nab-paclitaxel, paclitaxel, or gemcitabine plus carboplatin	Combined treatment with pembrolizumab and chemotherapy (nab-paclitaxel, paclitaxel, or gemcitabine plus carboplatin) can substantially improve PFS when compared with the placebo and chemotherapy combination in in patients with metastatic TNBC by increasing PD-L1 expression	Cortes et al. (2020)
		Doxorubicin	Combined treatment with pembrolizumab and doxorubicin can increase the number of CD8+ T cells in patients with metastatic TNBC	Egelston et al. (2023)
3	Tyrosine kinase	Tucatinib	Combined treatment with anti-PD-1 inhibitor and tucatinib improves the antitumor activity and survival in vivo by increasing CD8+ effector memory T cells	Li et al. (2023b)

CTLA-4 cytotoxic T-lymphocyte-associated protein 4, *PD-1* programmed cell death protein 1, *PDL-1* programmed death ligand 1, *PFS* progression-free survival, *TNBC* triple-negative breast cancer

Combining Immunotherapy with Chemotherapy

Chemotherapy, while traditionally viewed as immunosuppressive, can cause immunogenic cell death, making tumor cells more recognizable to the immune system. Combined with immunotherapies, chemotherapy can amplify the immune response against the tumor. In the phase III trial, atezolizumab, a PD-L1 inhibitor, and nab-paclitaxel increased the PFS in patients with metastatic TNBC (Schmid et al. 2018). In patients with metastatic TNBC, the combination of a PD-1 inhibitor (pembrolizumab) with various chemotherapy regimens (nab-paclitaxel, paclitaxel, or gemcitabine plus carboplatin) has resulted in substantial improvements in PFS rates, compared to those with a combination of placebo and chemotherapy (Cortes et al. 2020). In a phase I study combining pembrolizumab and doxorubicin, T cell responses were enhanced in patients with metastatic TNBC (Egelston et al. 2023).

Combining Immunotherapy with Tyrosine Kinase

Additionally, targeted therapies aimed at specific tumor-associated molecules or pathways can complement immunotherapies by weakening the tumor or rendering it more susceptible to immune attack. The combination of tucatinib, a selective HER2 inhibitor, with an anti-PD-1 inhibitor has been shown to enhance antitumor activity and improve survival in in vivo studies (Li et al. 2023b).

3 Integrative Bioinformatics Analysis for Overcoming Breast Cancer Resistance

Bioinformatics is an interdisciplinary field that integrates mathematical principles, information technology, and biological sciences to effectively manage, analyze, visualize, and preserve data pertaining to biological macromolecules (Mitra et al. 2022). The primary objective of bioinformatics is to utilize the generated information to address various biological inquiries (Patil and Gupta 2022). The field of bioinformatics provides valuable support to the discipline of molecular biology by employing computer-based tools to analyze and examine data pertaining to the functionality, regulation, and interconnected pathways of genes and proteins (Zafar et al. 2021). Bioinformatics methodologies for drug discovery commonly incorporate functional drug, gene, disease, and interaction network analyses (Li et al. 2020).

To date, millions of candidate anticancer agents have been suggested; however, only a limited number of these agents were found to both effective and safe (Stine et al. 2022). These select compounds have undergone successful development, subsequent clinical trials, and have been approved for patient treatment on a global scale annually (Stine et al. 2022). Furthermore, the research, development, and marketing of these drugs are expensive and time-consuming endeavors (Sleire et al. 2017). The development of information technology and the use of big data enable efficient and cost-effective drug development (Qian et al. 2019).

The advancement of technology has enhanced the discovery of regulatory genes implicated in cancer, as well as druggable genes, by utilizing genomic and proteomic methodologies (Abbasian et al. 2022). Additionally, the application of metabolomics techniques has facilitated the identification of cancer-associated metabolites and secondary metabolites derived from natural sources (Pang and Hu 2023). The use of omics-based technology and improvements in data storage, data mining, and machine learning have provided adequate sources of information regarding diseases, molecular mechanisms, and drugs (Mottini et al. 2019). Furthermore, advancements in the field of information technology, particularly in artificial intelligence and machine learning, could expedite the identification of protein targets, protein-protein interactions, drug candidates, and molecular mechanisms of anticancer agents

(Abbasian et al. 2022). This can be achieved by leveraging multiomics data sources, resulting in a more efficient and focused approach to discovery.

In the field of bioinformatics, the application of network analysis might facilitate the identification of therapeutic interventions that are tailored to the unique illness characteristics and cellular signaling pathways of individual patients (Infante et al. 2020). Integrative bioinformatics analysis is a comprehensive bioinformatics methodology that relies on the interaction between gene, target, and disease networks to gain a comprehensive understanding of the biological and pathophysiological mechanisms underlying diseases, with the ultimate aim of facilitating the identification and development of therapeutic modalities and diagnostic techniques (Rahman et al. 2021). This section explores the application of an integrative bioinformatics analysis approach in the realm of cancer research, specifically focusing on the identification and advancement of anticancer medications. In addition, we discuss numerous instances where the integrative bioinformatics analysis techniques were employed in research and drug discovery endeavors to address drug resistance observed in breast cancer.

3.1 Recent Integrative Bioinformatics Analysis for Biomarker Selection, Protein Target Identification, Drug Candidate Selection, and Prediction of Drug Sensitivity

Integrative bioinformatics analysis using machine learning/modeling for biomarker selection, protein target identification, drug candidate selection, and prediction of drug sensitivity have been conducted (Table 7). A machine learning methodology known as idTRAX was used to establish a connection between cell-based screening of small-molecule compounds and their kinase inhibition data (Gautam et al. 2019). This strategy was employed to immediately ascertain viable and easily targetable candidates for personalized therapy (Gautam et al. 2019).

By employing machine learning techniques, researchers can predict the response to paclitaxel using interpretable multi-variate classifiers (Bomane et al. 2019). Leveraging DNA methylation and miRNA data, the responsiveness of breast cancer cells to paclitaxel can be predicted to a certain extent (Bomane et al. 2019). Accordingly, these machine learning techniques and relevant data can offer valuable insights for the optimization of paclitaxel therapy in the context of clinical practice (Bomane et al. 2019).

A machine learning model was constructed to predict the inhibition of BCRP (Jiang et al. 2020). This model was shown to be a reliable and effective tool in distinguishing BCRP inhibitors from non-inhibitors within drug design and discovery pipelines (Jiang et al. 2020). The in vivo anticancer drug response was predicted using a comprehensive collection of patient-derived xenograft (PDX) pharmacogenomic data sets and by employing machine learning methods that reduce dimensionality (Nguyen et al. 2021).

Moreover, a multiple-layer cell line-drug response network (ML-CDN2) has been effectively constructed (Huang et al. 2021). This network was created by combining

Table 7 Machine learning and model for integrative bioinformatics analysis for disease, drug, and target identification to overcome breast cancer resistance

No	Machine learning and model	References
1	Machine learning with interpretable DNA methylation and miRNA data for paclitaxel response prediction	Bomane et al. (2019)
2	Machine learning methodology known as idTRAX to establish a connection between cell-based screening of small-molecule compounds and their kinase inhibition data	Gautam et al. (2019)
3	Machine learning model to forecast the inhibition of breast cancer resistance protein (BCRP) and to distinguish BCRP inhibitors from non-inhibitors within drug design and discovery pipelines	Jiang et al. (2020)
4	Machine learning methods using comprehensive collection of patient-derived xenograft (PDX) pharmacogenomic data to foresee the response of cancer drugs in vivo	Nguyen et al. (2021)
5	Multiple-layer cell line-drug response network (ML-CDN2) to forecast drug response in patient-derived breast cancer samples	Huang et al. (2021)
6	Comprehensive analysis of transcriptomic and proteomic data to identify a set of candidate genes correlated with cross-resistance to endocrine therapy in breast cancer	Li et al. (2022)
7	Proteomics and bioinformatics analysis to identify the proteins accountable for resistance to three chemotherapy drugs, namely doxorubicin, tamoxifen, and paclitaxel	Cha et al. (2022)
8	Machine learning techniques to identify probable substrate liability for both multi-drug resistance 1 (MDR1) and BCRP	Adachi et al. (2023)
9	In-depth proteomic investigation of breast cancer cells is conducted to pinpoint and rank potential treatments and provide valuable insights into the mechanisms underlying adaptive resistance, particularly in cases of triple-negative breast cancer	Sun et al. (2023)
10	Ligand- and structure-based methodologies on the modeling of transmembrane transporters in the development of drug candidates that exhibit enhanced efficacy against a specific target protein	Grandits and Ecker (2023)

a one-layer cell line similarity network, which relied on pathway activity profiles, with a three-layer drug similarity network, which considered drug structures, targets, and pan-cancer IC_{50} profiles (Huang et al. 2021). Consequently, the constructed model could forecast drug response in patient-derived breast cancer samples (Huang et al. 2021).

A comprehensive analysis of transcriptomic and proteomic data was undertaken to classify a group of candidate genes associated with cross-resistance to endocrine therapy in breast cancer (Li et al. 2022). The authors of the study highlighted the substantial clinical implications of their research, emphasizing that the identified candidate gene sets could play crucial roles in improving diagnoses, broadening the understanding of the underlying mechanisms and guiding the development of therapeutic strategies that address endocrine therapy resistance in breast cancer, encompassing both SERMs/SERDs and aromatase inhibitors (Li et al. 2022). Moreover, these findings hold promise for future advancements in the field of transcriptomics and proteomics (Li et al. 2022). A comprehensive proteomic and bioinformatic analysis was carried out to identify the proteins associated with resistance to three chemotherapy drugs, namely doxorubicin, tamoxifen, and paclitaxel, in breast cancer cells (Cha et al. 2022). The use of ensemble machine learning approaches, specifically utilizing molecular descriptors and graph convolutional networks, highlighted the advantages of employing sophisticated machine learning techniques to identify probable substrate liability for both MDR1 and BCRP (Adachi et al. 2023).

The application of a systematic approach involving multiomics techniques, combined with a comprehensive examination of the proteome of breast cancer cells through the utilization of data-independent acquisition, could effectively identify and prioritize therapeutics and offer valuable insights into the mechanisms of adaptive resistance, specifically in TNBC (Sun et al. 2023). The utilization of ligand- and structure-based methodologies in the modeling of transmembrane transporters enables the development of drug candidates that exhibit enhanced efficacy against a specific target protein. This, in turn, facilitates and reinforces subsequent synthetic endeavors (Grandits and Ecker 2023).

3.2 Recent Studies Employing Integrative Bioinformatics for Protein Target Compound Identification

This subsection discusses studies that explored potential protein targets of compounds to overcome breast cancer resistance, specifically using integrative bioinformatics analysis (Table 8). For example, ERα antagonists were effectively identified from an in-house natural product library using ligand-based machine learning techniques and structure-based molecular docking (Pang et al. 2018). The identified compounds, namely genistein, daidzein, phloretin, ellagic acid, ursolic acid, (−)-epigallocatechin-3-gallate, kaempferol, and naringenin, were subsequently validated in vitro, exhibiting varying degrees of antagonistic activity against ERα (Pang et al. 2018).

Using comprehensive bioinformatics techniques, ABL1, DNMT3B, and MLH1 were identified as prospective targets while also suggesting the involvement of the JAK-STAT cascade as a plausible molecular mechanism underlying the efficacy of hesperetin in combating breast cancer chemoresistance (Hermawan et al. 2020). A bioinformatics investigation identified CDK2, MDM2, KRAS, PIK3R1, CDH1, and ESR1 as specific targets of interest. Additionally, activation of the PI3K/Akt signaling pathway plays a crucial role in the molecular mechanism through which cinnamaldehyde can be used to overcome resistance to fulvestrant in breast cancer (Hermawan et al. 2021). Moreover, a genomic understanding of the role of fibroblast growth factor receptor 2 (FGFR2) as a paeoniflorin target is paramount in overcoming tamoxifen resistance in breast cancer (Wibowo et al. 2021). The creation of a cell line-specific network model for ER+ breast cancer has unveiled the potential mechanisms of resistance to PI3Kα inhibitors and medication combinations, providing valuable insights for combating drug resistance in cancer treatment (Gómez Tejeda Zañudo et al. 2021).

Table 8 Studies that implemented integrative bioinformatics analysis for drug target identification to overcome breast cancer resistance

No	Description of study	References
1	Ligand-based machine learning techniques and structure-based molecular docking to identify estrogen receptor (ER) α antagonists from natural product library	Pang et al. (2018)
2	Comprehensive bioinformatics techniques to identify prospective targets and plausible molecular mechanism underlying the efficacy of hesperetin in combating breast cancer chemoresistance	Hermawan et al. (2020)
3	Comprehensive bioinformatics investigation to explore molecular targets and mechanism of cinnamaldehyde to overcome breast cancer resistance to trastuzumab	Hermawan et al. (2021)
4	Bioinformatics analysis to identify paeoniflorin targets to overcome resistance to tamoxifen in breast cancer	Wibowo et al. (2021)
5	Utilizing a bioinformatic analysis with cell line-specific network models for ER+ breast cancer, we aim to identify a PI3Kα inhibitor that can effectively address resistance issues in breast cancer treatment	Gómez Tejeda Zañudo et al. (2021)
6	Bioinformatics analysis to explore targets and possible mechanism of glycyrrhizic acid to address resistance to tamoxifen in breast cancer cells	Azzahra et al. (2022)
7	Bioinformatics research and in vitro experimentation to establish the ability of honokiol to overcome resistance to tamoxifen in breast cancer cells	Hermawan et al. (2022)
8	Functional network analysis revealed p85 and PI3K as prospective gene targets and the underlying mechanism by which oleanolic acid overcomes tamoxifen resistance in breast cancer	Ibadurrahman et al. (2022)
9	Computational analysis discovers the targets of honokiol to understand the molecular mechanisms associated with cell cycle inhibition and immunotherapy in metastatic breast cancer stem cells	Skolastika et al. (2022)
10	Multiomics techniques and bioinformatics analysis to identify prospective targets for therapeutic interventions in TNBC	Ma et al. (2022)
11	Integrated bioinformatics analysis to identify the molecular pathways behind resistance to neoadjuvant chemotherapy	Saha Detroja et al. (2022)
12	Network-informed discovery to identify potential multidrug combinations for the treatment of ERα+/HER2−/PI3Kα-mutant breast cancer	Hany et al. (2023)
13	Machine learning–based virtual screening to identify novel and highly effective inhibitors targeting the EGFR protein for the treatment of breast cancer	Siddiqui et al. (2023)

EGFR epidermal growth factor receptor, *HER2* human epidermal growth factor receptor 2, *PI3Kα* phosphatidylinositol 3-kinase, isoform α

Using integrative bioinformatics, MDM2 was identified as a potential gene target for glycyrrhizic acid, while the PI3K-AKT signaling pathway holds promise as a possible mechanism for addressing resistance of tamoxifen in cells affected by breast cancer (Azzahra et al. 2022). Extensive bioinformatics research and in vitro experimentation demonstrated that honokiol could overcome tamoxifen resistance in breast cancer, potentially by selectively targeting *FGFR2*, *RET*, *ERBB4*, *MMP16*, *FN1*, and *SOX2* (Hermawan et al. 2022). Moreover, functional network analysis has shed light on p85 and PI3K as prospective gene targets, as well as the underlying mechanism through which oleanolic acid overcomes breast cancer resistance to tamoxifen (Ibadurrahman et al. 2022). According to in-depth computational analysis, CCND1, SIRT2, AURKB, VEGFA, HDAC1, CASP9, HSP90AA1, and HSP90AB1 were identified as targets, while the PI3K/Akt/mTOR and HIF1/NFkB pathways were identified as the molecular mechanisms through which honokiol exerts its effects on cell cycle inhibition and immunotherapy in metastatic breast cancer stem cells (Skolastika et al. 2022).

For TNBC, a prognostic signature comprising five genes was generated by utilizing multi-omics techniques and bioinformatics analysis (Ma et al. 2022), revealing possible prognostic markers for TNBC and identifying prospective targets for therapeutic interventions (Ma et al. 2022). Furthermore, integrated bioinformatics analysis was performed to effectively generate novel perspectives on the molecular pathways underlying resistance to neoadjuvant chemotherapy (Saha Detroja et al. 2022). Furthermore, the study has established a foundation for the implementation of medication-repurposing strategies and personalized treatment approaches aimed at overcoming resistance to neoadjuvant chemotherapy in breast cancer (Saha Detroja et al. 2022). A network-informed discovery was undertaken to identify potential multidrug combinations for the treatment of ERα+/HER2−/PI3Kα-mutant breast cancer (Hany et al. 2023). A comprehensive machine learning–based virtual screening approach, complemented by other computational approaches in conjunction with other computational methods, identified novel and highly effective inhibitors targeting the EGFR protein, with the ultimate goal of developing drugs for the treatment of breast cancer (Siddiqui et al. 2023). Collectively, extensive research has been conducted in the field of drug development, particularly during the identification phase, which is valuable in expediting the discovery of anticancer medications, particularly to address the challenge of resistance encountered in breast cancer treatment.

4 Conclusion

Recent studies have helped to elucidate the underlying mechanisms through which breast cancer cells acquire resistance to various targeted therapies and chemotherapeutic treatments. Clinical protocols for overcoming breast cancer resistance toward endocrine, targeted, chemo-, and immunotherapy have also been developed. In addition, advancements in the field of science and technology have facilitated the implementation of high throughput screening methodologies, including next-generation sequencing-based genomics, liquid chromatography with tandem mass spectrometry–based proteomics, and metabolomics. These techniques have enabled the examination of resistance mechanisms exhibited by breast cancer cells and the identification of possible protein targets for therapeutic compounds aimed at overcoming resistance. Furthermore, advances in information technology have fostered the growth of databases and software alongside concepts such as artificial intelligence and machine learning, which are utilized to identify potential gene targets and forecast their pharmacokinetic and toxicity profiles. These advancements will expedite the process of target identification and validation in the pursuit of discovering and developing anticancer medications capable of effectively combating the resistance of breast cancer cells toward targeted therapies and chemotherapy.

Acknowledgments The authors thank Ms. Ririn Widarti and Ms. Dian Anita for administrative support.

Competing Interests The authors declare no competing interests.

Ethics Approval and Consent to Participate Not applicable.

Funding Adam Hermawan was funded by the Penelitian Fundamental Reguler (PFR), contract number: 3132/UN1/DITLIT/Dit-Lit/PT.01.03/2023.

References

Abbasian MH, Ardekani AM, Sobhani N, Roudi R (2022) The role of genomics and proteomics in lung cancer early detection and treatment. Cancers (Basel) 14:5144

Abu Samaan TM, Samec M, Liskova A, Kubatka P, Büsselberg D (2019) Paclitaxel's mechanistic and clinical effects on breast cancer. Biomol Ther 9:789

Adachi A, Yamashita T, Kanaya S, Kosugi Y (2023) Ensemble machine learning approaches based on molecular descriptors and graph convolutional networks for predicting the efflux activities of MDR1 and BCRP transporters. AAPS J 25:88

Adrienne GW, Tanya EK, Tianyu L, Nabihah T, Gerburg MW, Edward TR III, Victoria A, Leilani A, Elizabeth AM, Beth O, Eric PW, Ian EK, Judith A, Eliezer MVA, Sara MT (2022) Phase Ib study of pembrolizumab in combination with trastuzumab emtansine for metastatic HER2-positive breast cancer. J Immunother Cancer 10:e005119

Ajabnoor GMA, Crook T, Coley HM (2012) Paclitaxel resistance is associated with switch from apoptotic to autophagic cell death in MCF-7 breast cancer cells. Cell Death Dis 3:e260–e260

Ali S, Rasool M, Chaoudhry H, Pushparaj PN, Jha P, Hafiz A, Mahfooz M, Abdus Sami G, Azhar Kamal M, Bashir S, Ali A, Sarwar Jamal M (2016) Molecular mechanisms and mode of tamoxifen resistance in breast cancer. Bioinformation 12:135–139

Aman NA, Doukoure B, Koffi KD, Koui BS, Traore ZC, Kouyate M, Effi AB (2019) HER2 overexpression and correlation with other significant clinicopathologic parameters in Ivorian breast cancer women. BMC Clin Pathol 19:1

Andrahennadi S, Sami A, Haider K, Chalchal HI, Le D, Ahmed O, Manna M, El-Gayed A, Wright P, Ahmed S (2021) Efficacy of Fulvestrant in women with hormone-resistant metastatic breast cancer (mBC): a Canadian Province experience. Cancers (Basel) 13:4163

André F, Ciruelos EM, Juric D, Loibl S, Campone M, Mayer IA, Rubovszky G, Yamashita T, Kaufman B, Lu YS, Inoue K, Pápai Z, Takahashi M, Ghaznawi F, Mills D, Kaper M, Miller M, Conte PF, Iwata H, Rugo HS (2021) Alpelisib plus fulvestrant for PIK3CA-mutated, hormone receptor-positive, human epidermal growth factor receptor-2-negative advanced breast cancer: final overall survival results from SOLAR-1. Ann Oncol 32:208–217

André F, Hee Park Y, Kim SB, Takano T, Im SA, Borges G, Lima JP, Aksoy S, Gavila Gregori J, De Laurentiis M, Bianchini G, Roylance R, Miyoshi Y, Armstrong A, Sinha R, Ruiz Borrego M, Lim E, Ettl J, Yerushalmi R, Zagouri F, Duhoux FP, Fehm T, Gambhire D, Cathcart J, Wu C, Chu C, Egorov A, Krop I (2023) Trastuzumab deruxtecan versus treatment of physician's choice in patients with HER2-positive metastatic breast cancer (DESTINY-Breast02): a randomised, open-label, multicentre, phase 3 trial. Lancet 401:1773–1785

Arcaro A, Guerreiro AS (2007) The phosphoinositide 3-kinase pathway in human cancer: genetic alterations and therapeutic implications. Curr Genomics 8:271–306

Arpino G, Wiechmann L, Osborne CK, Schiff R (2008) Crosstalk between the estrogen receptor and the HER tyrosine kinase receptor family: molecular mechanism and clinical implications for endocrine therapy resistance. Endocr Rev 29:217–233

Azzahra SNA, Hanif N, Hermawan A (2022) MDM2 is a potential target gene of glycyrrhizic acid for circumventing breast cancer resistance to tamoxifen: integrative bioinformatics analysis. Asian Pac J Cancer Prev 23:2341–2350

Bachelot T, Garcia-Saenz JA, Verma S, Gutierrez M, Pivot X, Kozloff MF, Prady C, Huang X, Khosravan R, Wang Z, Cesari R, Tassell V, Kern KA, Blay JY, Lluch A (2014) Sunitinib in combination with trastuzumab for the treatment of advanced breast cancer: activity and safety results from a phase II study. BMC Cancer 14:166

Baird RD, van Rossum AGJ, Oliveira M, Beelen K, Gao M, Schrier M, Mandjes IAM, Garcia-Corbacho J, Vallier AL, Dougall G, van Werkhoven E, Linossi C, Kumar S, van Tinteren H, Callari M, Beddowes E, Perez-Garcia JM, Rosing H, Platte E, Nederlof P, Schot M, de Vries SA, Bernards R, Saura C, Gallagher W, Cortès J, Caldas C, Linn SC (2019) POSEIDON trial phase 1b results: safety, efficacy and circulating tumor DNA response of the beta isoform-sparing PI3K inhibitor taselisib (GDC-0032) combined with tamoxifen in hormone receptor positive metastatic breast cancer patients. Clin Cancer Res 25:6598–6605

Ballhausen A, Wheler JJ, Karp DD, Piha-Paul SA, Fu S, Pant S, Tsimberidou AM, Hong DS, Subbiah V, Holley VR, Huang HJ, Brewster AM, Koenig KB, Ibrahim NK, Meric-Bernstam F, Janku F (2021) Phase I study of everolimus, letrozole, and trastuzumab in patients with hormone receptor-positive metastatic breast cancer or other solid tumors. Clin Cancer Res 27:1247–1255

Baselga J, Im S-A, Iwata H, Clemons M, Ito Y, Awada A, Chia S, Jagiełło-Gruszfeld A, Pistilli B, Tseng L-M (2015) PIK3CA status in circulating tumor DNA predicts efficacy of buparlisib plus fulvestrant in postmenopausal women with endocrine-resistant HR+/HER2–advanced breast cancer: first results from the randomized, phase III BELLE-2 trial. Proceedings of the 38th annual San Antonio breast cancer symposium, San Antonio

Baselga J, Im SA, Iwata H, Cortés J, De Laurentiis M, Jiang Z, Arteaga CL, Jonat W, Clemons M, Ito Y, Awada A, Chia S, Jagiełło-Gruszfeld A, Pistilli B, Tseng LM, Hurvitz S, Masuda N, Takahashi M, Vuylsteke P, Hachemi S, Dharan B, Di Tomaso E, Urban P, Massacesi C, Campone M (2017) Buparlisib plus fulvestrant versus placebo plus fulvestrant in postmenopausal, hormone receptor-positive, HER2-negative, advanced breast cancer (BELLE-2): a randomised, double-blind, placebo-controlled, phase 3 trial. Lancet Oncol 18:904–916

Basu A, Krishnamurthy S (2010) Cellular responses to cisplatin-induced DNA damage. J Nucleic Acids 2010:201367

Bellmunt J, Paz-Ares L, Cuello M, Cecere F, Albiol S, Guillem V, Gallardo E, Carles J, Mendez P, De la Cruz J (2007) Gene expression of ERCC1 as a novel prognostic marker in advanced bladder cancer patients receiving cisplatin-based chemotherapy. Ann Oncol 18:522–528

Blackwell KL, Zaman K, Qin S, Tkaczuk KHR, Campone M, Hunt D, Bryce R, Goldstein LJ (2019) Neratinib in combination with trastuzumab for the treatment of patients with advanced HER2-positive breast cancer: a phase I/II study. Clin Breast Cancer 19:97–104.e104

Bomane A, Gonçalves A, Ballester PJ (2019) Paclitaxel response can be predicted with interpretable multi-variate classifiers exploiting DNA-methylation and miRNA data. Front Genet 10: 1041

Bos MK, Lam SW, Motta G, Helmijr JCA, Beaufort CM, de Jonge E, Martens JWM, Boven E, Jansen M, Jager A, Sleijfer S (2023) Plasma ESR1 mutations and outcome to first-line paclitaxel and bevacizumab in patients with advanced ER-positive/HER2-negative breast cancer. Breast Cancer Res Treat 200:271–279

Bostner J, Karlsson E, Pandiyan MJ, Westman H, Skoog L, Fornander T, Nordenskjöld B, Stål O (2013) Activation of Akt, mTOR, and the estrogen receptor as a signature to predict tamoxifen treatment benefit. Breast Cancer Res Treat 137:397–406

Brabec V (2002) DNA modifications by antitumor platinum and ruthenium compounds: their recognition and repair. Prog Nucleic Acid Res Mol Biol 71:1

Braun L, Mietzsch F, Seibold P, Schneeweiss A, Schirmacher P, Chang-Claude J, Sinn HP, Aulmann S (2013) Intrinsic breast cancer subtypes defined by estrogen receptor signalling—prognostic relevance of progesterone receptor loss. Mod Pathol 26:1161–1171

Breugom AJ, Swets M, Bosset J-F, Collette L, Sainato A, Cionini L, Glynne-Jones R, Counsell N, Bastiaannet E, van den Broek CB (2015) Adjuvant chemotherapy after preoperative (chemo) radiotherapy and surgery for patients with rectal cancer: a systematic review and meta-analysis of individual patient data. Lancet Oncol 16:200–207

Brown A, Kumar S, Tchounwou PB (2019) Cisplatin-based chemotherapy of human cancers. J Cancer Sci Ther 11(4):97

Buchbinder EI, Desai A (2016) CTLA-4 and PD-1 pathways: similarities, differences, and implications of their inhibition. Am J Clin Oncol 39:98–106

Bydoun M, Marcato P, Dellaire G (2014) Breast cancer genomics. Elsevier, Cancer genomics, pp 213–232

Campone M, Im SA, Iwata H, Clemons M, Ito Y, Awada A, Chia S, Jagiełło-Gruszfeld A, Pistilli B, Tseng LM, Hurvitz S, Masuda N, Cortés J, De Laurentiis M, Arteaga CL, Jiang Z, Jonat W, Le Mouhaër S, Sankaran B, Bourdeau L, El-Hashimy M, Sellami D, Baselga J (2018) Buparlisib plus fulvestrant versus placebo plus fulvestrant for postmenopausal, hormone receptor-positive, human epidermal growth factor receptor 2-negative, advanced breast cancer: overall survival results from BELLE-2. Eur J Cancer 103:147–154

Cappetta D, De Angelis A, Sapio L, Prezioso L, Illiano M, Quaini F, Rossi F, Berrino L, Naviglio S, Urbanek K (2017) Oxidative stress and cellular response to doxorubicin: a common factor in the complex milieu of anthracycline cardiotoxicity. Oxidative Med Cell Longev 2017:1521020

Cha HK, Cheon S, Kim H, Lee KM, Ryu HS, Han D (2022) Discovery of proteins responsible for resistance to three chemotherapy drugs in breast cancer cells using proteomics and bioinformatics analysis. Molecules 27(6):1762

Chaganty BKR, Qiu S, Gest A, Lu Y, Ivan C, Calin GA, Weiner LM, Fan Z (2018) Trastuzumab upregulates PD-L1 as a potential mechanism of trastuzumab resistance through engagement of immune effector cells and stimulation of IFNγ secretion. Cancer Lett 430:47–56

Chia SK, Ellard SL, Mates M, Welch S, Mihalcioiu C, Miller WH Jr, Gelmon K, Lohrisch C, Kumar V, Taylor S, Hagerman L, Goodwin R, Wang T, Sakashita S, Tsao MS, Eisenhauer E, Bradbury P (2017) A phase-I study of lapatinib in combination with foretinib, a c-MET, AXL and vascular endothelial growth factor receptor inhibitor, in human epidermal growth factor receptor 2 (HER-2)-positive metastatic breast cancer. Breast Cancer Res 19:54

Chia SKL, Bedard PL, Hilton J, Amir E, Gelmon KA, Goodwin RA, Villa D, Cabanero M, Ritter H, Tu D, Tsao MS, Seymour L (2018) A phase I study of a PD-L1 antibody (Durvalumab) in combination with trastuzumab in HER-2 positive metastatic breast cancer (MBC) progressing on prior anti HER-2 therapies (CCTG IND.229)[NCT02649686]. J Clin Oncol 36:1029–1029

Cho HS, Mason K, Ramyar KX, Stanley AM, Gabelli SB, Denney DW Jr, Leahy DJ (2003) Structure of the extracellular region of HER2 alone and in complex with the Herceptin Fab. Nature 421:756–760

Christowitz C, Davis T, Isaacs A, van Niekerk G, Hattingh S, Engelbrecht AM (2019) Mechanisms of doxorubicin-induced drug resistance and drug resistant tumour growth in a murine breast tumour model. BMC Cancer 19:757

Ciruelos E, Villagrasa P, Pascual T, Oliveira M, Pernas S, Paré L, Escrivá-de-Romaní S, Manso L, Adamo B, Martínez E, Cortés J, Vazquez S, Perelló A, Garau I, Melé M, Martínez N, Montaño A, Bermejo B, Morales S, Echarri MJ, Vega E, González-Farré B, Martínez D, Galván P, Canes J, Nuciforo P, Gonzalez X, Prat A (2020) Palbociclib and trastuzumab in HER2-positive advanced breast cancer: results from the phase II SOLTI-1303 PATRICIA trial. Clin Cancer Res 26:5820–5829

Ciszewski WM, Tavecchio M, Dastych J, Curtin NJ (2014) DNA-PK inhibition by NU7441 sensitizes breast cancer cells to ionizing radiation and doxorubicin. Breast Cancer Res Treat 143:47–55

Clarke R, Tyson JJ, Dixon JM (2015) Endocrine resistance in breast cancer–an overview and update. Mol Cell Endocrinol 418:220–234

Coates AS, Winer EP, Goldhirsch A, Gelber RD, Gnant M, Piccart-Gebhart M, Thürlimann B, Senn H-J, Members P, André F (2015) Tailoring therapies—improving the management of early breast cancer: St Gallen International Expert Consensus on the primary therapy of early breast cancer 2015. Ann Oncol 26:1533–1546

Cortes J, Cescon DW, Rugo HS, Nowecki Z, Im SA, Yusof MM, Gallardo C, Lipatov O, Barrios CH, Holgado E, Iwata H, Masuda N, Otero MT, Gokmen E, Loi S, Guo Z, Zhao J, Aktan G, Karantza V, Schmid P, Investigators K (2020) Pembrolizumab plus chemotherapy versus placebo plus chemotherapy for previously untreated locally recurrent inoperable or metastatic triple-negative breast cancer (KEYNOTE-355): a randomised, placebo-controlled, double-blind, phase 3 clinical trial. Lancet 396:1817–1828

Cristofanilli M, Turner NC, Bondarenko I, Ro J, Im S-A, Masuda N, Colleoni M, DeMichele A, Loi S, Verma S (2016) Fulvestrant plus palbociclib versus fulvestrant plus placebo for treatment of hormone-receptor-positive, HER2-negative metastatic breast cancer that progressed on previous endocrine therapy (PALOMA-3): final analysis of the multicentre, double-blind, phase 3 randomised controlled trial. Lancet Oncol 17:425–439

Cui S, Feng J, Tang X, Lou S, Guo W, Xiao X, Li S, Chen X, Huan Y, Zhou Y, Xiao L (2023) The prognostic value of tumor mutation burden (TMB) and its relationship with immune infiltration in breast cancer patients. Eur J Med Res 28:90

Curigliano G, Dunton K, Rosenlund M, Janek M, Cathcart J, Liu Y, Fasching PA, Iwata H (2023) Patient-reported outcomes and hospitalization data in patients with HER2-positive metastatic breast cancer receiving trastuzumab deruxtecan or trastuzumab emtansine in the phase III DESTINY-Breast03 study. Ann Oncol 34:569–577

deGraffenried LA, Chandrasekar B, Friedrichs WE, Donzis E, Silva J, Hidalgo M, Freeman JW, Weiss GR (2004) NF-κB inhibition markedly enhances sensitivity of resistant breast cancer tumor cells to tamoxifen. Ann Oncol 15:885–890

Dickler MN, Saura C, Richards DA, Krop IE, Cervantes A, Bedard PL, Patel MR, Pusztai L, Oliveira M, Ware JA (2016) A phase II study of the PI3K inhibitor taselisib (GDC-0032) combined with fulvestrant (F) in patients (pts) with HER2-negative (HER2-), hormone receptor-positive (HR+) advanced breast cancer (BC). Am Soc Clin Oncol 34:520

DiMarco AV, Qin X, McKinney BJ, Garcia NMG, Van Alsten SC, Mendes EA, Force J, Hanks BA, Troester MA, Owzar K, Xie J, Alvarez JV (2022) APOBEC mutagenesis inhibits breast cancer growth through induction of T cell-mediated antitumor immune responses. Cancer Immunol Res 10:70–86

Droog M, Beelen K, Linn S, Zwart W (2013) Tamoxifen resistance: from bench to bedside. Eur J Pharmacol 717:47–57

Dusenbery AC, Maniaci JL, Hillerson ND, Dill EA, Bullock TN, Mills AM (2021) MHC class I loss in triple-negative breast cancer: a potential barrier to PD-1/PD-L1 checkpoint inhibitors. Am J Surg Pathol 45:701–707

Duz MB, Karatas OF (2021) Differential expression of ABCB1, ABCG2, and KLF4 as putative indicators for paclitaxel resistance in human epithelial type 2 cells. Mol Biol Rep 48:1393–1400

Dziadyk JM, Sui M, Zhu X, Fan W (2004) Paclitaxel-induced apoptosis may occur without a prior G2/M-phase arrest. Anticancer Res 24:27–36

Egelston CA, Guo W, Yost SE, Ge X, Lee JS, Frankel PH, Cui Y, Ruel C, Schmolze D, Murga M, Tang A, Martinez N, Karimi M, Somlo G, Lee PP, Waisman JR, Yuan Y (2023) Immunogenicity and efficacy of pembrolizumab and doxorubicin in a phase I trial for patients with metastatic triple-negative breast cancer. Cancer Immunol Immunother 72:3013–3027

Entwistle RA, Winefield RD, Foland TB, Lushington GH, Himes RH (2008) The paclitaxel site in tubulin probed by site-directed mutagenesis of Saccharomyces cerevisiae β-tubulin. FEBS Lett 582:2467–2470

Eustace AJ, Conlon NT, McDermott MSJ, Browne BC, O'Leary P, Holmes FA, Espina V, Liotta LA, O'Shaughnessy J, Gallagher C, O'Driscoll L, Rani S, Madden SF, O'Brien NA, Ginther C, Slamon D, Walsh N, Gallagher WM, Zagozdzon R, Watson WR, O'Donovan N, Crown J (2018) Development of acquired resistance to lapatinib may sensitise HER2-positive breast cancer cells to apoptosis induction by obatoclax and TRAIL. BMC Cancer 18:965

Fan W, Chang J, Fu P (2015) Endocrine therapy resistance in breast cancer: current status, possible mechanisms and overcoming strategies. Future Med Chem 7:1511–1519

Fang X, Cao J, Shen A (2020) Advances in anti-breast cancer drugs and the application of nano-drug delivery systems in breast cancer therapy. J Drug Deliv Sci Technol 57:101662

Fasching PA, Hein A, Kolberg HC, Häberle L, Uhrig S, Rübner M, Belleville E, Hack CC, Fehm TN, Janni W, Hartmann A, Erber R, Theuser AK, Brucker SY, Hartkopf AD, Untch M (2023) Pembrolizumab in combination with nab-paclitaxel for the treatment of patients with early-stage triple-negative breast cancer – a single-arm phase II trial (NeoImmunoboost, AGO-B-041). Eur J Cancer 184:1–9

Fattori S, Le Roy A, Houacine J, Robert L, Abes R, Gorvel L, Granjeaud S, Rouviere MS, Ben Amara A, Boucherit N, Tarpin C, Pakradouni J, Charafe-Jauffret E, Houvenaeghel G, Lambaudie E, Bertucci F, Rochigneux P, Goncalves A, Foussat A, Chretien AS, Olive D (2023) CD25high effector regulatory T cells hamper responses to PD-1 blockade in triple-negative breast cancer. Cancer Res 83:3026–3044

Feng Y, Spezia M, Huang S, Yuan C, Zeng Z, Zhang L, Ji X, Liu W, Huang B, Luo W (2018) Breast cancer development and progression: risk factors, cancer stem cells, signaling pathways, genomics, and molecular pathogenesis. Genes Dis 5:77–106

Fichtinger-Schepman AM, van der Veer JL, den Hartog JH, Lohman PH, Reedijk J (1985) Adducts of the antitumor drug cis-diamminedichloroplatinum(II) with DNA: formation, identification, and quantitation. Biochemistry 24:707–713

Fisusi FA, Akala EO (2019) Drug combinations in breast cancer therapy. Pharm Nanotechnol 7: 3–23

Fribbens C, O'Leary B, Kilburn L, Hrebien S, Garcia-Murillas I, Beaney M, Cristofanilli M, Andre F, Loi S, Loibl S (2016) Plasma ESR1 mutations and the treatment of estrogen receptor-positive advanced breast cancer

Fumoleau P, Koch KM, Brain E, Lokiec F, Rezai K, Awada A, Hayward L, Werutsky G, Bogaerts J, Marréaud S, Cardoso F (2014) A phase I pharmacokinetics study of lapatinib and tamoxifen in metastatic breast cancer (EORTC 10053 Lapatam study). Breast 23:663–669

Gadgeel SM, Lew DL, Synold TW, LoRusso P, Chung V, Christensen SD, Smith DC, Kingsbury L, Hoering A, Kurzrock R (2013) Phase I study evaluating the combination of lapatinib (a Her2/Neu and EGFR inhibitor) and everolimus (an mTOR inhibitor) in patients with advanced cancers: South West Oncology Group (SWOG) Study S0528. Cancer Chemother Pharmacol 72:1089–1096

Gajria D, Chandarlapaty S (2011) HER2-amplified breast cancer: mechanisms of trastuzumab resistance and novel targeted therapies. Expert Rev Anticancer Ther 11:263–275

Gala K, Chandarlapaty S (2014) Molecular pathways: HER3 targeted therapy. Clin Cancer Res 20: 1410–1416

Gao C, Li H, Liu C, Xu X, Zhuang J, Zhou C, Liu L, Feng F, Sun C (2021) Tumor mutation burden and immune invasion characteristics in triple negative breast cancer: genome high-throughput data analysis. Front Immunol 12:650491

Gautam P, Jaiswal A, Aittokallio T, Al-Ali H, Wennerberg K (2019) Phenotypic screening combined with machine learning for efficient identification of breast cancer-selective therapeutic targets. Cell Chem Biol 26:970–979.e974

Giessrigl B, Schmidt WM, Kalipciyan M, Jeitler M, Bilban M, Gollinger M, Krieger S, Jäger W, Mader RM, Krupitza G (2013) Fulvestrant induces resistance by modulating GPER and CDK6 expression: implication of methyltransferases, deacetylases and the hSWI/SNF chromatin remodelling complex. Br J Cancer 109:2751–2762

Giuliano M, Trivedi MV, Schiff R (2013) Bidirectional crosstalk between the estrogen receptor and human epidermal growth factor receptor 2 signaling pathways in breast cancer: molecular basis and clinical implications. Breast Care (Basel) 8:256–262

Giuliano M, Hu H, Wang YC, Fu X, Nardone A, Herrera S, Mao S, Contreras A, Gutierrez C, Wang T, Hilsenbeck SG, De Angelis C, Wang NJ, Heiser LM, Gray JW, Lopez-Tarruella S, Pavlick AC, Trivedi MV, Chamness GC, Chang JC, Osborne CK, Rimawi MF, Schiff R (2015) Upregulation of ER signaling as an adaptive mechanism of cell survival in HER2-positive breast tumors treated with anti-HER2 therapy. Clin Cancer Res 21:3995–4003

Gluz O, Liedtke C, Gottschalk N, Pusztai L, Nitz U, Harbeck N (2009) Triple-negative breast cancer—current status and future directions. Ann Oncol 20:1913–1927

Gluz O, Nitz UA, Christgen M, Kuemmel S, Holtschmidt J, Schumacher J, Hartkopf A, Potenberg J, Lüedtke-Heckenkamp K, Just M, Schem C, von Schumann R, Kolberg-Liedtke C, Eulenburg CZ, Schinköthe T, Graeser M, Wuerstlein R, Kates RE, Kreipe HH, Harbeck N (2023) Efficacy of endocrine therapy plus trastuzumab and pertuzumab vs de-escalated chemotherapy in patients with hormone receptor-positive/ERBB2-positive early breast cancer: the neoadjuvant WSG-TP-II randomized clinical trial. JAMA Oncol 9:946–954

Goetz MP, Okera M, Wildiers H, Campone M, Grischke EM, Manso L, André VAM, Chouaki N, San Antonio B, Toi M, Sledge GW Jr (2021) Safety and efficacy of abemaciclib plus endocrine therapy in older patients with hormone receptor-positive/human epidermal growth factor receptor 2-negative advanced breast cancer: an age-specific subgroup analysis of MONARCH 2 and 3 trials. Breast Cancer Res Treat 186:417–428

Goldenson B, Crispino JD (2015) The aurora kinases in cell cycle and leukemia. Oncogene 34: 537–545

Gomes I, de Almeida BP, Dâmaso S, Mansinho A, Correia I, Henriques S, Cruz-Duarte R, Vilhais G, Félix P, Alves P, Corredeira P, Barbosa-Morais NL, Costa L, Casimiro S (2020) Expression of receptor activator of NFkB (RANK) drives stemness and resistance to therapy in ER+HER2- breast cancer. Oncotarget 11:1714

Gómez Tejeda Zañudo J, Mao P, Alcon C, Kowalski K, Johnson GN, Xu G, Baselga J, Scaltriti M, Letai A, Montero J, Albert R, Wagle N (2021) Cell line-specific network models of ER(+) breast cancer identify potential PI3Kα inhibitor resistance mechanisms and drug combinations. Cancer Res 81:4603–4617

Graham PJ, Brar MS, Foster T, McCall M, Bouchard-Fortier A, Temple W, Quan ML (2015) Neoadjuvant chemotherapy for breast cancer, is practice changing? A population-based review of current surgical trends. Ann Surg Oncol 22:3376–3382

Grandits M, Ecker GF (2023) Ligand- and structure-based approaches for transmembrane transporter modeling. Curr Drug Res Rev 15

Guarneri V, Dieci MV, Griguolo G, Miglietta F, Girardi F, Bisagni G, Generali DG, Cagossi K, Sarti S, Frassoldati A, Gianni L, Cavanna L, Pinotti G, Musolino A, Piacentini F, Cinieri S, Prat A, Conte P (2021) Trastuzumab-lapatinib as neoadjuvant therapy for HER2-positive early breast cancer: survival analyses of the CHER-Lob trial. Eur J Cancer 153:133–141

Guo S, Sonenshein GE (2004) Forkhead box transcription factor FOXO3a regulates estrogen receptor alpha expression and is repressed by the her-2/neu/phosphatidylinositol 3-kinase/Akt signaling pathway. Mol Cell Biol 24:8681–8690

Gutierrez C, Schiff R (2011) HER2: biology, detection, and clinical implications. Arch Pathol Lab Med 135:55–62

Haddad TC, Suman VJ, D'Assoro AB, Carter JM, Giridhar KV, McMenomy BP, Santo K, Mayer EL, Karuturi MS, Morikawa A, Marcom PK, Isaacs CJ, Oh SY, Clark AS, Mayer IA, Keyomarsi K, Hobday TJ, Peethambaram PP, O'Sullivan CC, Leon-Ferre RA, Liu MC, Ingle JN, Goetz MP (2023) Evaluation of alisertib alone or combined with fulvestrant in patients with endocrine-resistant advanced breast cancer: the phase 2 TBCRC041 randomized clinical trial. JAMA Oncol 9:815–824

Haley B, Batra K, Sahoo S, Froehlich T, Klemow D, Unni N, Ahn C, Rodriguez M, Hullings M, Frankel AE (2021) A phase I/Ib trial of PD 0332991 (palbociclib) and T-DM1 in HER2-positive advanced breast cancer after trastuzumab and taxane therapy. Clin Breast Cancer 21:417–424

Hanna K, Mayden K (2021) Chemotherapy treatment considerations in metastatic breast cancer. J Adv Pract Oncol 12:6–12

Hanusch C, Schneeweiss A, Loibl S, Untch M, Paepke S, Kümmel S, Jackisch C, Huober J, Hilfrich J, Gerber B, Eidtmann H, Denkert C, Costa S, Blohmer JU, Engels K, Burchardi N, von Minckwitz G (2015) Dual blockade with AFatinib and trastuzumab as NEoadjuvant treatment for patients with locally advanced or operable breast cancer receiving taxane-anthracycline containing chemotherapy-DAFNE (GBG-70). Clin Cancer Res 21:2924–2931

Hany D, Zoetemelk M, Bhattacharya K, Nowak-Sliwinska P, Picard D (2023) Network-informed discovery of multidrug combinations for ERα+/HER2-/PI3Kα-mutant breast cancer. Cell Mol Life Sci 80:80

Hari M, Loganzo F, Annable T, Tan X, Musto S, Morilla DB, Nettles JH, Snyder JP, Greenberger LM (2006) Paclitaxel-resistant cells have a mutation in the paclitaxel-binding region of β-tubulin (Asp26Glu) and less stable microtubules. Mol Cancer Ther 5:270–278

Hermawan A, Wagner E, Roidl A (2016) Consecutive salinomycin treatment reduces doxorubicin resistance of breast tumor cells by diminishing drug efflux pump expression and activity. Oncol Rep 35:1732–1740

Hermawan A, Putri H, Utomo RY (2020) Comprehensive bioinformatics study reveals targets and molecular mechanism of hesperetin in overcoming breast cancer chemoresistance. Mol Divers 24:933–947

Hermawan A, Putri H, Utomo R (2021) Exploration of targets and molecular mechanisms of cinnamaldehyde in overcoming fulvestrant-resistant breast cancer: a bioinformatics study. Network Model Anal Health Inf Bioinf 10:30

Hermawan A, Putri H, Hanif N, Fatimah N, Prasetio HH (2022) Identification of potential target genes of honokiol in overcoming breast cancer resistance to tamoxifen. Front Oncol 12:1019025

Herrmann F, Lehr HA, Drexler I, Sutter G, Hengstler J, Wollscheid U, Seliger B (2004) HER-2/neu-mediated regulation of components of the MHC class I antigen-processing pathway. Cancer Res 64:215–220

Hickish T, Mehta A, Liu MC, Huang CS, Arora RS, Chang YC, Yang Y, Vladimirov V, Jain M, Tsang J, Pemberton K, Sadrolhefazi B, Jin X, Tseng LM (2022) Afatinib alone and in combination with vinorelbine or paclitaxel, in patients with HER2-positive breast cancer who failed or progressed on prior trastuzumab and/or lapatinib (LUX-Breast 2): an open-label, multicenter, phase II trial. Breast Cancer Res Treat 192:593–602

Holzer AK, Manorek GH, Howell SB (2006) Contribution of the major copper influx transporter CTR1 to the cellular accumulation of cisplatin, carboplatin, and oxaliplatin. Mol Pharmacol 70: 1390–1394

Howell SJ, Casbard A, Carucci M, Ingarfield K, Butler R, Morgan S, Meissner M, Bale C, Bezecny P, Moon S, Twelves C, Venkitaraman R, Waters S, de Bruin EC, Schiavon G, Foxley A, Jones RH (2022) Fulvestrant plus capivasertib versus placebo after relapse or progression on an aromatase inhibitor in metastatic, oestrogen receptor-positive, HER2-negative breast cancer (FAKTION): overall survival, updated progression-free survival, and expanded biomarker analysis from a randomised, phase 2 trial. Lancet Oncol 23:851–864

Hsu JL, Hung MC (2016) The role of HER2, EGFR, and other receptor tyrosine kinases in breast cancer. Cancer Metastasis Rev 35:575–588

Huang BE, Mulyasasmita W, Rajagopal G (2016) The path from big data to precision medicine. Expert Rev Precis Med Drug Dev 1:129–143
Huang D, Yang F, Wang Y, Guan X (2017) Mechanisms of resistance to selective estrogen receptor down-regulator in metastatic breast cancer. Biochimica et Biophysica Acta (BBA) – Reviews onCancer 1868:148–156
Huang JF, Wen CJ, Zhao GZ, Dai Y, Li Y, Wu LX, Zhou HH (2018) Overexpression of ABCB4 contributes to acquired doxorubicin resistance in breast cancer cells in vitro. Cancer Chemother Pharmacol 82:199–210
Huang D, Tang L, Yang F, Jin J, Guan X (2019) PIK3CA mutations contribute to fulvestrant resistance in ER-positive breast cancer. Am J Transl Res 11:6055–6065
Huang S, Hu P, Lakowski TM (2021) Predicting breast cancer drug response using a multiple-layer cell line drug response network model. BMC Cancer 21:648
Hudis CA (2007) Trastuzumab--mechanism of action and use in clinical practice. N Engl J Med 357:39–51
Huzil JT, Chen K, Kurgan L, Tuszynski JA (2007) The roles of beta-tubulin mutations and isotype expression in acquired drug resistance. Cancer Inform 3:159–181
Hwang S-Y, Park S, Kwon Y (2019) Recent therapeutic trends and promising targets in triple negative breast cancer. Pharmacol Ther 199:30–57
Ibadurrahman W, Hanif N, Hermawan A (2022) Functional network analysis of p85 and PI3K as potential gene targets and mechanism of oleanolic acid in overcoming breast cancer resistance to tamoxifen. J Genet Eng Biotechnol 20:66
Imyanitov E, Sokolenko A (2021) Mechanisms of acquired resistance of BRCA1/2-driven tumors to platinum compounds and PARP inhibitors. World J Clin Oncol 12:544
Infante T, Del Viscovo L, De Rimini ML, Padula S, Caso P, Napoli C (2020) Network medicine: a clinical approach for precision medicine and personalized therapy in coronary heart disease. J Atheroscler Thromb 27:279–302
Inoue M, Mimura K, Izawa S, Shiraishi K, Inoue A, Shiba S, Watanabe M, Maruyama T, Kawaguchi Y, Inoue S, Kawasaki T, Choudhury A, Katoh R, Fujii H, Kiessling R, Kono K (2012) Expression of MHC class I on breast cancer cells correlates inversely with HER2 expression. Onco Targets Ther 1:1104–1110
Jia Y, Zhang C, Zhou L, Xu H, Shi Y, Tong Z (2015) Micheliolide overcomes KLF4-mediated cisplatin resistance in breast cancer cells by downregulating glutathione. Onco Targets Ther 8: 2319–2327
Jiang D, Lei T, Wang Z, Shen C, Cao D, Hou T (2020) ADMET evaluation in drug discovery. 20. Prediction of breast cancer resistance protein inhibition through machine learning. J Cheminform 12:16
Johnston SRD, Hegg R, Im SA, Park IH, Burdaeva O, Kurteva G, Press MF, Tjulandin S, Iwata H, Simon SD, Kenny S, Sarp S, Izquierdo MA, Williams LS, Gradishar WJ (2021) Phase III, randomized study of dual human epidermal growth factor receptor 2 (HER2) blockade with lapatinib plus trastuzumab in combination with an aromatase inhibitor in postmenopausal women with HER2-positive, hormone receptor-positive metastatic breast cancer: updated results of ALTERNATIVE. J Clin Oncol 39:79–89
Juric D, Janku F, Rodón J, Burris HA, Mayer IA, Schuler M, Seggewiss-Bernhardt R, Gil-Martin M, Middleton MR, Baselga J, Bootle D, Demanse D, Blumenstein L, Schumacher K, Huang A, Quadt C, Rugo HS (2019) Alpelisib plus fulvestrant in PIK3CA-altered and PIK3CA-wild-type estrogen receptor-positive advanced breast cancer: a phase 1b clinical trial. JAMA Oncol 5:e184475
Kaklamani VG, Gradishar WJ (2003) Epirubicin versus doxorubicin: which is the anthracycline of choice for the treatment of breast cancer? Clin Breast Cancer 4:S26–S33
Kalluri R, Neilson EG (2003) Epithelial-mesenchymal transition and its implications for fibrosis. J Clin Invest 112:1776–1784

Kaminska K, Akrap N, Staaf J, Alves C, Ehinger A, Ebbesson A, Hedenfalk I, Beumers L, Veerla S, Harbst K, Ehmsen S, Borgquist S, Borg A, Perez-Fidalgo JA, Ditzel H, Bosch A, Honeth G (2021) Distinct mechanisms of resistance to fulvestrant treatment dictate level of ER independence and selective response to CDK inhibitors in metastatic breast cancer. Breast Cancer Res 23:26

Kaneko K, Ishigami S, Kijima Y, Funasako Y, Hirata M, Okumura H, Shinchi H, Koriyama C, Ueno S, Yoshinaka H, Natsugoe S (2011) Clinical implication of HLA class I expression in breast cancer. BMC Cancer 11:454

Kanthala S, Mill CP, Riese DJ 2nd, Jaiswal M, Jois S (2016) Expression and purification of HER2 extracellular domain proteins in Schneider2 insect cells. Protein Expr Purif 125:26–33

Kaplan BL, Norell H, Callender GG, Ohlum T, Kiessling R, Nishimura MI (2006) Interferon-gamma renders tumors that express low levels of her-2/neu sensitive to cytotoxic T cells. Cancer Immunol Immunother 55:653–662

Kasaian J, Mosaffa F, Behravan J, Masullo M, Piacente S, Iranshahi M (2016) Modulation of multidrug resistance protein 2 efflux in the cisplatin resistance human ovarian carcinoma cells A2780/RCIS by sesquiterpene coumarins. Phytother Res 30:84–89

Keenan TE, Li T, Vallius T, Guerriero JL, Tayob N, Kochupurakkal B, Davis J, Pastorello R, Tahara RK, Anderson L, Conway J, He MX, Shannon E, Godin RE, Sorger PK, D'Andrea A, Overmoyer B, Winer EP, Mittendorf EA, Van Allen EM, Shapiro GI, Tolaney SM (2021) Clinical efficacy and molecular response correlates of the WEE1 inhibitor adavosertib combined with cisplatin in patients with metastatic triple-negative breast cancer. Clin Cancer Res 27: 983–991

Kim HP, Yoon YK, Kim JW, Han SW, Hur HS, Park J, Lee JH, Oh DY, Im SA, Bang YJ, Kim TY (2009) Lapatinib, a dual EGFR and HER2 tyrosine kinase inhibitor, downregulates thymidylate synthase by inhibiting the nuclear translocation of EGFR and HER2. PLoS One 4:e5933

Koca E, Niravath PA, Ensor J, Patel TA, Li X, Hemati P, Wong H, Qian W, Boone T, Zhao J, Ramshesh PV, Cohen AL, Murthy A, Nair S, Darcourt JG, Belcheva A, Kaklamani VG, Chang JCN (2021) ANETT: PhAse II trial of NEoadjuvant TAK-228 plus tamoxifen in patients with hormone receptor-positive breast cancer. Breast Cancer Res Treat 188:433–439

Kornblum N, Manola J, Klein P, Ramaswamy B, Brufsky A, Stella P (2017) PrECOG 0102: a randomized, double-blind, phase II trial of fulvestrant plus everolimus or placebo in post-menopausal women with hormone receptor (HR)-positive, HER2-negative metastatic breast cancer (MBC) resistant to aromatase inhibitor (AI) therapy. Cancer Res 77:S1–02

Kumar R, George B, Campbell MR, Verma N, Paul AM, Melo-Alvim C, Ribeiro L, Pillai MR, da Costa LM, Moasser MM (2020) HER family in cancer progression: from discovery to 2020 and beyond. Adv Cancer Res 147:109–160

Kutuk O, Letai A (2008) Alteration of the mitochondrial apoptotic pathway is key to acquired paclitaxel resistance and can be reversed by ABT-737. Cancer Res 68:7985–7994

Latorre E, Tebaldi T, Viero G, Spartà AM, Quattrone A, Provenzani A (2012) Downregulation of HuR as a new mechanism of doxorubicin resistance in breast cancer cells. Mol Cancer 11:13

Lavasani SM, Somlo G, Yost SE, Frankel PH, Ruel C, Cui Y, Murga M, Tang A, Martinez N, Kruper L, Tumyan L, Schmolze D, Yeon C, Yuan Y, Waisman JR, Mortimer J (2023) Phase 2 prospective open label study of neoadjuvant nab-paclitaxel, trastuzumab, and pertuzumab in patients with HER2-positive primary breast cancer. Cancer 129:740–749

Le DT, Bui LT, Nguyen CV, Do KH, Tran GL, Do TA (2023) Neoadjuvant doxorubicin-paclitaxel combined chemotherapy in patients with inoperable stage III breast cancer: a retrospective cohort study with 10 years of follow-up in Vietnam. Oncol Ther 11:327–341

Ledys F, Kalfeist L, Galland L, Limagne E, Ladoire S (2021) Therapeutic associations comprising anti-PD-1/PD-L1 in breast cancer: clinical challenges and perspectives. Cancers (Basel) 13:5999

Leenhardt F, Alexandre M, Jacot W (2021) Alpelisib for the treatment of PIK3CA-mutated, hormone receptor-positive, HER2-negative metastatic breast cancer. Expert Opin Pharmacother 22:667–675

Li G, Zhao L, Li W, Fan K, Qian W, Hou S, Wang H, Dai J, Wei H, Guo Y (2014) Feedback activation of STAT3 mediates trastuzumab resistance via upregulation of MUC1 and MUC4 expression. Oncotarget 5:8317–8329

Li S, Li C, Jin S, Liu J, Xue X, Eltahan AS, Sun J, Tan J, Dong J, Liang X-J (2017) Overcoming resistance to cisplatin by inhibition of glutathione S-transferases (GSTs) with ethacraplatin micelles in vitro and in vivo. Biomaterials 144:119–129

Li K, Du Y, Li L, Wei D-Q (2020) Bioinformatics approaches for anti-cancer drug discovery. Curr Drug Targets 21:3–17

Li CL, Moi SH, Lin HS, Hou MF, Chen FM, Shih SL, Kan JY, Kao CN, Wu YC, Kao LC, Chen YH, Lee YC, Chiang CP (2022) Comprehensive transcriptomic and proteomic analyses identify a candidate gene set in cross-resistance for endocrine therapy in breast cancer. Int J Mol Sci 23(18):10539

Li Y, Qi L, Wang Y, Zhao X, Lv S, Feng Y, Liu C, Li P, Xiong B, Guo Y, Lv D, Liu Y, Mao T, Yuan K, Cheng X, Li Y, Wang X (2023a) Safety, tolerability, pharmacokinetics and immunogenicity of an antibody-drug conjugate (SHR-A1201) in patients with HER2-positive advanced breast cancer: an open, phase I dose-escalation study. Anti-Cancer Drugs 34:763–774

Li R, Sant S, Brown E, Caramia F, Nikolic B, Clarke K, Byrne A, Lara Gonzalez LE, Savas P, Luen SJ, Teo ZL, Virassamy B, Neeson PJ, Darcy PK, Loi S (2023b) Tucatinib promotes immune activation and synergizes with programmed cell death-1 and programmed cell death-ligand 1 inhibition in HER2-positive breast cancer. J Natl Cancer Inst 115:805–814

Liu PC, Liu X, Li Y, Covington M, Wynn R, Huber R, Hillman M, Yang G, Ellis D, Marando C, Katiyar K, Bradley J, Abremski K, Stow M, Rupar M, Zhuo J, Li YL, Lin Q, Burns D, Xu M, Zhang C, Qian DQ, He C, Sharief V, Weng L, Agrios C, Shi E, Metcalf B, Newton R, Friedman S, Yao W, Scherle P, Hollis G, Burn TC (2006) Identification of ADAM10 as a major source of HER2 ectodomain sheddase activity in HER2 overexpressing breast cancer cells. Cancer Biol Ther 5:657–664

Liu Z, Shi A, Song D, Han B, Zhang Z, Ma L, Liu D, Fan Z (2017) Resistin confers resistance to doxorubicin-induced apoptosis in human breast cancer cells through autophagy induction. Am J Cancer Res 7:574–583

Loi S, Giobbie-Hurder A, Gombos A, Bachelot T, Hui R, Curigliano G, Campone M, Biganzoli L, Bonnefoi H, Jerusalem G, Bartsch R, Rabaglio-Poretti M, Kammler R, Maibach R, Smyth MJ, Di Leo A, Colleoni M, Viale G, Regan MM, Andre F, International Breast Cancer Study G, the Breast International G (2019) Pembrolizumab plus trastuzumab in trastuzumab-resistant, advanced, HER2-positive breast cancer (PANACEA): a single-arm, multicentre, phase 1b-2 trial. Lancet Oncol 20:371–382

Loibl S, Turner NC, Ro J, Cristofanilli M, Iwata H, Im SA, Masuda N, Loi S, André F, Harbeck N, Verma S, Folkerd E, Puyana Theall K, Hoffman J, Zhang K, Bartlett CH, Dowsett M (2017) Palbociclib combined with fulvestrant in premenopausal women with advanced breast cancer and prior progression on endocrine therapy: PALOMA-3 results. Oncologist 22:1028–1038

Lollini PL, Nicoletti G, Landuzzi L, De Giovanni C, Rossi I, Di Carlo E, Musiani P, Muller WJ, Nanni P (1998) Down regulation of major histocompatibility complex class I expression in mammary carcinoma of HER-2/neu transgenic mice. Int J Cancer 77:937–941

Lovitt CJ, Shelper TB, Avery VM (2018) Doxorubicin resistance in breast cancer cells is mediated by extracellular matrix proteins. BMC Cancer 18:41

Lu YS, Im SA, Colleoni M, Franke F, Bardia A, Cardoso F, Harbeck N, Hurvitz S, Chow L, Sohn J, Lee KS, Campos-Gomez S, Villanueva Vazquez R, Jung KH, Babu KG, Wheatley-Price P, De Laurentiis M, Im YH, Kuemmel S, El-Saghir N, O'Regan R, Gasch C, Solovieff N, Wang C, Wang Y, Chakravartty A, Ji Y, Tripathy D (2022) Updated overall survival of ribociclib plus endocrine therapy versus endocrine therapy alone in pre- and perimenopausal patients with HR +/HER2- advanced breast cancer in MONALEESA-7: a phase III randomized clinical trial. Clin Cancer Res 28:851–859

Łukasiewicz S, Czeczelewski M, Forma A, Baj J, Sitarz R, Stanisławek A (2021) Breast cancer-epidemiology, risk factors, classification, prognostic markers, and current treatment strategies-an updated review. Cancers (Basel) 13:4287

Luque-Cabal M, García-Teijido P, Fernández-Pérez Y, Sánchez-Lorenzo L, Palacio-Vázquez I (2016) Mechanisms behind the resistance to trastuzumab in HER2-amplified breast cancer and strategies to overcome it. Clin Med Insights Oncol 10:21–30

Ma CX, Sanchez C, Gao F, Crowder R, Naughton M, Pluard T, Creekmore A, Guo Z, Hoog J, Lockhart AC, Doyle A, Erlichman C, Ellis MJ (2016) A phase I study of the AKT inhibitor MK-2206 in combination with hormonal therapy in postmenopausal women with estrogen receptor-positive metastatic breast cancer. Clin Cancer Res 22:2650–2658

Ma J, Chen C, Liu S, Ji J, Wu D, Huang P, Wei D, Fan Z, Ren L (2022) Identification of a five genes prognosis signature for triple-negative breast cancer using multi-omics methods and bioinformatics analysis. Cancer Gene Ther 29:1578–1589

Mackey JR, Martin M, Pienkowski T, Rolski J, Guastalla J-P, Sami A, Glaspy J, Juhos E, Wardley A, Fornander T (2013) Adjuvant docetaxel, doxorubicin, and cyclophosphamide in node-positive breast cancer: 10-year follow-up of the phase 3 randomised BCIRG 001 trial. Lancet Oncol 14:72–80

Mao Y, Lv M, Zhang Y, Nie G, Cui J, Wang Y, Wang Y, Cao W, Liu X, Wang X, Wang H (2020) APOBEC3B expression and its prognostic potential in breast cancer. Oncol Lett 19:3205–3214

Martini M, De Santis MC, Braccini L, Gulluni F, Hirsch E (2014) PI3K/AKT signaling pathway and cancer: an updated review. Ann Med 46:372–383

Matikas A, Johansson H, Grybäck P, Bjöhle J, Acs B, Boyaci C, Lekberg T, Fredholm H, Elinder E, Margolin S, Isaksson-Friman E, Bosch A, Lindman H, Adra J, Andersson A, Agartz S, Hellström M, Zerdes I, Hartman J, Bergh J, Hatschek T, Foukakis T (2023) Survival outcomes, digital TILs, and on-treatment PET/CT during neoadjuvant therapy for HER2-positive breast cancer: results from the randomized PREDIX HER2 trial. Clin Cancer Res 29:532–540

Maximov PY, Lee TM, Jordan VC (2013) The discovery and development of selective estrogen receptor modulators (SERMs) for clinical practice. Curr Clin Pharmacol 8:135–155

McCart Reed AE, Kalita-De Croft P, Kutasovic JR, Saunus JM, Lakhani SR (2019) Recent advances in breast cancer research impacting clinical diagnostic practice. J Pathol 247:552–562

Medina MA, Oza G, Sharma A, Arriaga L, Hernández Hernández JM, Rotello VM, Ramirez JT (2020) Triple-negative breast cancer: a review of conventional and advanced therapeutic strategies. Int J Environ Res Public Health 17:2078

Mercogliano MF, De Martino M, Venturutti L, Rivas MA, Proietti CJ, Inurrigarro G, Frahm I, Allemand DH, Deza EG, Ares S, Gercovich FG, Guzmán P, Roa JC, Elizalde PV, Schillaci R (2017) TNFα-induced mucin 4 expression elicits trastuzumab resistance in HER2-positive breast cancer. Clin Cancer Res 23:636–648

Meredith A-M, Dass CR (2016) Increasing role of the cancer chemotherapeutic doxorubicin in cellular metabolism. J Pharm Pharmacol 68:729–741

Miller TW, Balko JM, Arteaga CL (2011a) Phosphatidylinositol 3-kinase and antiestrogen resistance in breast cancer. J Clin Oncol 29:4452

Miller TW, Rexer BN, Garrett JT, Arteaga CL (2011b) Mutations in the phosphatidylinositol 3-kinase pathway: role in tumor progression and therapeutic implications in breast cancer. Breast Cancer Res 13:1–12

Mitra D, Mitra D, Bensaad MS, Sinha S, Pant K, Pant M, Priyadarshini A, Singh P, Dassamiour S, Hambaba L (2022) Evolution of bioinformatics and its impact on modern bio-science in the twenty-first century: special attention to pharmacology, plant science and drug discovery. Comput Toxicol 24:100248

Moerkens M, Zhang Y, Wester L, van de Water B, Meerman JH (2014) Epidermal growth factor receptor signalling in human breast cancer cells operates parallel to estrogen receptor α signalling and results in tamoxifen insensitive proliferation. BMC Cancer 14:283

Moo TA, Sanford R, Dang C, Morrow M (2018) Overview of breast cancer therapy. PET Clin 13: 339–354

Mottini C, Napolitano F, Li Z, Gao X, Cardone L (2019) Computer-aided drug repurposing for cancer therapy: approaches and opportunities to challenge anticancer targets, vol 68. Semin Cancer Biol, p 59

Muhammad J, Khan A, Ali A, Fang L, Yanjing W, Xu Q, Wei D-Q (2018) Network pharmacology: exploring the resources and methodologies. Curr Top Med Chem 18:949–964

Mukherjee A, Dhadda A, Shehata M, Chan S (2007) Lapatinib: a tyrosine kinase inhibitor with a clinical role in breast cancer. Expert Opin Pharmacother 8:2189–2204

Munzone E, Colleoni M (2018) Optimal management of luminal breast cancer: how much endocrine therapy is long enough? Ther Adv Med Oncol 10:1758835918777437

Murray S, Briasoulis E, Linardou H, Bafaloukos D, Papadimitriou C (2012) Taxane resistance in breast cancer: mechanisms, predictive biomarkers and circumvention strategies. Cancer Treat Rev 38:890–903

Musolino A, Boggiani D, Pellegrino B, Zanoni D, Sikokis A, Missale G, Silini EM, Maglietta G, Frassoldati A, Michiara M (2020) Role of innate and adaptive immunity in the efficacy of anti-HER2 monoclonal antibodies for HER2-positive breast cancer. Crit Rev Oncol Hematol 149: 102927

Nagata Y, Lan K-H, Zhou X, Tan M, Esteva FJ, Sahin AA, Klos KS, Li P, Monia BP, Nguyen NT, Hortobagyi GN, Hung M-C, Yu D (2004) PTEN activation contributes to tumor inhibition by trastuzumab, and loss of PTEN predicts trastuzumab resistance in patients. Cancer Cell 6: 117–127

Němcová-Fürstová V, Kopperová D, Balušíková K, Ehrlichová M, Brynychová V, Václavíková R, Daniel P, Souček P, Kovář J (2016) Characterization of acquired paclitaxel resistance of breast cancer cells and involvement of ABC transporters. Toxicol Appl Pharmacol 310:215–228

Nguyen LC, Naulaerts S, Bruna A, Ghislat G, Ballester PJ (2021) Predicting cancer drug response in vivo by learning an optimal feature selection of tumour molecular profiles. Biomedicines 9:1319

Nicolini A, Ferrari P, Kotlarova L, Rossi G, Biava M (2015) The PI3K-AKt-mTOR pathway and new tools to prevent acquired hormone resistance in breast cancer. Curr Pharm Biotechnol 16: 804–815

Nicolini A, Ferrari P, Duffy MJ (2018) Prognostic and predictive biomarkers in breast cancer: past, present and future, Seminars in cancer biology, vol 52. Elsevier, p 56

Nikolai BC, Lanz RB, York B, Dasgupta S, Mitsiades N, Creighton CJ, Tsimelzon A, Hilsenbeck SG, Lonard DM, Smith CL, O'Malley BW (2016) HER2 signaling drives DNA anabolism and proliferation through SRC-3 phosphorylation and E2F1-regulated genes. Cancer Res 76: 1463–1475

Niu B, Zhou Y, Liao K, Wen T, Lao S, Quan G, Pan X, Wu C (2022) "Pincer movement": reversing cisplatin resistance based on simultaneous glutathione depletion and glutathione S-transferases inhibition by redox-responsive degradable organosilica hybrid nanoparticles. Acta Pharm Sin B 12:2074–2088

Osborne CK, Schiff R (2011) Mechanisms of endocrine resistance in breast cancer. Annu Rev Med 62:233–247

Osborne CK, Neven P, Dirix LY, Mackey JR, Robert J, Underhill C, Schiff R, Gutierrez C, Migliaccio I, Anagnostou VK, Rimm DL, Magill P, Sellers M (2011) Gefitinib or placebo in combination with tamoxifen in patients with hormone receptor-positive metastatic breast cancer: a randomized phase II study. Clin Cancer Res 17:1147–1159

Osipo C, Meeke K, Cheng D, Weichel A, Bertucci A, Liu H, Jordan VC (2007) Role for HER2/neu and HER3 in fulvestrant-resistant breast cancer. Int J Oncol 30:509–520

Paananen J, Fortino V (2020) An omics perspective on drug target discovery platforms. Brief Bioinform 21:1937–1953

Palmieri D, Bronder JL, Herring JM, Yoneda T, Weil RJ, Stark AM, Kurek R, Vega-Valle E, Feigenbaum L, Halverson D, Vortmeyer AO, Steinberg SM, Aldape K, Steeg PS (2007) Her-2 overexpression increases the metastatic outgrowth of breast cancer cells in the brain. Cancer Res 67:4190–4198

Pandolfi PP (2004) Breast cancer—loss of PTEN predicts resistance to treatment. N Engl J Med 351:2337–2338

Pang H, Hu Z (2023) Metabolomics in drug research and development: the recent advances in technologies and applications. Acta Pharm Sin B 13:3238–3251

Pang X, Fu W, Wang J, Kang D, Xu L, Zhao Y, Liu AL, Du GH (2018) Identification of estrogen receptor α antagonists from natural products via in vitro and in silico approaches. Oxidative Med Cell Longev 2018:6040149

Park JH, Kang MJ, Ahn JH, Kim JE, Jung KH, Gong G, Lee HJ, Son BH, Ahn SH, Kim HH, Shin HJ, Moon DH, Kim SB (2016) Phase II trial of neoadjuvant letrozole and lapatinib in Asian postmenopausal women with estrogen receptor (ER) and human epidermal growth factor receptor 2 (HER2)-positive breast cancer [neo-ALL-IN]: highlighting the TILs, ER expressional change after neoadjuvant treatment, and FES-PET as potential significant biomarkers. Cancer Chemother Pharmacol 78:685–695

Park YH, Kim TY, Kim GM, Kang SY, Park IH, Kim JH, Lee KE, Ahn HK, Lee MH, Kim HJ, Kim HJ, Lee JI, Koh SJ, Kim JY, Lee KH, Sohn J, Kim SB, Ahn JS, Im YH, Jung KH, Im SA (2019) Palbociclib plus exemestane with gonadotropin-releasing hormone agonist versus capecitabine in premenopausal women with hormone receptor-positive, HER2-negative metastatic breast cancer (KCSG-BR15-10): a multicentre, open-label, randomised, phase 2 trial. Lancet Oncol 20:1750–1759

Patel L, Shukla T, Huang X, Ussery DW, Wang S (2020) Machine learning methods in drug discovery. Molecules 25:5277

Patil S, Gupta AD (2022) Bioinformatics and its application in computing biological data. Information retrieval in bioinformatics: a practical approach. Springer, pp 133–154

Perez EA (1998) Paclitaxel in breast cancer. Oncologist 3:373–389

Pircher M, Mlineritsch B, Fridrik MA, Dittrich C, Lang A, Petru E, Weltermann A, Thaler J, Hufnagl C, Gampenrieder SP, Rinnerthaler G, Ressler S, Ulmer H, Greil R (2015) Lapatinib-plus-pegylated liposomal doxorubicin in advanced HER2-positive breast cancer following trastuzumab: a phase II trial. Anticancer Res 35:517–521

Poornima P, Kumar JD, Zhao Q, Blunder M, Efferth T (2016) Network pharmacology of cancer: from understanding of complex interactomes to the design of multi-target specific therapeutics from nature. Pharmacol Res 111:290–302

Prat A, Pineda E, Adamo B, Galván P, Fernández A, Gaba L, Díez M, Viladot M, Arance A, Muñoz M (2015) Clinical implications of the intrinsic molecular subtypes of breast cancer. Breast 24: S26–S35

Qian T, Zhu S, Hoshida Y (2019) Use of big data in drug development for precision medicine: an update. Expert Rev Precis Med Drug Dev 4:189–200

Rago RP, Einstein A Jr, Lush R, Beer TM, Ko YJ, Henner WD, Bubley G, Merica EA, Garg V, Ette E, Harding MW, Dalton WS (2003) Safety and efficacy of the MDR inhibitor Incel (biricodar, VX-710) in combination with mitoxantrone and prednisone in hormone-refractory prostate cancer. Cancer Chemother Pharmacol 51:297–305

Rahman MH, Rana HK, Peng S, Kibria MG, Islam MZ, Mahmud SH, Moni MA (2021) Bioinformatics and system biology approaches to identify pathophysiological impact of COVID-19 to the progression and severity of neurological diseases. Comput Biol Med 138:104859

Reel PS, Reel S, Pearson E, Trucco E, Jefferson E (2021) Using machine learning approaches for multi-omics data analysis: a review. Biotechnol Adv 49:107739

Reslan L, Dalle S, Dumontet C (2009) Understanding and circumventing resistance to anticancer monoclonal antibodies. MAbs 1:222–229

Riggins RB, Zwart A, Nehra R, Clarke R (2005) The nuclear factor kappa B inhibitor parthenolide restores ICI 182,780 (Faslodex; fulvestrant)-induced apoptosis in antiestrogen-resistant breast cancer cells. Mol Cancer Ther 4:33–41

Rimawi MF, Niravath P, Wang T, Rexer BN, Forero A, Wolff AC, Nanda R, Storniolo AM, Krop I, Goetz MP, Nangia JR, Jiralerspong S, Pavlick A, Veeraraghavan J, De Angelis C, Gutierrez C, Schiff R, Hilsenbeck SG, Osborne CK (2020) TBCRC023: a randomized phase II neoadjuvant trial of lapatinib plus trastuzumab without chemotherapy for 12 versus 24 weeks in patients with HER2-positive breast cancer. Clin Cancer Res 26:821–827

Rinkenbaugh AL, Baldwin AS (2016) The NF-κB pathway and cancer stem cells. Cells 5:16

Roberts SA, Lawrence MS, Klimczak LJ, Grimm SA, Fargo D, Stojanov P, Kiezun A, Kryukov GV, Carter SL, Saksena G, Harris S, Shah RR, Resnick MA, Getz G, Gordenin DA (2013) An APOBEC cytidine deaminase mutagenesis pattern is widespread in human cancers. Nat Genet 45:970–976

Rocca A, Maltoni R, Bravaccini S, Donati C, Andreis D (2018) Clinical utility of fulvestrant in the treatment of breast cancer: a report on the emerging clinical evidence. Cancer Manag Res 10: 3083–3099

Rocha CRR, Silva MM, Quinet A, Cabral-Neto JB, Menck CFM (2018) DNA repair pathways and cisplatin resistance: an intimate relationship. Clinics 73:e478s

Rodler E, Sharma P, Barlow WE, Gralow JR, Puhalla SL, Anders CK, Goldstein L, Tripathy D, Brown-Glaberman UA, Huynh TT, Szyarto CS, Godwin AK, Pathak HB, Swisher EM, Radke MR, Timms KM, Lew DL, Miao J, Pusztai L, Hayes DF, Hortobagyi GN (2023) Cisplatin with veliparib or placebo in metastatic triple-negative breast cancer and BRCA mutation-associated breast cancer (S1416): a randomised, double-blind, placebo-controlled, phase 2 trial. Lancet Oncol 24:162–174

Røssevold AH, Andresen NK, Bjerre CA, Gilje B, Jakobsen EH, Raj SX, Falk RS, Russnes HG, Jahr T, Mathiesen RR, Lømo J, Garred Ø, Chauhan SK, Lereim RR, Dunn C, Naume B, Kyte JA (2022) Atezolizumab plus anthracycline-based chemotherapy in metastatic triple-negative breast cancer: the randomized, double-blind phase 2b ALICE trial. Nat Med 28:2573–2583

Rossi L, McCartney A, De Santo I, Risi E, Moretti E, Malorni L, Biganzoli L, Di Leo A (2019) The optimal duration of adjuvant endocrine therapy in early luminal breast cancer: a concise review. Cancer Treat Rev 74:29–34

Safaei R, Otani S, Larson BJ, Rasmussen ML, Howell SB (2008) Transport of cisplatin by the copper efflux transporter ATP7B. Mol Pharmacol 73:461–468

Saha Detroja T, Detroja R, Mukherjee S, Samson AO (2022) Identifying hub genes associated with neoadjuvant chemotherapy resistance in breast cancer and potential drug repurposing for the development of precision medicine. Int J Mol Sci 23(20):12628

Saha T, Lukong KE (2022) Breast cancer stem-like cells in drug resistance: a review of mechanisms and novel therapeutic strategies to overcome drug resistance. Front Oncol 12:856974

Sahu R, Pattanayak SP (2020) Strategic developments & future perspective on gene therapy for breast cancer: role of mTOR and Brk/PTK6 as molecular targets. Curr Gene Ther 20:237–258

Santa-Maria CA, Kato T, Park JH, Kiyotani K, Rademaker A, Shah AN, Gross L, Blanco LZ, Jain S, Flaum L, Tellez C, Stein R, Uthe R, Gradishar WJ, Cristofanilli M, Nakamura Y, Giles FJ (2018) A pilot study of durvalumab and tremelimumab and immunogenomic dynamics in metastatic breast cancer. Oncotarget 9:18985–18996

Saxon ML, Lee DC (1999) Mutagenesis reveals a role for epidermal growth factor receptor extracellular subdomain IV in ligand binding. J Biol Chem 274(40):28356–28362. https://doi.org/10.1074/jbc.274.40.28356

Scaltriti M, Rojo F, Ocaña A, Anido J, Guzman M, Cortes J, Di Cosimo S, Matias-Guiu X, Ramon y Cajal S, Arribas J, Baselga J (2007) Expression of p95HER2, a truncated form of the HER2 receptor, and response to anti-HER2 therapies in breast cancer. J Natl Cancer Inst 99:628–638

Scaltriti M, Eichhorn PJ, Cortés J, Prudkin L, Aura C, Jiménez J, Chandarlapaty S, Serra V, Prat A, Ibrahim YH, Guzmán M, Gili M, Rodríguez O, Rodríguez S, Pérez J, Green SR, Mai S, Rosen N, Hudis C, Baselga J (2011) Cyclin E amplification/overexpression is a mechanism of trastuzumab resistance in HER2+ breast cancer patients. Proc Natl Acad Sci USA 108: 3761–3766

Schärer OD (2005) DNA interstrand crosslinks: natural and drug-induced DNA adducts that induce unique cellular responses. Chembiochem 6:27–32

Schmid P, Adams S, Rugo HS, Schneeweiss A, Barrios CH, Iwata H, Dieras V, Hegg R, Im SA, Shaw Wright G, Henschel V, Molinero L, Chui SY, Funke R, Husain A, Winer EP, Loi S, Emens LA, Investigators IMT (2018) Atezolizumab and nab-paclitaxel in advanced triple-negative breast cancer. N Engl J Med 379:2108–2121

Schoeberl A, Gutmann M, Theiner S, Corte-Rodríguez M, Braun G, Vician P, Berger W, Koellensperger G (2022) The copper transporter CTR1 and cisplatin accumulation at the single-cell level by LA-ICP-TOFMS. Front Mol Biosci 9:1055356

Schöffski P, Cresta S, Mayer IA, Wildiers H, Damian S, Gendreau S, Rooney I, Morrissey KM, Spoerke JM, Ng VW, Singel SM, Winer E (2018) A phase Ib study of pictilisib (GDC-0941) in combination with paclitaxel, with and without bevacizumab or trastuzumab, and with letrozole in advanced breast cancer. Breast Cancer Res 20:109

Schoumacher M, Burbridge M (2017) Key roles of AXL and MER receptor tyrosine kinases in resistance to multiple anticancer therapies. Curr Oncol Rep 19:1–14

Serra V, Scaltriti M, Prudkin L, Eichhorn PJ, Ibrahim YH, Chandarlapaty S, Markman B, Rodriguez O, Guzman M, Rodriguez S, Gili M, Russillo M, Parra JL, Singh S, Arribas J, Rosen N, Baselga J (2011) PI3K inhibition results in enhanced HER signaling and acquired ERK dependency in HER2-overexpressing breast cancer. Oncogene 30:2547–2557

Shah D, Osipo C (2016) Cancer stem cells and HER2 positive breast cancer: the story so far. Genes Dis 3:114–123

Sharma P, Hu-Lieskovan S, Wargo JA, Ribas A (2017) Primary, adaptive, and acquired resistance to cancer immunotherapy. Cell 168:707–723

Shi X, Wang J, Lei Y, Cong C, Tan D, Zhou X (2019) Research progress on the PI3K/AKT signaling pathway in gynecological cancer (review). Mol Med Rep 19:4529–4535

Siddiqui AJ, Jahan S, Patel M, Abdelgadir A, Alturaiki W, Bardakci F, Sachidanandan M, Badraoui R, Snoussi M, Adnan M (2023) Identifying novel and potent inhibitors of EGFR protein for the drug development against the breast cancer. J Biomol Struct Dyn 41:1–13

Singh V, Afshan T, Tyagi P, Varadwaj PK, Sahoo AK (2023) Recent development of multi-targeted inhibitors of human topoisomerase II enzyme as potent cancer therapeutics. Int J Biol Macromol 226:473–484

Skolastika S, Hanif N, Ikawati M, Hermawan A (2022) Comprehensive computational analysis of honokiol targets for cell cycle inhibition and immunotherapy in metastatic breast cancer stem cells. Evid Based Complement Alternat Med 2022:4172531

Sleire L, Forde HE, Netland IA, Leiss L, Skeie BS, Enger PO (2017) Drug repurposing in cancer. Pharmacol Res 124:74–91

Smart E, Semina SE, Frasor J (2020) Update on the role of NFκB in promoting aggressive phenotypes of estrogen receptor-positive breast cancer. Endocrinology 161(10):bqaa152

Smith BN, Bhowmick NA (2016) Role of EMT in metastasis and therapy resistance. J Clin Med 5(2):17

Stefanski CD, Keffler K, McClintock S, Milac L, Prosperi JR (2019) APC loss affects DNA damage repair causing doxorubicin resistance in breast cancer cells. Neoplasia 21:1143–1150

Stine ZE, Schug ZT, Salvino JM, Dang CV (2022) Targeting cancer metabolism in the era of precision oncology. Nat Rev Drug Discov 21:141–162

Sui M, Huang Y, Park BH, Davidson NE, Fan W (2007) Estrogen receptor α mediates breast cancer cell resistance to paclitaxel through inhibition of apoptotic cell death. Cancer Res 67:5337–5344

Sun R, Ge W, Zhu Y, Sayad A, Luna A, Lyu M, Liang S, Tobalina L, Rajapakse VN, Yu C, Zhang H, Fang J, Wu F, Xie H, Saez-Rodriguez J, Ying H, Reinhold WC, Sander C, Pommier Y, Neel BG, Aebersold R, Guo T (2023) Proteomic dynamics of breast cancer cell lines identifies potential therapeutic protein targets. Mol Cell Proteomics 22:100602

Takeda K, Nakayama M, Hayakawa Y, Kojima Y, Ikeda H, Imai N, Ogasawara K, Okumura K, Thomas DM, Smyth MJ (2017) IFN-gamma is required for cytotoxic T cell-dependent cancer genome immunoediting. Nat Commun 8:14607

Taylor BC, Sun X, Gonzalez-Ericsson PI, Sanchez V, Sanders ME, Wescott EC, Opalenik SR, Hanna A, Chou ST, Van Kaer L, Gomez H, Isaacs C, Ballinger TJ, Santa-Maria CA, Shah PD, Dees EC, Lehmann BD, Abramson VG, Pietenpol JA, Balko JM (2023) NKG2A is a therapeutic vulnerability in immunotherapy resistant MHC-I heterogeneous triple negative breast cancer. Cancer Discov 14(2):290

Teven CM, Schmid DB, Sisco M, Ward J, Howard MA (2017) Systemic therapy for early-stage breast cancer: what the plastic surgeon should know. Eplasty 17:e7

Thery J-C, Spano J-P, Azria D, Raymond E, Llorca FP (2014) Resistance to human epidermal growth factor receptor type 2-targeted therapies. Eur J Cancer 50:892–901

Tolaney SM, Wardley AM, Zambelli S, Hilton JF, Troso-Sandoval TA, Ricci F, Im SA, Kim SB, Johnston SR, Chan A, Goel S, Catron K, Chapman SC, Price GL, Yang Z, Gainford MC, André F (2020) Abemaciclib plus trastuzumab with or without fulvestrant versus trastuzumab plus standard-of-care chemotherapy in women with hormone receptor-positive, HER2-positive advanced breast cancer (monarcHER): a randomised, open-label, phase 2 trial. Lancet Oncol 21:763–775

Treilleux I, Arnedos M, Cropet C, Wang Q, Ferrero JM, Abadie-Lacourtoisie S, Levy C, Legouffe E, Lortholary A, Pujade-Lauraine E, Bourcier AV, Eymard JC, Spaeth D, Bachelot T (2015) Translational studies within the TAMRAD randomized GINECO trial: evidence for mTORC1 activation marker as a predictive factor for everolimus efficacy in advanced breast cancer. Ann Oncol 26:120–125

Trowe T, Boukouvala S, Calkins K, Cutler RE Jr, Fong R, Funke R, Gendreau SB, Kim YD, Miller N, Woolfrey JR (2008) EXEL-7647 inhibits mutant forms of ErbB2 associated with lapatinib resistance and neoplastic transformation. Clin Cancer Res 14:2465–2475

Tsang RY, Sadeghi S, Finn RS (2011) Lapatinib, a dual-targeted small molecule inhibitor of EGFR and HER2, in HER2-amplified breast cancer: from bench to bedside. Clin Med Insights Ther 3: CMT.S3783

Tsimberidou AM, Shaw JV, Juric D, Verschraegen C, Weise AM, Sarantopoulos J, Lopes G, Nemunaitis J, Mita M, Park H, Ellers-Lenz B, Tian H, Xiong W, Kaleta R, Kurzrock R (2021) Phase 1 study of M2698, a p70S6K/AKT dual inhibitor, in patients with advanced cancer. J Hematol Oncol 14:127

Van Swearingen AED, Siegel MB, Deal AM, Sambade MJ, Hoyle A, Hayes DN, Jo H, Little P, Dees EC, Muss H, Jolly T, Zagar TM, Patel N, Miller CR, Parker JS, Smith JK, Fisher J, Shah N, Nabell L, Nanda R, Dillon P, Abramson V, Carey LA, Anders CK (2018) LCCC 1025: a phase II study of everolimus, trastuzumab, and vinorelbine to treat progressive HER2-positive breast cancer brain metastases. Breast Cancer Res Treat 171:637–648

Vassilopoulos A, Xiao C, Chisholm C, Chen W, Xu X, Lahusen TJ, Bewley C, Deng CX (2014) Synergistic therapeutic effect of cisplatin and phosphatidylinositol 3-kinase (PI3K) inhibitors in cancer growth and metastasis of Brca1 mutant tumors. J Biol Chem 289:24202–24214

Veglia F, Perego M, Gabrilovich D (2018) Myeloid-derived suppressor cells coming of age. Nat Immunol 19:108–119

Viedma-Rodríguez R, Baiza-Gutman L, Salamanca-Gómez F, Diaz-Zaragoza M, Martínez-Hernández G, Ruiz Esparza-Garrido R, Velázquez-Flores MA, Arenas-Aranda D (2014) Mechanisms associated with resistance to tamoxifen in estrogen receptor-positive breast cancer. Oncol Rep 32:3–15

Wahba HA, El-Hadaad HA (2015) Current approaches in treatment of triple-negative breast cancer. Cancer Biol Med 12:106–116

Wang H, Mao X (2020) Evaluation of the efficacy of neoadjuvant chemotherapy for breast cancer. Drug Des Devel Ther 14:2423–2433

Wang Y, Zhou BP (2011) Epithelial-mesenchymal transition in breast cancer progression and metastasis. Chin J Cancer 30:603–611

Wang SE, Narasanna A, Perez-Torres M, Xiang B, Wu FY, Yang S, Carpenter G, Gazdar AF, Muthuswamy SK, Arteaga CL (2006) HER2 kinase domain mutation results in constitutive phosphorylation and activation of HER2 and EGFR and resistance to EGFR tyrosine kinase inhibitors. Cancer Cell 10:25–38

Wang X, Hui R, Chen Y, Wang W, Chen Y, Gong X, Jin J (2019) Discovery of novel doxorubicin metabolites in MCF7 doxorubicin-resistant cells. Front Pharmacol 10:1434. https://doi.org/10.3389/fphar.2019.01434

Wang H, Guo S, Kim SJ, Shao F, Ho JWK, Wong KU, Miao Z, Hao D, Zhao M, Xu J, Zeng J, Wong KH, Di L, Wong AH, Xu X, Deng CX (2021) Cisplatin prevents breast cancer metastasis through blocking early EMT and retards cancer growth together with paclitaxel. Theranostics 11:2442–2459

Weiss A, Li T, Desai NV, Tung NM, Poorvu PD, Partridge AH, Nakhlis F, Dominici L, Sinclair N, Spring LM, Faggen M, Constantine M, Krop IE, DeMeo M, Wrabel E, Alberti J, Chikarmane S, Tayob N, King TA, Tolaney SM, Winer EP, Mittendorf EA, Waks AG (2023) Impact of neoadjuvant paclitaxel/trastuzumab/pertuzumab on breast tumor downsizing for patients with HER2+ breast cancer: single-arm prospective clinical trial. J Am Coll Surg 237:247–256

Wen C, Fu L, Huang J, Dai Y, Wang B, Xu G, Wu L, Zhou H (2019) Curcumin reverses doxorubicin resistance via inhibition the efflux function of ABCB4 in doxorubicin-resistant breast cancer cells. Mol Med Rep 19:5162–5168

Wibowo MA, Nugroho EP, Hermawan A (2021) Genomic understanding reveals the important role of FGFR2 as paeoniflorin target for circumventing breast cancer resistance to tamoxifen. Asian Pac J Cancer Prev 22:3949–3958

Wisinski KB, Tevaarwerk AJ, Burkard ME, Rampurwala M, Eickhoff J, Bell MC, Kolesar JM, Flynn C, Liu G (2016) Phase I study of an AKT inhibitor (MK-2206) combined with lapatinib in adult solid tumors followed by dose expansion in advanced HER2+ breast cancer. Clin Cancer Res 22:2659–2667

Xia W, Bisi J, Strum J, Liu L, Carrick K, Graham KM, Treece AL, Hardwicke MA, Dush M, Liao Q (2006) Regulation of survivin by ErbB2 signaling: therapeutic implications for ErbB2-overexpressing breast cancers. Cancer Res 66:1640–1647

Xie XF, Zhang QY, Huang JY, Chen LP, Lan XF, Bai X, Song L, Xiong SL, Guo SJ, Du CW (2023) Pyrotinib combined with trastuzumab and chemotherapy for the treatment of human epidermal growth factor receptor 2-positive metastatic breast cancer: a single-arm exploratory phase II trial. Breast Cancer Res Treat 197:93–101

Yakes FM, Chinratanalab W, Ritter CA, King W, Seelig S, Arteaga CL (2002) Herceptin-induced inhibition of phosphatidylinositol-3 kinase and Akt is required for antibody-mediated effects on p27, cyclin D1, and antitumor action. Cancer Res 62:4132–4141

Yam C, Mittendorf EA, Garber HR, Sun R, Damodaran S, Murthy RK, Ramirez D, Karuturi M, Layman RM, Ibrahim N, Rauch GM, Adrada BE, Candelaria RP, White JB, Ravenberg E, Clayborn A, Ding QQ, Symmans WF, Prabhakaran S, Thompson AM, Valero V, Tripathy D, Huo L, Moulder SL, Litton JK (2023) A phase II study of neoadjuvant atezolizumab and nab-paclitaxel in patients with anthracycline-resistant early-stage triple-negative breast cancer. Breast Cancer Res Treat 199:457–469

Yang F, Teves SS, Kemp CJ, Henikoff S (2014) Doxorubicin, DNA torsion, and chromatin dynamics. Biochimica et Biophysica Acta (BBA) – Reviews on Cancer 1845:84–89

Yao H, He G, Yan S, Chen C, Song L, Rosol TJ, Deng X (2017) Triple-negative breast cancer: is there a treatment on the horizon? Oncotarget 8:1913–1924

Yardley DA, Arrowsmith ER, Daniel BR, Eakle J, Brufsky A, Drosick DR, Kudrik F, Bosserman LD, Keaton MR, Goble SA, Bubis JA, Priego VM, Pendergrass K, Manalo Y, Bury M, Gravenor DS, Rodriguez GI, Inhorn RC, Young RR, Harwin WN, Silver C, Hainsworth JD, Burris HA 3rd (2017) TITAN: phase III study of doxorubicin/cyclophosphamide followed by ixabepilone or paclitaxel in early-stage triple-negative breast cancer. Breast Cancer Res Treat 164:649–658

Yin S, Bhattacharya R, Cabral F (2010) Human mutations that confer paclitaxel resistance. Mol Cancer Ther 9:327–335

Yin W, Xiang D, Wang T, Zhang Y, Pham CV, Zhou S, Jiang G, Hou Y, Zhu Y, Han Y (2021) The inhibition of ABCB1/MDR1 or ABCG2/BCRP enables doxorubicin to eliminate liver cancer stem cells. Sci Rep 11:10791

Yoshida R, Tazawa H, Hashimoto Y, Yano S, Onishi T, Sasaki T, Shirakawa Y, Kishimoto H, Uno F, Nishizaki M, Kagawa S, Fujiwara T (2012) Mechanism of resistance to trastuzumab and molecular sensitization via ADCC activation by exogenous expression of HER2-extracellular domain in human cancer cells. Cancer Immunol Immunother 61:1905–1916

Zafar I, Saba M, Raza MA, Rather MA, Rayan RA, Fatima M, Ain Q, Ishneiwra RM (2021) Roles of bioinformatics in reproductive science. In: Recent updates in molecular endocrinology and reproductive physiology of fish: an imperative step in aquaculture, pp 203–228

Zhang X, Diaz MR, Yee D (2013) Fulvestrant regulates epidermal growth factor (EGF) family ligands to activate EGF receptor (EGFR) signaling in breast cancer cells. Breast Cancer Res Treat 139:351–360

Zheng M (2022) Tumor mutation burden for predicting immune checkpoint blockade response: the more, the better. J Immunother Cancer 10:e003087

Index

© The Editor(s) (if applicable) and The Author(s), under exclusive license to Springer Nature Switzerland AG 2024
Interdisciplinary Cancer Research, https://doi.org/10.1007/978-3-031-65808-2

SPRINGER NATURE

GPSR Compliance

The European Union's (EU) General Product Safety Regulation (GPSR) is a set of rules that requires consumer products to be safe and our obligations to ensure this.

If you have any concerns about our products, you can contact us on ProductSafety@springernature.com

In case Publisher is established outside the EU, the EU authorized representative is:

Springer Nature Customer Service Center GmbH
Europaplatz 3
69115 Heidelberg, Germany

The manufacturer's authorised representative in the EU is Springer Nature Customer Service Centre GmbH, Europaplatz 3, 69115 Heidelberg, Germany. If you have any concerns regarding our products, please contact ProductSafety@springernature.com

Printed and bound by CPI Group (UK) Ltd, Croydon, CR0 4YY
28/09/2026
02224065-0008